Chemistry of Hazardous Materials

Chemistry of Hazardous Materials

Second Edition —

EUGENE MEYER

Consultant

 *A BRADY BOOK*

PRENTICE HALL BUILDING
Englewood Cliffs, New Jersey 07632

Library of Congress Cataloging in Publication Data

Meyer, Eugene
  Chemistry of hazardous materials.

  "A Brady book."
  Includes index.
  1. Hazardous substances—Fires and fire prevention.
  2. Hazardous substances. I. Title.
  TH9446.H38M48 1989   628.9'25   88-32252
  ISBN 0-89303-133-X

Editorial/production supervision: *Raeia Maes*
Manufacturing buyer: *Robert Anderson*

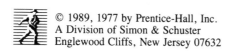 © 1989, 1977 by Prentice-Hall, Inc.
A Division of Simon & Schuster
Englewood Cliffs, New Jersey 07632

Printed in the United States of America

10 9 8 7 6 5 4 3 2 1

ISBN   0-89303-133-X

Prentice-Hall International (UK) Limited, *London*
Prentice-Hall of Australia Pty. Limited, *Sydney*
Prentice-Hall Canada Inc., *Toronto*
Prentice-Hall Hispanoamericana, S.A., *Mexico*
Prentice-Hall of India Private Limited, *New Delhi*
Prentice-Hall of Japan, Inc., *Tokyo*
Simon & Schuster Asia Pte. Ltd., *Singapore*
Editora Prentice-Hall do Brasil, Ltda., *Rio de Janeiro*

Contents

Preface

to the Second Edition

Many chemical substances are beneficial and essential for our existence. Nonetheless, certain industrially essential chemical substances, as well as some products made from them, are associated with an element of danger. They may be poisonous, flammable, water reactive, corrosive, explosive, or radioactive, or they may possess some other property that is cause for regarding them as dangerous. Using the jargon of emergency-response personnel, they are fittingly referred to as hazardous materials. These materials are used to produce the products of modern life, like televisions, automobiles, appliances, and home furnishings. Indeed, they are largely responsible for the relative affluence of the American life-style. Their widespread use has become a characteristic feature of the twentieth century.

While hazardous materials are routinely stored, handled, and transported without incident, mishaps do occur with sufficient regularity to require the attention of firefighters and others. When unintentionally released from their container or transport vehicle, hazardous materials often destroy property and wreak substantial health and environmental damage. Flammable and combustible materials burn, explosives detonate, poisonous materials kill, and other hazardous materials pose harmful risks in similar ways.

Since publication of the first edition, interest in the hazardous materials field has broadened significantly. This interest largely stems from the national impact that occupational, environmental, and transportation legislation has had on commerce and industry, as well as on the general public. Consequently, firefighters are now only one of several groups of individuals who require comprehensive training in the chemistry of hazardous materials. With this fact in mind, the second edition was prepared by primarily addressing the specific needs of firefighters; but the needs of individuals in other professions were also incorporated as they concern hazardous materials. For instance, more information has been provided in the second edition pertaining to procedures whereby hazardous materials may be safely handled and stored.

The resulting book naturally divides into two sections. The first, which comprises Chapters 1 through 4, is a review of elementary chemistry. The second part, Chapters 5 through 14, deals with hazardous materials regulations and illustrative examples from the major classes of hazardous materials.

Several individuals provided comments on the first edition, which I have attempted to incorporate herein. Some people remarked that the book should contain more chemistry, others requested less chemistry, and still others requested additional reference, whenever appropriate, to the regulatory aspects of hazardous materials. Chemistry plays such a pivotal role in understanding the properties of hazardous materials that its inclusion in a textbook of this type is absolutely essential. It is simply a matter of "how much?"

While preparing this second edition, I attempted to provide sufficient breadth and diversity so that instructors could choose the material they individually consider to be primarily important. The second edition provides more detailed discussions of chemical background topics. Additional examples of hazardous materials have been added, particularly several that have recently become newsworthy, such as asbestos, dioxin, polychlorinated biphenyls, and residential radon. Chapter 5, which is new to this edition, is a review of the federal hazardous materials transportation regulations.

By regulating the amount of time that is spent on specific chapters, an instructor may use this book for either a one- or two-semester course, with more or less "pure" chemistry. Students having had previous exposure to chemistry obviously require less time on the earlier chapters compared to the later ones. I recommend a general coverage of all chapters, but with a stronger emphasis on the chapters dealing with those hazardous materials most likely to be encountered by the group at interest.

Many individuals have helped generously with advise and assistance during preparation of the manuscript, for which I am extremely grateful. Among those deserving special mention are Capt. E. E. Comstock, U.S. Coast Guard (retired), who read the manuscripts of the first *and* second editions from beginning to end in their various stages and made innumerable, valuable suggestions; David W. Boykin (Delaware Technical Community College, Newark, Delaware) and William J. Vandevort (Program Coordinator, Field Operations, State Fire Marshall's Office, State of California, Sacramento, California), who read selected chapters and provided valuable critical reviews; and Matt McNearney, Raeia Maes, and others of the staff of Prentice Hall for their expert work and sound advice before and during production. Several individuals provided advice on certain sections that relate to their specific areas of expertise: Douglas E. Klapper and others of Pennwalt Corporation (organic peroxides), and Dr. John Riley of Ansul Corporation (reactive metals and Class D fires). For their comments, I am grateful.

Finally, I acknowledge the support of my wife, Phyllis, who has sustained me by her continuous encouragement throughout the ordeal of preparing the manuscript and proofreading the galleys. To her, I owe a special debt of gratitude, and to her, I lovingly dedicate this book.

Eugene Meyer

Chemistry
of
Hazardous
Materials

1

Introduction

Prior to the Industrial Revolution, disasters were primarily limited to those caused by uncontrollable natural events, like floods, earthquakes, volcanoes, hurricanes, and tornadoes. While disasters still result from such causes today, we are now confronted with others that are relatively unique to the twentieth century. One such cause directly relates to the materials used industrially to produce the amenities of modern life by the chemical, petroleum, and nuclear industries.

Consider just one product of the chemical industry, vinyl chloride. This raw material is used to make the synthetic plastic, polyvinyl chloride (PVC). Vinyl chloride is a highly flammable substance, so much so that the fire service considers it a severe fire and explosion hazard. When unleashed from its container, vinyl chloride poses a substantial threat to public health and safety. Bulk quantities of this substance are likely to be encountered in the workplace during storage or handling or during a train derailment or other transportation mishap.

Vinyl chloride is a single example of a substance that has come to be known as a *hazardous material*. There are thousands of such hazardous materials, more and more of which are now routinely encountered by firefighters and other emergency-response personnel. Fortunately, each such substance is typically encountered in only a limited number of ways. Nonetheless, hazardous materials may be found in a variety of places, even in the home. Some are found dispersed in the air we breathe; others are even in the food we eat and the water we drink. In truth, hazardous materials are likely to be found almost anywhere.

Long-established federal and state regulations exist that address the proper storage, handling, and transportation issues where hazardous materials are involved. While these regulations are unquestionably helpful, they represent only the first step toward maintaining public safety. More often than not, it is the performance of the emergency-response unit first arriving at the scene of a disaster that decides how well the public health and safety will be secured. This response action is unlike traditional firefighting in that personnel need to know the properties of the specific ma-

1

terials involved *before* they can safely approach a disaster scene or attempt a rescue mission.

What are these hazardous materials and what are their characteristics? How may we recognize them? How may we rapidly obtain technical information about them? These are some of the questions we shall address in this first chapter.

1-1 ROLE OF FIRE THROUGHOUT HISTORY

Fire is often the major concern at the site of an emergency disaster. Thus the fundamental question is how do we control fire. This question is answered for given fuel systems at appropriate points throughout this book; but it is pertinent to begin the study of hazardous materials by recognizing the role that fire has played in the world from prehistory to today.

According to modern geology, Earth has been continuously forming for some 3.6 billion or more years. Fire probably did not play a significant role in Earth's early evolution. However, with the passage of time, our planet slowly acquired an oxygenated atmosphere, probably similar to that we experience today. The oxygen resulted from plants, primarily algae, undergoing photosynthesis. It was only when atmospheric oxygen became sufficiently abundant that fire could first begin to manifest itself in nature. This occurred in several ways. For instance, volcanoes erupted hot, molten lava that kindled fires, and electrical thunderstorms generated lightning, which kindled other fires.

By the time hominid species are known to have first arrived on Earth, perhaps some 4 to 8 million years ago, fires had been most likely already burning from time to time. The majesty and destructive force of such fires were probably viewed by primeval humans, but this is a matter upon which we can only speculate.

As civilization developed, so did the ingenuity of the early members of the human race. First, relatively crude methods of kindling fires were learned, which did not involve the intervention of nature: Rocks were struck together to create sparks, or wooden sticks were twisted in holes until the friction caused the wood to kindle. Later, with the passage of time, our early ancestors learned to use fire for warmth, to cook food, to shape tools and weapons, to ward off animals, and to furnish light.

In ancient times, fire was regarded as a sacred and powerful force, which was often feared, honored, and worshipped. When provoked, dragons were believed to exhale fire; the sun was perceived as a giant ball of fire; and virtually every early culture associated a supernatural character with fire. The ancient Greeks, for example, believed the god Prometheus stole fire from the heavens and brought it to Earth. The event was depicted by Rubens in the early 1600s in a famous painting (Fig. 1-1). For this bad deed, Prometheus was sentenced to be bound in chains to a mountain, with a vulture picking away at his liver until some mortal consented to die in his place.

Fire was not perceived too differently until around 500 B.C., when the Greek

Figure 1-1 According to Greek mythology, the human race first received fire when Prometheus stole burning flames from the heavens and brought them to earth. The event is depicted by Rubens in the famous work of art, *Prometeo (boceto)*. (Courtesy of Museo del Prado, Madrid)

philosopher Empedocles gave it the distinction of being one of the fundamental constituents out of which all materials are made. Empedocles carefully observed wood as it burned, noting that smoke and flames are produced. Holding a cool surface near the flames, Empedocles further noted the condensation of water vapor. When the wood was entirely burned, he noted that ashes were the only solid residue. Hence, he concluded, wood must consist of four basic components: earth (the ashes), fire, air (the smoke), and water. Empedocles generalized these observations, identifying them as the four constituents of all materials.

This view, that fire was a constituent of all materials, was generally accepted until the early 1700s. Then, two German scientists, Becker and Stahl, proposed an alternative theory concerning the structural nature of materials. This theory was also concerned with the burning process. According to these scientists, combustible materials contained a basic component, called *phlogiston*, which caused combustion and escaped as materials burned. Other than as a manifestation of combustion, fire itself did not play a particularly unique role in this phlogiston theory.

It was not until 1774 that Antoine Laurent Lavoisier, a French chemist, reexamined the nature of combustion and provided us with a new insight into how materials burn. Lavoisier was one of the first scientists to recognize the importance of oxygen in the combustion process. He noted that when a material burns the component parts of that material combine with oxygen; that is, the material is not actually consumed or destroyed, but changes into new forms. His fundamental experiments on combustion earned Lavoisier respect and a leading position among scientists of the 1700s. Even today, the scientific information first established by Lavoisier is still helping us to control fire, as well as to use it to our advantage.

The control of fire has played a major role in the advance of civilization and has profoundly influenced the development of science and technology. Today, fire still remains the world's major source of warmth, high temperature, and power. Systems employing fire are now commonplace throughout all major parts of the civilized world: bunsen flames, arcs, industrial and home heaters, internal combustion engines, jet engines, rockets and numerous chemically different combustion systems employing solid, liquid, and gaseous propellants.

Yet fire has also been one of our worst enemies. Countless lives have been lost due to incidents caused by fire, as Table 1-1 illustrates. The first disastrous fire in recorded history occurred in Rome in 64 A.D. during the reign of the emperor Nero. History tells us that Nero was notoriously cruel and depraved; having ordered the burning of Rome, he is said to have enjoyed the spectacle. In modern times, the most well-known fire in the United States is probably the infamous Chicago fire of 1871, triggered perhaps by Mrs. O'Leary's cow kicking over a lantern. This conflagration is now recalled each October through activities conducted during Fire Prevention Week, proclaimed annually in the United States and Canada.

It was much earlier, during America's colonial period, that the first firefighting tactic was developed, the bucket brigade. Early settlers passed water in buckets from

TABLE 1-1   SOME CATASTROPHIC FIRES

Year	Place	Deaths
64	Rome	Unknown
1666	London	Thousands
1871	Peshtigo, Wisconsin (forest fire)	1152
1871	Chicago	250
1892	St. John's, Newfoundland	600
1903	Chicago (Iroquois Theater)	600
1904	New York Harbor (steamboat fire)	1021
1906	San Francisco	452
1922	Izmir (Smyrna), Turkey	1000
1929	Cleveland, Ohio (clinic)	125
1942	Boston, Massachusetts (Coconut Grove Night Club)	492
1949	Chungking, China	1700
1956	Marcinelle, Belgium (coal mine fire)	262
1963	Fitchville, Ohio (nursing home)	63
1965	Salt Lake City, Utah (727 aircraft crash)	43
1970	Marietta, Ohio (Harmer House Nursing Home)	22
1972	Sunshine Mine disaster	92
1973	Isle of Mann resort	50
1987	San Juan, Puerto Rico (Du Pont Plaza Hotel)	97

a water supply to the fire and back again. These people were probably also America's first volunteer firefighters; but the first known regular group of organized volunteer firefighters was directed by Benjamin Franklin in 1736. The first municipal fire department was established in Boston in 1679, but most American cities continued to use volunteer fire service until the early 1900s.

Since the early 1950s, we have also experienced a change in the manner by which we control the spread of fire. As an industrialized society, we have come to regularly use petroleum, plastics, explosives, and rocket fuels. We simply cannot hope to control fires involving such materials without the occasional use of special firefighting tactics. Such tactics sometimes involve the use of fire extinguishers other than water, such as carbon dioxide, Halon agents and "alcohol" and "protein" foams.

In contemporary times, it takes the combined effort of architects, scientists, engineers, fire technologists, firefighters, transportation officials, and others to solve the problems associated with the potential for fire in given situations, as well as to respond to its actuality. This effort has resulted in the promulgation of related codes and regulations designed to promote safety. In modern high-rise buildings, for example, we now have the required use of fire-resistant materials, the installation of automatic sprinkler systems, and the construction of accessible safety exits. It is this combination of technical resources that has helped to save lives and property in the latter quarter of this century more effectively than ever before.

This does not mean that all problems involving the control of fire have been solved. Just the opposite is true! Even the nature of fire in given systems is not clearly understood. Furthermore, effective tactics have not been developed for fighting fire in every situation that is likely to be encountered. Yet it is through the continuation of this effort of using scientific data and experience that we can hope to improve our ability of controlling fire and promoting safety to the benefit of humankind.

1-2 WHY STUDY CHEMISTRY?

We are normally too engrossed with the activities of our daily lives to consider what life would be like without chemistry. If we were to ponder this point, however, it would soon be apparent that chemistry affects everything we do. Not a single moment of time goes by during which we are not affected somehow by a chemical substance or chemical process.

Chemistry is regarded as one of the *natural sciences*; that is, it is concerned with a study of natural phenomena. Specifically, chemistry deals with the composition of substances, as well as the changes that occur in them. However, this includes not only substances that occur naturally, but also synthetic substances. By knowing the composition of substances and by understanding how changes occur in them, we are in a better position to control and use them to our benefit.

Chemistry is broadly divided into two branches: *organic chemistry* and *inor-*

ganic chemistry. At one time, it was thought that certain substances originated only in living things—plants and animals. These substances were called organic, since it was felt that their origin required the vital force of life itself.

However, in 1828, Friedrich Wöhler disproved this hypothesis when he prepared urea, a substance isolated from urine, entirely from other substances that were unconnected with either plants or animals. The idea that a vital force was needed to produce organic substances like urea was abandoned; however, the name of the branch of chemistry, organic chemistry, was retained. Since Wöhler's research, organic chemistry has come to mean the study of substances that contain carbon in their chemical composition. While some of these organic substances are actually found in living organisms, many others have been synthesized that have no known natural counterpart. The study of organic substances that affect the life process is a subdivision of organic chemistry called *biochemistry*.

By contrast, the study of substances that do not contain carbon in their composition is known as *inorganic chemistry*. The substances themselves are often called inorganic; they include aluminum, iron, sulfur, oxygen, and many others.

To be an effective firefighter, is it really necessary to study organic and inorganic chemistry? Fortunately, not in great detail! It would be incorrect to give the impression that the study of classical chemistry is absolutely essential for success in the firefighting profession. However, some aspects of organic and inorganic chemistry relate directly to the study of hazardous materials and thus affect a person's performance as an effective firefighter in the industrialized world of today.

Consider a few basic questions like the following: Why is water ineffective in extinguishing a fire of burning magnesium? Why does a red-brown cloud of smoke result from a fire involving items made from polyurethane? Why is lime effective in neutralizing a spill of sulfuric acid? What causes spontaneous combustion? None of these questions can be properly answered without using chemistry. Yet what the average firefighter needs is not a *great* knowledge of chemistry, but rather an understanding of a relatively few simple principles. These are the principles of chemistry we shall study.

1-3 CHARACTERISTICS OF HAZARDOUS MATERIALS

From the perspective of fire science, a hazardous material is often regarded as any substance or mixture that, if improperly handled, may be damaging to our health and well-being or to the environment. Based on this viewpoint, there are seven basic classes of hazardous materials. While we shall note the features of these classes here, in Chapter 5 we will also review the manner by which the U.S. Department of Transportation classifies hazardous materials. The latter system is much more inclusive than the traditional approach sometimes followed in the fire service. Nevertheless, the traditional approach is useful as an introduction to the characteristics that are likely to cause a substance to be regarded as a hazardous material.

F, S

The following are the seven classes of hazardous materials conventionally noted in fire science:

Fla

1. *Flammable materials.* These are any solid, liquid, vapor, or gaseous materials that ignite easily and burn rapidly when exposed to an ignition source. Examples of flammable materials within this broad definition include certain solvents like benzene and ethanol, dusts like flour and certain finely dispersed powders like aluminum, and gases like hydrogen and methane.

Spon

Ex

Oxi

2. *Spontaneously ignitable materials.* These are solid or liquid materials that ignite spontaneously without an ignition source, usually but not necessarily due to the dangerous buildup of heat during storage caused by oxidation or microbiological action. An example of a substance that ignites spontaneously without an ignition source is white phosphorus; examples of substances that spontaneously ignite due to the buildup of heat are fishmeal and grass.

3. *Explosives.* These chemical substances detonate, usually as the result of shock, heat, or some other initiating mechanism. Examples are dynamite and trinitrotoluene (TNT).

Oxi

4. *Oxidizers.* These are substances that evolve or generate oxygen, either at ambient conditions or when exposed to heat. Examples are ammonium nitrate and benzoyl peroxide.

Corro

5. *Corrosive materials.* These are solids or liquid materials, like battery acid, that burn or otherwise damage skin tissue at the site of contact.

Toy

6. *Toxic materials.* Broadly, these are poisons that in small doses either kill or cause adverse health effects. Examples of toxic materials of primary concern to firefighters are carbon monoxide and hydrogen cyanide.

Rad

7. *Radioactive materials.* These materials are characterized by transformations occurring in their atomic nuclei. Uranium hexafluoride is an example of a radioactive material.

Not only are these materials individually hazardous, but they may pose an even more severe hazard when mixed. For instance, while a flammable liquid is hazardous, it poses an even greater threat to health and the environment if it has been mixed with an explosive, toxic, or radioactive material.

In addition, chemical substances may interact to produce new substances that are likely to be hazardous materials. The following examples, illustrated in Fig. 1-2, typify the manner by which a chemical mixture of two or more substances may result in production of a hazardous material.

Substances may interact to form a toxic material.   Suppose we mix the acid found in some commercial rust removers or toilet bowl cleaners with laundry bleach or swimming pool disinfectant. A chemical reaction occurs spontaneously, resulting in production of chlorine gas. Chlorine is highly toxic by inhalation; the human body can tolerate no more than one part per million in air.

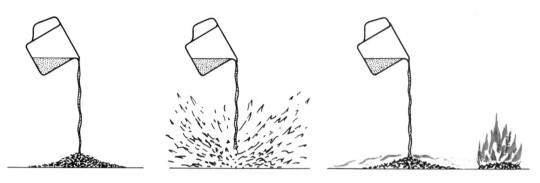

Figure 1-2 Mixing two substances may be dangerous in several ways: A toxic gas may evolve, the mixture may explode or burst into flame, or the heat of chemical reaction may cause other nearby materials to self-ignite.

Substances may interact to form a flammable or explosive material. Important examples of such substances to firefighters are those that are water reactive. For instance, metallic sodium reacts with water to form flammable hydrogen gas. Sufficient heat is generated during this chemical reaction to cause the hydrogen to self-ignite.

Substances may interact with the simultaneous evolution of heat. Oxidizers are examples of hazardous materials that may readily react with flammable materials, causing their self-ignition. For instance, when concentrated nitric acid, an oxidizer, is cautiously poured on finely dispersed sawdust, the resulting chemical reaction evolves enough heat to cause the spontaneous ignition of the sawdust.

1-4 CLASSES OF FIRE

Fire is often associated with hazardous materials incidents, and is divided into four distinct classes: class A, class B, class C and class D.

Class A fires. Class A fires result from the combustion of ordinary cellulosic materials, such as wood and paper, and similar natural and synthetic materials, like rubber and plastics. Figure 1-3 illustrates some common combustible materials that often make up the fuel of class A fires. Such fires usually leave embers or ashes; water is generally an effective means of extinguishing them.

Class B fires. Class B fires result from the combustion of flammable gases and flammable and combustible liquids, several of which are illustrated in Fig. 1-4. Carbon dioxide, dry chemical extinguishers, or foam are suitable extinguishers of class B fires, but not water.

Class C fires. Class C fires result from the combustion of materials occur-

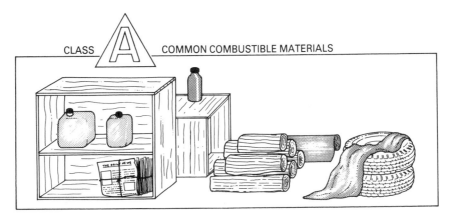

Figure 1-3   Class A fires result from the ignition of materials made from cellulose (for example, wood and cotton), plastics, and rubber.

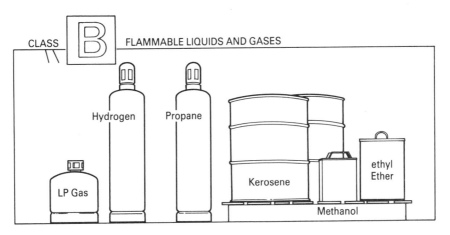

Figure 1-4   Class B fires result from the ignition of flammable gases and flammable and combustible liquids.

ring in or originating from energized electrical circuits, such as those noted in Fig. 1-5. The use of carbon dioxide or dry chemical extinguishers is generally recommended for extinguishing ordinary class C fires.

Class D fires.   Class D fires result from the combustion of certain metals that possess relatively unique chemical reactivities. Examples of such metals are noted in Fig. 1-6; they include titanium, magnesium, zirconium, aluminum, and sodium. Class D fires are often extinguished with special fire extinguishers, like

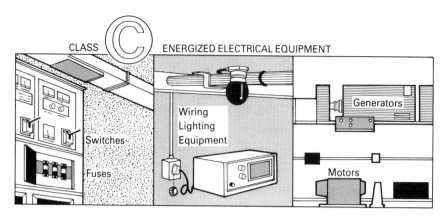

Figure 1-5   Class C fires originate in energized electrical equipment.

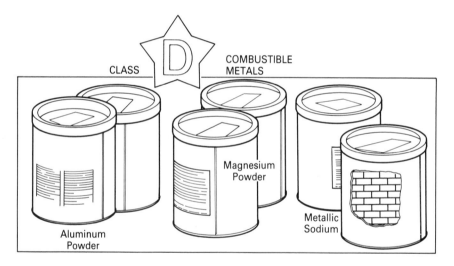

Figure 1-6   Class D fires result from the ignition of certain reactive metals.

graphite or sodium chloride (ordinary table salt); water should never be used as an extinguisher of a class D fire.

1-5 FEDERAL STATUTES AFFECTING HAZARDOUS MATERIALS

Over 20 major federal statutes affect hazardous materials control in one manner or another, from production, marketing, transportation, use, and exposure to their ultimate disposal. Table 1-2 presents a brief overview of these statutes, the federal agency responsible for their administration, and the type of hazardous materials control that each provides. The statutes define hazardous materials, hazardous substances, toxic pollutants, and other similar terms. We shall review the purpose of these statutes to acquire an initial perception of what is meant by describing a material as hazardous.

Contents and Labeling of Household Products

Five of the statutes in Table 1-2 affect the identification of the contents of products that are sold to consumers for home use. The *Federal Hazardous Substances Act* establishes labeling requirements for certain substances likely to be found in household products; these substances are called hazardous substances. The Consumer Product Safety Commission is permitted to issue an outright ban of a hazardous substance in household products when the substance "possesses such a degree or nature of hazard that adequate cautionary labeling cannot be written, and the public health and safety can be served only by keeping such articles out of intrastate commerce." Under this statute, a hazardous substance is specifically one that is highly toxic, toxic, corrosive, irritating, a strong sensitizer, extremely flammable, flammable, combustible, or capable of generating pressure, "if such substance . . . may cause substantial personal injury or substantial illness during . . . any customary or reasonably foreseeable handling or use." A degree of hazard is implied by terms like "extremely flammable" as opposed to "flammable," but these distinctions will not concern us here.

Products containing hazardous substances must bear a label of a specific size that provides appropriate initial advisory information, such as DANGER, WARNING, CAUTION, or POISON. The label must also include a description of the principal hazards involved in the use of the product, such as "Flammable" or "Vapor Harmful"; it must also include a statement of measures that the user should take to avoid the hazard, such as "Use Only in a Well-Ventilated Area." It must list the common names of the hazardous ingredients; provide first-aid instructions; identify the name and location of the manufacturer, distributor, or repacker; and state "Keep Out of Reach of Children" or some other appropriate equivalent statement. Figure 1-7 illustrates an example of proper labeling on a container of aerosol cleaning fluid intended to be used on home kitchen appliances.

The Consumer Product Safety Commission has banned the following from

DANGER: Keep from heat or flame; Do not puncture, incinerate or store above 120°F. KEEP OUT OF REACH OF CHILDREN. CONTAINS PETROLEUM DISTILLATES. HARMFUL OR FATAL IF SWALLOWED. If swallowed, do not induce vomiting. Call a physician immediately. Spray only in well-ventilated area. Avoid prolonged breathing of vapors. In case of eye contact, flush with copious amounts of water.

Figure 1-7 The Federal Hazardous Substances Act requires that products be properly labeled to warn users of their potential hazards. This label identifies the potential hazards of an aerosol cleaning fluid intended for use in the home.

TABLE 1-2 FEDERAL STATUTES CONCERNED WITH HAZARDOUS MATERIALS

Statute	Year enacted	Responsible agency[a]	Sources covered
Toxic Substances Control Act	1976	USEPA	Requires premarket evaluation of all new chemical substances other than food additives, drugs, pesticides, alcohol, and tobacco; allows USEPA to regulate existing chemical hazards not covered by other federal laws dealing with toxic substances
Clean Air Act	1970, amended 1977	USEPA	Hazardous air pollutants
Federal Water Pollution Control Act	1972, amended 1977	USEPA	Toxic pollutants; hazardous substances
Safe Drinking Water Act	1974, amended 1977	USEPA	Priority pollutants
Federal Insecticide, Fungicide, and Rodenticide Act	1948, amended 1972, 1975, 1978, 1980	USEPA	Pesticides
Act of July 22, 1954, codified as 346(a) of the Food, Drug and Cosmetic Act	1954, amended 1972	USEPA	Tolerances for pesticide residues in food
Resource Conservation and Recovery Act	1976, amended 1980	USEPA	Hazardous wastes
Marine Protection, Research, and Sanctuaries Act	1972	USEPA	Ocean dumping
Comprehensive Environmental Response, Compensation, and Liability Act ("Superfund")	1980, amended 1986	USEPA	Cleanup of sites where hazardous substances are improperly disposed; restoration and replacement of natural resources when damaged or lost
Superfund Amendments and Reauthorization Act	1986	USEPA	Same as above, but adds right-to-know provisions intended to bring to the general public more information about and a better understanding of local chemical related operations
Federal Food, Drug and Cosmetic Act	1938	USFDA	Basic coverage of food, drugs, and cosmetics

Act	Year	Agency	Description
Food additives amendment	1958	USFDA	Food additives
Color additive amendments	1960	USFDA	Color additives
New drug amendments	1962	USFDA	Drugs
New animal drug amendments	1968	USFDA	Animal drugs and feed additives
Medical device amendments	1976	USFDA	Medical devices
Wholesome Meat Act	1967	USDA	Food, feed and color additives; pesticide residues in meat and poultry
Wholesome Poultry Products Act	1968	USDA	
Occupational Safety and Health Act	1970	OSHA	Workplace hazardous substances
Federal Hazardous Substances Act	1966, amended 1981	CPSC	Hazardous substances in household (consumer) products
Consumer Product Safety Act	1972, amended 1981	CPSC	Dangerous consumer products
Poison Prevention Packaging Act	1970, amended 1981	CPSC	Packaging of products that could lead to the poisoning of children
Hazardous Materials Transportation Act	1970	USDOT[b]	Transportation of hazardous materials (generally)
Federal Railroad Safety Act	1970	USDOT[c]	Transportation of hazardous materials by railway
Ports and Waterways Safety Act	1972	USDOT[d]	Transportation of hazardous materials by water
Dangerous Cargo Act	1952	USDOT	Shipment of certain groups of hazardous materials by water

[a] USEPA, U.S. Environmental Protection Agency
FDA, U.S. Food and Drug Administration
USDA, U.S. Department of Agriculture
OSHA, Occupational Safety and Health Administration
CPSC, Consumer Products Safety Commission
USDOT, U.S. Department of Transportation

[b] Enforced by the Materials Transportation Bureau
[c] Enforced by the Federal Railroad Administration
[d] Enforced by the U.S. Coast Guard

household products: carbon tetrachloride; extremely flammable water repellants; certain fireworks; liquid drain cleaners containing 10% or more potassium hydroxide; products containing soluble cyanides; paint with a lead content above 0.06% by dry weight (except artist's paint); general use garments, tape joint compounds and artificial fireplace embers containing asbestos; and aerosols containing vinyl chloride.

The *Federal Food, Drug, and Cosmetic Act* directs the Food and Drug Administration to ensure that foods are safe to eat, that drugs are safe and effective, and that packaging and labeling of foods and drugs are truthful and informative. This statute affects toxicants insofar as the Food and Drug Administration sets tolerance limits for poisonous substances in food, assesses the safety of food additives, and prohibits the addition to food of any substance known to induce cancer in humans or animals; the latter prohibition is commonly called the *Delaney Clause*. The administration monitors and assesses the health risk of all types of chemical contaminants.

The *Federal Insecticide, Fungicide, and Rodenticide Act* regulates the manufacture, use, and disposal of pesticides for agricultural, forestry, household, and other activities. It requires registration of all pesticides, classifies pesticides for general as opposed to restricted use, requires "informative and accurate labeling," and provides for the suspension or cancellation of registration for "imminent hazard" or "unreasonable adverse effects on the environment." Under this statute, a *pesticide* is "any substance or mixture of substances intended for preventing, destroying, repelling or mitigating" insects, rodents, fungi, weeds, and other forms of plant and animal life that are "injurious to health or the environment."

Pesticide containers must be labeled with the following information, prominently placed with such conspicuousness and in such terms as to render it likely to be read and understood by the ordinary individual: the name, brand, or trademark; directions for use that are "necessary for effecting the purpose for which the product is intended" and adequate if complied with "to protect health and the environment"; an ingredient statement, giving the name and percentage of each active ingredient and the percentage of all inert ingredients; a statement of the use classification under which it is registered; the name and address of the producer; and the appropriate warning or caution statements.

Finally, the *Toxic Substances Control Act* authorizes the U.S. Environmental Protection Agency (USEPA) to obtain production and test data from industry on *toxic substances*. The latter are defined as commercial chemical substances, except pesticides and drugs, whose manufacture, processing, distribution, use, or disposal may present an unreasonable risk of injury or damage to the environment. This comprehensive program includes the testing, registration, and, if necessary, limitation or prohibitions on manufacture and use of new chemical substances; and inventory, testing, and reevaluation of existing chemical substances for use-registration. Four substances have been banned or severely restricted in use under this statute: polychlorinated biphenyls (PCBs); chlorofluorocarbons in aerosols; 2,3,7,8-tetrachlorodibenzo-p-dioxin (TCDD); and asbestos in numerous materials.

Treatment, Storage, and Disposal of Pollutants and Wastes

The USEPA has been charged under several federal statutes with the responsibility of assuring protection of human health and the environment from the treatment, storage, and disposal of certain substances and mixtures, either as raw materials or industrial chemical wastes, that could potentially inflict harm. Specifically, these environmental statutes direct USEPA to regulate the discharge of such materials to air, water, and soil within the boundaries of the United States and its territories.

The modern approach to air pollution regulation was initiated with passage of the *Clean Air Act* in 1963, followed in 1970 by the Clean Air Act Amendments. Enacted in these amendments were provisions that directed the federal regulation of *hazardous air pollutants*. Accordingly, USEPA established standards for each pollutant. Hazardous air pollutants affect our health and well-being: When inhaled or absorbed in the body, they may cause adverse health effects, like respiratory ailments. USEPA regulates the air emission of these hazardous air pollutants by a series of technical criteria known as *National Ambient Air Quality Standards*. Each standard states the maximum allowable concentration of the pollutant in ambient air. Each standard is composed of two parts: a primary standard, designed to protect human health, and a secondary standard, designed to protect vegetation and other aspects of the environment, as well as personal comfort and well-being.

The federal water pollution control effort began seriously with the passage of the *Federal Water Pollution Control Act* in 1948, but a comprehensive regulatory scheme for controlling water pollution discharges was first set by USEPA with passage of the Federal Water Pollution Control Act Amendments of 1972. The primary regulatory approach of these amendments now consists of USEPA-promulgated, industry by industry, technology-based effluent limitations, which apply to the discharge of *toxic pollutants* in all waters within the United States. Approximately 150 substances are currently listed as toxic pollutants. In addition, there are conventional water pollutants designated as biochemical oxygen demand,* total suspended soils, pH, fecal coliform, and oil and grease. A *toxic pollutant* is defined in this statute as a substance that "after discharge and upon exposure, ingestion, inhalation or assimilation into any organism . . . , will, on the basis of information . . . , cause death, disease, behavioral abnormalities, cancer, genetic mutations, physiological malfunctions (including malfunctions in reproduction) or physical deformations, in such organisms or their offspring."

Under the Federal Water Pollution Control Act, USEPA also regulates the discharge of *hazardous substances* to navigable waters of the United States. This includes not only direct discharges to water bodies, but also indirect discharges, such as into storm drains or sewers, that have the potential to enter navigable waters. The

---

* Biochemical oxygen demand (BOD) is a standardized means of estimating the degree of contamination of water supplies, particulary those potentially contaminated by sewage and industrial wastes.

U.S. Coast Guard has jurisdiction over discharges in marine waters, while USEPA and the individual state counterparts have jurisdiction in fresh waters; practically speaking, all three agencies receive notification of spills and other forms of discharge and respond accordingly. Approximately 300 hazardous substances are designated under the Federal Water Pollution Control Act.

USEPA has also been directed to assure the existence of supplies of safe drinking water for human consumption; this mandate comes from the *Safe Drinking Water Act*. The regulatory scheme is encompassed in *National Primary Drinking Water Regulations*, which impose specific monitoring, analysis, contaminant level, and report requirements on coliform bacteria, turbidity, certain organic and inorganic chemicals, and radioactivity. USEPA has also prepared criteria documents on 129 substances, called *priority pollutants*, that specify the maximum concentrations of these substances in ambient water regarded as acceptable for the protection of aquatic organisms and human health.

In 1976, Congress passed another environmental statute, called the *Resource Conservation and Recovery Act* (RCRA), and in 1980, USEPA promulgated regulations applicable to facilities that generate, transport, treat, store, and/or dispose of *hazardous waste*. The statute defines *hazardous* waste as a material

> "which because of its quantity, concentration, or physical, chemical, or infectious characteristics may:
> **(a)** cause, or significantly contribute to an increase in mortality or an increase in serious irreversible, or incapacitating, reversible, illness; or
> **(b)** pose a substantial present or potential hazard to human health or the environment when improperly treated, stored or disposed of, or otherwise managed."

The USEPA-promulgated regulations list hazardous waste in four categories. They also establish four characteristics that cause a material to be regarded as a hazardous waste under RCRA: ignitability, corrosivity, reactivity, and EP toxicity (extraction-procedure toxicity). These characteristics have specific regulatory definitions, which require more chemical background; hence, they shall be discussed in more detail at appropriate points in later chapters.

In 1980, Congress passed another environmental law that attempts to correct past mistakes resulting from the improper disposal of *hazardous substances*. This law is the *Comprehensive Environmental Response, Compensation, and Liability Act* (CERCLA), known more commonly as *Superfund*. CERCLA addresses the prevention and cleanup of releases of hazardous substances and the restoration and replacement of natural resources that are damaged or lost because of such releases. As defined in CERCLA, hazardous substances refer broadly to materials defined largely in several other federal environmental statutes, as follows:

1. Any substance designated as a *hazardous substance* under the Federal Water Pollution Control Act.

2. Any substance specifically designated as a *hazardous substance* under CERCLA.

3. Any *hazardous waste* listed under RCRA or having the characteristics defined in RCRA.

4. Any substance designated as a *toxic pollutant* under the Federal Water Pollution Control Act.

5. Any substance designated as a *hazardous air pollutant* under the Clean Air Act.

6. Any *imminently hazardous substance* designated under the Toxic Substances Control Act.

No specific designation of a hazardous substance under CERCLA has been made, nor has any substance been designated as an imminently hazardous substance under the Toxic Substances Control Act.

From an environmental perspective, the definition of a hazardous substance under CERCLA conveys the broadest interpretation of "hazardousness," as it encompasses the meaning established by four other environmental statutes.

Hazardous Substances in the Workplace

The *Occupational Safety and Health Act* is a federal statute that attempts to protect employees in the workplace from occupational illness and injuries caused by exposure to hazardous substances. This act empowers the Occupational Safety and Health Administration (OSHA) to regulate certain aspects of the workplace to reduce the incidence of chemically induced occupational illnesses and injuries. For instance, OSHA establishes permissible exposure limits to certain chemical substances when encountered in the workplace. Sometimes, the National Institute for Occupational Safety and Health (NIOSH) recommends a revision of these limits based on their own research efforts. It is the responsibility of OSHA and NIOSH to ensure, insofar as possible, that every working individual is provided with a safe and healthful working environment.

In 1983, OSHA promulgated a standard that sets minimum requirements to which employers must adhere for communicating hazard information to workers; it is often referred to as the *right-to-know law*. In brief, its intent is to assure workers of their right to know about hazards associated with substances to which they are being exposed in their place of employment.

This federal standard requires chemical manufacturers and importers to assess the hazards of the chemicals that they produce or import and to transmit the hazard information to users of the chemicals. This information must be sufficiently comprehensive to allow user-employers to devise appropriate employee protection programs and to give employees the information that they personally need to protect themselves against potential incidents associated with these hazards. The regulation applies to *all* chemical substances known to be present in the workplace to which employees may be exposed under normal working conditions and to those substances to which they could be exposed in a foreseeable emergency.

Under the OSHA regulation, hazard information is conveyed to workers in the following ways:

1. *Labels and other forms of warning.* Every manufacturer and importer of a chemical substance must ensure that each container of a hazardous substance is labeled, tagged, or marked with the identity of the product, appropriate hazard warnings, and the name and address of the manufacturer, importer, or other responsible party. Labels may use symbols, pictures, and/or words to present their message. A warning label used by J. T. Baker Chemical Company on containers of hydrochloric acid is illustrated in Fig. 1-8. Note that the pictorial information on this label includes the use of a number system that conveys information about the product regarding health, flammability, reactivity, and contact. This system of conveying information closely parallels that employed by the National Fire Protection Association (NFPA), which we shall review in Sec. 1-6.

2. *Material Safety Data Sheets (MSDS).* An MSDS is a technical bulletin containing detailed information about a hazardous substance. OSHA requires that manufacturers of these substances also prepare the MSDS. An example of an MSDS for hydrochloric acid is illustrated in Fig. 1-9. Note that the information on the label (Fig. 1–8) is keyed to the MSDS, which contains more extensive information about hydrochloric acid. The MSDS must accompany each sample or order of hydrochloric acid shipped to a location for the first time.

The OSHA regulation stipulates that the MSDS must be designed as the most comprehensive source of written information for the employee. The following minimum information must be provided: the identity of the product as used on the container label; the chemical and common name of all ingredients having known health hazards present in concentrations greater than 1% and, for cancer-causing substances, if present at 0.1% or more; the physical and chemical characteristics of the hazardous components; the physical and health hazards, including signs and symptoms of exposure and prior and/or existing medical conditions that may warn against exposure; the primary routes of entry; any known exposure limits; whether the hazardous substances are listed as carcinogens (cancer-causing substances) or potential carcinogens; the precautions for safe handling and use and procedures for spill or leak cleanup; control measures; the emergency first-aid procedures; the date of preparation; and the name, address, and telephone number of the company or the responsible employee distributing the MSDS.

Training is the critical link in the hazard-communication program, serving to explain and to reinforce the information presented on warning labels and MSDSs. Specifically, the OSHA standard requires that employers train workers in how to recognize hazardous substances that may be released in the workplace (for example, by odor or appearance) and how to protect themselves from these hazards.

Figure 1-8 The label used on containers of hydrochloric acid distributed by J. T. Baker, Inc. Note that the label provides the following information: (a) identification of the name of the product; (b) appropriate potential hazard warnings; and (c) the name and address of the responsible party. Users of this substance must ensure that this information is similarly conveyed to workers in the workplace. (Courtesy of J. T. Baker, Inc., Phillipsburg, New Jersey)

J. T. Baker Chemical Co.

222 Red School Lane   Phillipsburg, N.J. 08865
24-Hour Emergency Telephone -- (201) 859-2151
Chemtrec # (800) 424-9300
National Response Center # (800) 424-8802

MATERIAL
SAFETY DATA
SHEET

```
H3880 -02                   Hydrochloric Acid                    Page: 1
Effective: 08/07/86                                       Issued: 10/19/87
==========================================================================
                    SECTION I - PRODUCT IDENTIFICATION
==========================================================================
Product Name:    Hydrochloric Acid
Formula:         HCl
Formula Wt:         36.46
CAS No.:         7647-01-0
NIOSH/RTECS No.: MW4025000
Common Synonyms: Muriatic Acid;  Chlorohydric Acid;  Hydrochloride
Product Codes:   9543,9539,9535,9534,9544,9529,9542,4800,9549,9530,9548,9540
                 5537,9547,9546,9537,5367
==========================================================================
                    PRECAUTIONARY LABELLING
==========================================================================
```

BAKER SAF-T-DATA™ System

HEALTH	FLAMMABILITY	REACTIVITY	CONTACT
3	0	2	3
SEVERE	NONE	MODERATE	SEVERE

Laboratory Protective Equipment

GOGGLES & SHIELD	LAB COAT & APRON	VENT HOOD	PROPER GLOVES

Precautionary Label Statements

POISON! DANGER!
CAUSES SEVERE BURNS
MAY BE FATAL IF SWALLOWED OR INHALED
Do not get in eyes, on skin, on clothing.
Do not breathe vapor. Causes damage to Respiratory system (lungs),
eyes and skin. Keep in tightly closed container. Loosen closure cautiously.
Use with adequate ventilation. Wash thoroughly after handling. In case
of spill neutralize with soda ash or lime and place in dry container.

```
==========================================================================
                    SECTION II - HAZARDOUS COMPONENTS
==========================================================================
                Component                              %     CAS No.

Hydrochloric Acid  (23° Baume)                       35-40  7647-01-0
==========================================================================
                    SECTION III - PHYSICAL DATA
==========================================================================
Boiling Point:      110°C (  230°F)          Vapor Pressure(mmHg):  N/A
```

Continued on Page: 2

Figure 1-9 The Material Safety Data Sheet (MSDS) for hydrochloric acid distributed by
J. T. Baker, Inc. Other manufacturers, importers, or distributors of hydrochloric acid may use
a different format, but the information noted on this MSDS must be minimally provided to
individuals who work with hydrochloric acid. (Courtesy of J. T. Baker, Inc., Phillipsburg,
New Jersey)

J. T. Baker Chemical Co.
222 Red School Lane   Phillipsburg, N.J. 08865
24-Hour Emergency Telephone -- (201) 859-2151
Chemtrec # (800) 424-9300
National Response Center # (800) 424-8802

MATERIAL SAFETY DATA SHEET

```
H3880 -02                  Hydrochloric Acid                  Page: 2
Effective: 08/07/86                               Issued: 10/19/87
==============================================================================
                    SECTION III - PHYSICAL DATA (Continued)
==============================================================================
```

Melting Point: -25°C (-13°F)   Vapor Density(air=1): 1.3

Specific Gravity: 1.19   Evaporation Rate: N/A
   ($H_2O=1$)   (Butyl Acetate=1)

Solubility(H_2O):   Complete (in all proportions) % Volatiles by Volume: 100

Appearance & Odor: Clear, colorless or slightly yellow, pungent, fuming liquid.
```
==============================================================================
                SECTION IV - FIRE AND EXPLOSION HAZARD DATA
==============================================================================
```
Flash Point: N/A   NFPA 704M Rating: 3-0-0

Flammable Limits: Upper - N/A %   Lower - N/A %

Fire Extinguishing Media
    Use extinguishing media appropriate for surrounding fire.

Special Fire-Fighting Procedures
    Firefighters should wear proper protective equipment and self-contained
    breathing apparatus with full facepiece operated in positive pressure mode.
    Move containers from fire area if it can be done without risk. Use water
    to keep fire-exposed containers cool.
    Do not get water inside containers.

Unusual Fire & Explosion Hazards
    May emit hydrogen gas upon contact with metal.

Toxic Gases Produced
    hydrogen chloride, hydrogen gas
```
==============================================================================
                    SECTION V - HEALTH HAZARD DATA
==============================================================================
```
PEL and TLV listed denote ceiling limit.

Threshold Limit Value (TLV/TWA):   7 mg/m^3 (5   ppm)

Permissible Exposure Limit (PEL):   7 mg/m^3 (5   ppm)

Toxicity: LD_{50} (oral-rabbit)(mg/kg)   - 900
             LD_{50} (ipr-mouse)(mg/kg)   - 40
             LC_{50} (inhl-rat-1H) (ppm)   - 3124
```

Continued on Page:  3

**Figure 1-9 (continued)**
```

J. T. Baker Chemical Co.

222 Red School Lane   Phillipsburg, N.J. 08865
24-Hour Emergency Telephone -- (201) 859-2151

Chemtrec # (800) 424-9300
National Response Center # (800) 424-8802

**MATERIAL
SAFETY DATA
SHEET**

H3880 -02   Hydrochloric Acid   Page: 3
Effective: 08/07/86   Issued: 10/19/87
===
                    SECTION V - HEALTH HAZARD DATA (Continued)
===

Carcinogenicity: NTP: No   IARC: No   Z List: No   OSHA req: No

Effects of Overexposure
     Inhalation of vapors may cause pulmonary edema, circulatory system
     collapse, damage to upper respiratory system, collapse.
     Inhalation of vapors may cause coughing and difficult breathing.
     Liquid may cause severe burns to skin and eyes.
     Ingestion is harmful and may be fatal.
     Ingestion may cause severe burning of mouth and stomach.
     Ingestion may cause nausea and vomiting.

Target Organs
     respiratory system, eyes, skin

Medical Conditions Generally Aggravated By Exposure
     None Identified

Routes Of Entry
     ingestion, inhalation, skin contact, eye contact

Emergency and First Aid Procedures
     CALL A PHYSICIAN.
     If swallowed, do NOT induce vomiting; if conscious, give water, milk, or
     milk of magnesia.
     If inhaled, remove to fresh air. If not breathing, give artificial
     respiration. If breathing is difficult, give oxygen.
     In case of contact, immediately flush eyes or skin with plenty of water for
     at least 15 minutes while removing contaminated clothing and shoes.
     Wash clothing before re-use.
===
                         SECTION VI - REACTIVITY DATA
===
Stability: Stable   Hazardous Polymerization: Will not occur

Conditions to Avoid: heat, moisture

Incompatibles: most common metals, water, amines, metal oxides,
                         acetic anhydride, propiolactone, vinyl acetate,
                         mercuric sulfate, calcium phosphide, formaldehyde,
                         alkalies, carbonates, strong bases,
                         sulfuric acid, chlorosulfonic acid

Decomposition Products: hydrogen chloride, hydrogen, chlorine
===
                    SECTION VII - SPILL AND DISPOSAL PROCEDURES
===
Steps to be taken in the event of a spill or discharge
     Wear self-contained breathing apparatus and full protective clothing. Stop

Continued on Page: 4

Figure 1-9 (continued)

J. T. Baker Chemical Co.

222 Red School Lane   Phillipsburg, N.J. 08865
24-Hour Emergency Telephone -- (201) 859-2151
Chemtrec # (800) 424-9300
National Response Center # (800) 424-8802

MATERIAL SAFETY DATA SHEET

```
H3880 -02                    Hydrochloric Acid                    Page: 4
Effective: 08/07/86                                        Issued: 10/19/87
================================================================================
              SECTION VII - SPILL AND DISPOSAL PROCEDURES (Continued)
================================================================================
```

leak if you can do so without risk. Ventilate area. Neutralize spill with
soda ash or lime. With clean shovel, carefully place material into clean,
dry container and cover; remove from area. Flush spill area with water.

J. T. Baker Neutrasorb^R or Neutrasol^R "Low Na+" acid neutralizers
are recommended for spills of this product.

Disposal Procedure
    Dispose in accordance with all applicable federal, state, and local
    environmental regulations.

EPA Hazardous Waste Number: D002 (Corrosive Waste)
```
================================================================================
              SECTION VIII - INDUSTRIAL PROTECTIVE EQUIPMENT
================================================================================
```

Ventilation: Use general or local exhaust ventilation to meet
                          TLV requirements.

Respiratory Protection: Respiratory protection required if airborne
                          concentration exceeds TLV. At concentrations up
                          to 100 ppm, a chemical cartridge respirator with
                          acid cartridge is recommended. Above this level,
                          a self-contained breathing apparatus is advised.

Eye/Skin Protection: Safety goggles and face shield, uniform,
                          protective suit, acid-resistant gloves are
                          recommended.
```
================================================================================
              SECTION IX - STORAGE AND HANDLING PRECAUTIONS
================================================================================
```

SAF-T-DATA^{TM} Storage Color Code: White (corrosive)

Special Precautions
    Keep container tightly closed. Store in corrosion-proof area.
    Isolate from incompatible materials.
    Do not store near oxidizing materials.
```
================================================================================
              SECTION X - TRANSPORTATION DATA AND ADDITIONAL INFORMATION
================================================================================
```
DOMESTIC (D.O.T.)

Proper Shipping Name   Hydrochloric acid
Hazard Class   Corrosive material (liquid)
UN/NA   UN1789
Labels   CORROSIVE
Reportable Quantity   5000 LBS.

Continued on Page: 5

Figure 1-9 (continued)

J. T. Baker Chemical Co.

222 Red School Lane   Phillipsburg, N.J. 08865
24-Hour Emergency Telephone -- (201) 859-2151
Chemtrec # (800) 424-9300
National Response Center # (800) 424-8802

**MATERIAL
SAFETY DATA
SHEET**

```
H3880 -02                    Hydrochloric Acid                      Page: 5
Effective: 08/07/86                                        Issued: 10/19/87
==============================================================================
     SECTION X - TRANSPORTATION DATA AND ADDITIONAL INFORMATION (Continued)
==============================================================================

INTERNATIONAL (I.M.O.)

Proper Shipping Name    Hydrochloric acid, solution
Hazard Class            8
UN/NA                   UN1789
Labels                  CORROSIVE
==============================================================================
N/A = Not Applicable or Not Available
---
The information published in this Material Safety Data Sheet has been compiled
from our experience and data presented in various technical publications. It is
the user's responsibility to determine the suitability of this information for
the adoption of necessary safety precautions. We reserve the right to revise
Material Safety Data Sheets periodically as new information becomes available.
J. T. Baker makes no warranty or representation about the accuracy or complete-
ness nor fitness for purpose of the information contained herein.
COPYRIGHT 1987 J.T.BAKER INC.




                          --  LAST PAGE  --
```

Figure 1-9 (continued)

While the OSHA standard is not intended to apply to the firefighting profession, labels and MSDSs can also provide information to firefighters and other emergency-response forces. They are likely to be useful to fire inspectors attempting to ascertain the identity of chemical substances at a given facility during an on-site inspection, as well as to identify their associated hazards. Information of this type is often useful for the protection of lives and property if the facility becomes involved in a fire or other emergency mishap.

Hazardous Materials in Transit

The *U.S. Department of Transportation* (USDOT) regulates carriers who offer or accept hazardous materials for intrastate, interstate, or international transportation through certain marking, labeling, placarding, packaging, and other requirements. The information conveyed by warning labels, placards, and shipping papers is so important to emergency-response units that Chapter 5 is devoted to reviewing some aspects of these requirements in some detail.

Hazardous Substances within Communities

When CERCLA was reauthorized in 1986, a free-standing act was also enacted: the *Emergency Planning and Community Right-To-Know Act*. To better prepare communities with information concerning the chemical substances within their boundaries, Congress included two right-to-know reporting requirements. The first stipulates that facilities handling chemical substances must submit copies of appropriate MSDSs to the local emergency planning committee, the state emergency response commission, and the local fire department. The second reporting requirement stipulates that facilities must submit an emergency and hazardous chemical inventory form to the same groups, identifying their hazardous chemical substances and quantities. This combined information ensures citizens that they have sufficient information to prepare beforehand for an emergency involving chemical substances within their communities.

1-6 NFPA 704 SYSTEM OF IDENTIFYING POTENTIAL HAZARDS

At the scene of an emergency, how may the potential hazards associated with a given material be rapidly identified? One way is by recognizing certain markings on hazardous materials storage tanks and at other locations resulting from use of the *704 system*. This system was developed by the National Fire Protection Association (NFPA). It is now frequently used by many manufacturing and process industries, hospitals, and virtually any other facility that stores hazardous materials. It is also used throughout a noteworthy publication of the NFPA titled *Fire Protection Guide on Hazardous Materials*.* This handbook is extremely useful for rapidly identifying

*Available from the National Fire Protection Association, 470 Atlantic Avenue, Boston, Massachusetts 02210.

TABLE 1-3 THE NFPA SIGNAL CODES FOR HEALTH, FLAMMABILITY AND REACTIVITY

Identification of health		Identification of flammability		Identification of reactivity (stability) color	
Hazard color code: BLUE		Hazard color code: RED		Hazard color code: YELLOW	
Signal	Type of Possible Injury	Signal	Susceptibility of Materials to Burning	Signal	Susceptibility to Release of Energy
4	Materials that on very short exposure could cause death or major residual injury even though prompt medical treatment was given.	4	Materials that will rapidly or completely vaporize at atmospheric pressure and normal ambient temperature, or that are readily dispersed in air, and will burn readily.	4	Materials that in themselves are readily capable of detonation or of explosive decomposition or reaction at normal temperatures and pressures.
3	Materials that on short exposure could cause serious temporary or residual injury even though prompt medical treatment was given.	3	Liquids and solids that can be ignited under almost all ambient temperature conditions.	3	Materials that in themselves are capable of detonation or explosive reaction but require a strong initiating source or that must be heated under confinement before initiation or that react explosively with water.
2	Materials that on intense or continued exposure could cause temporary incapacitation or possible residual injury unless prompt medical treatment was given.	2	Materials that must be moderately heated or exposed to relatively high ambient tempertures before ignition can occur.	2	Materials that in themselves are normally unstable and readily undergo violent chemical change but do not detonate. Also, materials that may react violently with water or may form potentially explosive mixtures with water.
1	Materials that on exposure would cause irritation but only minor residual injury even if no treatment was given.	1	Materials that must be preheated before ignition can occur.	1	Materials that in themselves are normally stable, but which can become unstable at elevated temperatures and pressures or which may react with water with some release of energy but not violently.

TABLE 1-3 (continued)

	Identification of health		Identification of flammability		Identification of reactivity (stability) color
Hazard color code: BLUE		Hazard color code: RED		Hazard color code: YELLOW	
Signal	Type of Possible Injury	Signal	Susceptibility of Materials to Burning	Signal	Susceptibility to Release of Energy
0	Materials that on exposure under fire conditions would offer no hazard beyond that of ordinary combustible material.	0	Materials that will not burn.	0	Materials that in themselves are normally stable, even under fire exposure conditions, and which are not reactive with water.

certain hazardous features of materials that are likely to be encountered during emergencies.

The 704 system identifies the relative degree of three hazards associated with a given material: health, flammability, and chemical reactivity. This is accomplished through the use of one of five numbers, zero through four, for each property. Appropriate numbers are displayed in each of the top three quadrants of a diamond-shaped diagram, as shown by the example in Fig. 1-10. NFPA recommends color-coding each quadrant: blue for health hazard, red for fire hazard, and yellow for chemical reactivity hazard, beginning with the left-hand quadrant and proceeding clockwise. The significance of the numbers as it relates to each property is provided in Table 1-3. Note that the severity of a given property increases from 0 to 4. The number 0 means there is essentially no hazard, whereas the number 4 denotes the highest degree of hazard.

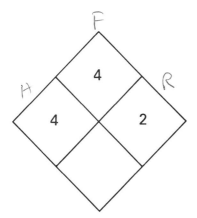

Figure 1-10 The potential hazards of a substance may be identified by use of a color-coded numeral system on this diamond-shaped diagram. The relative degree of severity is expressed by numbers ranging from 0 (no hazard) to 4 (maximum hazard). The substance illustrated by the number 4 (left) indicates that exposure may rapidly cause death or major residual illness; the 4 (top) indicates the material readily vaporizes at ambient conditions and burns if ignited; and the 2 (right) indicates the material is normally unstable, but does not detonate.

Four special symbols are also employed in the 704 system, each of which is displayed in the bottom quadrant, when appropriate, which is colorless: a radiation hazard symbol resembling a propeller, which tells us the material at issue is radioactive; the letter W with a line drawn through its center to caution against the application of water; the letters OXY to indicate the material is an oxidizer; and the letter P, which indicates there is a potential for the material to undergo autopolymerization.

Today, we often find such NFPA ratings as components of information on Material Safety Data Sheets. We most often find 704 diamonds displayed on fixed tanks used to store bulk quantities of hazardous materials. But can industry always be relied on to supply this information without exception to emergency-response forces? Unfortunately, the answer to this question is no, although such information generally is supplied in regions where local ordinances require the use of the 704 system.

As we study individual hazardous materials beginning in Chapter 6, the appropriate 704 diamond will be displayed in the page margin near the first point at which a discussion of each material begins.

1-7 CHEMTREC

CHEMTREC stands for Chemical Transporation Emergency Center, a public service of the Manufacturing Chemists Association with offices in Washington, D.C. It provides immediate advice for those at the scene of emergencies and then promptly contacts the shipper of the hazardous materials involved for more detailed assistance and appropriate follow-up. CHEMTREC operates around the clock, 24 hours a day, 7 days a week, to receive direct-dial, toll-free calls from any point in the continental United States through a wide-area service (WATS) telephone number.* Shippers, often member companies of the Manufacturing Chemists Association, are notified through preestablished phone contacts, providing 24-hour accessibility. As circumstances warrant, the National Transportation Safety Board, appropriate offices of the U.S. Environmental Protection Agency, the U.S. Coast Guard, and others may be notified.

CHEMTREC is not intended, and is not equipped, to function as a general information source, but by design is confined to dealing with chemical transportation

---

*In the 48 contiguous states, telephone 800–424–9300. For calls originating within the District of Columbia, Alaska, or Hawaii, telephone 202–483–7616. In Canada, help may be summoned through any control center of an analogous service called TEAP, the *Transportation Emergency Assistance Plan* of the Canadian Chemical Producers Association. The Canadian regions and the telephone numbers to call for assistance are as follows:

Atlantic Provinces	819–537–1123
and eastern/central Quebec	(Shawinigan, Quebec
Southwestern Quebec	514–373–8330
	(Valleyfield, Quebec)

emergencies. Shipping documents of participating companies include the following: "In the event of any emergency concerning the chemicals in this shipment, call the toll-free number (800) 424–9300, day or night." CHEMTREC's number has also been widely circulated in professional literature distributed to emergency service personnel, carriers, and the chemical industry, and has been further circulated in bulletins of governmental agencies, trade associations, and similar groups.

An emergency reported to CHEMTREC is received by the communicator on duty, who records details in writing and by tape recorder. The communicator attempts to learn what happened, where, and when; the hazardous materials involved; the type and condition of containers; shipper and shipping point; carrier, consignee, and destination; general nature and extent of injuries to people, property, and the environment, if any; prevailing weather; composition of the surrounding area; who the caller is and where the caller is located; and, of utmost importance, how and where telephone contact can be reestablished with the caller or another responsible party at the scene.

With the caller remaining on the line, the communicator draws on the best available information on the hazardous materials reported to be involved, thereby giving specific indication of the hazards and what to do (as well as what not to do) in case of spills, fire, or exposure as the immediate first steps in controlling the emergency. Information on each chemical, furnished by its producers, is within arm's length. Trade names and synonyms of chemical names are cross-referenced for ready identification by whatever name is given.

CHEMTREC's communicators are not usually scientists. They are chosen for their ability to remain calm under emergency conditions. To preclude unfounded personal speculation regarding the specific features of a reported emergency, they are under instructions to abide strictly by the information prepared by technical experts for their use.

Having advised the caller, the communicator proceeds immediately to notify the shipper by telephone. The known particulars of the emergency thus relayed, responsibility for further guidance, including dispatching personnel to the scene or whatever seems warranted, passes to the shipper.

Although proceeding to the second stage of assistance becomes more difficult

---

Eastern Ontario	613–348–3616 (Maitland, Ontario)
Central Ontario	416–356–8310 (Niagara Falls, Ontario)
Southwestern Ontario	519–339–3711 (Sarnia, Ontario)
Northern/Western Ontario	705–682–2881 (Copper Cliff, Ontario)
Manitoba, Saskatchewan, and Alberta	403–477–8339 (Edmonton, Alberta)
British Columbia	604–929–3441 (Vancouver, British Columbia)

when the shipper is unknown, communicators are armed with other resources upon which they can rely. For instance, CHEMTREC can contact the U.S. Nuclear Regulatory Commission in situations involving radioactive materials.

CHEMTREC does not seek to displace specialized programs, but rather to collaborate with them and enhance their effectiveness. One telephone number to CHEMTREC can readily afford this opportunity when the need arises.

REVIEW EXERCISES

Classification of Hazardous Materials

1.1. Categorize each of the following materials into one or more of the hazardous material classes noted in Sec. 1-3:

    (a) Wet hay   **(b)** Octane   **(c)** Insecticide
    (d) Type C fireworks   **(e)** Sulfur dioxide   **(f)** Lye
    (g) Rocket warheads   **(h)** Kerosene   **(i)** Natural gas
    (j) Zinc peroxide   **(k)** *Botulinus* toxin

Federal Statutes

1.2. Identify the federal statute that has responsibility for effecting control in each of the following situations and the federal agency that oversees the statute's implementation:

    (a) Manufacture of an insecticide that kills mites
    (b) Production of a medication to counteract muscular aches
    (c) Sale of a disinfectant that contains phenol for use in the home
    (d) Nonintentional discharge of 5000 lb of carbon tetrachloride into the Mississippi River
    (e) Occupational exposure to asbestos
    (f) Transportation of hydrochloric acid by rail tank car from Los Angeles to Portland
    (g) Treatment of an arsenic-bearing industrial waste

1.3. The ABC Chemical Company develops a furniture polish for use in the home. The product contains a constituent that has been shown in research studies to kill more than 50% of a population of rats who were orally given the substance in a dose of 50 milligrams per kilogram (50 mg/kg). Devise a warning label, consistent with the requirements promulgated under the Federal Hazardous Substances Act, to be used on containers of this product.

1.4. Fire inspectors are often assigned to determine what hazardous materials are regularly used at industrial facilities within the jurisdiction of their fire departments. To help accomplish this aim, what documents required under the Occupational Safety and Health Act should they request when visiting these facilities?

NFPA 704 Diamonds

1.5. Assign numbers to each of the appropriate quadrants of a 704 diamond characterizing lithium hydride, based only on the following information:

(a) Large doses of substances containing lithium cause dizziness and prostration in humans.

(b) When lithium hydride contacts droplets of atmospheric moisture, hydrogen forms, which self-ignites; with larger amounts of water, the reaction may occur explosively.

1.6. Assign numbers to each of the appropriate quadrants of a 704 diamond characterizing chloral (trichloroacetaldehyde), based only on the following information:

(a) Both the liquid and vapor of chloral are extremely hazardous to handle, as inhalation of the substance may cause serious injury to the lungs on short exposure.

(b) Liquid chloral dissolves in water, but does not react violently with water.

(c) A mixture of chloral vapor with air does not burn when exposed to an ignition source.

1.7. For each of the following groups of 704 diamonds, indicate which symbolizes the greater degree of:

(a) Fire hazard:

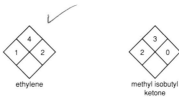

ethylene   methyl isobutyl
                           ketone

(b) Health hazard:

epichlorohydrin   methyl formate

(c) Chemical reactivity hazard:

peracetic acid   nitroethane

(d) Combined fire and health hazard:

1, 2-butylene oxide   allyl chloride

CHEMTREC

1.8. Prior to placing a telephone call to CHEMTREC in an emergency, what general information should the caller either know or be in the process of obtaining?

1.9. Arriving at his place of employment, a director of environmental affairs notices a valve has been mistakenly left opened on a 5000-gal holding tank, thus allowing hundreds of gallons of tributylamine to escape into a diked area surrounding the tank. The following marking is stenciled on the side of the tank:

(a) Why is it most likely unsafe for the employee to enter the area and turn off the valve without the use of self-contained breathing apparatus?

(b) How may the employee ascertain from the marking whether tributylamine will ignite if exposed to an ignition source? F - 2

(c) Discovering he cannot turn off the valve without help from others, the employee first telephones the fire department and then decides to also place a call to CHEMTREC. What information concerning this incident should the employee have available to inform CHEMTREC? I.D. & AmT

(d) What information should the employee be prepared to immediately convey to the arriving response team?

ID AmT

EYE- DEE -AmT

2

Some Features of Matter and Energy

At first glance, the huge number of hazardous materials is certain to overwhelm the average nonscientist. How is it possible to learn the individual properties of so many materials and recall them under the disordered conditions that often prevail when lives and property are in jeopardy?

Fortunately, even for professional scientists, the characteristics of many chemical substances have been effectively correlated and systematized. Furthermore, many properties of substances can be related to a feature of their state of matter. For instance, all gases are known to possess certain common behavioral properties. So when the chemistry of gases is studied, we first learn these common properties, and turn our attention later to the features that cause individual gases to be regarded as unique substances.

In this chapter, we shall learn some common properties of matter and energy and note how these properties influence certain phenomena connected with chemical hazards. Also, as we review a feature of matter or energy, we will learn how it relates to an issue dealing with firefighting. Specifically, in this chapter we learn how the modes of heat transfer contribute to the propagation of fire, the reason why liquid water often effectively serves as a good fire extinguisher, and why confined gases are likely to explode when excessively heated.

2-1 MATTER DEFINED

Each day countless types of matter are encountered by everyone. Air, water, metals, stones, dirt, animals, and plants—the materials of which the world is made—are all different kinds of *matter*. When we look for common features among all its forms, we note that matter possesses mass and occupies space. In other words, matter is distinguishable from empty space by its presence in it. *Mass* is closely related to the concept of *weight*. In our universe, every form of matter is attracted to all other forms by a force called *gravity*. On Earth, the weight of matter is a measure of the force with which it is pulled by gravity toward the Earth's center. As we leave Earth's surface, the gravitational pull decreases, eventually becoming virtually insignificant, while the weight of matter accordingly reduces to zero. Yet the matter still possesses the same amount of mass. Hence, the mass and weight of matter are proportional to each other.

Since matter occupies space, a given form of matter is also associated with a definite volume. Space should not be confused with air, since air is itself a form of matter. *Volume* refers to the actual amount of space that a given form of matter occupies.

As usually experienced, matter has three different forms or states of aggregation: gas, liquid, and solid. These are called the *physical states of matter* and are distinguishable from one another by means of two general features, shape and volume.

Solids

A *solid* is a form of matter that possesses a rigid state that is independent of the size and shape of its container. Consider this book. It retains its shape regardless of its position in space and does not need to be placed in a container to retain that shape. Left to itself, it will never spontaneously assume a shape different from what it has now.

Solids also possess a fairly definite volume at a given temperature and pressure. We can squeeze most solids with all our might or heat or cool them, but their total volume change is relatively small.

Liquids

A *liquid* is a form of matter that does not possess a characteristic shape; rather, its shape depends on the shape of the container it occupies. Consider a glass of water. The liquid water takes the shape of the glass up to the level it occupies. If we pour the water into a cup, the water takes the shape of the cup; or, if we pour it into a bowl, the water takes the shape of the bowl. Of course, sufficient space must be available for the water; otherwise, the liquid overflows. But, assuming that space is available, any liquid assumes whatever shape its container possesses.

Like solids, liquids possess a fairly definite volume at a given temperature and pressure, and they tend to maintain this volume when they are exposed to a change in either of these conditions. Liquids and solids are often considered incompressible; their volume barely changes at all with application of pressure. However, liquids do expand when heated, more than solids do, but not nearly to the degree that gases expand. The expansion of heated liquids is a topic we shall revisit in Sec. 2-10.

Gases

A *gas* is another form of matter that does not possess a characteristic shape, but takes the shape of its container. A gas or mixture of gases, like air, can be put into a balloon, and it will take the shape of the balloon; or it can be put into a tire, and it will assume the shape of the tire.

A gas is also identified by its lack of a characteristic volume. When confined to a container with nonrigid, flexible walls, the volume that a confined gas occupies depends on its temperature and pressure. When confined to a balloon, for instance, the gas's volume expands and contracts depending on the prevailing temperature and pressure. When confined to a container with rigid walls, however, the volume of the gas is forced to remain constant. This feature of gases may cause them to explode from their containers, which we shall discuss further in Sec. 2-12.

These various properties of solids, liquids, and gases may now be used to formally define them. Matter in the *solid* state possesses a definite volume and a definite shape; matter in the *liquid* state possesses a definite volume, but lacks a definite shape; and matter in the *gaseous* state possesses neither a definite volume nor a definite shape.

2-2 UNITS OF MEASUREMENT

The necessity to measure, and to measure with accuracy, is essential to any kind of scientific or technological endeavor. "To measure" means to find the number of units in a sample of something. For instance, when we measure the distance from one point to another along a wall, we determine how many feet, yards, or meters are between the points. The foot, yard, and meter are examples of common units of length.

Two systems of units have survived the test of time. In the United States, we frequently use a system formerly called the *English system*, although even the English no longer use it. We continue to use an array of units that have no obvious interrelationship, like inches (in.), feet (ft), yards (yd), miles, ounces (oz), pounds (lb), tons, pints (pt), quarts (qt), gallons (gal), bushels, pecks, teaspoons (tsp), tablespoons (tbsp), and so forth, all of which are still called *English units of measurement*.

Most of the rest of the world and all the scientific community use a system of measurement called the *metric system*. In relatively recent times, the metric system has been modified so that today we actually use the *SI* system, based on the official French name, Le Systeme International d'Unites. The SI system encourages the use of certain fundamental units from which all other measurements are constructed. These fundamental units are called the *SI base units*. We shall be concerned with two SI base units, the meter and kilogram. They are the SI base units of length and mass, respectively.

Certain prefixes are used in the SI system to denote multiples and fractions of the units of measurement. Each prefix is a fraction or multiple of the number 10. For example, we noted earlier that the SI base unit of length is the meter. When we wish to refer to 1000 meters, we use the word *kilometer*. The prefix *kilo* means 1000 times the base unit. Only four prefixes are commonly used in studying the chemistry of hazardous materials, which are listed in Table 2-1. These particular prefixes should be committed to memory.

There are three types of measurements with which we need to become familiar: length, mass, and volume. It is appropriate to discuss each type separately.

TABLE 2-1 COMMON PREFIXES USED IN THE SI SYSTEM

Prefix	Meaning
kilo-	One thousand times the SI base unit[a]
centi-	One-hundredth of the SI base unit
milli-	One-thousandth of the SI base unit
micro-	One-millionth of the SI base unit

[a] See text for an exception in the case of the SI base unit of mass.

Length

We noted earlier that the SI unit of length is the *meter* (m). One meter is slightly longer than one yard. Specifically, 39.37 inches equals 1 meter. We measure length in the metric system with a metric ruler. One meter is equivalent to 100 *centimeters* (cm) and to 1000 *millimeters* (mm).

$$1 \text{ m} = 100 \text{ cm} = 1000 \text{ mm}$$

For very large lengths, we use the *kilometer* (km). Once again, 1000 meters is equivalent to 1 kilometer.

$$1 \text{ km} = 1000 \text{ m}$$

Figure 2-1 illustrates the relationship between the centimeter and inch.

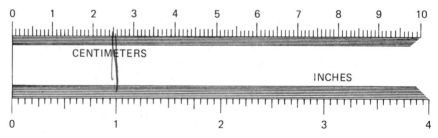

Figure 2-1 The relationship between the inch and centimeter. Note that 1 inch (in.) equals 2.54 centimeters (cm).

Mass

The SI unit of mass is the *kilogram*. One kilogram is equivalent to 2.2 lb. A bit of attention needs to be given to constructing multiples and fractions of mass measurements. This is due to the fact that the unit of mass is the only one that already contains a prefix (kilo-). The names of the various multiples and fractions of the unit of mass are constructed by attaching the appropriate prefix to the word, *gram*, not kilogram. In other words, the gram is used as though it is the SI unit of mass, even though it actually is not.

Thus, one one-thousandth of a gram is called a *milligram* (mg), and one one-millionth of a gram is called a *microgram* (μg). One thousand grams is one *kilogram* (kg).

$$1 \text{ mg} = 0.001 \text{ g}$$

$$1 \text{ } \mu\text{g} = 0.000001 \text{ g}$$

$$1 \text{ kg} = 1000 \text{ g}$$

Volume

The approved SI unit of volume is the *cubic meter* (m^3). This is not an SI base unit; rather, it is a unit that can be derived directly from the SI unit of length. To accomplish this, consider a cube, like the one pictured in Fig. 2-2, measuring 1 meter to each edge. The volume of this cube is determined by taking the product of the length, width, and height.

$$\text{Volume} = 1 \text{ m} \times 1 \text{ m} \times 1 \text{ m} = 1 \text{ m}^3$$

Hence, the volume of this cube is 1 cubic meter. Since it is possible to derive the SI unit of volume directly from a unit previously defined (the meter), it is senseless to define some other unit to express volume.

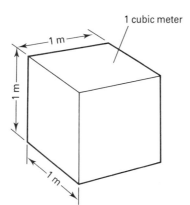

Figure 2-2   A cube that measures 1 meter to each edge. The volume occupied by this cube is 1 *cubic meter* (m^3), the approved SI unit of volume.

Although the cubic meter is the approved unit of volume in the SI system, another unit has been used for years to measure volume. This unit is the *liter* (L). One liter is slightly larger than one quart. Chemists continue to use the liter for measuring volume, because its size is so convenient for laboratory scale measurements. By comparison, the cubic meter is often too large.

One liter is equivalent to one one-thousandth of a cubic meter.

$$1 \text{ L} = 0.001 \text{ m}^3$$

If we construct a cube measuring 1 meter to each edge, and then divide the resulting volume into 1000 equally sized cubes, the volume of the small cubes is 1 liter. This is illustrated in Fig. 2-3. Imagine further dividing each of the small cubes into another 1000 equally sized cubes. Each of these cubes has a volume of 1 *milliliter* (mL). (Remember the prefix milli- means one one-thousandth of a unit.)

$$1 \text{ L} = 1000 \text{ mL}$$

Formerly, the milliliter was known as a *cubic centimeter* (cm^3). The cubic centimeter is still occasionally used in chemistry.

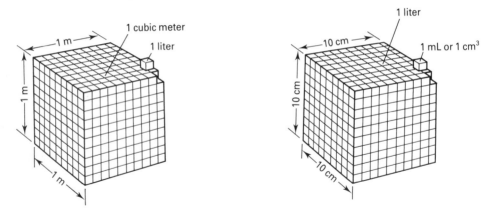

Figure 2-3 The cube on the left measures 1 meter to each edge and has been divided into 1000 equally sized cubes. The volume occupied by each of the smaller cubes is 1 liter (L). The cube on the right measures 10 cm to each edge and has been divided into 1000 equally sized cubes. The volume of the larger cube is 1 liter and the volume of each smaller cube is 1 milliliter (mL) (or 1 cubic centimeter).

2-3 CONVERTING BETWEEN UNITS OF THE SAME KIND

Suppose we wish to convert between grams and kilograms, pounds and grams, or liters and gallons. How would we accomplish this task? Problems like these can best be solved by using a simple procedure called the *factor-unit method*. Briefly, it consists of the following key steps:

1. Identify the unit that is desired.
2. Choose the proper conversion factor(s).
3. Multiply the given measurement by the conversion factor(s), being certain to multiply, divide, and cancel equal units, just like equal numbers.

The conversion factor, mentioned in the second point, is crucial to obtaining the correct answer to a problem. It is always a fraction numerically equal to 1 and relates the amount of one quantity (the numerator) to another quantity (the denominator). For example, we know there are 1000 m in 1 kilometer. Either of the following fractions could serve as a conversion factor:

$$\frac{1000 \text{ m}}{1 \text{ km}} \quad \text{or} \quad \frac{1 \text{ km}}{1000 \text{ m}}$$

The first factor is read as follows: 1000 meters per kilometer, the second as one kilometer per 1000 meters.

Suppose we know the height of the Sears Building in Chicago to be 443.2 m, measured from ground level. What is its height in kilometers? Following the factor-unit method, we multiply the given measurement, 443.2 m, by a conversion factor that relates meters to kilometers.

$$443.2 \text{ m} \times \frac{1 \text{ km}}{1000 \text{ m}} = 0.4432 \text{ km}$$

Note that the "m" in 443.2 m cancels with the "m" in 1000 m. The arithmetic involves dividing 443.2 km by 1000, which yields 0.4432 km.

People who are practiced in using the metric system simply move decimal points from left to right, or vice versa, as the need requires.

Let's next convert between a metric measurement and its equivalent in the English system. Some of the more frequently used metric units and their equivalent English counterparts are given in Table 2-2. These interrelationships permit us to write conversion factors.

TABLE 2-2   COMMON SI (METRIC) UNITS
AND THEIR ENGLISH EQUIVALENTS

1 m = 39.37 in.
2.54 cm = 1 in.
1 kg = 2.2 lb
454 g = 1 lb
946 mL = 1 qt

Suppose we know that a particular fixed storage tank has a capacity of 10,000 gal and desire to know its equivalent in liters. Table 2-2 does not directly provide an interrelationship between gallons and liters. However, the table does indicate that 946 mL equals 1 quart, and we know 1 quart equals 4 gallons and 1 liter equals 1000 milliliters. Hence, we have three conversion factors:

$$\frac{946 \text{ mL}}{1 \text{ qt}} \quad \text{and} \quad \frac{1 \text{ qt}}{4 \text{ gal}} \quad \text{and} \quad \frac{1 \text{ L}}{1000 \text{ mL}}$$

Ten thousand gallons can thus be converted into its equivalent in liters through the use of the factor-unit method as follows:

$$10,000 \text{ gal} \times \frac{4 \text{ qt}}{\text{gal}} \times \frac{946 \text{ mL}}{1 \text{ qt}} \times \frac{1 \text{ L}}{1000 \text{ mL}} = 37,840 \text{ L}$$

Note that "gal" in 10,000 gal cancels with "gal" in 1 gal; "qt" in 4 qt cancels with "qt" in 1 qt; and "mL" in 946 mL cancels with "mL" in 1000 mL.

2-4 DENSITY OF MATTER

If the particular volume of a given mass of a substance is known, we can readily compute its mass per unit volume. This property is called the *density* of that substance.

$$\text{Density} = \frac{\text{mass}}{\text{volume}}$$

For instance, suppose we weigh exactly 50.00 lb of water on a scale and then transfer the water to a graduated cylinder from which we can establish the volume that it occupies. When this simple exercise is performed, we learn that 50.00 lb of water occupies a volume of 6.00 gal. Hence, the density of water can be computed as 8.33 lb/gal, as follows:

$$\text{Density} = \frac{50.00 \text{ lb}}{6.00 \text{ gal}} = 8.33 \text{ lb/gal}$$

In the English system of units, densities of solids and liquids are usually determined in pounds per gallon (lb/gal) or pounds per cubic foot (lb/ft³); densities of gases are normally determined in pounds per cubic foot. In the metric system, densities of solids and liquids are normally determined in grams per milliliter (g/mL); normally, the densities of gases are determined in grams per milliliter or milligrams per cubic meter (mg/m³). The densities of some common substances are noted in Table 2-3. Notice that the density of water when determined in metric units is 1.00 g/mL. This is due to the fact that water was originally used in the metric system as the reference standard for establishing the units of mass and volume.

Often the mass of liquids and solids is compared to the mass of an equal volume of water. This yields an abstract or dimensionless number called the *specific gravity* of the substance. For instance, if we weigh 1 gallon of sulfuric acid, we find it weighs 15.33 lb. Earlier we determined the density of water, from which we learned that 1 gallon of water weighs 8.33 lb. From this combination of information,

TABLE 2-3 DENSITIES OF SOME COMMON LIQUIDS AND SOLIDS

	Density	
Substance	g/mL at 20°C	lb/gal at 68°F
Acetone	0.792	6.60
Aluminum	2.70	22.5
Benzene	0.879	7.33
Carbon disulfide	1.274	10.62
Carbon tetrachloride	1.595	13.29
Chloroform	1.489	12.41
Diethyl ether	0.730	6.08
Ethyl alcohol	0.791	6.59
Gasoline	0.66–0.69	5.5–5.7
Kerosene	0.82	6.83
Lead	11.34	94.5
Mercury	13.6	113
Silver	10.5	87.5
Sulfur	2.07	17.3
Turpentine	0.87	7.25
Water (4°C)	1.00	8.33

we can compute the specific gravity of sulfuric acid as follows:

$$\text{Specific gravity} = \frac{15.33 \text{ lb/gal}}{8.33 \text{ lb/gal}} = 1.84$$

This computation tells us that sulfuric acid is 1.84 times heavier than water.

Specific gravities of liquids can also be directly determined through the use of a *hydrometer*, a cylindrical glass stem containing a bulb weighted with shot so that it floats upright in a liquid. The liquid is generally poured into a tall jar, like that shown in Fig. 2-4, and the hydrometer is lowered into the liquid until it floats. To determine the specific gravity of the liquid, a graduated scale inside the stem is read at the point where the surface of the liquid touches the stem of the hydrometer. Thus, for pure water, the hydrometer reads 1.0. The specific gravities of some other common liquids are noted in Table 2-4.

Many substances are soluble in water to some degree. But there are also many substances that are relatively insoluble in water. When a liquid does not appreciably dissolve in water, we say that the two liquids are mutually *immiscible*. It is characteristic of two immiscible liquids that they coexist as two separate and distinct phases, one on top of the other. For instance, oil and water are so insoluble in each other that they are considered immiscible liquids.

When liquids are immiscible with water, their specific gravities tell us how they behave when mixed with water. Liquids that are immiscible with water *float* on water if they possess specific gravities less than 1.0; they *sink* below water if they possess specific gravities greater than 1.0. For instance, all grades of oil have

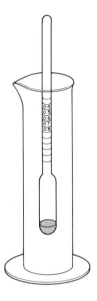

Figure 2-4 A hydrometer is a device that is often employed to determine the specific gravity of a liquid. The specific gravity of this liquid is 0.68. The hydrometer operates on the principle that an object floats "high" in a liquid of relatively high specific gravity (greater than 1.0) and "low" in a liquid of relatively low specific gravity (less than 1.0).

TABLE 2-4 SPECIFIC GRAVITIES OF SOME LIQUIDS[a]

Substance	Specific gravity at 20°C (68°F)
Acetic acid	1.05
Allyl chloride[a]	0.94
Chlorobenzene	1.11
Cyclopentanone[a]	0.95
2-Ethylhexanol[a]	0.83
Heptane[a]	0.68
Hydrochloric acid	1.19
Methyl acetate	0.97
Nitric acid	1.50
Sulfuric acid	1.84
Tetraethyl lead[a]	1.66
Trichlorofluoromethane[a]	1.49

[a] Immiscible with water and flammable, except that trichlorofluoromethane is nonflammable.

specific gravities less than 1.0. If an oil tanker ruptures at sea, any oil that spills from that tanker floats on the water, forming an oil slick.

Knowing the specific gravity of a liquid can be useful for deciding whether to use water as the fire extinguisher if the liquid catches fire. Suppose a liquid immiscible with water having a specific gravity less than 1.0 is confined to a drum or tank. If this liquid is ignited, the addition of water will normally be ineffective at extinguishing the fire. This ineffectiveness arises from the fact that the applied water sinks below the surface of this burning liquid. If the drum or tank overflows, the main constituent of the overflowing liquid will be the burning material, not water. Hence, application of water in such instances may contribute to the spread of fire by means of the normal fluid motion of the liquid to adjoining areas.

On the other hand, consider a liquid immiscible with water having a specific gravity greater than 1.0. If this liquid is ignited, the addition of water effectively extinguishes the fire. In this instance, the water floats on the heavier liquid and smothers the fire. These two situations are dramatized further in Fig. 2-5.

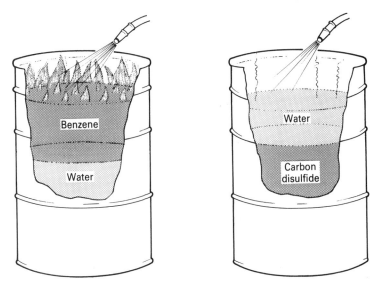

Figure 2-5   On the left, water is added to a drum containing burning benzene (specific gravity = 0.879). Water and benzene are immiscible liquids. The water settles below the benzene and does not ordinarily extinguish the fire. On the right, water is added to a drum containing burning carbon disulfide (specific gravity = 1.274). Water and carbon disulfide are also immiscible liquids. The water floats on the carbon disulfide, which prevents further contact of the fuel with atmospheric oxygen. In this case, the addition of water effectively extinguishes the fire.

Often, the mass of gases and vapors is compared to the mass of an equal volume of air or other reference gas. This yields another dimensionless property called the *vapor density*. Dry air, which is itself a mixture of several gases, has a density of 1.29 g/L, or 0.011 lb/gal.

Suppose we consider the vapor density of oxygen. Table 2-3 tells us that 1 liter of oxygen weighs 1.43 g. We may use this information to compute the vapor density of oxygen as follows:

$$\text{Vapor density of oxygen} = \frac{1.43 \text{ g/L}}{1.29 \text{ g/L}} = 1.11$$

The vapor densities of all gases and vapors may be determined in this fashion. Table 2-5 lists the vapor densities for some substances.

TABLE 2-5 VAPOR DENSITIES OF
SOME COMMON GASES

Substance	Vapor density (air = 1)
Acetylene	0.899
Ammonia	0.589
Carbon dioxide	1.52
Carbon monoxide	0.969
Chlorine	2.46
Fluorine	1.7
Hydrogen	0.07
Hydrogen chloride	1.26
Hydrogen cyanide	0.938
Hydrogen sulfide	1.18
Methane	0.553
Nitrogen	0.969
Oxygen	1.11
Ozone	1.66
Propane	1.52
Sulfur dioxide	2.22

An awareness of the vapor densities of gases and vapors is useful for evaluating the potentially hazardous features of these substances. If the vapor density of a gas or vapor is greater than 1.0, an arbitrary volume of the gas or vapor is heavier than the same volume of air. A leak of such a substance from its container tends to displace air and concentrate near low areas before ultimately dissipating into the atmosphere. This can be cause for concern if the substance is flammable, since flammable gases and the vapors of flammable liquids constitute serious fire hazards. For example, a volume of gasoline vapor is heavier than the same volume of air; hence, it accumulates in low areas like the bilges in boats, where it poses a risk of fire and explosion. Gases and vapors that are denser than air can also be cause for concern when the substances are toxic; in this case, their accumulation in a confined area could pose a serious health hazard. On the other hand, if the vapor density of a gas or vapor is less than 1.0, the substance tends to rise in relatively calm atmospheres. Such a gas or vapor typically disperses quite rapidly. When released inside buildings, it often escapes through open windows and doors.

Sometimes an advantageous use can be made of vapor densities. Consider carbon dioxide, a nonflammable gas whose vapor density is 1.52. When discharged at the base of a fire, carbon dioxide displaces the air. Due to its vapor density, the carbon dioxide blankets the fire and shuts out or dilutes the air that sustains the combustion.

Vapor densities may be useful for examining situations involving gases and vapors, just as specific gravities may be useful for examining situations involving certain liquids and solids. An important distinction, however, is that while *some* liquids and solids are immiscible with water, all gases and vapors are totally miscible with air. Given adequate time, gases and vapors become completely mixed with the components of the atmosphere; we say that they are *infinitely miscible* with air.

2-5 ENERGY

Energy is defined as the *capacity to do work*. Energy is thus proportional to work. In a broad sense, when we expend an effort in order to accomplish something, the activity is referred to as work. Scientists define work in terms of some simple type of labor, like a push or pull of matter that results in moving it.

Energy is found about us in a variety of forms, although these forms are typically abstract: radiant energy (light), thermal energy (heat), acoustical energy (sound), mechanical energy, chemical energy, atomic energy, and electricity. Many of these forms of energy play a great role in the study of chemistry. Let's briefly consider just one such form, chemical energy. All forms of matter have energy stored in them because of their chemical nature. Dynamite, for example, has a certain amount of chemical energy. When it explodes, some of this chemical energy is released in the form of light, heat, and sound. All hazardous materials have chemical energy stored in them as well, and this energy may play a role in their hazardous nature.

Units of energy that we often encounter in the United States are the British thermal unit (Btu) and the calorie. The SI unit of energy is the *joule*. One joule (J) is the energy possessed by a 2-kilogram mass moving at a velocity of 1 meter per second. We shall revisit the British thermal unit and calorie in Sec. 2-8.

Matter and energy are closely interrelated. Whereas neither can be created nor destroyed, energy may be developed from matter and turned into matter, and vice versa. These observations are summarized in an important law of nature known as the *law of conservation of mass and energy*: Regardless of which transformations occur, the total amount of mass and energy in the universe remains constant.

2-6 MEASUREMENT OF TEMPERATURE

Temperature is a property of matter associated with its degree of hotness or coldness. Hot matter is associated with high temperatures, and cold matter is associated with low temperatures. To say that something is hot or cold, however, merely points out a relative condition. A temperature tells us just how hot or cold matter is. Hence, a temperature is an indication of a condition of matter at some point, just as size is an indication of how large or small it is.

The temperature of substances near room conditions is most commonly determined through the use of the simple mercury-in-glass column thermometer. Such a thermometer is a sealed glass capillary tube that has been partially filled with mercury and then calibrated following a prescribed procedure. When inserted into a substance, the temperature of the glass and mercury rise or fall as the mercury expands or contracts, respectively, until it comes to the temperature of the substance itself. By observing the height of the column of mercury, we may then measure the temperature of this substance.

Suppose we take an unmarked capillary tube containing mercury and place it in steam above boiling water at sea level. The mercury level rises and remains stationary. This height can now be etched on the glass. It serves as the first reference point, called the *steam point*. If we place the same tube in ice water, the mercury level drops and again comes to a point where it remains stationary. This height of the mercury column serves as the second reference point, called the *ice point*.

On the *Fahrenheit scale*, the ice point and steam point are assigned the values of 32 and 212 degrees, respectively. There are 180 equally spaced divisions between these two reference points, each division called one degree Fahrenheit (1°F). A thermometer calibrated in this fashion is called a *Fahrenheit thermometer*.

On the Celsius scale, the ice point and steam point are assigned the values of 0 and 100 degrees, respectively. There are 100 equally spaced divisions between these two reference points; each division is called one degree Celsius (1°C). This is called a *Celsius thermometer*. Note that the Fahrenheit and Celsius thermometers are alike in that they use the freezing and boiling points of water to define each reference point, but they differ from one another by the numbers that are assigned to define them.

Since the same temperature interval is divided into 180° on the Fahrenheit scale and 100° on the Celsius scale, the temperature range corresponding to 1 Celsius degree is 180/100, or 9/5, as great as that corresponding to 1 Fahrenheit degree. Thus, the Celsius and Fahrenheit scales are related to each other by formulas such as the following:

$$t(°F) = \tfrac{9}{5} t(°C) + 32$$

$$t(°C) = \tfrac{5}{9} [t(°F) - 32]$$

When using either formula, it is essential to perform the arithmetic exactly as ₊ne formulas are written. To determine a Fahrenheit reading, nine-fifths of a Celsius reading is taken first and then 32 degrees are added. Suppose we desire to convert 46°C to its equivalent temperature reading on the Fahrenheit scale. We take nine-fifths of 46, which equals 83, and add 32:

$$t(°F) = [\tfrac{9}{5} \times 46] + 32 = 83 + 32 = 115°F$$

To determine a Celsius reading, 32 is subtracted from the Fahrenheit reading, and then five-ninths of this difference is taken. Suppose we desire to convert 115°F to its equivalent temperature on the Celsius scale. We subtract 32 from 115, which equals 83, and then take five-ninths of it:

$$t(°C) = \tfrac{5}{9} \times [115 - 32] = \tfrac{5}{9} \times 83 = 46°C$$

Subdivision of the Fahrenheit and Celsius scales can be continued indefinitely above and below the two reference points. When a thermometer is read below the zero mark, the divisions are read as *minus* degrees, or as degrees Celsius or Fahrenheit *below zero*. When a temperature reading below zero on one of these scales is converted to a temperature on the other scale, it is important to account for the negative numbers algebraically.

While an upper temperature limit has never been identified, scientific theory and experiment do show that there is a limit below which it is impossible to cool matter. This coldest temperature is $-273.15°C$ or $-459.69°F$, each called *absolute zero*. Two additional temperature scales have been defined by referring to these temperatures as zero degrees: the Kelvin and Rankine temperature scales.

The Kelvin temperature scale is defined as follows:

$$T(K) = t(°C) + 273.15°C$$

Thus, to obtain a Kelvin temperature, we merely add 273.15 to a Celsius reading. The *kelvin* (without the degree sign) has been adopted as the SI unit of temperature.

The Rankine temperature scale is defined as follows:

$$T(°R) = t(°F) + 459.69°F$$

In other words, to obtain a Rankine temperature, we add 459.69 to the Fahrenheit reading.

Since 0 K, or 0°R, represents the lowest attainable temperature, we do not encounter the use of negative numbers on either the Kelvin or Rankine temperature scales.

Figure 2-6 summarizes the interrelationships between the four temperature scales discussed here.

Temperature values are often useful in chemistry. For instance, chemical substances can be characterized by temperatures that are unique to them, such as their melting and boiling points.

In fire science, temperature values are frequently useful, too. For instance, arson investigators are sometimes required to estimate the temperature that was achieved during a fire. This may be accomplished by examining the condition of objects remaining after the fire. Knowledge of the melting points of metals, plastics, and other building materials assists investigators in establishing the approximate temperature of the fire. Table 2-6 lists approximate melting points of some materials that are likely to be encountered after a fire in a home. Suppose that the investigator identifies a necklace that melted during the fire and subsequently solidified. If this jewelry is gold, the investigator can estimate that the fire minimally achieved a temperature of nearly 2000°F.

2-7 MEASUREMENT OF PRESSURE

Pressure is defined as a force applied to an area; hence, it is expressed as a unit of force per unit of area, like pounds per square inch (psi). In the SI system, the unit of force is the newton (N); it is the force that accelerates a 1-kilogram body 1 meter per

$$F = 1 \, kg \left(\frac{m}{sec^2} \right)$$

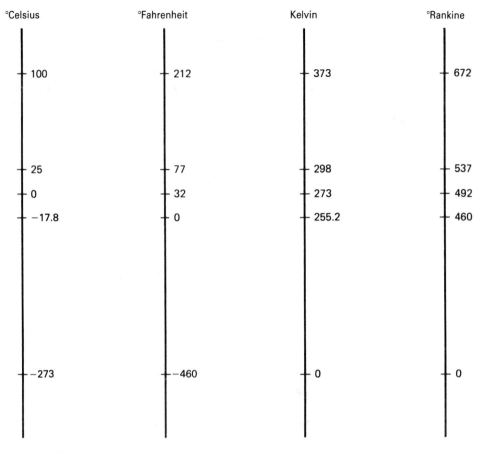

°Celsius	°Fahrenheit	Kelvin	°Rankine
100	212	373	672
25	77	298	537
0	32	273	492
−17.8	0	255.2	460
−273	−460	0	0

Figure 2-6 The interrelationship between four temperature scales: Celsius (C), Fahrenheit (F), Kelvin (K), and Rankine (R). Note that 0°C = 32°F and 100°C = 212°F. Absolute zero is approximately −273°C or −460°F, which equal zero kelvin (0 K) and zero degrees Rankine (0°R), respectively. Average room temperature is also illustrated: 25°C = 77°F = 298 K = 537°R.

TABLE 2-6 APPROXIMATE MELTING POINTS OF SOME MATERIALS LIKELY TO BE ENCOUNTERED AFTER A FIRE

Material	Temperature °F	Temperature °C
Solder	361	183
Tin	449	232
Lead	618	326
Zinc	878	471
Magnesium	1202	650
Aluminum	1220	661
Silver	1761	960
Gold	1945	1060
Copper	1981	1082
Iron	2781	1527
Chromium	3407	1875

second for each second. Hence, the SI unit of pressure is the newton per square meter (N/m²), which is called 1 pascal (Pa).

The force resulting from the weight of the overlying air produces the phenomenon of *atmospheric pressure.* The Earth's gravity is sufficiently powerful to give the atmosphere an average downward force of 14.7 psi at sea level; that is, on the average, 14.7 lb of air bears down on each square inch of the Earth's surface. The pressure exerted by the atmosphere at sea level also supports on the average a column of mercury 760 mm high (written as 760 mm Hg). This pressure is called 1 *atmosphere* (atm).

$$1 \text{ atm} = 760 \text{ mm Hg} = 14.7 \text{ psi} = 101{,}325 \text{ Pa}$$

Atmospheric pressure is normally recorded on a barometer. When the atmospheric pressure decreases below the average value of 760 mm Hg, the mercury level is said to fall; if it increases above the average value, the mercury level is said to rise.

Gases and vapors are normally confined to cylinders or tanks to which a pressure gauge has been attached, like that shown in Fig. 2-7. When the gauge is read, the pressure exerted by the contents of the vessel can be determined. Since atmospheric pressure is constantly applied to all forms of matter on Earth, pressure gauges usually measure the amount of pressure by which the gas exceeds atmospheric pressure. This is called the *gauge pressure;* units of gauge pressure note this by the inclusion of a "g" with the unit, as psig.

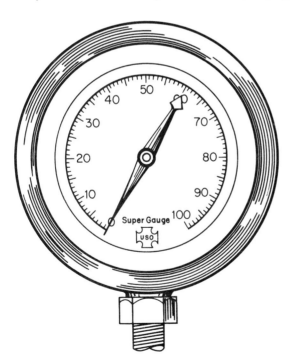

Figure 2-7 A pressure gauge. The difference between the actual pressure that a substance exerts and the pressure of the atmosphere is called the *gauge pressure.* Thus, when the needle reads zero, as illustrated, the gauge pressure is zero, which refers to an *absolute pressure* of 14.7 psi (or 101.3 kilopascals).

The true pressure, called the *absolute pressure,* is the sum of the gauge pressure and the atmospheric pressure. In the English system of units,

$$P\,(\text{absolute}) = P\,(\text{gauge}) + 14.7 \text{ psi}$$

Units of absolute pressure are sometimes noted by inclusion of an "a" with the unit, as psia. A pressure of 19.4 psig corresponds to an absolute pressure of 34.1 psia.

If a liquid is confined to a closed container at a given temperature, some of it evaporates into the space above the liquid until equilibrium arises between the liquid and its vapor. Equilibrium is characterized by two opposing changes that simultaneously occur. Some of the confined liquid evaporates, while its vapors condense back into the liquid phase. The pressure exerted by the vapor that is in equilibrium with the liquid at a given temperature is the *vapor pressure* of the liquid at that temperature. It is a measure of the liquid's ability to evaporate or give off vapors.

While the vapor pressure of liquids could be expressed in any unit of pressure, it is normally noted in pounds per square inch or millimeters of mercury (that is, the height of a column of mercury that the vapor supports).

The vapor pressure is a characteristic property of all liquids, as well as some solids, and varies with the applied temperature. As the temperature increases, more and more of the liquid or solid becomes vapor; the vapor pressure accordingly increases. This is illustrated by the vapor pressure data in Table 2-7 and Fig. 2-8 for several common substances as a function of temperature.

TABLE 2-7   VAPOR PRESSURES OF SOME COMMON LIQUIDS

Temperature		Water (mm Hg)	Ethyl alcohol (mm Hg)	Benzene (mm Hg)
°F	°C			
14	−10	2.1	5.6	15
32	0	4.6	12.2	27
50	10	9.2	23.6	45
68	20	17.5	43.9	74
86	30	31.8	78.8	118
122	50	92.5	222.2	271
167	75	289.1	666.1	643
212	100	760.0	1693.3	1360

Flammable liquids with relatively high vapor pressures, like ether as noted in Fig. 2-8, may pose unique problems when shipped. The structural features of their nonbulk containers and bulk transport vehicles are regulated by USDOT. During the course of shipping such liquids almost anywhere, they are likely to absorb heat from the surroundings and increase in temperature. Considerable vapor may form within the shipping containers or tanks. This vapor fills the void space and exerts pressure on the walls of the vessels.

USDOT requires the DOME placard noted in Fig. 2-9 to be affixed to rail tank cars, in addition to the FLAMMABLE placard, when flammable liquids are shipped

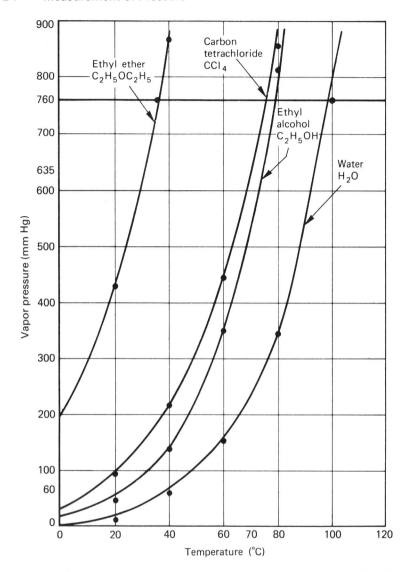

Figure 2-8 The variance of the vapor pressure of some common liquids with temperature. When the vapor pressure equals atmospheric pressure (760 mm Hg), the liquid boils; the corresponding temperature is called the *boiling point* of the liquid. Note that the boiling point of water is 100°C at atmospheric pressure.

having either of the following properties: a vapor pressure exceeding 16 psia at 100°F but not exceeding 27 psia at 100°F, or a vapor pressure exceeding 27 psia at 100°F but not exceeding 40 psia at 100°F. A DOME placard is affixed to each side of the dome, and one is affixed on the top near the manhole in line with the ladders.

AVOID ACCIDENTS	CAUTION Unscrew This Bung SLOWLY

AVOID ACCIDENTS

DO NOT REMOVE THIS DOME COVER
WHILE GAS PRESSURE EXISTS IN TANK

KEEP LIGHTED LANTERNS AWAY

CAUTION   Unscrew This Bung
SLOWLY

Do not unscrew entirely until all interior pressure has escaped through the loosened threads.

REMOVE BUNG IN OPEN AIR. Keep all open flame lights and fires away. Inclosed Electric Lights are safe.

DOME placard   BUNG label

Figure 2-9   The DOME placard and BUNG label. USDOT requires their application, in addition to the FLAMMABLE placard or FLAMMABLE LIQUID label, as appropriate, on tanks, metal drums, or barrels used for transporting certain flammable liquids having relatively high vapor pressures. The printing on both the placard and label is black on a white background. DOME placards are required to be applied one on each side of the tank dome and one on the top near the manhole in line with the ladders. The BUNG label is applied on metal drums or barrels near the bung hole.

USDOT also requires the use of a unique label when flammable liquids are shipped in metal drums if the liquid has either of the vapor pressures previously noted. This is the BUNG label, also shown in Fig. 2-9, which must be affixed to the metal drum near its bung. The rationale for requiring the use of the DOME placard and the BUNG label is directly connected with the potential buildup of internal pressure in these vessels when they become heated.

Under most ordinary conditions, it is the vapor of substances that burns. It is also the vapor of substances that often causes adverse health effects by inhalation or absorption through the skin. Consequently, when containers or tanks of liquid hazardous materials are emptied, they may still pose a health, fire, or explosion hazard. This is caused by the presence of a vapor in "empty" vessels that arises from the evaporation of residual liquid. Even though the liquid contents of such vessels may have been removed, the vessels are not truly empty.

USDOT regulates this potentially hazardous situation in the case of empty rail tank cars by requiring notification that the tanks have been emptied of their contents. One such requirement is the use of precautionary wording on waybills. For instance, waybills referring to the transportation of an empty tank that previously contained gasoline read as follows: "EMPTY: Gasoline, 3, Flammable liquid, UN1203, PG II, Placarded" or "EMPTY: Last contained Gasoline, 3, Flammable liquid, UN1203, PG II, Placarded." Another pertinent requirement, implied in the preceding wording, is the use of placards that clearly specify that the tank has been emptied. An example of this placard is noted in Fig. 2-10. These regulations pertain to empty rail tank cars that contained any form of hazardous materials, not just those whose hazard class is "flammable liquid." (See Chapter 5 for additional information.)

Figure 2-10 USDOT requires this FLAMMABLE—EMPTY placard to be displayed on a rail tank car that has been "emptied" of its contents, but still contains sufficient residue to pose a risk of fire and explosion. The triangular section at the top of the placard is black; the word EMPTY is white; but the other features are identical to those of the FLAMMABLE placard [see Figure 6-8(b)]. Similar placards must be displayed on rail tank cars emptied of hazardous materials other than flammable liquids.

2-8 TRANSMISSION OF HEAT

Heat is defined as a form of energy caused by the motion of molecules, small particles of which most matter is composed. Heat is manifested as a result of three phenomena: (1) a change in temperature, (2) a change in the physical state of matter, or (3) a change in the chemical identity of a substance. Heat is either absorbed or evolved as these processes occur. A process resulting in the absorption of heat is called *endothermic;* a process resulting in the release of heat to the surroundings is called *exothermic*.

Substances lose their chemical identity when chemical reactions occur. One substance changes to another substance. The thermal energy accompanying the chemical reaction is called its *heat of reaction*. When a substance burns, the heat of reaction is called the *heat of combustion*. When a substance dissolves, the heat of reaction is called the *heat of solution*.

The proper control of heat is of utmost importance in a great many manufacturing processes. The utilization of heat evolved from combustion processes such as the burning of coal, gasoline, fuel oils, wood, natural gas, and other materials is of enormous industrial importance.

The proper control of heat evolved from combustion and other chemical reactions is also extremely important in fire control. Fires continue as self-sustained phenomena only when sufficient heat has evolved from the exothermic combustion reaction to substitute for the input energy provided from a burning match or similar ignition source. Similarly, many fires cannot be extinguished until some means is used to reduce or remove the heat.

In the United States, two units are commonly employed for measuring heat: the *British thermal unit* (Btu) and the *calorie* (cal). One British thermal unit represents the heat that must be supplied to raise 1 pound of water 1 degree Fahrenheit, specified at the temperature of water's maximum density (4°C or 39°F). The calorie is defined as the amount of heat required to raise the temperature of 1 gram of water 1 degree Celsius from 14.5° to 15.5°C. One calorie is the same amount of energy as 4.184 joules. One British thermal unit equals approximately 252 calories.

$$1 \text{ cal} = 4.184 \text{ J}$$

$$1 \text{ Btu} = 252 \text{ cal}$$

Heat is transmitted from material to material or from one spot to another by three independent modes: conduction, convection, and radiation. These three modes are very important for understanding the spread of fire.

Conduction

The handle of a silver spoon placed in hot coffee rapidly gets hot itself, even though the part of the handle that is touched is not directly in the hot liquid. This transfer of heat between two or more objects—from the coffee to the spoon and then to our fingers—is called *conduction*.

Every material conducts heat to some extent, but there is a wide variation in the ability of substances to conduct heat. Metals conduct heat best, like silver, copper, iron, and aluminum. Nonmetals, like glass and air, do not conduct heat well.

Convection

Heat may also be transferred from spot to spot within a given substance or even between two or more substances by the mixing of their component parts. This happens when we add cold cream to hot coffee or when we warm soup on a stove. The transfer of heat within a substance or between substances by means of mixing is known as *convection*.

Convection is responsible for the circulation of warm air from a heat vent about a relatively cool room. Maybe you've heard that heat rises? What is really meant is that hot air rises. As air issues from a heat vent, it rises toward the ceiling since it is less dense than the cooler surrounding air. Cool air descends to take the place of the warm air, is heated, and then it, too, rises toward the ceiling. Warm and cool air may exchange in this fashion, thus establishing air currents.

The origin of the convection phenomenon can be traced to the impact that gravity has on matter. When zero gravity exists, as in outer space, convection cannot occur unless it is artificially induced. When a flame burns on Earth, convection enables its combustion products to move away from the flame and to dissipate into the surrounding atmosphere; but in zero gravity the combustion products remain in the immediate area of the combustion zone where they quickly extinguish the flame.

Radiation

Imagine a 200-watt light bulb hanging from a ceiling. When the light is turned on, heat can be felt when we hold our hands *under* the bulb. This transfer of heat from the bulb to our hands cannot be due to convection, since hot air currents rise upward. It also cannot be due to conduction, since the air between the bulb and our hands is a nonmetal, and thus a poor conductor of heat. This third means by which heat may be transmitted is called *radiation*.

Unlike conduction and convection, radiation may occur even when there is no material contact between objects. It is the mechanism responsible for the warming of Earth by the sun, whereby heat travels through space.

At elevated temperatures, all matter radiates heat. When the temperature of a heated object exceeds approximately 500°C, the radiation becomes visible. Burning flames, glowing coals, and molten metal are examples of matter so hot that they radiate visible light. Hot objects also radiate heat at temperatures less than 500°C, but it is generally infrared light, which we are incapable of seeing.

Conduction, convection, and radiation may contribute to the sustenance and spread of freely burning fires, as Fig. 2-11 illustrates; but convection and radiation are principally responsible. Convection contributes by naturally mixing hot and cold air throughout the area. As hot air rises, the heat is transmitted to adjoining materials. When these materials become hot enough, they also ignite. Simultaneously, cool fresh air flows inward into the combustion site to replace hot air. This movement is turbulent and is strongly affected by the density difference between the hot and cold gases. Radiation affects the spread of fire by transmitting heat, often laterally, from the immediate combustion site to nearby materials.

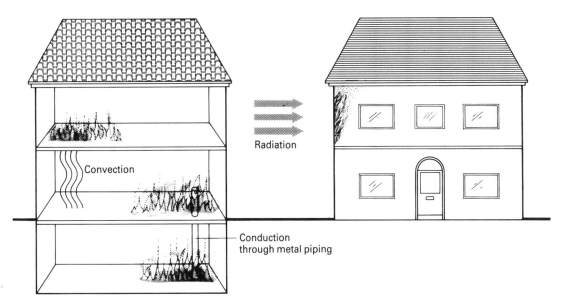

Figure 2-11 Conduction, convection, and radiation contribute to the spread of fire. In the building on the left, heat is first conducted through metal piping from a basement fire to the first floor. Heat is then transmitted to the second floor by convection and to the nearby adjacent building on the right by radiation.

The transfer of heat may also affect the performance of fire personnel. Hot air inhaled by firefighters causes fatigue, thereby threatening a firefighter's ability to carry out routine physical duties. Inhalation of hot air also diminishes an individual's capability to make responsible decisions. In some instances, the inhalation of hot air may even cause lung damage and thus be responsible for long-lasting, adverse health effects and, possibly, even death.

2-9 CALCULATION OF HEAT

Two or more substances differ from one another by the quantity of heat required to produce a given elevation of temperature in the same mass. This quantity of heat is called the *heat capacity* of the substance or, alternatively, its *heat content*. It is expressed in the units of Btu/lb °F or cal/g °C. Heat capacities are listed for some common liquids in Table 2-8.

TABLE 2-8   HEAT CAPACITIES OF SOME COMMON LIQUIDS

Liquid	Heat capacity (cal/g °C or Btu/lb °F)
Acetone	0.506
Benzene	0.406
Carbon tetrachloride	0.201
Chloroform	0.234
Diethyl ether	0.547
Ethyl alcohol	0.581
Methyl alcohol	0.600
Turpentine	0.411
Water (from 0° to 100°C)	1.000
Water (below 0° and above 100°C)	0.5

Values of heat capacity vary with temperature. The heat capacity for water, for instance, is 1.00 Btu/lb °F between 32° and 212°F, whereas it becomes 0.5 Btu/lb °F below 32°F and above 212°F.

The ratio of the heat capacity of a substance to the heat capacity of water at the same temperature is a dimensionless number called its *specific heat*. Most common substances have specific heat values of less than 1.

Heat may be transferred to a substance, causing a change in its physical state or simply causing a change in its temperature. Heat transferred to a substance that causes a change in temperature, but no change in its physical state, is sometimes called the *sensible heat*. It is determined from the following equation:

$$Q = m \times C \times \Delta T$$

Q is the heat, m is the mass of the substance, C its heat capacity, and ΔT (delta T)

the difference between the final and initial temperatures. Suppose we wish to determine the sensible heat when 100 g of copper is heated from 30° to 100°C, given the heat capacity of copper.

$$\Delta T = 100° - 30° = 70°C$$

$$Q = 100 \text{ g} \times 0.093 \text{ cal/g } °C \times 70°C = 651 \text{ cal}$$

The change in the physical state of matter occurs without a change in temperature. If relatively warm water is exposed to a cold winter temperature, the temperature of the water falls until it reaches 32°F. But then the temperature remains fixed until the entire liquid mass freezes to solid ice. This is called the *freezing point* of water. The quantity of heat per unit mass required to change any liquid to a solid at its freezing point is called its *latent heat of fusion*. Following solidification at 32°F, the temperature decreases until it reaches the ambient temperature of the surroundings. The freezing of liquid water to solid ice is accompanied by the liberation of 144 Btu/lb or 80 cal/g. *Melting* is the reverse of freezing; hence, the same amount of heat is absorbed from the surroundings as ice melts.

We can similarly take a mass of water and heat it to 212°F. But then the temperature remains fixed while the liquid water converts to steam. This is the temperature at which the vapor pressure of water equals the atmospheric pressure of 1 atm. It is called the *boiling point* of water. The amount of heat that must be supplied to any material at its boiling point to convert it from a liquid to a vapor is called its *latent heat of vaporization*. The heat of vaporization of water is relatively high when compared to other substances, 970 Btu/lb or 540 cal/g. *Condensation* is the reverse of boiling; hence, the same amount of heat is liberated when steam condenses to water vapor.

Figure 2-12 illustrates the amount of heat liberated to the surroundings when 1 lb of water undergoes a temperature change from −20°F to 300°F. Note that the largest amount of heat is liberated over this temperature range when the water evaporates, due to the relatively high value of water's latent heat of vaporization. This explains why water is such an effective fire extinguisher. When applied to a fire, water draws heat from the burning material and ultimately vaporizes, extracting 970 Btu for each pound that vaporizes. As heat is drawn from the burning material, its temperature ultimately drops to the ambient temperature of the surroundings, which is normally insufficient to reinitiate combustion; the fire subsequently dies out. Water is commonly employed as a fire extinguisher because it is nonflammable and generally plentiful and inexpensive; but the primary reason water effectively extinguishes fire is due to its ability to remove heat from the burning material. This is true only because nature provides water with a relatively large heat of vaporization.

As we see from Table 2-8, liquid water also possesses a relatively larger heat capacity than other substances. Because of this relatively large heat capacity, water has the ability to absorb more heat within the same fixed temperature range than most liquids of the same mass.

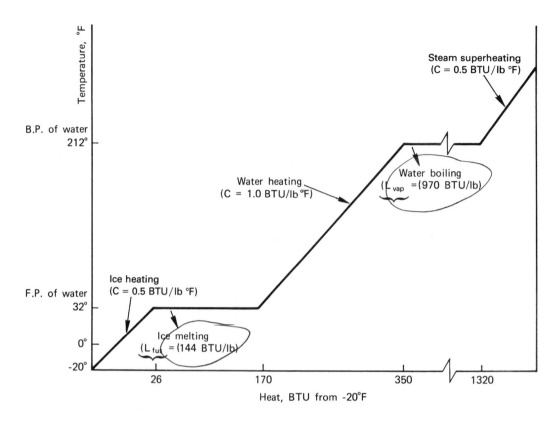

Figure 2-12 A plot of Fahrenheit temperature versus the heat in BTU for 1 pound of pure water at atmospheric pressure (14.7 psia or 101.3 kilopascals) (not to scale).

2-10 EXPANSION OF LIQUIDS

With few exceptions, the dimensions of materials expand when they are heated and contract when cooled. Gaseous matter expands most, as we will note further in Sec. 2-12; liquids expand less, and solids even less. Water is the common exception to this rule: Water begins to expand when it is cooled below 4°C.

The volume to which liquids and solids expand is determined by means of the following formula:

$$V_2 = V_1(1 + \alpha \Delta T) \quad \text{or} \quad V_2 = V_1 + V_1 \alpha \Delta T$$

Here V_1 and V_2 are the initial and final volumes of a material, respectively, ΔT is the difference in temperature, and the Greek letter alpha (α) is a measure of the change in volume per unit of original volume per degree change in temperature. It is called the *coefficient of volume expansion*. Values of this parameter are listed in Table 2-9 for some common liquids.

TABLE 2-9 COEFFICIENT OF VOLUME EXPANSION
FOR SOME COMMON LIQUIDS

Liquid	$\alpha/°F$	$\alpha/°C$
Acetic acid	0.000594	0.001071
Acetone	0.000825	0.001487
Benzene	0.000686	0.001237
Carbon tetrachloride	0.000687	0.001236
Diethyl ether	0.000919	0.001656
Ethyl alcohol	0.000622	0.001120
Gasoline	0.000599	0.001080
Glycerine	0.000280	0.000505
Methyl alcohol	0.000665	0.001199
Pentane	0.000892	0.001608
Turpentine	0.000541	0.000973
Water	0.000115	0.000207

Tanks, metal drums, and other containers filled with liquids to the brim must either overflow or rupture upon expansion of the liquid contents. This statement is always true unless the container itself is capable of either withstanding the increased pressure or of expanding to compensate for the increase in volume. The amount of material by which a container of liquid falls short of being full is called its *outage* or *ullage*. It is generally expressed in percentage by volume.

While manufacturers of chemical products are normally cognizant of outage, the allowance may be inadequate to compensate for the volume increase that is experienced by liquid materials during fires. This may be illustrated by considering a 55-gal steel drum filled to the brim with benzene at 68°F, which is subsequently heated to 170°F during a fire.

$$V_2 = 55 \text{ gal} + \left(55 \text{ gal} \times \frac{0.000686}{°F} \times 102°F\right) = 59 \text{ gal}$$

This calculation points out that 55 gal of benzene expands when heated from 68° to 170°F by 4 gal, or 7.3% by volume. If the benzene has been heated in a nearly constant volume steel container, the increase in internal pressure may cause the container to rupture.

USDOT requires containers, rail tank cars, and cargo tanks of flammable liquids and corrosive liquids to be incompletely filled. When offered or accepted for transportation, sufficient outage must always be provided in containers of 110 gal or less so that the containers are not liquid full at 130°F (55°C).

USDOT requires that flammable liquids may not be loaded into the domes of rail tank cars. When the dome of a tank car does not itself provide sufficient outage, vacant space must be left in the shell to provide the required outage. No cargo tank or compartment used for the transportation of any flammable liquid should be liquid full. According to USDOT regulations, the outage in a cargo tank or compartment used in the transportation of flammable liquids should be not less than 1% by volume; sufficient space should be left vacant to prevent leakage from, or distortion of, the tank or compartment by expansion of the contents due to a rise in temperature. For many specific flammable liquids, USDOT specifies a minimum outage when the liquids are transported in rail tank cars.

When corrosive liquids are transported in rail tank cars, USDOT also requires that sufficient outage be provided in the dome of the tank car. When this is impossible, vacant space must be left in the shell to allow for the required outage. This required outage is usually not less than 2% by volume, except for certain USDOT-specified rail tank cars, for which the outage must not be less than 1% by volume.

2-11 FLAMMABILITY

Flammable gases and the vapors of flammable liquids generally ignite readily when they are mixed with air and exposed to a source of ignition. However, for every flammable and combustible substance, there is a concentration in air above and below which it does not burn. The minimum concentration of gas or vapor in air below which a substance does not burn when exposed to an ignition source is called the *lower explosive limit;* the maximum concentration above which ignition does not occur is called the *upper explosive limit*. These limits are normally expressed in percentage by volume of gas or vapor in air.

For instance, the lower explosive limit of xylene is 1.1% by volume in air, whereas the upper explosive limit is 7.0% in air. These measurements mean that a mixture of xylene vapor and air having a concentration of xylene less than 1.1% by volume is too *lean* in xylene vapor to burn; similarly, a mixture containing more than 7.0% xylene in air is too *rich* in xylene to burn. Xylene is said to possess a flammable range of 5.9. The *flammable range* of a substance, sometimes called its *explosive range,* is the numerical difference between the upper and lower explosive limits. Thus all concentrations by volume of xylene vapor in air between 1.1% and 7.0% are in the flammable or explosive range.

Generally, liquids and solids do not actually burn as the liquid and solid states of matter. Rather, they give off vapors that ignite only when a combustible mixture with air has been attained. Some flammable liquids give off insufficient vapor at their ambient temperatures to burn. The liquid must be heated to a temperature at which the concentration of the vapor is within its flammable range; then it can ignite.

There is a minimum temperature at which a liquid gives off sufficient vapor to first form an ignitable mixture with the air near the surface of the liquid or within a test vessel; this temperature is called the *flash point*. A substance does not continuously burn at its flash point; the ignitable mixture only momentarily flashes. USDOT defines the flash point as "the minimum temperature at which a substance gives off flammable vapors which, in contact with sparks or flame, will ignite." Flash points are determined in a specialized scientific apparatus, like that illustrated in Fig. 2-13.

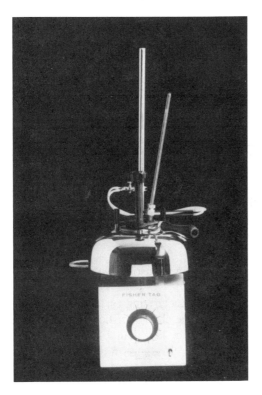

Figure 2-13 The flash point of a flammable or combustible liquid is experimentally determined by using specialized equipment, like this Pensky-Martens closed cup tester. A sample of the liquid is contained in the test cup and slowly heated. An ignition source is periodically applied in the vapor space above the liquid. When a flash of fire is momentarily observed, the temperature on the thermometer is recorded. This is the flash point of the liquid. (Courtesy of Fisher Scientific Company, Pittsburgh, Pennsylvania)

While the concept of a substance's flash point is useful for describing the flammability of liquids, the term does not ordinarily have meaning when applied to flammable gases or solids. At ambient conditions, most flammable gases do not need to be heated in order to ignite. However, it may be necessary to heat flammable solids to a minimum temperature before they begin burning when exposed to an ignition source; for solids this temperature is called the *kindling point*.

As the temperature of a flammable liquid is increased beyond its flash point, a temperature is reached at which self-sustained combustion occurs. This temperature is called the *fire point*. The fire point of many liquids is 30° to 50°F higher than their flash points. As the temperature of a flammable liquid is further increased, a minimum temperature is attained at which self-sustained combustion occurs in the *absence* of an ignition source; this is called the *autoignition point*. For xylene, the fire point is 111.2°F and the autoignition point is 924°F.

The interrelationship between flammability and flash point is apparent in certain NFPA definitions. For instance, a liquid having a flash point below 100°F and having a vapor pressure not exceeding 40 psia at 100°F is referred to by NFPA as a *class I flammable liquid*. There are three subclasses of class I flammable liquids:

1. Class IA. These are liquids with flash points below 73°F and boiling points below 100°F. An example of a class IA flammable liquid is *n*-pentane.

2. Class IB. These are liquids with flash points below 73°F and boiling points at or above 100°F. Examples of class IB flammable liquids are benzene, gasoline, and acetone.

3. Class IC. These are liquids with flash points at or above 73°F and below 100°F. Examples of Class IC flammable liquids are turpentine and *n*-butyl acetate.

NFPA refers to a *combustible liquid* as one having a flash point at or above 100°F. Combustible liquids are subdivided further into classes as follows:

1. Class II. These are liquids with flash points at or above 100°F, but below 140°F. Examples of class II combustible liquids are kerosene and camphor oil.

2. Class IIIA. These are liquids with flash points at or above 140°F, but below 200°F. Examples of class IIIA combustible liquids are creosote oils and phenol.

3. Class IIIB. These are liquids having flash points at or above 200°F. Ethylene glycol is an example of a class IIIB combustible liquid.

The flash point is also used for characterizing certain liquid hazardous wastes (Sec. 1-5). A hazardous waste exhibits the characteristic of ignitability if it is a liquid and possesses a flash point equal to or less than 140°F (60°C). Other features pertaining to this characteristic are noted elsewhere in this book. Hazardous wastes that possess the characteristic of ignitability are assigned the EPA hazardous waste number D001.

2-12 GENERAL HAZARDS ACCOMPANYING THE GASEOUS STATE

Of the three states of matter, only the gaseous state is capable of being described in comparatively simple terms. The description consists of the relationship between the volume of a gas and its temperature and pressure.

Robert Boyle, an Irish physicist, demonstrated experimentally that, if the temperature remains fixed, the volume of a confined mass of a gas varies inversely with its *absolute* pressure. Constant-temperature experiments performed by Boyle show that the product of the volume of a gas and its absolute pressure is always constant. This statement is commonly called *Boyle's law*, which is expressed mathematically in either of the following equivalent forms:

$$P \times V = \text{constant}$$

$$P_1 \times V_1 = P_2 \times V_2$$

Here P and V represent pressure and volume, respectively. P_1 and V_1 are the initial pressure and volume, and P_2 and V_2 are the final pressure and volume.

Suppose we wish to know what volume 40 cubic feet (40 ft³) of an arbitrary gas occupies if its pressure is changed from 14.7 to 450 psia, but its temperature remains fixed. Since we know that an increase in pressure causes the volume to decrease, the new volume must be smaller than 40 ft³. Using Boyle's law, we can readily solve this problem as follows:

$$V_2 = 40 \text{ ft}^3 \times \frac{14.7 \text{ psia}}{450 \text{ psia}} = 1.3 \text{ ft}^3$$

Jacques Charles and Joseph Gay-Lussac, two French scientists, independently demonstrated that, if the pressure of a gas remains constant, its volume increases with an increase in the temperature. Charles first showed that the volume of a gas is directly proportional to the *absolute* temperature. This statement became known as *Charles's law*. Expressed mathematically, either of the following equivalent statements is Charles's law:

$$\frac{V}{T} = \text{constant}$$

$$\frac{V_1}{V_2} = \frac{V_2}{T_2}$$

Here V and T represent volume and absolute temperature, respectively. V_1 and T_1 are initial volume and temperature, and V_2 and T_2 are final volume and temperature.

Suppose an arbitrary gas occupies a volume of 300 mL at 0°C. What volume does it occupy at 100°C if the pressure remains fixed? We must first remember to convert the Celsius temperature readings to absolute temperatures. Hence, the initial temperature is 273.15 K, and the final temperature is 373.15 K. Now we can determine the new volume as follows:

$$V_2 = 300 \text{ mL} \times \frac{373.15 \text{ K}}{273.15 \text{ K}} = 410 \text{ mL}$$

Boyle's and Charles's laws may also be combined into the following expression, which applies when neither the temperature nor pressure is fixed:

$$\frac{V_1}{V_2} = \frac{T_1}{T_2} \times \frac{P_2}{P_1}$$

This is usually called the *combined gas law*.

The volume of a gas may be forced to remain fixed, too, as when it is confined to a steel cylinder or other storage container. Then the ratio of initial and final volumes, V_1/V_2, equals 1, and the combined gas law takes on the following form:

$$\frac{P_1}{T_1} = \frac{P_2}{T_2}$$

This relationship is sometimes called *Amontons's law*. When a gas enclosed in a steel storage cylinder is heated, its pressure accordingly increases. The magnitude of the pressure increase is dramatized in Fig. 2-14. The first storage container in this figure confines a gas at 1000 psia at normal room conditions, 70°F. The second container

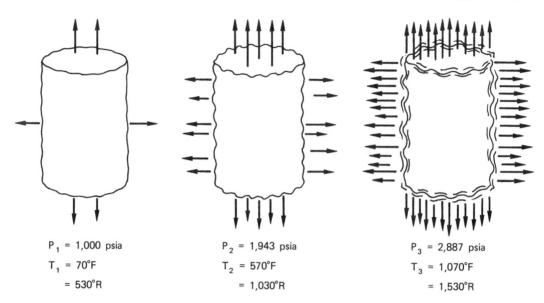

P$_1$ = 1,000 psia

T$_1$ = 70°F

   = 530°R

P$_2$ = 1,943 psia

T$_2$ = 570°F

   = 1,030°R

P$_3$ = 2,887 psia

T$_3$ = 1,070°F

   = 1,530°R

Figure 2-14 The effect of applied temperature on a gas confined to a constant-volume container, such as a metal drum or barrel. An increase in the temperature of the gas by 500°F causes the internal pressure to nearly double, which often causes the container to rupture.

confines the same mass of gas heated to 570°F; its pressure nearly doubles. The third container confines the same mass of gas heated to 1070°F; its pressure nearly triples the original value. These are not abnormal temperatures encountered during fires. To relieve the strain caused by the overly excessive internal pressure on the walls of these containers at elevated temperatures, the containers often rupture.

Oxygen, hydrogen, nitrogen, and air are among the many compressed gases that are typically contained in steel cylinders. Under general conditions, they should be stored in an upright position, strapped or chained as shown in Fig. 2-15, and protected from high temperatures. Such compressed gases are generally safely stored under such conditions. However, if heated to approximately 130°F (54°C), gas cylinders are likely to burst, normally near their valves for the reason just cited. If unsecured, they could jettison like rockets, perhaps for hundreds of yards.

USDOT requires many gas cylinders to be marked with a *service pressure*. This is a designation of the authorized pressure applicable to a specific type of USDOT-approved gas cylinder. For instance, for cylinders marked DOT 3A1800, the service pressure is 1800 psig. To assure safety, the pressure of the containers' contents at 70°F must not exceed the service pressure for which the particular cylinders are designed. Furthermore, USDOT requires that the actual pressure in containers at 130°F shall not exceed five-fourths the service pressure, except for containers that have been charged with acetylene, liquefied nitrous oxide, or liquefied carbon dioxide.

When exposed to temperatures exceeding 130°F, gas cylinders should be treated as potential bombs, since their contents are beyond the specification temper-

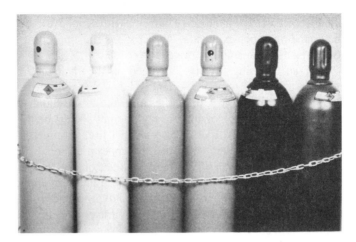

Figure 2-15   To assure stability, cylinders of compressed gases should be securely strapped or chained to a wall or other permanent structure. When not in use, a good practice is to keep them capped. Care should always be taken to keep compressed gas cylinders away from sources of heat or ignition. (Courtesy of Compressed Gas Association, Arlington, Virginia)

ature authorized by USDOT for the containers. If a gas cylinder ruptures, the concentration of a flammable gas issuing from the container is usually within its flammable range. As it mixes with air, the gas spontaneously ignites. The entire area often is immediately engulfed in flames. It is for this reason that the explosion of gas cylinders represents one of the more serious potential hazards encountered during firefighting.

Even the explosion of cylinders containing nonflammable gases can be extremely dangerous. Although such gases cannot burn, the explosion of any gas cylinder may readily cause further destruction in the immediate area by the force exerted when the contents expand.

An analogous hazardous situation exists when rail tank cars containing highly flammable gases or vapors under pressure are exposed to the direct flames of a fire. Such situations may occur during rail and truck accidents, as when a leak of a flammable gas or vapor from a punctured tank car ignites and then impinges on the shell of another nearby tank car containing a flammable gas under pressure. Gases are often shipped in rail tank cars under pressure so that they actually exist in two phases. A heavier liquid phase settles to the bottom of the tank car, while a gaseous phase coexists in the space above the liquid. Contact of fire with the shell of the tank car causes a simultaneous loss of strength in the metal and a rapid buildup of internal pressure in the vapor space above the liquid. Although a venting mechanism is necessarily built into the structure of USDOT-approved rail tank cars, it is normally incapable of relieving the rapid buildup of internal pressure caused by the excess pressure.

The combination of a weakened shell structure and the buildup of internal pressure results in an instantaneous release and ignition of vapor. The phenomenon is called a BLEVE (pronounced "blevey" with a long e), an acronym for *b*oiling *l*iquid *e*xpanding *v*apor *e*xplosion. A BLEVE is illustrated in Fig. 2-16; note its size by comparison to the standing water tower.

After an exposing fire has impinged on the shell of a tank car, a BLEVE is likely to occur at any time. Experience shows that failure in the metal structure is likely to result within 10 to 30 minutes after the initial flame contact unless the tank car is cooled. Whenever possible, every attempt should be taken to prevent the actual occurrence of the BLEVE. The principal firefighting tactic is to direct streams of water at the main vulnerable points: the uppermost area of the tank car (the vapor phase area) and the actual site of flame contact on the tank car shell. If cooling of the tank car is to be effected, 500 gal/min is a minimum application rate.

The risk to firefighters, spectators, and property from a BLEVE cannot be overexaggerated. Fragments of the tank may travel like missiles for hundreds of feet in any direction, potentially causing death or dismemberment. Nearby structures are often destroyed by these rocketing shells and from secondary fires originating from radiation. Since the risk to fire personnel is great, the use of unmanned monitors is essential for discharging water whenever the potential for a BLEVE exists.

Figure 2-16 A BLEVE results when a flammable liquid is rapidly heated to relatively high temperatures above its boiling point. The venting mechanisms of cargo tanks are generally incapable of releasing the huge buildup of internal pressure caused by the expansion of the vapor. Structural failure results, causing a massive explosion with an accompanying fireball.

2-13 GENERAL HAZARDS OF CRYOGENS

All gases can be reduced eventually to liquids and the liquids to solids by appropriately decreasing their temperature and/or increasing their pressure. However, for all gases there is a temperature above which pressure alone cannot condense them to liquids. This temperature is called the *critical temperature*. For instance, carbon dioxide readily liquefies when sufficient pressure is applied while the gas is cooled below 88°F. But no amount of applied pressure can cause liquefaction of the gas above 88°F.

The pressure required to liquefy a gas at its critical temperature is called the *critical pressure*. The critical pressure of carbon dioxide is 73 atm; that is, the application of 73 atm causes carbon dioxide at 88°F to liquefy.

Table 2-10 lists the critical temperature and pressure for several gases.

TABLE 2-10   CRITICAL TEMPERATURE AND CRITICAL PRESSURE OF SOME
SELECTED SUBSTANCES

	Critical Temperature		Critical Pressure	
Substance	°C	°F	atm	psi
Ammonia	130	266	115	1690
Butane	152	305.6	38	558
Carbon dioxide	31.1	88	73	1073
Diethyl ether	197	387	35.8	526
Hydrogen	−234.5	−390	20	294
Nitrogen	−146	−230.8	33	485
Oxygen	−118	−180	50	735
Propane	96.7	206	42	617
Sulfur dioxide	155.4	311.7	78.9	1160

Gases that liquefy at reduced temperatures are called *cryogenic liquids,* or *cryogens,* a word adapted from the Greek word *kyros,* meaning icy cold. USDOT defines a cryogenic liquid as a refrigerated liquefied gas having a boiling point colder than −130°F (−90°C) at one atmosphere, absolute. The science concerned with the physical and chemical phenomena of these icy cold substances is called *cryogenics*.

The handling and storage of cryogenic liquids require special attention. When handling a cryogenic liquid, a shield should be worn to protect the eyes and gloves to protect the hands, as shown in Fig. 2-17. If a mishap occurs during which the liquid mistakenly flows inside the gloves, they should be removed immediately to prevent frostbite.

USDOT also regulates the design of vessels (cylinders, cargo tanks, and rail tank cars) used for transporting cryogenic liquids. These regulations specify requirements for the installation of pressure control systems, pressure relief devises, valves, and piping.

Figure 2-17 Cryogenic fluids are generally contained in a Dewar flask, a double-walled glass vessel with a silvered interior. When transferring a cryogen from one vessel to another, it is necessary to protect the eyes and skin from possible contact by splattering.

Due to their extremely cold nature, cryogenic liquids are associated with the following three common hazardous features:

1. *High expansion rate on vaporization.* To illustrate this point, consider cryogenic methane. When it vaporizes, it expands to approximately 630 times its initial volume; that is, 1 ft^3 of liquid expands to 630 ft^3 of the gas. As we have observed elsewhere in this chapter, the internal pressure developed by the vaporization of liquids confined to containers is likely to result in rupture of the container. This expansion may also cause a health hazard by displacing the air needed for respiration, thereby reducing oxygen to a concentration incapable of supporting life.

2. *Ability to liquefy other gases.* Extremely cold substances liquefy or solidify many other substances that they contact. Even air solidifies when exposed to several cryogenic liquids. The solidification of air poses a major hazard in the handling of cryogenic liquids, since the "ice" may block the passageway in venting tubes, which prevents the release of internal pressure.

3. *Potential for damaging living tissue.* When tissue is exposed to cryogenic liquids for an adequate time, the tissue solidifies. Pain does not typically accompany this freezing, but pain is usually experienced when the tissue subsequently thaws. The solidification of tissue causes a local arrest in the circulation of blood in the affected area. But as the tissue melts, the circulation of blood is restored, during which the formation of blood clots is highly probable. Widespread cellular damage may also occur in affected tissues, since they are highly vulnerable to bacterial infection. Serious skin burns may result from exposure to cryogenic liquids, especially when the specific heat of the liquid is relatively high, as in the case of liquid oxygen. Such burns resemble first-, second- or third-degree thermal burns, depending on the depth to which tissues have frozen.

Unless transportation to a medical facility warrants a delay, tissues that have been frozen by exposure to cryogenic liquids should be restored to normal body temperature as soon as practical. Rapid rewarming of frozen tissues minimizes the potential for further tissue damage. This is best accomplished by flushing the affected area with large volumes of tepid water.

REVIEW EXERCISES

States of Matter

2.1. In which state of matter does each of the following materials exist at room conditions (70°F and 1 atm): **(a)** concrete; **(b)** asbestos; **(c)** rainwater; **(d)** aspirin tablets; **(e)** mercury; **(f)** neon; **(g)** air?

2.2. Identify the following characteristics as those exhibited by a gas, liquid, or solid, or any combination of them: **(a)** possess a characteristic shape; **(b)** possess a characteristic volume; **(c)** lack a characteristic volume; **(d)** lack a characteristic shape; **(e)** possess a characteristic shape *and* volume; **(f)** possess a characteristic volume, but lack a characteristic shape; **(g)** lack a characteristic volume and shape.

Conversions Between Units of Measurement

2.3. The standard length of fire hose used in the United States is 50 ft. What is its equivalent length in meters?

2.4. Portable tanks used to haul water to regions unequipped with fire hydrants normally carry from 1000 to 3000 gal of water.
(a) What is the minimum volume of water these tanks carry when expressed in liters?
(b) What is the maximum volume of water these tanks carry when expressed in cubic meters?

2.5. To control spillage from aboveground flammable liquid storage tanks, NFPA recommends in the *Flammable and Combustible Liquids Code* (NFPA 30) the construction of a remote impounding area into which, if necessary, the liquid may drain. The capacity of the impounding area must not be less than that of the largest tank that can drain into it. Consider a single tank that can hold 100,000 gal of gasoline. Is a remote impounding area measuring 250 m² in area and 1.5 m high adequate to contain the maximum capacity of this storage tank?

2.6. When called to an emergency, a fire truck uses 1 liter of gasoline to traverse a distance of 5 kilometers. How many miles per gallon does the fire truck achieve?

2.7. To prevent eye irritation and possible impairment of lung function, the state of California recommends for nitrogen dioxide an upper tolerance level of 200 $\mu g/m^3$ in air. What is the equivalent concentration in milligrams per liter (mg/L)?

2.8. An inspector desires to cordon off an area 15.0 ft × 12.0 ft in order to isolate and preserve potential arson evidence. What size tarpaulin measured in square meters would completely cover the area?

Density of Matter

2.9. The density of natural gas used for domestic and commercial heating purposes is 0.83 kg/m^3 at a given temperature and pressure.

    (a) What volume in cubic meters does 25 kg of natural gas occupy at this temperature and pressure?

    (b) What is the density of natural gas when expressed in grams per liter (g/L)?

2.10. A piece of metal alleged to be aluminum weighs 13.21 g and occupies 4.89 mL. Use Table 2-3 to determine if the metal is actually aluminum.

2.11. The density of kerosene is 0.88 g/mL. Suppose kerosene ignites in an open-head, 55-gal drum half-filled with the fuel.

    (a) Why is the use of water as a fire extinguisher generally ineffective for this type of fire?

    (b) If excess water is added to the drum, which liquid overflows from it, kerosene or water?

2.12. *o*-Nitrotoluene, a flammable liquid, is immiscible with water and has a specific gravity of 1.16. Consider an incident in which 500 gal of *o*-nitrotoluene leaks from its storage tank into a diked enclosure surrounding the tank. A 704 diamond is not visibly identifiable on the tank.

    (a) Would the use of water effectively extinguish burning *o*-nitrotoluene?

    (b) If excess water is added to burning *o*-nitrotoluene in the diked enclosure, which liquid overtops it, *o*-nitrotoluene or water?

2.13. When propane escapes from a bottled gas cylinder on a relatively calm day, does it tend to rise in the atmosphere or settle in low spots?

2.14. An accident involves two automobiles and a section of a train with two rail tank cars placarded POISON GAS and marked CHLORINE, both of which have ruptured and are leaking chlorine gas to the atmosphere. One side of the nearby railroad track slopes to a valley, while the other inclines to a hill. The emergency-response personnel responding to this incident are informed to move the victims of this incident to fresh air as a first-aid measure. In a relatively calm atmosphere, where should members of this emergency-response crew move victims of this incident, into the valley or up the hill?

Temperature Conversions

2.15. The surface of an iron can burn skin at about 65°C, a temperature at which pain can be felt. This temperature is called the *threshold for pain*. At what equivalent Fahrenheit temperature is pain felt?

2.16. After approximately 8 min, a typical fire causes room temperatures to increase nearly 800°C in high-rise buildings [B. T. Zinn and others, *Fire Technology*, Vol. 10 (1974), p. 35]. What is the equivalent increase in temperature in degrees Fahrenheit?

2.17. Bromine has a narrow range in which it exists in the liquid state of matter, 20° to 138°F. In what range of temperature expressed in Celsius degrees is bromine liquid?

2.18. Liquid air boils at −318°F. Express the boiling point of air on each of the following absolute temperature scales: **(a)** Rankine; **(b)** Kelvin.

Pressure

2.19. Liquid storage tanks built to Underwriter Laboratories, Inc., requirements (UL 142–1972) may be used for operating pressures not exceeding 1 psig, and are limited to 2.5 psig under emergency venting conditions. What operating and emergency venting pressures expressed in kilopascals (kPa) are required by Underwriters Laboratories for these tanks?

2.20. The vapor pressure of ethyl mercaptan at 20°F is 442 mm Hg, whereas the vapor pressure of ethylamine is 1.18 atm at the same temperature. Both substances are flammable liquids. Assuming other conditions to be the same, which substance poses the greater fire hazard?

2.21. Which of the following incidents most likely poses the greatest hazard, based on the degree of damage that they could cause to the nearby environment: **(a)** a container of water ruptures at 300 psig; **(b)** a container of air ruptures at 200 psig; or **(c)** a container of air ruptures at 300 psig?

Transmission of Heat

2.22. Which of the mechanisms of heat transfer is involved in each of the following situations:
(a) An icepack is applied to a bruise.
(b) An air-conditioning unit is used to cool a room.
(c) Body heat is rapidly lost when wearing lightweight, noninsulating clothing.
(d) An overhead heat lamp is used outside a shower.

2.23. When firefighters wish to ventilate a burning building, they often open a large hole at the highest point on its roof. Why is such a point selected?

2.24. Why does a fire that originates on the fifth floor of a building spread more rapidly to the ninth floor than it does to the first floor?

2.25. Upon first entering a hot room, why do firefighters generally aim a stream of water at the ceiling rather than toward the floor?

Calculation of Heat

2.26. How many Celsius degrees will the temperature of 900 g of liquid water be raised by 4000 cal of heat?

2.27. What amount of heat in Btu is removed from a burning pile of rubble when 100 gal of water is applied as a fire extinguisher, all of which subsequently evaporates to the atmosphere?

2.28. Since carbon tetrachloride does not burn in air, it is potentially useful as a fire extinguisher. Compare the abilities of equal masses of carbon tetrachloride and water to remove heat from separate but otherwise identical fires. (*Hint:* Assume 50 lb of each extinguisher is applied to separate fires, each of which experiences a temperature difference of 10°F.)

Expansion of Liquids

2.29. Whizzo Solvents Co., Inc., manufactures methyl alcohol and wishes to transport this flammable liquid in 55-gal, closed-head drums. If the drums are filled to a capacity of 54 gal at 68°F, has the manufacturer allowed sufficient outage so that the drums will not be completely full at 130°F?

Flammability

2.30. Other factors being equal, which flammable liquid poses the greater fire hazard: methyl acetate, whose flash point is 15°F, or ethanol, whose flash point is 54°F?

2.31. When flammable and combustible liquids are stored in containers outdoors (such as in multiple 55-gal drums), NFPA 30, the *Flammable and Combustible Liquids Code*, recommends the confinement of these containers to separate piles having no more than the maximum total gallonage noted:

Class IA, 1100 gal
Class IB, 2200 gal
Class IC, 4400 gal
Class II, 8800 gal
Class III, 22,000 gal

What maximum gallonage is recommended by NFPA in each such pile for the outdoors storage of 30,000 gal of each of the following flammable or combustible liquids?
(a) 2-Pinene, whose flash point is 91°F and whose boiling point is 313°F.
(b) Chlorobenzene, whose flash point is 84°F and whose boiling point is 270°F.
(c) Dipentene, whose flash point is 45°C and whose boiling point is 178°C.
(d) Ethanethiol, whose flash point is <27°C and whose boiling point is 37°C.

Gaseous State

2.32. The contents of an aerosol container of furniture polish are packed under pressure. The container is marked with a warning statement, which reads as follows: "Do not store at temperatures above 120°F."
(a) What hazardous feature of pressurized gases is illustrated by this statement?
(b) Does the container lose this hazardous feature when the furniture polish has been largely removed?

2.33. Oxygen is stored in a gas cylinder 5.0 ft long and 8.0 in. in internal diameter under a pressure of 225 psia. To what volume, in cubic feet (ft³), does the oxygen expand when it is suddenly released from its container at its original temperature, but to a pressure of 14.7 psia?

2.34. A gas cylinder is filled with hydrogen, a flammable gas, at a pressure of 235 psig when the ambient temperature is 65°F. During a fire, the temperature of the cylinder and its contents is raised to 350°F.
(a) What gauge pressure is experienced by the hydrogen at 350°F?
(b) What is likely to happen to the gas cylinder at this elevated temperature?

Cryogenic Liquids

2.35. Will gaseous ammonia at 250°F liquefy when 1000 psi of pressure is applied to it?

2.36. When responding to emergency incidents involving cryogenic liquids, why are members of emergency-response forces warned to prevent such materials from entering their boots?

3

Chemical Forms of Matter

Consider commercial products like paint, pesticide formulations, ammunition, gasoline, and other petroleum fuels. Each of these products is a hazardous material, comprised of a mixture of distinctly different components, each separable from one another. For instance, if we take a sample of paint into a chemical laboratory and subject it to chemical analysis, we find that it is a mixture of some resin or other film-forming compound, a solvent or thinner, and an organic or inorganic pigment. All such mixtures are characterized in a similar fashion as a blend of different components, dispersed either uniformly or nonuniformly, but capable of separation from one another.

When any mixture has been separated into its components, we note that these components are unique forms of matter. We shall identify these forms of matter in this chapter, as well as some of their general properties. In particular, we shall observe that any substance is either an element or a compound and possesses a characteristic set of physical and chemical properties. We shall also observe how chemists describe the structure of such substances in terms of atoms, molecules, and ions. Finally, we shall learn how chemists name unique substances and write their chemical formulas.

3-1 ELEMENTS AND COMPOUNDS

A material that has been separated from all other materials is called a *pure substance* or, more simply, a *substance*. Examples of such substances are aluminum metal, copper metal, oxygen, distilled water, and table sugar. No matter what procedures are used to purify them or what their origin is, samples of the same substance are indistinguishable from one another. All samples of distilled water are alike and indistinguishable from all other samples; that is, water is the same substance whether it has been distilled from rainwater, well water, seawater, or water from any other source.

A substance is characterized as a material having a fixed composition, usually expressed in terms of percentage by mass. Distilled water is a pure substance consisting of 11.2% hydrogen and 88.8% oxygen by mass. By contrast, coal is not a pure substance; its carbon content alone may vary anywhere from 35% to 90% by mass. Materials that are not pure substances, like coal, are *mixtures*.

Suppose we take a pure substance like limestone, chemically named calcium carbonate, and heat it, as shown in Fig. 3-1. The limestone ultimately crumbles to a white powder. Careful examination shows that carbon dioxide evolves from the limestone while it is heated. Substances like limestone that can be broken down into two or more simpler substances are called *compound substances,* or simply *compounds.* Heating is a common way of decomposing compounds, but other forms of energy may often be used as well.

Substances that resist attempts to decompose them into simpler forms of matter are called *elements.* The elements are the fundamental substances of which all matter is composed. There are only 106 known elements, but there are over one million known compounds. Of the 106 elements, only 88 are present in detectable amounts on Earth, and many of these 88 are very rare. Table 3-1 shows that only ten ele-

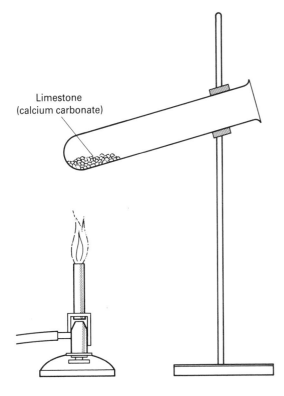

Limestone
(calcium carbonate)

Figure 3-1 The heating of limestone (calcium carbonate) in a test tube. The limestone chemically decomposes, evolving carbon dioxide. This illustrates that calcium carbonate is a compound and not an element.

$$CaCO_3 (s) \xrightarrow{\Delta} CO_2 (g) + CaO (s) \xrightarrow{H_2O} Ca(OH)_2 (s)$$

TABLE 3-1   NATURAL ABUNDANCE OF THE ELEMENTS (EARTH'S CRUST, OCEANS, AND THE ATMOSPHERE)

Oxygen	49.5%	Chlorine	0.19%
Silicon	25.7%	Phosphorus	0.12%
Aluminum	7.5%	Manganese	0.09%
Iron	4.7%	Carbon	0.08%
Calcium	3.4%	Sulfur	0.06%
Sodium	2.6%	Barium	0.04%
Potassium	2.4%	Chromium	0.033%
Magnesium	1.9%	Nitrogen	0.030%
Hydrogen	1.9%	Fluorine	0.027%
Titanium	0.58%	Zirconium	0.023%
		All others	<0.1%

ments make up approximately 99% by mass of the Earth's crust, including the atmosphere, the surface layer, and the bodies of water. We observe from this table that the most abundant element on Earth is oxygen, which is found in the free state in the atmosphere, as well as in combined form with other elements in numerous minerals and ores.

Each element has a specific name and symbol. Some common elements are listed in Table 3-2. The symbols of the elements consist of either one or two letters,

TABLE 3-2   SYMBOLS OF THE MOST COMMON ELEMENTS

Element	Symbol	Element	Symbol
Aluminum	Al	Lithium	Li
Antimony	Sb	Magnesium	Mg
Argon	Ar	Manganese	Mn
Arsenic	As	Mercury	Hg
Barium	Ba	Neon	Ne
Bismuth	Bi	Nickel	Ni
Boron	B	Nitrogen	N
Bromine	Br	Oxygen	O
Cadmium	Cd	Phosphorus	P
Calcium	Ca	Platinum	Pt
Carbon	C	Potassium	K
Chlorine	Cl	Radium	Ra
Chromium	Cr	Silicon	Si
Cobalt	Co	Silver	Ag
Copper	Cu	Sodium	Na
Fluorine	F	Strontium	Sr
Gold	Au	Sulfur	S
Helium	He	Tin	Sn
Hydrogen	H	Titanium	Ti
Iodine	I	Tungsten	W
Iron	Fe	Uranium	U
Lead	Pb	Zinc	Zn

with the first letter capitalized. It is best to memorize the symbols of the elements listed in Table 3-2, since they are often encountered in a study of hazardous materials.

A complete listing of all known elements is provided on the inside back cover of this book.

3-2 METALS, NONMETALS, AND METALLOIDS

Each element may be classified as a metal, nonmetal, or metalloid. *Metals* are elements that usually possess properties like the following: they conduct heat and electricity well, generally melt and boil at very high temperatures, possess relatively high densities, are normally malleable (able to be hammered into sheets) and ductile (able to be drawn into a wire), and display a brilliant luster. Examples of metals are iron, platinum, and silver. Almost all metals are solids at room temperature (68°F or 20°C). Mercury is the only common liquid metal at room temperature; none is gaseous.

Elements that do not possess the general physical properties just mentioned are called *nonmetals*. These elements generally melt and boil at relatively low temperatures, do not possess a luster, are less dense than metals, and hardly conduct heat and electricity, if at all. At room temperature, most nonmetals are either solids or gases. Bromine is the only liquid nonmetal at room temperature. Oxygen, fluorine, and nitrogen are examples of gaseous nonmetals, while carbon, sulfur, and phosphorus are examples of solid nonmetals.

There are exceptions to these general properties of metals and nonmetals. For instance, carbon conducts heat and electricity well, and one form of carbon melts at 2550°C. Yet chemists classify it as a nonmetal.

Several elements have properties resembling both metals and nonmetals. They are called *metalloids*. The metalloids are boron, silicon, germanium, arsenic, antimony, tellurium, and polonium. Silicon and germanium have the luster associated with metals, but do not conduct heat and electricity well.

3-3 PHYSICAL AND CHEMICAL CHANGES

The constant composition associated with a given substance is maintained by internal linkages among its units, such as between one atom and another. These linkages are called *chemical bonds*. When a particular process occurs that involves the making and breaking of these bonds, we say that a *chemical reaction* or a *chemical change* has occurred. Combustion and corrosion are common examples of chemical changes associated with some hazardous materials.

Let's briefly consider combustion. When a substance burns, it normally combines with atmospheric oxygen. The resulting products of combustion are compounds containing oxygen, called *oxides*. For instance, many synthetic gasolines are mixtures of several substances, including one called *octane*. When octane burns completely, it becomes carbon dioxide gas and water vapor; carbon dioxide is an oxide of carbon as its name implies, while water is an oxide of hydrogen. Carbon dioxide and water vapor are unlike octane or other gasoline components in many ways. They have different properties and different compositions. This conversion of octane to carbon dioxide and water is typical of a chemical change in any substance.

By contrast, a substance may undergo a change in which its compostion remains the same. Such changes are called *physical changes*. Let's consider octane again. When exposed to the ambient environment, it evaporates, but its composition remains unchanged. Alterations in the physical state of a substance, as here from a liquid to a vapor, are considered physical changes. Other examples of physical changes are melting, freezing, boiling, crushing, and pulverizing. Figure 3-2 illustrates the physical and chemical changes discussed here for octane.

Figure 3-2 An example of a physical and chemical change in octane, a component of gasoline. On the left the substance evaporates (that is, it changes its physical state from the liquid to vapor); this is a physical change. On the right a spill of octane ignites and burns, thereby changing to carbon dioxide and water vapor; this is a chemical change.

The types of behavior that a substance exhibits when undergoing chemical changes are called its *chemical properties*. The characteristics that do not involve changes in the chemical identity of a substance are called its *physical properties*. All substances may be distinguished from one another by these properties, in much the same way as certain features (fingerprints, for example) distinguish one human being from another. The study of hazardous materials is concerned to a great extent with learning the chemical and physical properties of many substances, some examples of which are listed in Table 3-3.

TABLE 3-3 CHARACTERISTICS OF SOME SUBSTANCES

Substance	Physical properties	Chemical properties
Oxygen, an element	Odorless, colorless gas; does not conduct heat or electricity; density = 1.43 g/L; becomes liquid at −297°F (−183°C)	Combines readily with many elements (a chemical reaction called oxidation)
Phosphorus, an element	White or red solid; does not conduct heat or electricity; density = 1.82 g/mL (white) and 2.34 g/mL (red)	Readily combines with oxygen, chlorine, and fluorine; white form spontaneously ignites in dry air
Carbon dioxide, a compound	Odorless and colorless gas; solidifies at −83°F (−67°C) under pressure, forming Dry Ice; soluble in water under pressure	Does not burn; reacts with water-soluble metal compounds, forming metallic carbonates
Hydrogen chloride, a compound	Strong-smelling, colorless gas; density = 1.20 g/mL; soluble in water, forming hydrochloric acid	Reacts with many minerals, forming water-soluble products; reacts with ammonia, forming ammonium chloride

3-4 INTERNAL STRUCTURE OF ATOMS

If a small piece of an element, say aluminum, is hypothetically divided and subdivided into smaller and smaller pieces, until its subdivision is no longer possible, the result is one particle of aluminum. This smallest particle of the element, which is still representative of the element, is called an *atom*, from the Greek word *atomos*, meaning indivisible.

An atom is infinitesimally small. Yet it is composed of particles, principally electrons, protons, and neutrons. *Electrons are* negatively charged particles and are responsible for the chemical reactivity of the atoms of a given element. *Protons* are positively charged particles, and *neutrons* are neutral particles. Electrons and protons bear the same magnitude of charge, but of opposite sign. For convenience, the charge of electron is assigned to be −1, and of the proton +1.

Protons are relatively massive; they are 1836 times more massive than electrons. Neutrons are even slightly more massive than protons. These fundamental characteristics of electrons, protons, and neutrons are summarized in Table 3-4.

The protons and neutrons of an atom reside in a central area called its *nucleus*. Electrons reside primarily in designated regions of space surrounding the nucleus, called *atomic orbitals*. There are several types of atomic orbitals, some of which are relatively close to the nucleus, while others are relatively remote from it. Only a prescribed number of electrons may reside in a given type of atomic orbital. Two electrons are always close to the nucleus, in an atom's innermost atomic orbital (except

TABLE 3-4 SOME BASIC ATOMIC PARTICLES

Particle	Proton	Electron	Neutron
Symbol	p^+	e^-	n
Relative charge	+1	−1	0
Relative mass	1	About 0^a	1

[a] The mass of an electron is 1/1836 the mass of the proton.

in hydrogen atoms, which have only one electron). In most atoms, other electrons are located in atomic orbitals some distance from the nucleus.

When atoms of the same element are neutral, they contain the same number of protons and electrons. The number is called the *atomic number*. We often use atomic numbers in the study of chemistry to determine the number of electrons possessed by neutral atoms of an element. Atoms of hydrogen have one electron, those of helium have two, those of lithium have three, and so forth.

While neutral atoms of the same element have an identical number of protons and electrons, they may differ by the number of neutrons in their nuclei. Atoms of the same element having different numbers of neutrons are called *isotopes* of that element.

Thus, carbon is an element composed of carbon atoms, and all carbon atoms exhibit nearly the same physical and chemical properties. Some may have slightly different masses due to different numbers of neutrons in their nuclei, but these carbon atoms all act the same when they are involved in chemical changes. Similarly, oxygen is an element composed of oxygen atoms, and all oxygen atoms possess nearly the same properties, too. But, carbon and oxygen atoms are not alike, since they have different numbers of elementary particles.

Over the past two centuries, scientists have been able to determine the relative masses of the atoms of all known elements. These relative masses are called *atomic weights*. Note that they are not absolute masses, but rather masses that have been measured relative to the mass of a particular type of atom selected as a standard. Since they are relative parameters, atomic weights are unitless. A specific isotope of carbon, called *carbon-12*, has been selected as the atom whose mass serves as a reference standard. This carbon isotope consists of six electrons, six protons, and six neutrons and is assigned a mass of exactly 12.

Scientists experimentally establish the atomic weights of all elements by taking into account their naturally occurring isotopes and natural abundances. In other words, the atomic weight of any element is obtained from the average relative mass of its naturally occurring isotopes, weighted according to their natural abundances. This means that the atomic weight of a given element is a number that tells us how the mass of an *average atom* of that element compares with the mass of a carbon-12 atom.

Use of the carbon-12 standard results in an atomic weight for natural carbon itself of 12.011, because of the slight difference involved in averaging the masses of its natural isotopic forms. The atomic numbers and atomic weights of the elements have been provided with the listing on the inside back cover of this book.

3-5 PERIODIC CLASSIFICATION OF THE ELEMENTS

During the last half of the nineteenth century, several scientists first noted that the chemical properties of any given element were similar to the chemical properties of certain other elements. Let's consider sodium metal. Sodium reacts explosively with water and burns spontaneously in air. When these two chemical properties are compared with the properties of other elements, we find that potassium is among the metals that also explodes on contact with water and burns spontaneously with air.

Another example of the repetition of properties is provided by helium. As far as we know, helium is chemically inert. When we look to see if any other element is chemically inert, neon and argon are found to be among the elements that lack chemical activity.

These are just two of many examples that illustrate that the chemical properties of the elements repeat themselves. Chemists summarize all such observations in the *periodic law*: The properties of the elements vary periodically with their atomic numbers.

Suppose we write the name of each element in a square and then arrange the 106 squares by order of increasing atomic number. This means that the total number of electrons possessed by each element increases in this arrangement, one at a time, as we move from one square to the next. Then, let's further arrange them into rows and columns in such a manner that when elements are put into the same column, they possess similar properties. Of course, one would need to know a great deal of chemistry to accomplish this feat. A similar exercise was first performed over a hundred years ago, when many elements had not even been discovered.

Such an arrangement of the chemical elements into a chart designed to represent the periodic law is called a *periodic table*. A modern version is illustrated in Fig. 3-3. Elements positioned within the same column of a periodic table are called the members of a *family* of elements. Each family is identified by a Roman numeral and a capital letter at the top of the column, such as IA, IIA, and so on. Thus, for example, helium, neon, argon, krypton, xenon, and radon belong to the same family, identified by VIIIA. Elements that are positioned in the same row of a periodic table belong to the same *period*. In Fig. 3-3, the periods are numbered on the far left of the table from 1 to 7. There is one period of two elements, two periods of eight elements each, two more of eighteen elements each, one period of thirty-two elements, and a final period, which presently has twenty elements.

The periodic table is an important aid for learning chemistry, since it tabulates a variety of information in one spot. We observe immediately that the atomic numbers of the elements are tabulated on the periodic table. (On larger tables, such as those found in chemistry classrooms, the periodic table also contains a listing of atomic weights.) Another useful feature is that it allows us to readily identify which

METALS

NONMETALS

Period	Group IA	IIA	IIIB	IVB	VB	VIB	VIIB	VIIIB			IB	IIB	IIIA	IVA	VA	VIA	VIIA	VIIIA
1	1 Hydrogen H																1 Hydrogen H	2 Helium He
2	3 Lithium Li	4 Beryllium Be											5 Boron B	6 Carbon C	7 Nitrogen N	8 Oxygen O	9 Fluorine F	10 Neon Ne
3	11 Sodium Na	12 Magnesium Mg											13 Aluminum Al	14 Silicon Si	15 Phosphorus P	16 Sulfur S	17 Chlorine Cl	18 Argon Ar
4	19 Potassium K	20 Calcium Ca	21 Scandium Sc	22 Titanium Ti	23 Vanadium V	24 Chromium Cr	25 Manganese Mn	26 Iron Fe	27 Cobalt Co	28 Nickel Ni	29 Copper Cu	30 Zinc Zn	31 Gallium Ga	32 Germanium Ge	33 Arsenic As	34 Selenium Se	35 Bromine Br	36 Krypton Kr
5	37 Rubidium Rb	38 Strontium Sr	39 Yttrium Y	40 Zirconium Zr	41 Niobium Nb	42 Molybdenum Mo	43 Technetium Tc	44 Ruthenium Ru	45 Rhodium Rh	46 Palladium Pd	47 Silver Ag	48 Cadmium Cd	49 Indium In	50 Tin Sn	51 Antimony Sb	52 Tellurium Te	53 Iodine I	54 Xenon Xe
6	55 Cesium Cs	56 Barium Ba	* 57 Lanthanum La	72 Hafnium Hf	73 Tantalum Ta	74 Tungsten W	75 Rhenium Re	76 Osmium Os	77 Iridium Ir	78 Platinum Pt	79 Gold Au	80 Mercury Hg	81 Thallium Tl	82 Lead Pb	83 Bismuth Bi	84 Polonium Po	85 Astatine At	86 Radon Rn
7	87 Francium Fr	88 Radium Ra	** 89 Actinium Ac	104 Kurcha-tovium Ku	105 Hahnium Ha	106												

* Lanthanide Series

6	58 Cerium Ce	59 Prase-odymium Pr	60 Neodymium Nd	61 Promethium Pm	62 Samarium Sm	63 Europium Eu	64 Gadolinium Gd	65 Terbium Tb	66 Dysprosium Dy	67 Holmium Ho	68 Erbium Er	69 Thulium Tm	70 Ytterbium Yb	71 Lutetium Lu

** Actinide Series

7	90 Thorium Th	91 Protactinium Pa	92 Uranium U	93 Neptunium Np	94 Plutonium Pu	95 Americium Am	96 Curium Cm	97 Berkelium Bk	98 Californium Cf	99 Einsteinium Es	100 Fermium Fm	101 Mendelevium Md	102 Nobelium No	103 Lawrencium Lr

Key: 11 Atomic Number — Sodium Name — Na Symbol

Figure 3-3 A version of the periodic table of the elements. The zigzag solid line separates metals (on the left) from nonmetals (on the right). The elements that lie to the immediate right or left of the zigzag line are metalloids.

elements are metals, nonmetals, and metalloids. The bold, zigzag line separates metals from nonmetals, while those elements lying to each immediate side of the line are the metalloids. Metals fall to the left of the line, and nonmetals fall to the right of it.

Another useful feature of the periodic table is evident from its design. When we observe elements as the members of a family, we know that these elements possess similar chemical properties. Four families deserve special recognition in this regard. Group IA is called the *alkali metal family*; its members are lithium, sodium, potassium, rubidium, cesium, and francium. We noted earlier that each of them is a metal that reacts explosively with water and ignites on exposure to the air.

Group IIA is called the *alkaline earth family*; its members are beryllium, magnesium, calcium, strontium, barium, and radium. These elements are also chemically reactive, but not nearly as reactive as the alkali metals. They cause water to decompose, but slowly. They ignite in air, but only after they have been heated or exposed to an ignition source.

Group VIIA is called the *halogen family*; its members are fluorine, chlorine, bromine, iodine, and astatine. These elements are nonmetals, each of which is particularly reactive. Fluorine ranks as the most chemically reactive of all elements.

Group VIIIA is called the *noble gas family*. Its six members are also nonmetals: helium, neon, argon, krypton, xenon, and radon. Chemists originally thought that these gases were totally inert to chemical combination and called them *inert gases*. However, some of these elements are now known to form compounds, disproving the idea that they are totally inert. Yet, from a chemical viewpoint, the noble gases uniquely stand out as a group of elements lacking the chemical reactivity associated with all other elements.

Hydrogen is a unique element in that it occupies two positions on the periodic table, once in IA and then again in VIIA. This is because hydrogen has some properties of the alkali metals and some properties of the halogens. Although positioned in each family, hydrogen is considered neither an alkali metal nor a halogen; rather, chemists think of hydrogen as an element by itself, to illustrate its uniqueness and individuality.

3-6 MOLECULES AND IONS

While the smallest representative particle of an element is the atom, not all uncombined elements exist as single atoms. In fact, only six elements actually exist as single atoms. These are the noble gases. We say these elements are *monatomic*.

When elements other than the noble gases exist in either the gaseous or liquid state of matter at room conditions, they consist of units containing pairs of like atoms. These units are called *molecules*. For example, hydrogen, oxygen, nitrogen, and chlorine are gases as we generally encounter them. Each exists as a molecule having two atoms. These molecules are said to be *diatomic* and are symbolized by the notations H_2, O_2, N_2 and Cl_2, respectively. They are illustrated in Fig. 3-4.

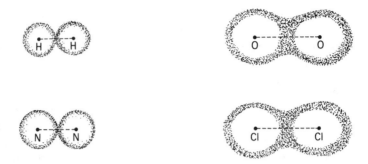

Figure 3-4   Some simple diatomic molecules (not to scale).

The smallest particle of many compounds is also the molecule. Molecules of compounds contain atoms of two or more elements. For example, the water molecule consists of two atoms of hydrogen and one atom of oxygen. The methane molecule consists of one carbon atom and four hydrogen atoms. Chemists denote these molecules as H_2O and CH_4, respectively.

Not all compounds occur naturally as molecules. Many of them occur as aggregates of oppositely charged atoms or groups of atoms called *ions*. Atoms become charged by gaining or losing some of their electrons. In general, metal atoms lose electrons, while nonmetal atoms gain electrons. Atoms of metals that lose their electrons become positively charged; atoms of nonmetals that gain electrons become negatively charged.

Let's consider the difference between the sodium atom and the sodium ion. The atomic number of sodium is 11, which we establish from either Fig. 3-3 or the listing on the inside back cover. Thus, the neutral sodium atom has 11 electrons and 11 protons. If the sodium atom somehow loses an electron, it then has only 10 left, although it still has its 11 protons. By losing the electron, the sodium atom becomes a sodium ion. Its net charge is $+1$; that is, $+11 + (-10) = +1$.

Consider magnesium as a second example. Magnesium has an atomic number of 12; thus, the neutral magnesium atom has 12 electrons and 12 protons. Suppose the magnesium atom somehow loses two electrons. When the magnesium atom loses two electrons, it becomes a magnesium ion. It has 10 electrons and 12 protons, and its net charge is $+2$.

These processes by which metals become positive ions may be represented as follows:

$$Na \longrightarrow Na^+ + e^-$$

$$Mg \longrightarrow Mg^{2+} + 2e^-$$

Here, e^- represents an electron; writing e^- to the right of the arrow means an electron is lost from the atom of the metal whose symbol appears to the left of the arrow. Na^+ and Mg^{2+} denote sodium and magnesium ions, respectively.

As noted earlier, atoms of nonmetals tend to *gain* electrons. Consider atoms of fluorine. The atomic number of fluorine is 9. Hence, a fluorine atom possesses 9 electrons and 9 protons. If a fluorine atom somehow attains another electron, it would then have 10 electrons, but still only 9 protons. Its net charge would be -1; that is, $+9 + (-10) = -1$. Fluorine atoms occur naturally as diatomic molecules (F_2). Hence, we represent this ionization process as follows:

$$F_2 + 2e^- \longrightarrow 2F^-$$

Atoms of metals that have lost one or more electrons to become positively charged ions are named with the name of the metal. As noted earlier, Na^+ and Mg^{2+} are named the sodium and magnesium ions, respectively. Atoms of nonmetals that have gained one or more electrons to become negatively charged ions are named by modifying the name of the nonmetal so that it ends in *-ide*; F^- is thus named the fluoride ion.

Frequently, two or more nonmetal atoms unite to form a *polyatomic ion*. For instance, one sulfur atom and four oxygen atoms often unite to form an ion with a net charge of -2; it is called the *sulfate ion* and is symbolized as SO_4^{2-}. We shall observe other examples of polyatomic ions in Sec. 3-10.

3-7 NATURE OF CHEMICAL BONDING

We noted earlier that, as compounds form, the atoms of one element become attached to, or associated with, atoms of other elements by forces called *chemical bonds*. But how do these chemical bonds form? The answer to this question is one upon which chemists have speculated for centuries.

In Sec. 3-5, we observed that the noble gases are relatively inert. This observation implies that their atomic structures are associated with a special electronic stability. With the exception of helium, atoms of the noble gases have eight electrons in their outermost atomic orbitals, that is, those farthest from their nuclei. It is this presence of eight electrons in the outermost atomic orbitals of most noble gases that gives them their unique electronic stability. An atom of helium has only two electrons, both in its innermost atomic orbital, and these two electrons give helium its special electronic stability.

Molecules and ions often form when their constituent atoms are capable of acquiring the electronic stability of the noble gases. A neutral atom has only a prescribed number of electrons; hence, it must somehow interact with the electrons from other atoms to acquire the structure of a noble gas.

Bonding occurs between atoms by either of two mechanisms, called *ionic bonding* and *covalent bonding*, which we shall discuss independently. These two procedures for describing the bonding capabilities of atoms represent extreme situations, but, in actuality, some degree of each type of bonding exists in all substances.

3-8 LEWIS SYMBOLS

While an atom of any element possesses a definite number of electrons, only some of them are normally involved in chemical bonding. These electrons are called the *valence electrons* of an atom. They are the electrons in an atom's outermost atomic orbital. Electrons that do not participate in bonding are called *nonbonding electrons.*

We display valence electrons by means of a *Lewis symbol*, named after the American chemist, G. N. Lewis. A Lewis symbol consists of a symbol of an element together with a certain number of dots, which represent the element's valence electrons. For example, Na· is the Lewis symbol for the sodium atom. The sodium atom has only one electron that participates in chemical bonding.

Table 3-5 lists the Lewis symbols of some representative elements important to the study of hazardous materials. Note that a simple way exists for determining the number of valence electrons for any element of an A family. The number that identifies the A family on the periodic table is also the number of valence electrons possessed by elements in that family. For instance, we observe from the periodic table that the halogens are located in the family denoted as VIIA, and we note from Table 3-5 that each halogen atom has seven valence electrons.

TABLE 3-5   LEWIS SYMBOLS OF SOME REPRESENTATIVE ELEMENTS

Family	IA	IIA	IIIA	IVA	VA	VIA	VII A	
	Li·	·Be·	·B·	·C·	·N·	·O·	:F·	H·
	Na·	·Mg·	·Al·	·Si·	·P·	·S·	:Cl·	
	K·	·Ca·	·Ga·	·Ge·	·As·	·Se·	:Br·	
	Rb·	·Sr·				·Te·	:I·	

3-9 IONIC BONDING

Electrons may be *transferred* from an atom of one element to an atom of another element to form positive and negative *ions*. This phenomenon generally occurs between the atoms of metals and nonmetals. The ions that form are attracted to each other by virtue of their opposite charges, an electrostatic force of attraction that chemists call an *ionic bond*. This is based on a fundamental law of nature by which forms of matter with like charges (+/+ or −/−) repel each other, whereas forms of matter with unlike charges (+/−) attract.

Many atoms of the elements transfer or accept just the number of electrons that gives them the same number as the noble gas nearest to the elements in atomic number. This allows these atoms to attain the electronic stability of the noble gas.

For illustration, consider the ionic bonding in sodium fluoride. Table 3-5 indicates that the Lewis symbols of sodium and fluorine are Na· and :F̈· , respectively. When these two elements combine at the atomic level, an atom of sodium transfers its single electron to a fluorine atom. By transferring an electron, the sodium atom electronically resembles neon. By accepting it, the fluorine atom electronically resembles neon. The atoms correspondingly become charged; that is they become ions. The process can be written schematically as follows:

$$\text{Na·} + \text{:F̈·} \longrightarrow \text{Na}^+ \text{:F̈:}^-$$

The attraction between these oppositely charged ions constitutes the ionic bond that binds them together.

3-10 COVALENT BONDING

Electrons may be *shared* by atoms of identical or different elements to form molecules of elements or compounds. This sharing of electrons is usually between atoms of the nonmetals. Atoms of the same nonmetal bond to form molecules of an element; atoms of different nonmetals bond to form molecules of a compound.

Atoms of nonmetals acquire their electronic stability by sharing electrons in a manner that permits them to resemble atoms of the noble gases nearest to them in atomic number. For all atoms except the hydrogen atom, this means acquiring a total of eight electrons in their outermost atomic orbital, that is, the atom's valence electrons plus those it shares with another atom. For hydrogen atoms, it means sharing two electrons. One shared pair of electrons between any two atoms is called a *covalent bond*.

Let's observe what happens when two atoms of hydrogen combine to form a molecule of this element. H· is the Lewis symbol for hydrogen. To achieve the electronic structure of the nearest noble gas, helium, one hydrogen atom shares its only electron with the single electron from a second hydrogen atom. We may represent this process as follows:

$$\text{H·} + \text{H·} \longrightarrow \text{H:H}$$

The pair of electrons shared between these atoms of hydrogen is the covalent bond. This manner of representing the hydrogen molecule, that is, as H:H, is called its *Lewis structure*.

Two chlorine atoms combine to form a molecule of chlorine. The Lewis symbol of chlorine is :C̈l· . To achieve the electronic structure of the noble gas nearest to it in atomic number (argon), each chlorine atom shares its unpaired electron. This formation of the chlorine molecule from two chlorine atoms may be represented as follows:

$$\text{:C̈l·} + \text{:C̈l·} \longrightarrow \text{:C̈l:C̈l:}$$

Let's consider next the combination of hydrogen and chlorine. A hydrogen atom and a chlorine atom may share an electron pair to form one molecule of the substance called *hydrogen chloride*. Formation of the hydrogen chloride molecule from hydrogen and chlorine atoms may be represented as follows:

$$H \cdot + \; :\ddot{C}l \cdot \; \longrightarrow \; H:\ddot{C}l:$$

The hydrogen atom shares an electron pair, so its electronic structure resembles that of the helium atom; the chlorine atom shares an electron pair, so it electronically resembles the argon atom.

An atom may also form more than one covalent bond by simply sharing more than one pair of electrons. For instance, consider the formation of the methane molecule, which consists of one carbon atom and four hydrogen atoms. Lewis symbols for carbon and hydrogen are $\cdot \dot{C} \cdot$ and $H \cdot$, respectively. In methane, the carbon atom shares each of its four electrons with the electrons from four hydrogen atoms, as follows:

$$\cdot \dot{C} \cdot + 4H \cdot \; \longrightarrow \; \begin{matrix} H \\ H:\ddot{C}:H \\ H \end{matrix}$$

The sharing of electrons in the methane molecule results in an electronic arrangement like neon for the carbon atom and an electronic arrangement like helium for each of the four hydrogen atoms.

Finally, there are some compounds in which two nonmetallic atoms share more than one pair of electrons. This behavior is particularly characteristic of carbon atoms. It results in the formation of *multiple bonds*. Two types of multiple bonds exist: double and triple bonds. A *double bond* consists of the sharing of two pairs of electrons (::), while a *triple bond* consists of the sharing of three pairs of electrons (:::).

Carbon dioxide and carbon monoxide molecules exemplify the formation of double and triple bonds, respectively. The carbon dioxide molecule consists of one atom of carbon and two atoms of oxygen. Lewis symbols for carbon and oxygen are $\cdot \dot{C} \cdot$ and $\cdot \dot{O} \cdot$, respectively. The formation of a carbon dioxide molecule from carbon and oxygen atoms may be represented as follows:

$$\cdot \ddot{O} \cdot + \cdot \dot{C} \cdot + \cdot \ddot{O} \cdot \; \longrightarrow \; \ddot{O}::C::\ddot{O}$$

Note the existence of two pairs of electrons on each side of the carbon atom. These two shared pairs of electrons constitute a double bond. By sharing electrons in this fashion, the carbon atom and two oxygen atoms achieve electronic arrangements like atoms of neon. Chemists symbolize the carbon dioxide molecule as CO_2, that is, one atom of carbon and two atoms of oxygen.

The formation of the carbon monoxide molecule from carbon and oxygen atoms may be represented as follows:

$$\cdot \dot{C} \cdot + \cdot \ddot{O} \cdot \; \longrightarrow \; :C:::O:$$

The three shared pairs of electrons between the carbon and oxygen atoms constitute a triple bond. Once again, the carbon and oxygen atoms achieve the electronic stability of the neon atom by sharing electrons between them. Since the carbon monoxide molecule consists of one carbon atom and one oxygen atom, we symbolize it as CO.

Figure 3-5 illustrates the structures of several of these molecules.

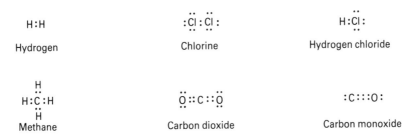

Figure 3-5   The Lewis structures of some simple molecules.

For the sake of simplicity, Lewis structures are usually written by using a long dash, or bond, for a shared pair of electrons and omitting the dots representing any other electrons. Using bonds to represent a shared pair of electrons, the compounds previously discussed may be represented as follows:

$$H-H \qquad Cl-Cl \qquad H-Cl \qquad H-\overset{\displaystyle H}{\underset{\displaystyle H}{|}}C-H \qquad O=C=O \qquad C=O$$

hydrogen   chlorine   hydrogen   methane   carbon   carbon
chloride   dioxide   monoxide

3-11 IONIC AND COVALENT COMPOUNDS

Chemical compounds are often classified into either of two groups based on the nature of the bonding between their atoms. Chemical compounds consisting of atoms bonded together by means of ionic bonds are called *ionic compounds*. Compounds whose atoms are bonded together by covalent bonds are called *covalent compounds*.

Ionic and covalent compounds are associated with a number of general properties, which are summarized in Table 3-6. Note from this table that ionic compounds usually boil and melt at much higher temperatures than do covalent compounds. Most of them dissolve in water and conduct an electric current when they are melted or dissolved in solution, while the opposite is true of covalent compounds. These are generalizations only and reflect only the characteristics of most ionic and covalent compounds.

TABLE 3-6 CONTRASTING PROPERTIES BETWEEN IONIC AND COVALENT
COMPOUNDS

Ionic compounds	Covalent compounds
High melting points	Low melting points
High boiling points	Low boiling points
High solubility in water	Low solubility in water
Nonflammable	Flammable
Molten substances and their water solutions conduct an electric current	Molten substances do not conduct an electric current
Exist predominantly as solids at room temperature	Exist as gases, liquids, and solids at room temperature

3-12 CHEMICAL FORMULAS

Chemists condense information regarding the chemical composition of substances by
writing a *chemical formula* for them. In the case of substances that consist of
molecules, the chemical formula indicates the kinds of atoms present in each
molecule and the actual number of them. For instance, we observed earlier in this
chapter that hydrogen and oxygen are symbolized as H_2 and O_2, respectively. These
are the chemical formulas for hydrogen and oxygen. The subscript 2 means that each
molecule of these elements contains two atoms. These chemical formulas are read
verbally as "H-two" and "O-two," respectively.

We have also observed that the chemical formulas of hydrogen chloride and
methane are HCl and CH_4, respectively. These formulas represent the composition
of a hydrogen chloride molecule and a methane molecule. One molecule of hydrogen chloride consists of one hydrogen atom and one chlorine atom, while a molecule
of methane consists of one carbon atom and four hydrogen atoms. We verbally read
these chemical formulas as "HCl" and "CH-four", respectively.

Figure 3-6 interprets the nature of two additional chemical formulas.

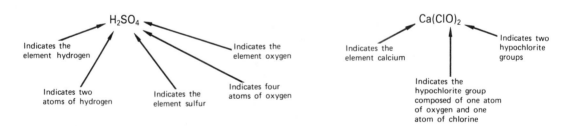

Figure 3-6 Explanation of the chemical formulas H_2SO_4 and $Ca(ClO)_2$. Sulfuric acid is denoted by the formula H_2SO_4; it is a corrosive liquid used to produce rayon, certain explosives,
dyes, other acids and detergents. Calcium hypochlorite is denoted by $Ca(ClO)_2$; it is a constituent of some solid bleaching agents and disinfecting agents.

On the other hand, substances are not always composed of molecules. For instance, sodium fluoride, an ionic compound, is composed of ions held together by ionic bonds. Experimental evidence indicates that the components of ionic compounds are not molecules, but rather aggregates of positive and negative ions. When we analyze a sample of solium fluoride, we learn that it is composed of an equal number of sodium and fluoride ions; that is, sodium fluoride exists as aggregates of Na^+ and F^- ions. Chemists write its chemical formula as Na^+F^-, representing the lowest number of formula units. In common practice, the net charges are omitted and we simply write NaF.

When responding to incidents involving a hazardous material, knowing what the substance actually is from its name or chemical formula is extremely valuable. The identity of the unique hazardous material in a package, container, or transport vehicle is irreplaceable information. When a substance's identity is known, the response to an emergency mishap can be more properly directed by accounting for the unique characteristics of the hazardous material at issue.

During emergency-response actions, the names and formulas of chemical commodities may be determined in several ways. They are provided on Material Data Safety Sheets (MSDSs) when the commodity is a single substance and not a mixture. When hazardous materials are transported, specific names may often be identified by reviewing the information provided on shipping papers. Furthermore, the "proper shipping name" of a hazardous material is also required to be marked on nonbulk packages (for example, boxes and steel drums) as well as cargo tanks, tank cars, and multiunit rail tank cars containing hazardous materials. The proper shipping name and the true chemical name of a substance are frequently synonymous. (See Chapter 5 for details.)

3-13 WRITING CHEMICAL FORMULAS AND NAMING IONIC COMPOUNDS

The chemical formulas of ionic compounds are obtained from the symbols of the ions that make up a given substance. The most commom ions are listed in Table 3-7. The names and symbols of these ions should be memorized. There are several simple rules by which these ions are named, which makes the memorization less tedious.

Positive ions

1. Alkali metals, alkaline earth metals, aluminum, zinc, and hydrogen form only one positive ion. They take the name of the metal from which they are derived. The net charge on hydrogen and each alkali metal is $+1$, and the net charge on each alkaline earth metal ion is $+2$.

Na^+	sodium ion	Ba^{2+}	barium ion
Al^{3+}	aluminum ion	H^+	hydrogen ion

TABLE 3-7 SOME COMMON IONS

Positive ions	Negative ions
Ammonium, NH_4^+	Acetate, $C_2H_3O_2^-$
Copper(I) or cuprous, Cu^+	Bromide, Br^-
Hydrogen, H^+	Chloride, Cl^-
Silver, Ag^+	Chlorate, ClO_3^-
Sodium, Na^+	Chlorite, ClO_2^-
Potassium, K^+	Cyanide, CN^-
	Fluoride, F^-
Barium, Ba^{2+}	Hydrogen carbonate or bicarbonate, HCO_3^-
Cadmium, Cd^{2+}	Hydrogen sulfate or bisulfate, HSO_4^-
Calcium, Ca^{2+}	Hydrogen sulfite or bisulfite, HSO_3^-
Cobalt(II) or cobaltous, Co^{2+}	Hydroxide, OH^-
Copper(II) or cupric, Cu^{2+}	Hypochlorite, ClO^-
Iron(II) or ferrous, Fe^{2+}	Iodide, I^-
Lead(II) or plumbous, Pb^{2+}	Nitrate, NO_3^-
Magnesium, Mg^{2+}	Nitrate, NO_2^-
Manganese(II) or manganous, Mn^{2+}	Perchlorate, ClO_4^-
Mercury(I) or mercurous, Hg_2^{2+}	Permanganate, MnO_4^-
Strontium, Sr^{2+}	
Tin(II) or stannous, Sn^{2+}	Carbonate, CO_3^{2-}
Zinc, Zn^{2+}	Oxide, O^{2-}
	Peroxide, O_2^{2-}
Chromium(III) or chromic, Cr^{3+}	Sulfate, SO_4^{2-}
Iron(III) or ferric, Fe^{3+}	Sulfide, S^{2-}
	Sulfite, SO_3^{2-}
Tin(IV) or stannic, Sn^{4+}	
	Phosphate, PO_4^{3-}

2. If a metal forms more than one positive ion, the ion is named in either of two
   ways:

   a. The ion takes the English name of the metal from which it is derived, im-
      mediately followed by a Roman numeral written in parentheses that indi-
      cates the net charge on the ion. This system of naming ionic compounds is
      called the *Stock system*.

   Cu^+ copper(I) ion Sn^{2+} tin(II) ion

   Cu^{2+} copper(II) ion Sn^{4+} tin(IV) ion

   b. The ion takes the Latin name of the metal from which it is derived, to-
      gether with one of the suffixes -*ous* or *ic*, representing the lower and higher
      net charge, respectively. We call this system of naming ionic compounds
      the *older system*.

   Cu^+ cuprous ion Sn^{2+} stannous ion

   Cu^{2+} cupric ion Sn^{4+} stannic ion

3. There are only two common polyatomic positive ions:

NH_4^+ ammonium ion Hg_2^{2+} mercury(I) or
                                                  mercurous ion

Negative Ions

1. Monatomic negative ions are named by adding the suffix *-ide* to the stem of the name of the nonmetals from which they are derived.

F^- fluoride ion Cl^- chloride ion

O^{2-} oxide ion S^{2-} sulfide ion

2. Polyatomic negative ions are named according to the following system:
   a. Two polyatomic negative ions are named with the suffix *-ide*:

CN^- cyanide ion OH^- hydroxide ion

   b. The polyatomic negative ions containing carbon are named uniquely. Two are especially important, the carbonate and acetate ions.

CO_3^{2-} carbonate ion $C_2H_3O_2$ acetate ion

   c. When a nonmetal forms two different negative ions containing oxygen, the suffix *-ite* is used to name the ion with the lesser number of oxygen atoms, and the suffix *-ate* is used to name the ion with the higher number of oxygen atoms.

NO_2^- nitrite ion SO_3^{2-} sulfite ion

NO_3^- nitrate ion SO_4^{2-} sulfate ion

   d. When a nonmetal forms more than two different negative ions containing oxygen, the prefix *hypo* (meaning lower than usual) or *per* (meaning higher than usual) is used together with the suffixes *-ite* and *-ate*, in order of increasing number of oxygen atoms, as follows: hypo_____ite, _____ite, _____ate, and per_____ate. Thus, the polyatomic ions of chlorine and oxygen are named as follows:

ClO^- hypochlorite ion

ClO_2^- chlorite ion

ClO_3^- chlorate ion

ClO_4^- perchlorate ion

   e. Certain negative ions containing hydrogen and oxygen atoms (exceptions are the hydroxide and acetate ions) are named hydrogen _____, or alternatively bi_____.

$$HCO_3^-$$ hydrogen carbonate, or bicarbonate

$$HSO_4^-$$ hydrogen sulfate, or bisulfate

When we know the chemical formulas of substances, we name the associated compounds by simply listing the names of the positive and negative ions in that order. For instance, NaCl is sodium chloride, MgS is magnesium sulfide, and NH_4NO_3 is ammonium nitrate.

Frequently, we encounter chemical formulas containing parentheses, such as $Fe(ClO_4)_3$. It is sometimes best to rewrite such formulas to indicate the net charges on the constituent ions. In this case, we have $Fe^{3+}(ClO_4^-)_3$. This shows that the $+3$ charge on the iron(III) ion balances the -3 charge on three perchlorate ions. In other words, the total positive and negative charges of the ions are equal ($+3$ and -3), since the compound is neutral. The compound is named iron(III) perchlorate by the Stock system and ferric perchlorate by the older system.

Before proceeding further, it is best to note the following examples:

$AlCl_3$ aluminum chloride

$(NH_4)_2CO_3$ ammonium carbonate

K_2SO_4 potassium sulfate

SnF_2 stannous fluoride or tin(II) fluoride

$Ca_3(PO_4)_2$ calcium phosphate

Suppose next that we are interested in writing a chemical formula when we know the name of the substance. It is again best to follow two simple rules, as follows:

1. If the net charges on the positive and negative ions are equal (but opposite in sign), the formula of a substance is obtained by simply listing the symbol of the positive ion first, and then the symbol of the negative ion. For instance, NaF, CaS, ZnO, KOH, $NaClO_3$, $BaSO_4$, and $AlPO_4$ are the chemical formulas of compounds whose component ions have numerically equal charges.

2. If the net charges on the positive and negative ions are not numerically equal, first write the symbol of the positive and negative ions together with their respective charges. Then, simply cross the numbers that represent the charge (but not the sign) to the opposite ion, as shown in the following examples:

Potassium sulfate: $K_2^+ \ SO_4^{2-}$,   or K_2SO_4 (1 is not written)

Aluminum oxide: $Al_2^{3+} \ O_3^{2-}$,   or Al_2O_3

Barium iodide: $Ba^{2+} \ I_2^-$,   or BaI_2 (1 is not written)

Ammonium sulfate: $NH_4^+ \quad SO_4^{2-}$,   or $(NH_4)_2SO_4$ (1 is not written)

In the use of the second rule, one special case should be noted. When the charges are multiples of one another, such as $+4$ and -2, the lowest common multiple should be used. For instance, the chemical formula of tin(IV) oxide is SnO_2, not Sn_2O_4. In the study of hazardous materials, this exception arises only for compounds containing the tin(IV) ion.

3-14 WRITING SOME CHEMICAL FORMULAS AND NAMING COVALENT COMPOUNDS

It is generally more difficult to name covalent compounds and to write their chemical formulas. In this book, we shall learn many of them as the occasions present themselves. Nevertheless, a great many covalent compounds have acquired common names through the passage of time, which are still used as part of the language of chemistry. Table 3-8 lists the common names and formulas of some of these covalent compounds that are named nonsystematically. These names and formulas should be memorized.

TABLE 3-8   COMMON NAMES FOR SOME
SIMPLE COVALENT COMPOUNDS

Formula of compound	Common name
C_2H_2	Acetylene
NH_3	Ammonia
C_6H_6	Benzene
C_2H_6	Ethane
N_2H_4	Hydrazine
H_2S	Hydrogen sulfide
CH_4	Methane
PH_3	Phosphine
C_3H_8	Propane
H_2O	Water

A number of covalent compounds contain only two elements, and these substances can be named in a simple fashion. First, we name the element that is written first in a chemical formula, preceded by a Greek prefix that tells us how many atoms of the element are in the compound; then we name the other element, preceding it also with a similar prefix indicating the number of atoms of this second element. The name of the second element is modified so that it ends in -ide. Table 3-9 lists the Greek prefixes used for naming these simple compounds.

We have encountered the following chemical formulas of several simple covalent compounds containing only two elements: HCl, CO, and CO_2. Following our

TABLE 3-9 GREEK PREFIXES USED IN NAMING SIMPLE COVALENT COMPOUNDS

Prefix		Compound
Mono-	1	Carbon monoxide, CO^a
Di-	2	Carbon dioxide, CO_2
Tri-	3	Phosphorus trichloride, PCl_3
Tetra-	4	Carbon tetrachloride, CCl_4
Penta-	5	Phosphorus pentachloride, PCl_5
Hexa-	6	Sulfur hexafluoride, SF_6
Hepta-	7	Dichlorine heptoxide, $Cl_2O_7{}^a$
Octa-	8	Dichlorine octoxide, Cl_2O_8
Ennea-[b]	9	Tetraiodine enneaoxide, I_4O_9
Deca-	10	Tetraphosphorus decoxide, $P_4O_{10}{}^a$

[a] When two vowels appear next to one another, as *oo* in monooxide and *ao* in heptaoxide, the vowel from the Greek prefix is dropped for the sake of euphony.
[b] The Latin prefix *nona-* is also used to denote nine atoms. Thus, I_4O_9 may also be correctly named tetraiodine nonoxide.

earlier discussion, the compounds represented by these formulas are named hydrogen chloride, carbon *mono*xide and carbon *di*oxide, respectively. Other examples of naming covalent compounds are provided in Table 3-9.

3-15 NAMING ACIDS

Acids are a class of chemical compounds that we shall study more extensively in Chapter 7. For now, we shall consider acids as compounds that give hydrogen ions when dissolved in water. This means that all acids contain hydrogen in their chemical structures.

Binary acids contain hydrogen and one other nonmetal. This general chemical formula of a binary acid is HX or H_2X, where X is the symbol of some nonmetal. These acids are named by placing the prefix *hydro-* before the stem of the name of an identifying nonmetal, attaching the suffix *-ic* to this stem, and adding the word acid. There are three binary acids whose chemical formulas should be noted:

HF   hydrofluoric acid

HCl   hydrochloric acid

H_2S hydrosulfuric acid

There is also one special case of a ternary acid (one containing hydrogen and two other elements), which is named as if it is a binary acid.

HCN   hydrocyanic acid

It is important to note that these four names are applicable only to the substances when they have been dissolved in water. As single substances, they are named as compounds of hydrogen: hydrogen fluoride, hydrogen chloride, hydrogen sulfide, and hydrogen cyanide, respectively.

Another important group of acids are the *oxyacids*. These acids contain one or more oxygen atoms in their structure, in addition to hydrogen and the identifying nonmetal. The general chemical formula of the oxyacids is H_mXO_n, where X is the symbol of the identifying nonmetal, and m and n are numbers. There are only ten oxyacids important to the study of hazardous materials. Their names and chemical formulas are listed in Table 3-10. They should be memorized.

TABLE 3-10 NAMES AND CHEMICAL
FORMULAS OF SOME OXYACIDS

Name	Chemical formula
Sulfurous acid	H_2SO_3
Sulfuric acid	H_2SO_4
Nitrous acid	HNO_2
Nitric acid	HNO_3
Phosphorous acid	H_3PO_3
Phosphoric acid	H_3PO_4
Hypochlorous acid	$HClO$
Chlorous acid	$HClO_2$
Chloric acid	$HClO_3$
Perchloric acid	$HClO_4$

The names of oxyacids correlate directly with the names of their associated polyatomic ions. Polyatomic ions whose names end in *-ate* are derived from acids whose names end in *-ic*. For example, the nitrate ion is derived from nitric acid, and the perchlorate ion is derived from perchloric acid. Polyatomic ions whose names end in *-ite* are derived from acids whose names end in *-ous*. For example, the nitrite ion is derived from nitrous acid, and the hypochlorite ion is derived from hypochlorous acid.

3-16 MOLECULAR WEIGHTS, FORMULA WEIGHTS, AND THE MOLE

The relative weight of a compound that occurs as molecules is called the *molecular weight*. It is the sum of the atomic weights of each atom that comprises the molecule. Consider the water molecule. Its molecular weight is determined as follows (to five significant figures):

$$2 \text{ hydrogen atoms} = 2 \times 1.008 = 2.016$$
$$1 \text{ oxygen atom} = 1 \times 15.999 = \underline{15.999}$$
$$\text{Molecular weight of } H_2O = 18.015$$

The relative weight of a compound that occurs as formula units is called the *formula weight*. It is the sum of the atomic weights of all atoms that comprise one formula unit. Consider sodium fluoride. Its formula weight is determined as follows (to five significant figures):

$$1 \text{ sodium ion} = 22.990$$
$$1 \text{ fluoride ion} = \underline{18.998}$$
$$\text{Formula weight of NaF} = 41.988$$

Chemists also frequently make use of a unit quantity called the *mole* (mol). The mole is the approved SI unit for the amount of any substance. As a unit, it is written as *mol* (without the -e). One mole of atoms, molecules, or formula units represents the amount of substance that has a mass in grams equal to its atomic weight, molecular weight, or formula weight. Thus 1 mol of carbon is an amount of carbon that weighs 12.011 g, or 1 mol of water is an amount of water that weighs 18.105 g. Since the concept of the mole applies to any type of particle, it is important to point out just what the particle is. For instance, 1 mol of hydrogen *atoms* is an amount that has a mass of 1.0079 g, but 1 mol of hydrogen *molecules* is an amount that has a mass of 2.016 g.

Finally, 1 mole of atoms, molecules, or formula units contains a definite number of these units: 6.022×10^{23}. This exponential notation refers to a number in which the decimal point has been moved 23 places to the right. When we write this number without using an exponent, we get 602,200,000,000,000,000,000,000. The mole is analogous to units like the dozen or ream, which means 12 and 500 of something, respectively. One mole of hydrogen atoms is 6.022×10^{23} H atoms, and 1 mol of hydrogen molecules is 6.022×10^{23} H$_2$ molecules. The number is called the *Avogadro number*, in honor of the scientist who first suggested its existence.

The Avogadro number is not just huge—it is so enormous that it is difficult to appreciate how large it is. Imagine that you wished to count this number of jelly beans at the rate of 3 beans per second. A single person would have to count jelly beans at this rate for 6×10^{15} years! This is a million times older than the age of the Earth. The enormous magnitude of this number reflects the minute dimensions of atoms and molecules on the scale of "ordinary" measurements.

REVIEW EXERCISES

Elements and Compounds

3.1 Some characteristic information about a particular substance is indicated next. Decide whether the substance is an element or compound, and give the reasoning for your selection.

Substance X is a yellow solid that does not dissolve in water, but does dissolve in the petrochemical solvent toluene. When a solution of toluene and substance X is allowed to evaporate, only substance X remains in a form indistinguishable from the original substance.

Substance X also reacts chemically with a variety of metals and nonmetals forming new substances unlike X itself.

Finally, attempts to decompose X through the use of light, heat, or an electric current are unsuccessful, although X may undergo a phase change during these experimental studies.

3.2. Without consulting a tabulation, give the name of the element that corresponds to each of these symbols: Cu, Zn, Al, S, Cl, O, P, Fe, Sn, Ag, Mg, Mn, Si, Na, and He.

3.3. Without consulting a tabulation, give the symbols of each of the following elements:
- **(a)** Titanium
- **(b)** Antimony
- **(c)** Potassium
- **(d)** Calcium
- **(e)** Fluorine
- **(f)** Uranium
- **(g)** Lead
- **(h)** Neon
- **(i)** Strontium
- **(j)** Hydrogen
- **(k)** Barium
- **(l)** Cobalt

Physical and Chemical Changes

3.4. Classify each of the following observations as a physical or chemical change:
- **(a)** Methyl ethyl ketone, a liquid solvent at 68°F (20°C), evaporates at higher temperatures.
- **(b)** Charcoal burns, becoming carbon dioxide and water vapor.
- **(c)** Sodium bicarbonate (baking soda) is added to an acid spill, causing neutralization of the acid.
- **(d)** The lead plate of an automobile battery becomes coated with white, insoluble lead sulfate during discharging.

3.5. Lye, chemically known as sodium hydroxide, is sometimes used to clean clogged drains. Some characterists of lye are noted next. Classify them as either physical or chemical properties.

Lye is a white, solid substance whose density is 2.13 g/mL. It dissolves in water, evolving a relatively large amount of heat. Concentrated water solutions of lye attack water-insoluble silicate minerals found in glass and porcelain, converting them into water-soluble silicates.

Lye also reacts with the chemical substances found in animal fats to form soap. In addition, it reacts with metallic aluminum, forming hydrogen; it is for this reason that the label on containers of lye warns the user not to mix lye with water in an aluminum vessel.

The Periodic Table

3.6. Which of the following elements most closely resembles calcium: sulfur, bromine, phosphorus, or strontium?

3.7. Classify each of the following elements as a metal, nonmetal, or metalloid:
    (a) Chlorine   (b) Sulfur   (c) Copper
    (d) Oxygen   (e) Phosphorus   (f) Bromine
    (g) Antimony   (h) Magnesium   (i) Lead

3.8. Which of the following elements are alkali metals: potassium, fluorine, aluminum, zinc, bromine, titanium, sodium, or barium?

3.9. Which of the following elements are halogens: potassium, fluorine, aluminum, zinc, bromine, titanium, sodium, or barium?

Atoms, Molecules, and Ions

3.10. Why do sodium atoms have a net charge of 0, while sodium ions have a net charge of $+1$?

3.11. Determine the net charge of the following ions: (a) potassium; (b) chloride; (c) oxide.

3.12. The poisonous gas sulfur dioxide exists as molecular units consisting of one sulfur atom and two oxygen atoms. How is a molecule of sulfur dioxide symbolized in chemistry?

Chemical Bonding

3.13. Ordinary table salt is known chemically as sodium chloride. Illustrate the manner by which sodium chloride forms from its constituent elements, sodium and chlorine.

3.14. Fluorite is a mineral commonly used in making opalescent glass and in enamel cooking utensils. The principal chemical compound found in fluorite is calcium fluoride.
    (a) Describe how calcium fluoride forms from its constituent elements, calcium and fluorine.
    (b) What is the chemical formula for calcium fluoride?

3.15. Give the Lewis structures of molecules of the following elements:
    (a) hydrogen; (b) fluorine; (c) nitrogen; (d) oxygen.

3.16. Nitrogen trichloride (NCl_3) is a yellow liquid often considered an explosion hazard, since even a slight shock may cause it to decompose violently. Write the Lewis structure of the nitrogen trichloride molecule.

3.17. Carbon disulfide (CS_2) is one of the few commercial solvents capable of dissolving sulfur and certain plastic substances. Since its flash point is $-22°F$ ($-30°C$), carbon disulfide is often considered a fire and explosion risk. Write the Lewis structure of the carbon disulfide molecule.

3.18 Formaldehyde, whose chemical formula is HCHO is probably best known in the form of its water solution, called *formalin*. This solution is widely used as a disinfecting, sterilizing, and embalming agent. Write the Lewis structure of the formaldehyde molecule.

Chemical Formulas

3.19. The substance whose chemical formula is $BaSO_4$ is commonly used to increase contrast in X-ray photographs of the stomach and intestinal tract. What is the chemical name of this substance?

3.20. Wood consists mainly of cellulose and a binding agent called *lignin*. In the paper industry, lignin is removed by treatment of pulverized wood with a substance whose chemical formula is $Ca(HSO_3)_2$. What is the chemical name of this substance?

3.21. A red variety of mercuric sulfide is used in artist's paint and is known by the common name *vermillion*. What is its chemical formula?

3.22. Lead acetate is sometimes known as "sugar of lead" since it has a sweet taste. (*Caution*: Don't taste it; all lead compounds are poisonous!) What is its chemical formula?

3.23. Name each substance with the following chemical formulas:

(a) $CdSO_4$	(b) $KHSO_4$	(c) $AgCl$
(d) $CaBr_2$	(e) Na_2S	(f) NiF_2
(g) CCl_4	(h) BaO_2	(i) H_2
(j) $Sr(NO_3)_2$	(k) $(NH_4)_2CO_3$	(l) CO
(m) $KClO_4$	(n) $FeSO_3$	(o) NH_3
(p) $Mg(HCO_3)_2$	(q) KOH	(r) PCl_3
(s) $AgC_2H_3O_2$	(t) $NaClO$	(u) $Pb(ClO_3)_2$
(v) Na_3PO_4	(w) $Cu_3(PO_4)_2$	(x) C_2H_2
(y) ZnS	(z) Fe_2O_3	

3.24. Give the chemical formula for each of the following substances:

(a) Nitrogen	(b) Carbon dioxide
(c) Hydrogen chloride	(d) Lead oxide
(e) Iron(II) sulfate	(f) Nickel hydroxide
(g) Magnesium carbonate	(h) Cadmium nitrate
(i) Dinitrogen pentoxide	(j) Ammonium chloride
(k) Barium chlorate	(l) Chromium(III) sulfide
(m) Oxygen	(n) Hydrazine
(o) Calcium hypochlorite	(p) Aluminum nitrite
(q) Sodium oxide	(r) Zinc phosphate
(s) Nickel fluoride	(t) Copper(I) sulfate
(u) Ferrous carbonate	(v) Potassium bisulfite
(w) Hydrogen sulfide	(x) Silver bromide
(y) Potassium cyanide	(z) Diphosphorus pentoxide

Molecular Weights, Formula Weights, and the Mole

3.25. Potassium permanganate is widely used as a bleaching and disinfecting agent. What is the formula weight of potassium permanganate?

3.26. Tetraethyl lead, $Pb(C_2H_5)_4$, is a colorless, volatile liquid soluble in gasoline. It is used as an antiknock agent for preventing the premature explosion of gasoline–air mixtures in engines when the mixture is compressed. What is the molecular weight of tetraethyl lead?

3.27. α-Chloroacetophenone, C_8H_7OCl, is a relatively harmless, but very potent, lachrymator used as tear gas for law enforcement. What is its molecular weight?

3.28. About four-fifths of the air is nitrogen, a colorless, odorless, and tasteless gas. Nitrogen is often used industrially for many purposes. For instance, it is used as a blanketing agent in missiles to counteract the possibility of fire. What number of moles of nitrogen molecules is contained in 250 g of nitrogen?

4

Principles of Chemical Reactions

Having observed how to name chemical substances and how to write their chemical formulas, we are now in a position to examine how such substances interact. Actually, the reactions of individual hazardous materials will be examined throughout the remainder of this textbook, but in this chapter we shall examine the general features that apply to all chemical reactions. This underscores the necessity for acquiring an ability to write and balance equations and learning how certain factors affect the speed of chemical reactions.

The general type of chemical reaction of utmost concern in fire science is combustion. Hence, combustion will be reviewed here in detail. Not only shall we observe what occurs when a fuel burns, but also what happens when certain substances are used to extinguish their fires.

4-1 CHEMICAL REACTIONS

A substance that has undergone a chemical change is no longer the same substance. In other words, it became one or more new substances. When this chemical change is described, it is common to indicate that the substance *reacted* in a particular manner. For instance, we say that dynamite exploded, hydrogen burned, acid corroded, or some other similar expression.

Each of the previous statements relates to a chemical change, more often called a *chemical reaction*. In chemistry, it is commonplace to summarize the result of a given reaction in the form of a *chemical equation*. An equation is a shorthand method for expressing a reaction in terms of written chemical formulas. For instance, if we wish to describe the combustion of elemental carbon, the following is one way of writing an equation that illustrates this chemical change:

$$C + O_2 \longrightarrow CO_2$$

On the left of the arrow is written the chemical formulas of the substances that enter the chemical reaction; they are called *reactants*. On the right of the arrow is written the formulas of the substances formed as a consequence of the reaction; they are called *products*. The arrow itself is read as yields, produces, forms, or gives. Consequently, one way to read the previous equation is "carbon plus oxygen yields carbon dioxide."

Chemical equations are the basic language of chemistry. They should contain as much information as possible concerning the specific chemical change under consideration. Not only may we list the formulas of the substances reacting and forming, but we may also list their physical states under the temperature and pressure conditions of the reaction. Physical states are indicated in parentheses next to the chemical formulas of the reactants and formulas by using italicized letters such as *s*, *l*, *g* and *aq*. These symbols mean solid, liquid, gas, and aqueous solution, respectively. Some examples are provided in Table 4-1. The combustion of elemental carbon could also be written in the following form:

$$C(s) + O_2(g) \longrightarrow CO_2(g)$$

Before a chemical equation can be written, we first need to know the chemical formulas of the reactants and products. To write the more complete form of an equation, we must also know the physical state of the reactants and products. In this book, we shall list the physical states of reactants and products when equations are written, but students are not normally expected to know this information. They usually find physical states in reference books or through laboratory experimentation.

TABLE 4-1   SYMBOLS USED IN CHEMICAL EQUATIONS

Symbol	Meaning	Examples
$(g)^a$	Gaseous reactant or product	$H_2(g)$, $CO_2(g)$
(l)	Liquid reactant or product	$H_2O(l)$, $Br_2(l)$
$(s)^b$	Solid reactant or product	$Fe(s)$, $S_8(s)$
$(aq)^c$	Reactant or product dissolved in water	$NaCl(aq)$, $KNO_3(aq)$
$(conc)^d$	Reactant undiluted with water	$HCl(conc)$

[a] An arrow pointing upward (↑) is also used when the gas is a product of a reaction.

[b] An arrow pointing downward (↓) is also used when a solid precipitates from solution.

[c] Meaning *aqueous*.

[d] Meaning *concentrated*.

4-2 BALANCING SIMPLE EQUATIONS

Not only must an equation summarize what occurs qualitatively during a chemical change, but it also must be accountable for other, more fundamental observations. In particular, each equation must be written so that the chemical change at issue adheres to the law of conservation of mass and energy (Sec. 2-5). This law summarizes the observation that in "ordinary" chemical reactions, there is no apparent change in mass. Ordinary chemical reactions are limited to those occurring at the molecular or ionic level, as opposed to the reactions of atomic nuclei.

Conservation of mass requires that during a given chemical change the atoms of any element are neither created nor destroyed. This means that the same number of atoms remain after a reaction as there were before the reaction occurred. Hence, when chemical equations are written, we make certain that an *equal* number of atoms of each type of element occurs on both sides of the arrow. Such an equation is then said to be *balanced*. The equation written in Sec. 4-1 is balanced, since it has one carbon atom and two oxygen atoms on each side of the arrow.

Not all equations are directly balanced after writing the chemical formulas for their reactants and products. In fact, they are generally unbalanced at this point. Thus, we must select a proper coefficient to place *in front of* the appropriate formula so that the equation becomes balanced. The correct formula of a substance must never be changed when we balance an equation. Furthermore, coefficients are never written in the middle of a formula, such as H_23O or some similar concoction.

Most simple equations may be balanced by inspection. While there are no absolute rules for balancing equations by inspection, the following points are useful when first learning the process:

1. Write the correct formula for each reactant and product, and separate the reactants from the products with an arrow.
2. Write the physical state of the reactants and products in parentheses after each formula, when this information is known.
3. Choose the formula of the substance containing the greatest number of atoms of an arbitrary element. Insert a number in front of one or both formulas so that the numbers of atoms for this particular element are balanced.
4. If the formulas for polyatomic ions appear in an equation, balance them as single units only when they retain their identity on both sides of the arrow.
5. Balance any remaining atoms or ions.

Let's consider an example. Suppose you wish to write the balanced chemical equation for the reaction that occurs when methane burns in air to form carbon dioxide and water vapor. (Methane is the principal component of natural gas.) This is an ordinary combustion reaction involving the chemical union of methane and atmospheric oxygen. First, we write the chemical formulas of the reactants and products. The formula of methane is CH_4 (from Table 3-8), oxygen is O_2, carbon dioxide is CO_2, and water is H_2O. Under the reaction conditions, each is a gas or vapor. Hence, we initially write the following:

$$CH_4(g) + O_2(g) \longrightarrow CO_2(g) + H_2O(g)$$

Next, we note that this unbalanced equation contains more hydrogen atoms (in CH_4) than any other type of atom. We balance the hydrogen atoms by inserting a 2 in front of the formula for water. The equation now looks like the following:

$$CH_4(g) + O_2(g) \longrightarrow CO_2(g) + 2H_2O(g)$$

No ions are in this equation, so we next balance the oxygen atoms. This is accomplished by inserting a 2 in front of O_2. The balanced equation takes on the following form:

$$CH_4(g) + 2O_2(g) \longrightarrow CO_2(g) + 2H_2O(g)$$

When such exercises are performed, it is usually best to make one final check: one carbon atom, four hydrogen atoms, and four oxygen atoms on each side of the arrow.

4-3 TYPES OF CHEMICAL REACTIONS

By now, one point should be apparent: Equations illustrating the chemistry of a hazardous material cannot be written if we do not know the products of a reaction. Determining the products that form is not always a simple feat, but many reactions in which we are interested can be classified as one of four types. These types are reviewed next with illustrative examples.

Combination or Synthesis Reaction

In this type of chemical reaction, two or more simpler substances combine to form a more complex substance. Some examples of combination reactions are illustrated by the following equations:

$$\underset{\text{hydrogen}}{2H_2(g)} + \underset{\text{oxygen}}{O_2(g)} \longrightarrow \underset{\text{water}}{2H_2O(l)}$$

$$\underset{\text{sodium}}{2Na(s)} + \underset{\text{chlorine}}{Cl_2(g)} \longrightarrow \underset{\text{sodium chloride}}{2NaCl(s)}$$

$$\underset{\text{carbon}}{C(s)} + \underset{\text{oxygen}}{O_2(g)} \longrightarrow \underset{\text{carbon dioxide}}{CO_2(g)}$$

Decomposition Reaction

In a decomposition reaction, a relatively complex substance is broken down into several simpler substances. Some examples of decomposition reactions are illustrated by the following equations:

$$\underset{\text{water}}{2H_2O(l)} \longrightarrow \underset{\text{hydrogen}}{2H_2(g)} + \underset{\text{oxygen}}{O_2(g)}$$

$$\underset{\text{sodium carbonate}}{Na_2CO_3(s)} \longrightarrow \underset{\text{sodium oxide}}{Na_2O(s)} + \underset{\text{carbon dioxide}}{CO_2(g)}$$

$$\underset{\text{ammonium dichromate}}{(NH_4)_2Cr_2O_7(s)} \longrightarrow \underset{\text{chromic oxide}}{Cr_2O_3(s)} + \underset{\text{nitrogen}}{N_2(g)} + \underset{\text{water}}{4H_2O(g)}$$

Single Replacement or Single Displacement Reaction

In this type of chemical reaction, an element and a compound react so that the free element replaces or displaces an element in the compound. Some examples of such reactions are illustrated by the following equations:

$$\underset{\text{magnesium}}{Mg(s)} + \underset{\text{hydrochloric acid}}{2HCl(aq)} \longrightarrow \underset{\text{magnesium chloride}}{MgCl_2(aq)} + \underset{\text{hydrogen}}{H_2(g)}$$

$$\underset{\text{copper}}{Cu(s)} + \underset{\text{silver nitrate}}{2AgNO_3(aq)} \longrightarrow \underset{\text{silver}}{2Ag(s)} + \underset{\text{cupric nitrate}}{Cu(NO_3)_2(aq)}$$

$$\underset{\text{potassium iodide}}{2KI(aq)} + \underset{\text{chlorine}}{Cl_2(g)} \longrightarrow \underset{\text{potassium chloride}}{KCl(aq)} + \underset{\text{iodine}}{I_2(s)}$$

Double Replacement or Double Displacement Reaction

In a double replacement reaction, there is an exchange of the positively charged ions in two compounds. Some examples of this type of chemical reaction are illustrated by the following equations:

$$\underset{\text{sodium cyanide}}{2NaCN(s)} + \underset{\text{sulfuric acid}}{H_2SO_4(aq)} \longrightarrow \underset{\text{sodium sulfate}}{Na_2SO_4(aq)} + \underset{\text{hydrogen cyanide}}{2HCN(g)}$$

$$\underset{\text{lead(II) sulfide}}{PbS(s)} + \underset{\text{hydrochloric acid}}{2HCl(aq)} \longrightarrow \underset{\text{lead(II) chloride}}{PbCl_2(aq)} + \underset{\text{hydrogen sulfide}}{H_2S(q)}$$

$$\underset{\text{sodium sulfate}}{Na_2SO_4(aq)} + \underset{\text{barium chloride}}{BaCl_2(aq)} \longrightarrow \underset{\text{barium sulfate}}{BaSO_4(s)} + \underset{\text{sodium chloride}}{2NaCl(aq)}$$

4-4 OXIDATION–REDUCTION REACTIONS

Another way that chemists classify chemical reactions is based on oxidation–reduction phenomena, frequently called *redox* reactions. For classification purposes, chemical phenomena either involve oxidation–reduction or they do not. For instance, combination, decomposition, and simple replacement reactions involve oxidation–reduction processes, whereas double replacement reactions do not. We shall study the oxidation–reduction phenomenon in depth in Chapter 10. But at this point it is well to be introduced to the subject, since a great many processes are concerned with it.

Oxidation is a chemical process by which a system becomes less associated with its electrons. For ionic substances, this is directly accomplished by the complete loss of one or more electrons. For example, consider the following equations:

$$Na(s) \longrightarrow Na^+(aq) + e^-$$

$$Mg(s) \longrightarrow Mg^{2+}(aq) + 2e^-$$

$$Cu(s) \longrightarrow Cu^{2+}(aq) + 2e^-$$

$$Fe^{2+}(aq) \longrightarrow Fe^{3+}(aq) + e^-$$

$$2Cl^-(aq) \longrightarrow Cl_2(g) + 2e^-$$

These equations illustrate oxidation. In the first three examples, neutral atoms of sodium, magnesium, and copper lose one or more electrons and become positively charged ions. In the fourth example, the iron(II) ion loses an electron and becomes the iron(III) ion. In the fifth example, negative chloride ions lose electrons to form a neutral molecule of chlorine.

These few examples serve to define oxidation in ionic systems: *Oxidation* is a chemical process in which electrons are lost. In the previous examples, we say that the atoms and ions on the left of the arrow have been *oxidized*.

Oxidation is always accompanied by another process called *reduction*. For ionic systems, reduction is associated with the complete gain of electrons. This is illustrated by the following equations:

$$Cl_2(g) + 2e^- \longrightarrow 2Cl^-(aq)$$

$$S_8(s) + 16e^- \longrightarrow 8S^{2-}(aq)$$

$$Fe^{3+}(aq) + e^- \longrightarrow Fe^{2+}(aq)$$

$$Fe^{2+}(aq) + 2e^- \longrightarrow Fe(s)$$

In the first two examples, neutral elements gain electrons and form negative ions. In the third example, the iron(III) ion gains an electron and becomes the iron(II) ion. In the final example, the iron(II) ion gains two electrons and becomes neutral elemental iron. The molecules and ions on the left of the arrow are *reduced*.

Oxidation may also occur in completely covalent systems, too, but a complete transfer of electrons does not occur. For instance, consider the chemical reaction represented by the following equation:

$$H_2(g) + Cl_2(g) \longrightarrow 2HCl(g)$$

In the hydrogen and chlorine molecules, the electron pairs that constitute the covalent bonds are shared equally by their respective atoms. In hydrogen chloride, however, the sharing of the electron pair between the hydrogen and chlorine atom is unequal, as illustrated in Fig. 4-1. In the case of the hydrogen chloride molecule, the chlorine atom has a greater share in the pair of bonding electrons that does the hydrogen atom.

Although a complete transfer of electrons has not occurred, a partial transfer of electrons has, in the sense that the electron distribution is not symmetrical about the atoms in hydrogen chloride. This unsymmetrical distribution of electrons is typical of oxidation in covalent systems. In other words, hydrogen has been oxidized and chlorine has been reduced, but a complete transfer of electrons has not occurred.

In either ionic or covalent systems, the substances that are oxidized are called *reducing agents*. The substances that are reduced are called *oxidizing agents*.

Let's consider an example. When a camera flashbulb sends out its brilliant blaze, a chemical reaction occurs in which magnesium burns to form magnesium

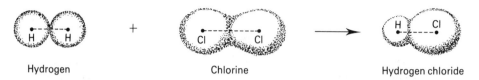

Hydrogen   Chlorine   Hydrogen chloride

Figure 4-1 When the oxidation–reduction phenomenon occurs between covalently bonded substances, electrons are not completely transferred from one reacting species to the next. In the hydrogen and chlorine molecules, shown to the left of
the arrow, the electron pairs are mutually shared between the like atoms; but in the
hydrogen chloride molecule, shown to the right of the arrow, the electronic distribution is asymmetric about the center. This partial loss and gain of electron density
is typical of the oxidation–reduction phenomenon between all covalent substances.

oxide. This reaction may be illustrated by the following equation:

$$2Mg(s) + O_2(g) \longrightarrow 2MgO(s)$$

In this example, a magnesium atom loses two electrons to become a magnesium ion;
hence, it is oxidized. Each atom of an oxygen molecule gains two electrons to become oxide ions; hence, they are reduced. Magnesium is the reducing agent, and
oxygen is the oxidizing agent.

4-5 FACTORS AFFECTING THE RATE OF REACTION

Each chemical reaction occurs at a definite speed, called its *rate of reaction*. Sometimes the rate of reaction is referenced to a given chemical phenomenon, as the combustion rate, corrosion rate, or explosion rate. Chemists establish these rates of reaction by experimentally noting the change in concentration of a reactant or product
with time.

   The speed at which a given substance undergoes a chemical change is often associated with its hazardous nature. For instance, consider the explosion of nitroglycerin. Several grams of nitroglycerin completely decompose within a millionth of a
second. Fortunately, all chemical reactions do not occur this rapidly. Otherwise, we
would have even greater problems when responding to emergencies involving chemical mishaps.

   The rate of reaction depends on at least seven factors, each of which will be
discussed independently. When appropriate, the influence of each factor will be
noted as it bears on the rate of combustion.

Nature of the Material

When a piece of white phosphorus is exposed to air, it ignites spontaneously, even
without exposure to an ignition source. But, more commonly, most combustible substances that we encounter, like wood, paper, hydrogen, magnesium, and sulfur, do
not begin burning until they are first exposed to a spark, flame, or other ignition

source. On the other hand, some substances do not appear to burn at all under any conditions, like water, carbon dioxide, nitrogen, and the noble gases.

The rates of combustion of these particular substances vary from zero to some finite value. It is their individual chemical nature that causes some substances to burn spontaneously, others to burn only when kindled, and others not to burn at all.

Subdivision of the Reactants

Logs do not burn readily in a fireplace. Ordinarily, they must first be kindled, usually by the heat generated from the burning of smaller pieces of wood. Yet a dispersion of sawdust from the same type of wood may ignite spontaneously or when exposed to an ignition source and burn with explosive violence. In fact, the dust of practically any combustible material often burns explosively when mixed with air and exposed to a spark or flame.

In this instance, it is the subdivision of the reactant that affects the reaction rate. In general, anytime its surface area is increased, a substance reacts faster. This is illustrated in Fig. 4-2, which shows that a flammable liquid burns fastest in the vessel which provides the greatest surface area.

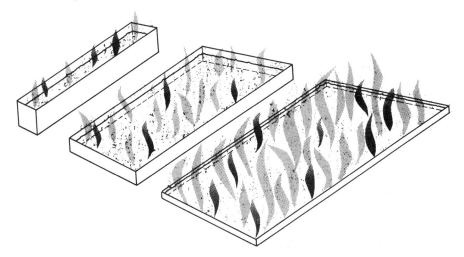

Figure 4-2 The same amount of a flammable liquid has been added to three containers of progressively increasing size. When ignited at the same instant, the liquid burns at the fastest rate in the container that provides the largest surface area.

State of Aggregation

The rate at which a substance reacts is frequently affected by its physical state of matter. This is particularly true of the rate of combustion. Generally, reactions between gases occur faster than those between liquids, and reactions between either gases or liquids occur faster than those between solids.

These differences in reaction rates are due to the structure of the gaseous, liq-uid, and solid states of matter. In gases, the particles of the substance are relatively far apart and exert small attractive forces on each other. This permits diffusion to occur readily. But in the liquid and solid states of matter, the particles of a substance at issue are in contact and held together tightly within its bulk. This hinders contact between the particles, which reduces their reaction rate.

Concentration of Reactants

Before any chemical reaction occurs, the particles that make up the structure of the reactants must contact each other. (On the other hand, particle contact does not nec-essarily mean that reaction will occur.) The probability of particles contacting each other increases as the number of particles in a given volume increases. In other words, if provided with the same reactants in two vessels, the rate of their reaction is generally faster in the vessel containing the greater concentration of reactants.

Imagine that we have four containers of equal volume holding different num-bers of two hypothetical molecules, A and B, as pictured in Fig. 4-3. What is the relative number of collisions between unlike molecules? (Collisions between like molecules are ignored, as A contacting A or B contacting B, since these collisions do not cause a chemical reaction.)

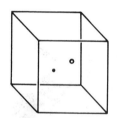 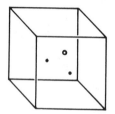 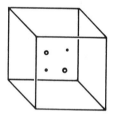 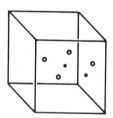

Figure 4-3 The probability of a chemical reaction increases as the number of reactant molecules confined to a container increases. The molecules of the reactants A and B are des-ignated as dots and open circles, respectively.

Let's consider each container separately. The first container holds one mole-cule of A and one molecule of B, while the second container holds two molecules of A and one molecule of B. It follows that the likelihood of an A molecule colliding with a B molecule is twice as great in the second container as compared with the first one. In the third container, which holds two molecules of A and two molecules of B, the probability of collision between unlike molecules is increased to four times the possibility in the first container and is two times as great as in the second one. Finally, in the fourth container, which holds two molecules of A and four molecules of B, the probability of collision of unlike molecules is increased to eight times that observed in the first container. This illustrates that an increase in the concentration of the reactants causes the rate of a given reaction to increase.

We also noted in Sec. 2-11 that the concentration of reactants has an important bearing on the rate of combustion. Unless the concentration of flammable gases or the vapors of flammable liquids is in the *flammable range* of a given material, the material does not burn.

The concentration of oxygen also has an effect on combustion. In clean air at sea level, the concentration of oxygen is only about 21% by volume. Most of the air, about 78% by volume, consists of nitrogen. Hence, atmospheric nitrogen serves as a diluent during the combustion of materials in air and actually retards their combustion rate. In an atmosphere of pure oxygen, combustion always occurs with a greatly increased intensity. Some substances that are ordinarily stable in air even burn spontaneously in an atmosphere of pure oxygen.

Activation Energy

While some substances react spontaneously upon contact, most substances need to have some minimum amount of energy supplied to them before they chemically react. For instance, pieces of combustible solids lying exposed to atmospheric oxygen do not just suddenly begin burning. However, if an ignition source is brought near them, they are likely to burn quite readily. This minimum amount of energy that must be supplied to initiate chemical reactions is called the *activation energy*.

In exothermic reactions, the activation energy does not need to be continually supplied as the reaction proceeds. The activation energy is replaced or substituted for by the energy that is released during the process. Thus, once combustion reactions have been initiated, they continue until either the fuel or oxygen is exhausted.

An amount of activation energy must also be supplied to initiate endothermic reactions. However, we must not only supply the activation energy, but enough additional energy to replace the energy absorbed by the reactants. In the case of endothermic reactions, the phenomenon ceases if the source of energy has been removed. Consider electrolysis, the process of decomposing a substance through the use of an electric current. Electrolysis is an endothermic phenomenon. When we disconnect the current, electrolysis stops.

Temperature

Heating a mixture of reactants causes its particles to move more rapidly, and this increases the probability that the particles will collide. As the speed of the particles increases, the temperature accordingly rises. But the increase in reaction rate at this higher temperature does not depend on these additional collisions as much as it does on the additional number of particles that are now sufficiently activated to chemically react. Hence, a reaction rate increases with a rise in temperature, because proportionately more molecules become sufficiently activated at higher temperatures.

According to a classic rule in chemistry, the rate of reaction doubles for every 10°C (or 18°F) rise. Thus, twice as many molecules are initiated to react when the temperature of the reacting system has been increased by 10°C. Four times as many are activated when the temperature is increased 20°C, and eight times as many for a 30°C increase in temperature. While the same number of reactant particles are present in each instance, the fraction of them that attain enough energy to chemically react is greater for those at elevated temperatures.

Catalysis

A *catalyst* is any substance of which a fractionally small percentage affects the rate of reaction. Thus, following a given reaction, a catalyst appears to remain unconsumed. Usually, catalysts accelerate the reaction rate, but some retard it.

A familiar example of catalysis is associated with the rusting of iron, which is catalyzed by atmospheric water vapor. Iron exposed to moisture corrodes much faster than iron in a dry atmosphere. Another example of catalysis is provided by the oxidation of hydrogen. A mixture of hydrogen and oxygen may be kept in a vessel for years without appreciably reacting. However, if a small amount of platinum is introduced into this vessel, the mixture explodes.

During a given reaction, the catalyst itself may actually undergo a chemical change, but then react at least once again so that its overall chemical identity is not lost. However, the catalyst does not always undergo a chemical change. It is often altered physically by particles of one or more reactants adhering to its surface.

Catalysis is extremely important for regulating the speed of biological reactions in living systems. In this case the catalysts are called *enzymes*.

4-6 COMBUSTION AND THE ENERGETICS OF CHEMICAL REACTIONS

Every chemical reaction is somehow associated with energy. Yet the role that energy plays in a given chemical reaction is not always apparent.

Combustion is probably the ideal example for illustrating the relationship between energy and a chemical reaction. Combustion is a chemical reaction that releases energy as heat and usually, light. When combustion is referred to in everyday practice, we generally indicate that something "is burning" or "on fire." Hence, combustion and burning are essentially equivalent terms. The light or luminosity that normally accompanies burning is called a *flame*.

In Sec. 2-5, we noted that chemical energy is present in every substance. This chemical energy has two components. Some chemical energy arises in connection with the motion of the atoms and molecules that make up the substance. Chemical energy is also stored in a substance as the result of its unique structure. This latter contribution to the chemical energy of a substance is associated with the strength of its chemical bonds and is called the *bond energy*. During any chemical reaction, some bonds break and new ones form. Any excess energy is either absorbed or released to the environment. This is the source of the energy released during combustion.

The terms combustion and oxidation are often used synonymously, but, technically, there is a subtle difference between them. During combustion, two or more substances chemically unite. In practice, one of them is almost always atmospheric oxygen, but combustion reactions are known in which oxygen is not one of the reactants. For instance, the combustion of some substances occurs in an atmosphere of chlorine. In Sec. 4-4, we noted that oxidation is a chemical process associated with the loss of electrons. Combustion manifests itself as fire when the oxidation occurs relatively fast. Thus, combustion is sometimes described as a *rapid oxidation*.

The substance that actually burns is the *fuel*. We noted earlier that as a fuel burns, its atoms are never destroyed. Instead, in ordinary combustion, they unite with atoms of atmospheric oxygen and enter into new chemical forms. In most cases, they become oxides. Oxygen itself does not burn. It is said to *support* combustion.

Many flammable and combustible substances contain carbon in their molecular framework. When these substances burn in air, the constituent carbon atoms unite with oxygen atoms, normally forming either carbon monoxide or carbon dioxide. When carbon dioxide forms, the combustion is said to be *complete*. When carbon monoxide forms, we say it is *incomplete*. But sometimes the carbon atoms do not unite with oxygen. When this occurs, elemental carbon is a product of combustion. During a fire, it is manifested as black smoke. The carbon is dispersed as tiny particles in air, together with other combustion products.

The energy released during combustion is called the *heat of combustion*. Each fuel has a characteristic heat of combustion, some examples of which are noted in Table 4-2. Heats of combustion are generally noted in an energy unit per gram, pound, or mole of substance burned, like Btu/lb, kcal/g, or kJ/mol. Materials having relatively high heats of combustion are good fuels for heating and cooking purposes.

The ordinary combustion of methane illustrates the burning process. The following equation may be written to describe the overall phenomenon:

$$CH_4(g) + 2O_2(g) \longrightarrow CO_2(g) + 2H_2O(g) + 213 \text{ kcal/mol}$$

First, we note that methane is the fuel. Second, we observe that, as a consequence of

TABLE 4-2 APPROXIMATE AVERAGE HEATS OF COMBUSTION OF SOME COMMON MATERIALS

Fuel	Heat of combustion	
	Btu/lb	kcal/g
Methane[a]	23,000	13.3
Gasoline	20,000	11.0
Fuel oils	19,000	10.5
Wood/paper	8,000	4.4
Hydrogen	60,000	28.7

[a] Methane is the principal component of natural gas.

burning, the carbon and hydrogen atoms comprising the molecular structure of methane become carbon dioxide and water vapor. Third, we note that the heat of combustion of methane is 213 kcal/mol; that is, 213 kcal of energy is evolved as heat and light per mole of methane burned.

4-7 SPONTANEOUS COMBUSTION

Some substances undergo oxidation very slowly, initially such that their oxidation is virtually imperceptible. When such substances slowly oxidize in confined spaces where the circulation of air is poor, they may absorb their own heat of reaction. The absorbed heat raises the temperature of these substances to their auto-ignition or kindling temperatures at which the substances self-ignite. Since a source of ignition is unnecessary in these situations, burning initiated by the accumulation of heat from slow oxidation processes is called *spontaneous combustion*.

Many flammable and combustible animal and vegetable oils, like linseed oil, oxidize slowly. Figure 4-4 illustrates that these substances should always be regarded as potential sources of spontaneous combustion. A commonly encountered incident associated with this phenomenon involves turpentine-soaked rags that were improperly disposed following their use. When the rags are negligently tossed into a pile in a broom closet, for instance, the heat cannot dissipate to the surroundings. The heat

Figure 4-4 Oily rags and improperly stored linseed oil, varnishes, lacquers, and other oil-based paint products are among the materials likely to undergo spontaneous combustion. Each year, millions of dollars in property are needlessly lost because of fires that originated from such sources.

concentrates in the rags and the temperature rises to the kindling point of the cloth. Soon thereafter, the rags burn.

Spontaneous combustion may also occur in undried agricultural products, such as damp hay. Microorganisms proliferate in these damp materials. Their physiological activity evolves heat; its production is called *thermogenesis*. This biological activity is supplemented by chemical oxidation until the temperature of the product rises to approximately 160°F. At this elevated temperature, the microorganisms can no longer survive, but chemical oxidation continues to occur. When heat evolves from this oxidation faster than it dissipates, the product burns.

Thick, black smoke usually accompanies the fires of animal and vegetable oils initiated by spontaneous combustion. Billows of white smoke accompany the spontaneous combustion of hay, grass, and other agricultural products.

4-8 FIRE TETRAHEDRON

In the past, the components of fire were often visualized in terms of a triangle, called the *fire triangle,* shown in Fig. 4-5. The three sides of the fire triangle represent the essential constituents of an ordinary combustion reaction: fuel, oxygen, and heat.

We still concur with the basic concept of the fire triangle. But as scientists learned more about the intricacies of the combustion process, a fourth component was added to it, requiring that we now represent combustion by a four-sided figure. This is the fire tetrahedron illustrated in Fig. 4-6. The fourth component needed to

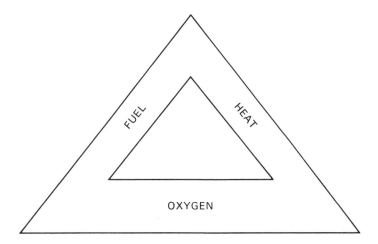

Figure 4-5   The fire triangle. The self-sustenance of an ordinary fire was once considered possible only when ample fuel, oxygen, and heat simultaneously existed, the three components of this triangle. Today, however, this concept has been broadened into the fire tetrahedron shown in Fig. 4-6, which includes a fourth component, free radicals.

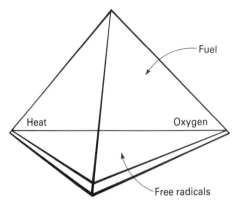

Figure 4-6 The fire tetrahedron, the four components of which are fuel, oxygen, heat, and free radicals. An ordinary fire is self-sustained only when there is simultaneously ample fuel, oxygen, heat, and free radical propagation. Furthermore, fires are extinguished by removing any one of these components from the other three.

totally describe combustion is a reactive species called a *free radical*. A free radical is a molecular fragment having one or more unpaired electrons. Consider the methane molecule. If one of its four chemical bonds is broken or split, the molecular structure of methane must be represented as follows:

$$\text{H}-\overset{\displaystyle\overset{\text{H}}{|}}{\underset{\displaystyle\underset{\text{H}}{|}}{\text{C}}}\cdot$$

The dot represents an unpaired electron. The species represented here is the *methyl free radical*.

The unpaired electron causes free radicals to be highly reactive. Thus, free radicals are short-lived, having only a transient existence. Nevertheless, free radicals are capable of initiating many kinds of chemical reactions, including combustion. They accomplish this by reacting with other available free radicals, atoms, and molecules. Hence, ordinary combustion actually requires the interrelationship of four components: fuel, oxygen, heat, and free radicals. Unless all four of these components are simultaneously present in the proper proportion, a fire cannot occur.

To illustrate the importance of free radicals in combustion, let's consider the burning of methane in more detail. Chemists now view the combustion of methane as a series of individual reactions, each of which contributes to a *chain reaction*. In other words, the products of one reaction activate additional molecules, which then interact in new chemical reactions. This sequence of individual steps is called the *mechanism* of the reaction. The mechanism whereby methane burns is highly complex. Nonetheless, some aspects of the combustion may be summarized by means of the sequential processes noted as follows:

$$CH_4(g) \longrightarrow CH_3\cdot(g) \longrightarrow CH_2O(g) \longrightarrow$$
$$HCO\cdot(g) \longrightarrow CO(g) \longrightarrow CO_2(g)$$

The first mechanistic step associated with the combustion of methane involves production of a methyl free radical and hydrogen atom. This step occurs as a methane molecule first absorbs energy of activation from its source of ignition. This initiation step is represented as follows:

$$CH_4(g) \longrightarrow CH_3 \cdot (g) + H \cdot (g)$$

Then hydrogen atoms react with molecular oxygen, forming a hydroxyl radical ($\cdot OH$) and an oxygen atom ($O \cdot$).

$$H \cdot (g) + O_2(g) \longrightarrow \cdot OH(g) + O \cdot (g)$$

This is called the *chain branching* step of the mechanism. This is the most important mechanistic step in the combustion of methane.

Next, a series of reactions occurs in which free radicals and molecules multiply in number. This combination of reactions is called the *propagation* step of the mechanism, which is illustrated by the following group of equations:

$$CH_4(g) + \cdot OH(g) \longrightarrow H_2O(g) + CH_3 \cdot (g)$$
$$CH_4(g) + H \cdot (g) \longrightarrow CH_3 \cdot (g) + H_2(g)$$
$$CH_3 \cdot (g) + O \cdot (g) \longrightarrow CH_2O(g) + H \cdot (g)$$
$$CH_2O(g) + CH_3 \cdot (g) \longrightarrow CHO \cdot (g) + CH_4(g)$$
$$CH_2O(g) + \cdot OH(g) \longrightarrow CHO \cdot (g) + H_2O(g)$$
$$CH_2O(g) + H \cdot (g) \longrightarrow CHO \cdot (g) + H_2(g)$$
$$CH_2O(g) + O \cdot (g) \longrightarrow CHO \cdot (g) + \cdot OH(g)$$
$$CHO \cdot (g) \longrightarrow CO(g) + H \cdot (g)$$
$$CO(g) + \cdot OH(g) \longrightarrow CO_2(g) + H \cdot (g)$$

Not all reactions illustrated by these equations are equally important during propagation, but each reaction may contribute to the formation of radicals and molecules as indicated.

Finally, the mechanism is associated with a *termination* step, in which free radicals combine. The following equation illustrates an example of termination:

$$CH_3 \cdot (g) + H \cdot (g) \longrightarrow CH_4(g)$$

It is curious to note that, while molecular oxygen is a major reactant during ordinary combustion, it participates only in a single step of the mechanism: the chain branching step. Furthermore, conversion of carbon monoxide to carbon dioxide is dominated by the elementary step represented by the following equation:

$$CO(g) + \cdot OH(g) \longrightarrow CO_2(g) + H \cdot (g)$$

This equation illustrates the second most important reaction in the combustion of methane.

4-9 CHEMISTRY OF FIRE EXTINGUISHMENT

The concept of the fire tetrahedron allows us to easily identify the components that contribute to an ordinary fire. By the same token, this concept assists in understanding the chemistry associated with the effectiveness of certain substances as fire extinguishing agents. All mechanisms of fire extinguishment reduce to the process of removing one component of the fire tetrahedron from the other three.

An essential feature of any fire extinguishing agent is its incapability of burning. Any flammable or combustible substance is either totally useless or has only a limited utility at extinguishing fires. There are four common types of fire extinguishing agents which effectively extinguish one or more of the classes of fire previously identified in Sec. 1-4. Each type is discussed independently.

Water

Water is an oxide of hydrogen and does not readily combine with more oxygen. Hence, it is a nonflammable substance. In Sec. 2-9, we noted that the effectiveness of water as a fire extinguishing agent is mainly associated with its ability to extract heat from a fire as it vaporizes. By removing heat, only three components of the fire tetrahedron remain; hence, a fire cannot continue.

Water is generally effective as a fire extinguishing agent only on class A fires. However, we noted in Sec. 2-4 that water may sometimes be used to extinguish class B liquid fires when water is immiscible with the liquid and floats on the burning liquid. Water can also be effective as a fire extinguishing agent for certain class B fires in which it serves as a diluent. When flammable or combustible liquids dissolve in sufficient water, the mixture becomes nonflammable. Burning acetone is an example of a fire that water may extinguish by dilution. Of course, not all liquids are water soluble. Hence, the success of water at extinguishing flammable liquid fires depends considerably on the extent to which water dissolves in the burning liquid.

Water is often selected to extinguish fires because it is usually available in relatively large quantities. Yet water is not a good, all-purpose fire extinguishing agent. Considerable damage often results from using water to extinguish fires. In fact, the damage resulting from the use of water frequently surpasses that caused directly by the fire. While water is generally effective at extinguishing class A fires, it is often more dense than liquids that burn as class B fires and is immiscible with them. Hence, it sinks below the surface of such fuels, where it is incapable of extinguishing fires (Sec. 2-4). Water also ruins delicate electrical circuitry and thus is not recommended as an extinguishing agent of class C fires. Water containing dissolved mineral salts conducts electricity, which puts firefighters at risk of being electrocuted. Finally, water may react with the burning metals that constitute class D fires and may actually aid in sustaining combustion, rather than extinguishing the fire.

Water also has the disadvantageous property that in many parts of the world it can freeze during use. Antifreeze agents, like ethylene glycol, may be added to water to keep it liquid at temperatures as low as $-60°F$ ($-51°C$). However, these water solutions are frequently corrosive when in contact with metals and require special pressure mechanisms for their release. Furthermore, if insufficient water and antifreeze are applied to a fire, the water evaporates, after which the antifreeze may burn.

The use of water on fires is also inefficient, due to water's fluidity. Most water applied to a fire runs off to areas where it does not interact with the fire and thus does not vaporize. This disadvantageous feature of water can be altered by changing its method of application. Figure 4-7 illustrates that the efficiency of water can be markedly improved by using it as a spray or fog. The increased surface area of the particulates of fog leads to more rapid vaporization and thus to faster removal of heat.

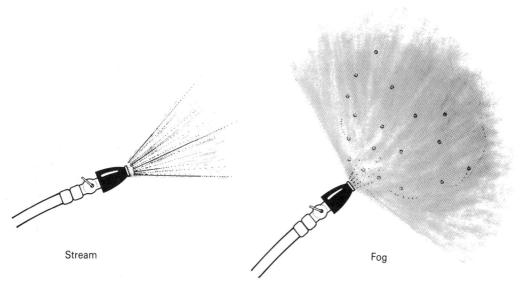

Stream   Fog

Figure 4-7   A comparison between the application of water to a fire as a solid stream and as a fog. In the latter case, the droplets of water present more surface area to the heat source and thus are capable of absorbing heat more effectively than the solid stream.

The use of water is also more efficient when it has been sealed into the structure of a gel or foam. Several film-forming foams can be produced that retain water. The principal advantage in their use as fire extinguishing agents is flexibility. The foam can be varied in fluidity by adjusting the concentration of the foaming agent when it is mixed with water. The result is a foam that can be as viscous as molasses or as mobile as water itself. The main disadvantage in using foam on class B fires is the possibility that it will saturate the burning fuel. Saturation results when the foam is applied by plunging it deep into the fuel, rather than across its surface.

While several such foams are available commercially as fire extinguishing agents, we shall briefly note only two of them here. One is called *alcohol foam*. It is often used for extinguishing the class B fires of flammable liquids that are immiscible with water and less dense than water. It is also recommended as an extinguishing agent for water-soluble, flammable liquid fires.

Another popular foaming agent is called *protein foam*. This material is prepared from natural protein materials like fishmeal, horn and hoof meal, or feather meal. Protein foam typically contains from 3% to 6% by mass of a protein concentrate. Protein foam has a lubricating nature. It is for this reason that protein foam is often used on runways to assist disabled aircraft in landing.

All foams are immediately effective as fire extinguishing agents, since they act as blanketing agents over fires. In other words, they form a barrier between the fuel and the atmosphere, which starves a fire of oxygen.

Carbon Dioxide

Carbon dioxide is a nonflammable oxide of carbon. It is often effective at extinguishing class B and class C fires and even some class A fires, but it is never effective at extinguishing class D fires. The effectiveness of carbon dioxide as a fire extinguishing agent is mainly associated with its vapor density. The vapor density of carbon dioxide is 1.52 (Sec. 2-4), making it 1.5 times more dense than air. Thus, carbon dioxide smothers a fire, preventing the fuel from contacting atmospheric oxygen.

When intended for use as a fire extinguishing agent, carbon dioxide is often encountered as a compressed gas in a portable extinguisher like that illustrated in Fig. 4-8. Portable hand extinguishers containing carbon dioxide are available commercially in capacities from 2 to 25 lb (0.9 to 11 kg). Carbon dioxide is sometimes discharged from vessels in which it has been confined in the liquid state. In such situations it is much colder than the surrounding atmosphere. The cold carbon dioxide aids in fire extinguishment by cooling burning materials below the minimum temperature at which they can burn.

By contrast with water, carbon dioxide does not ordinarily damage the fire area. It is said to be a "clean" fire extinguishing agent; being gaseous at normal temperature and pressure conditions, it dissipates into the surrounding atmosphere after having been discharged from its containing vessel. But this feature of carbon dioxide may also be associated with a disadvantage in its use. It essentially limits the use of carbon dioxide to areas where the atmosphere is calm, as inside buildings. Even when used within buildings, firefighters need to apply a sufficient quantity of carbon dioxide so that the burning material has an opportunity to cool. Otherwise, once the gas dissipates into the atmosphere and the fuel again contacts atmospheric oxygen, the fire may begin burning again.

The principal disadvantage in the use of carbon dioxide is associated with the fact that it does not support the life process. Since it is 1.5 times more dense than air, the total amount of carbon dioxide applied for extinguishment often replaces the available oxygen needed to sustain life in an area. Then carbon dioxide may cause death by suffocation. For this reason, it is often necessary to first evacuate the area before flooding it with carbon dioxide.

Figure 4-8   A portable carbon dioxide fire extinguisher. This extinguisher delivers a quick smothering action to flames, suffocating the fire of atmospheric oxygen. Its operation is simple: remove the locking pin, aim the horn at the base of the fire, squeeze the lever to discharge carbon dioxide, and release the lever to stop the discharge. (Courtesy of Walter Kidde, Mebane, North Carolina)

The physical properties of carbon dioxide at room temperature are relatively unique. For instance, most solids liquefy before they vaporize. However, solid carbon dioxide, called *Dry Ice,* passes directly from the solid state of matter to the vapor state, without passing through the liquid state. This transformation is called *sublimation*. Figure 4-9 shows the pressure–temperature relationship for carbon dioxide. The curves divide the plane into three regions, which provide the temperature and pressure conditions in which carbon dioxide exists. At atmospheric pressure, carbon dioxide can exist only as a solid or vapor. Hence, Dry Ice sublimes unless its pressure exceeds 5.11 atm, or 75.1 psi.

Carbon dioxide may be stored as a liquid under a pressure of at least 56.6 atm (830 psi) at room conditions. In fact, carbon dioxide is available as the liquid for use in portable extinguishers and total flooding systems as a fire extinguishing agent. Total flooding means that the affected area is literally flooded on demand with carbon dioxide. Figure 4-10 illustrates the use of carbon dioxide for fire extinguishment by means of total flooding.

Two systems involving the use of carbon dioxide are available for total

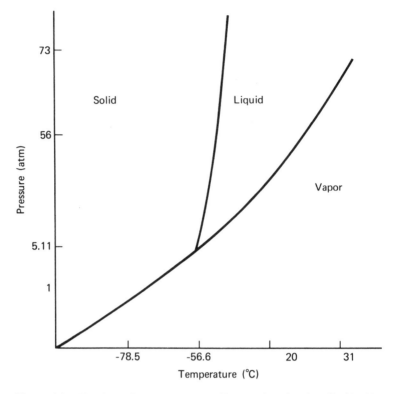

Figure 4-9 The phase diagram (not to a uniform scale) of carbon dioxide. Note that at 1 atm carbon dioxide only exists as either its vapor or solid (Dry Ice). Liquid carbon dioxide exists only at pressures equal to or greater than 75.1 psi (5.11 atm) and temperatures between $-69.9°F$ ($-56.6°C$) and approximately $88°F$ ($31°C$).

flooding. The first type is known as a *high-pressure* system, in which liquid carbon dioxide is stored in cylinders under 850 psi. The second type is the *low-pressure* system, in which liquid carbon dioxide is maintained at $0°F$ ($-18°C$) by means of refrigeration under 300 psi.

Some fire extinguishers take advantage of the production of carbon dioxide by chemical action. The operation of the once popular soda-acid fire extinguisher is due to a chemical reaction that produces carbon dioxide. Soda is the common name of the substance whose chemical name is sodium bicarbonate. A solution of soda and a solution of sulfuric acid are arranged as shown in Fig. 4-11; when the assembly is inverted, the two solutions mix. The resulting chemical reaction produces carbon dioxide, as illustrated by the following equation:

$$2NaHCO_3(aq) + H_2SO_4(aq) \longrightarrow Na_2SO_4(aq) + 2H_2O(l) + 2CO_2(g)$$

Carbon dioxide produced in the soda-acid fire extinguisher generates pressure that forces water out the nozzle. Hence, it is actually water, not carbon dioxide, that plays the role of extinguishing fires when soda-acid extinguishers are used. The stan-

Figure 4-10 Carbon dioxide is discharged in a total flooding system over a foundary quench tank without interfering with normal plant operations. (Courtesy of Chemetron Corporation, Chicago, Illinois)

dard 2.5 gallon soda-acid fire extinguisher provides 2.5 gallons of water. Sodium bicarbonate solutions slowly deteriorate with time. Hence, soda-acid extinguishers need to be periodically recharged.

Carbon dioxide may also be contained as bubbles in a *chemical foam*. Here, sodium bicarbonate and aluminum sulfate are stored as separate water solutions, or as powders, within an extinguisher. When mixed, they form carbon dioxide, as il-

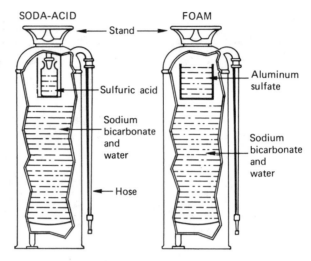

Figure 4-11 On the left is the soda-acid fire extinguisher, now relatively obsolete, containing sodium bicarbonate and sulfuric acid solutions. When inverted, the acid and bicarbonate solutions mix and chemically react, producing carbon dioxide, which forces the solution mixture out of the nozzle. On the right is a modification of this fire extinguisher, which uses solutions of aluminum sulfate and soldium bicarbonate to produce the carbon dioxide.

lustrated by the following equation:

$$6NaHCO_3(aq) + Al_2(SO_4)_3(aq) \longrightarrow 3Na_2SO_4(aq) + 2Al(OH)_3(s) + 6CO_2(g)$$

Sometimes, an extract of licorice root is mixed with these solutions. Together with aluminum hydroxide, they form a tough coating around each bubble of carbon dioxide. The entire mass issues as a foam and acts as a wet blanket on a fire, preventing the burning material from contacting atmospheric oxygen.

Some large petroleum refineries and/or their outlying tank farms employ a two-solution, wet-foam system for smothering fires that may occur in petroleum storage tanks. Solutions of aluminum sulfate and sodium bicarbonate are stored in separate tanks. These tanks are equipped with foam-mixing chambers, usually located at the top ring or roof of the tank, into which the solutions mix and form the chemical foam. When spread on the surface of a burning petroleum product, the foam starves the fire by preventing its contact with atmospheric oxygen.

Halon Agents

Many halogen-containing covalent compounds are nonflammable and thus are potentially useful as fire extinguishing agents. They are examples of substances known as *fluorocarbons* (Sec. 11-7), some of which have been commercially valuable in fire extinguishers. The fluorocarbons that are useful as fire extinguishing agents are substances whose molecules have one or two carbon atoms and from four to six atoms of fluorine, chlorine, or bromine. In other words, the molecular structures of these compounds resemble methane and ethane, in which certain hydrogen atoms have been substituted with halogen atoms. When used to extinguish fires, they are called *Halon agents*.

The first of such compounds discovered to have firefighting properties was carbon tetrachloride, a totally nonflammable liquid. It has a potential for use on class B fires as long as water is strictly absent. Applied as a liquid, carbon tetrachloride readily evaporates, forming a heavy cloud of vapor that smothers the fire. Other features that make it potentially useful as a fire extinguisher are its boiling point of 169°F (78°C) and freezing point of −3.1°F (−19.5°C).

However, carbon tetrachloride has disadvantages that far outweigh its strong points. When inhaled, its vapor may cause liver and kidney damage. But another disadvantage in the use of carbon tetrachloride during fire situations is that it may be converted chemically to the toxic gas, phosgene, whose formula is $COCl_2$. This chemical reaction occurs when carbon tetrachloride combines with moisture in the atmosphere, as illustrated by the following equation:

$$CCl_4(g) + H_2O(g) \longrightarrow COCl_2(g) + 2HCl(g)$$

The formation of phosgene is even possible in the total absence of fire, as, for instance, when carbon tetrachloride is spilled on hot metallic surfaces. Due to the potential formation of phosgene, the use of carbon tetrachloride has been strongly discouraged as a fire extinguisher. The use of carbon tetrachloride as a fire extinguishing agent has been illegal for decades in the United States with one exception: Utility companies may use carbon tetrachloride in small units intended to extinguish fires on pole tops only.*

The firefighting properties of several other Halon agents have been known since the 1940s. The common ones are listed with their physical properties in Table 4-3. The technical names of these substances are based on a unique system of nomenclature, in which a three- or four-digit number is used following the word Halon. Each digit in the number corresponds to the number of carbon and halogen atoms in the substance, as follows:

> First digit: number of carbon atoms
> Second digit: number of fluorine atoms
> Third digit: number of chlorine atoms
> Fourth digit: number of bromine atoms
> Fifth digit: number of iodine atoms

For instance, carbon tetrachloride is called Halon 104; bromochlorodifluoromethane is called Halon 1211.

*Although carbon tetrachloride has restricted use as a fire extinguishing agent, whenever it becomes necessary throughout the remainder of this book to discuss the chemistry of Halon agents, carbon tetrachloride will be selected as a "model" Halon agent for illustrative purposes only, because its molecular structure is much simpler than those of the other Halon agents.

TABLE 4-3   SOME PHYSICAL PROPERTIES OF THE COMMON HALON AGENTS

Fire Extinguishing Agent	Chemical Formula	Halon No.	Boiling Point	Melting Point	Specific Gravity of Liquid at 68°F (20°C)
Bromochloromethane	CH_2BrCl	1011	151°F(66°C)	−124°F(−191°C)	1.93
Dibromodifluoromethane	CBr_2F_2	1202	76°F(24.5°C)	−223°F(−142°C)	2.28
1,2-dibromo-1,1,2,2-tetrafluoroethane	$F_2BrC-CBrF_2$	2402	117°F(243°C)	−167°F(−111°C)	2.17
Bromotrifluoromethane	$CBrF_3$	1301	−72°F(−58°C)	−270°F(−168°C)	1.57
Dichlorodifluoromethane	CCl_2F_2	122	−22°F(−30°C)	−252°F(−158°C)	1.31
Bromochlorodifluoromethane	$CBrClF_2$	1211	25°F(−3.9°C)	−257°F(−161°C)	1.83
1,2-dichloro-1,1,2,2-tetrafluoroethane	$F_2ClC-CClF_2$	242	39°F(3.9°C)	−137°F(−94°C)	1.44

Like carbon dioxide, the Halon agents are collectively considered clean fire extinguishing agents. This factor is highly desirable for the protection of delicate electronic equipment, computer circuitry, and aircraft interiors. For this reason, Halon agents are ideally suitable when applied to fires in vaults, museums, libraries, hospitals, and similar areas. They may be discharged on class B or class C fires and on some class A fires. They are often considered more desirable than either water or carbon dioxide for the same purpose.

However, Halon agents also have several disadvantages. Halon agents should never be applied to class D fires, since they may do more harm than good. In addition, each Halon agent may also cause an undesirable physiological response when inhaled. Inhalation of elevated concentrations has been known to cause respiratory stimulation, tremors, convulsions, depression, lethargy, and unconsciousness. Halon agents decompose under fire conditions and may react with the fuel to form hydrogen fluoride, hydrogen chloride, and/or hydrogen bromide. These latter gases are very toxic and may achieve intolerable concentrations during fires.

Halon agents have one other undesirable feature. Following their use as fire extinguishing agents, Halon agents disperse in the atmosphere, finally rising to an area of the stratosphere known as the ozone layer (Secs. 6-1 and 11-7). Here these fluorocarbons react with ozone. Such reactions are believed to be associated with the precipitous drop in the concentration of stratospheric ozone first detected during 1987. This reduction in ozone interferes with vital processes essential for protecting the Earth from the effects of ultraviolet radiation. To inhibit further damage to the ozone layer, the production and use of all fluorocarbons, including Halon agents, is to be sharply curtailed throughout the civilized world by 1999. Because of a worldwide agreement to accomplish this feat, the use of Halon agents as fire extinguishers will presumably diminish in popularity.

In the United States, the most popular Halon agents are Halon 1301 (bromotrifluoromethane) and Halon 1211 (bromochlorodifluoromethane). They are readily available in fire extinguishers with convenient sizes for immediate discharge, such as 2.5, 5, and 9 lb (1.1, 2, and 4 kg), one of which is illustrated in Fig. 4-12.

At room temperature, the Halon agents are more dense than air. For example, Halon 1301 is a gas having a density about five times that of air. This physical property is one reason Halon agents are often capable of smothering fires. However, the principal reason they effectively extinguish fires is associated with their ability to form free radicals and atoms under fire conditions. These radicals and atoms react with the intermediates that form during the combustion of a fuel. Then certain mechanistic steps in combustion cannot occur, and this causes the fire to cease. The free radicals and atoms formed from the decomposition of Halon 1301 and Halon 1211 have a greater affinity for these intermediates than the intermediates do for each other.

Halon agents containing bromine, like either Halon 1301 or Halon 1211, are particularly effective as free radical scavengers. This is due to the ability of these agents to form bromine atoms when they are exposed to high temperatures. For instance, under fire conditions, Halon 1301 forms bromine atoms as illustrated by the

Figure 4-12 A portable Halon 1211 fire extinguisher. This extinguisher contains bromochlorodifluoromethane as the fire extinguishing agent. It may be used to effectively suppress class A, B, and C fires. Since the agent evaporates directly into the atmosphere following its application, no residue remains. This is a particularly desirable feature in certain environments, like data-processing centers. (Courtesy of Walter Kidde, Mebane, North Carolina)

following equation:

$$\text{CF}_3\text{Br}(g) \longrightarrow \underset{\underset{\text{F}}{|}}{\overset{\overset{\text{F}}{|}}{\text{F}-\text{C}\cdot}}(g) + \text{Br}\cdot(g)$$

Bromine atoms scavenge free radicals produced during the combustion of petroleum fuels. Consider once again the combustion of methane. Some pertinent reactions between combustion intermediates and bromine atoms are represented by the following equations:

$$\text{CH}_4(g) + \text{Br}\cdot(g) \longrightarrow \text{CH}_3\cdot(g) + \text{HBr}(g)$$

$$\text{HBr}(g) + \text{H}\cdot(g) \longrightarrow \text{H}_2(g) + \text{Br}\cdot(g)$$

$$\text{HBr}(g) + \cdot\text{OH}(g) \longrightarrow \text{H}_2(g) + \text{Br}\cdot(g)$$

In other words, bromine atoms react with the intermediates normally formed during combustion, thereby removing atoms and free radicals from the burning system, so that combustion cannot continue.

Dry Chemicals

There are five chemical substances generally used in the solid state as dry chemical fire extinguishing agents: sodium bicarbonate, potassium bicarbonate, sodium chloride, potassium chloride, and monoammonium phosphate. These substances are called *dry chemicals*. Certain of the dry chemical fire extinguishing agents effectively suppress class B and C fires, others are useful on class D fires, and some extinguish class A, B, and C fires. The substances that effectively suppress class D fires are also forms of fire extinguishers called *dry powder.* However, some dry powder formulations contain graphite. These graphite-based formulations are *not* dry chemical fire extinguishing agents.

Sodium bicarbonate and potassium bicarbonate are sometimes referred to as "regular" dry chemical fire extinguishing agents. Potassium bicarbonate is widely known by the trademark, *Purple K*. The name of this extinguishing agent is presumably related to the purple color imparted to a flame by potassium atoms. Both of these alkali metal bicarbonates effectively suppress and extinguish flammable liquid fires. In addition, they are often used as an initial tactic against high-pressure gas fires. Their use in such instances is only a temporary measure. Fires involving flammable gases are only fully extinguished when the gas leaks have been stopped, as by turning off a valve or plugging a hole.

The effectiveness of the alkali metal bicarbonates as fire extinguishing agents is due in part to the fact that heat causes them to decompose, forming carbon dioxide. As noted earlier, carbon dioxide smothers fire. The following equation illustrates the thermal decomposition of sodium bicarbonate:

$$2NaHCO_3(s) \longrightarrow Na_2CO_3(s) + CO_2(g) + H_2O(g)$$

However, there is a more fundamental reason why dry chemicals extinguish fires when they contain an alkali metal in their chemical structure. This is true not only of the alkali metal bicarbonates, but also the alkali metal chlorides. The latter are constituents of dry chemical fire extinguishers intended for use on alkali metal fires; for instance, sodium chloride is a fire extinguishing agent used to suppress some class D fires. When exposed to the high temperatures accompanying a major fire, the alkali metal bicarbonates and chlorides decompose, forming alkali metal atoms and chlorine atoms.

One theory that has been proposed to explain the effectiveness of such compounds as fire extinguishing agents deals with the presence of these alkali metal atoms. These atoms may react with other reactive chemical species to form water. For instance, in fires involving petroleum fuels, atoms of the alkali metals react with hydroxyl radicals as noted by the following equations:

$$Na \cdot (g) + \cdot OH(g) \longrightarrow NaOH(g)$$
$$NaOH(g) + \cdot H(g) \longrightarrow H_2O(l) + Na \cdot (g)$$

Figure 4-13 A portable, multipurpose, dry chemical fire extinguisher. This extinguisher contains monoammonium phosphate as the fire extinguishing agent. It may be used to effectively suppress most class A, B, and C fires. (Courtesy of Walter Kidde, Mebane, North Carolina)

In this case, the overall phenomenon involves the reaction of hydrogen atoms and hydroxyl radicals to form water.

$$H \cdot (g) + \cdot OH(g) \longrightarrow H_2O(g)$$

Another dry chemical is *monoammonium phosphate,* known more properly in chemistry as ammonium dihydrogen phosphate. This fire extinguishing agent is employed in the *multipurpose ABC* fire extinguisher, like that illustrated in Fig. 4-13. When properly used, its effectiveness as a fire extinguishing agent is twofold. Applied to a burning material, it decomposes endothermically as noted by the following equation:

$$2NH_4H_2PO_4(s) \longrightarrow P_2O_5(s) + 2NH_3(g) + 3H_2O(g)$$

Since it absorbs heat as it decomposes, the fuel is cooled below the minimum temperature at which it can burn. Under such conditions, the fire ceases. In addition, the ammonia generated by the decomposition of monoammonium phosphate may react with hydroxyl radicals. Hence, when this substance is applied to fires involving fuels that generate hydroxyl radicals, it functions by removing two components of the fire tetrahedron: heat and free radicals.

REVIEW EXERCISES

Writing Chemical Equations

4.1. In the Apollo 13 spacecraft, canisters containing solid lithium hydroxide were used to absorb carbon dioxide exhaled by the astronauts. Absorption results in the formation of solid lithium carbonate and water. Write the balanced equation for this chemical reaction.

4.2. An emergency oxygen supply can be obtained in high-flying aircraft by means of a chemical reaction involving the decomposition of sodium chlorate. This substance decomposes to oxygen and sodium chloride. Write the balanced chemical equation for this reaction.

4.3. Convert the following word equations into formula equations and balance them:
(a) Ammonia + oxygen → nitrogen monoxide + water
(b) Copper(II) hydroxide + sulfuric acid → copper(II) sulfate + water
(c) Cobalt(II) chloride + potassium hydroxide → cobalt(II) hydroxide + potassium chloride
(d) Silicon tetrachloride + calcium carbonate → silicon dioxide + calcium chloride + carbon dioxide
(e) Zinc sulfide + oxygen → zinc oxide + sulfur dioxide
(f) Iodine + hydrogen sulfide → hydrogen iodide + sulfur
(g) Ferric nitrate + stannous nitrate → ferrous nitrate + stannic nitrate
(h) Calcium hypochlorite → calcium chloride + oxygen
(i) Zinc sulfate + sodium sulfide → zinc sulfide + sodium sulfate
(j) Ammonium sulfate + sodium hydroxide → ammonia + sodium sulfate + water
(k) Ammonium acetate + silver nitrate → silver acetate + ammonium nitrate

Balancing Equations

4.4. Balance each of the following equations:

(a) $C_4H_{10}(g) + O_2(g) \rightarrow CO_2(g) + H_2O(l)$

(b) $As_2O_3(s) + C(s) \rightarrow As(s) + CO_2(g)$

(c) $Na(s) + H_2O(l) \rightarrow NaOH(aq) + H_2(g)$

(d) $CS_2(l) + Cl_2(g) \rightarrow CCl_4(l) + S_2Cl_2(l)$

(e) $MnO_2(s) + HCl(conc) \rightarrow MnCl_2(aq) + H_2O(l) + Cl_2(g)$

(f) $H_3PO_4(aq) + BaCl_2(aq) \rightarrow Ba_3(PO_4)_2(s) + HCl(aq)$

(g) $Ag_2S(s) + H_2O(l) \rightarrow Ag(s) + H_2S(g) + O_2(g)$

(h) $Ca_3P_2(s) + H_2O(l) \rightarrow Ca(OH)_2(s) + PH_3(g)$

(i) $PbCl_2(s) + Cl_2(g) + H_2O(l) \rightarrow PbO_2(s) + HCl(g)$

(j) $CaSO_4(s) + C(s) \rightarrow CaO(s) + CO_2(g) + SO_2(g)$

(k) $KClO_3(s) \rightarrow KClO_4(s) + KCl(s)$

(l) $Mg(s) + AgCl(s) \rightarrow Ag(s) + MgCl_2(aq)$

(m) $Mg_3N_2(s)\ H_2O(l) \rightarrow NH_3(g) + Mg(OH)_2(s)$

(n) $H_2S(g) + O_2(g) \rightarrow H_2O(l) + SO_2(g)$

Types of Chemical Reactions

4.5. Identify whether each of the following equations illustrates a combination, decomposition, single replacement, or double replacement reaction:

(a) $N_2O_5(g) + H_2O(l) \rightarrow 2HNO_3(aq)$

(b) $FeCO_3(s) \rightarrow FeO(s) + CO_2(g)$

(c) $Pb(C_2H_3O_2)_2(aq) + H_2S(g) \rightarrow 2HC_2H_3O_2(aq) + PbS(s)$

(d) $Ca(s) + 2HCl(aq) \rightarrow CaCl_2(aq) + H_2(g)$

(e) $Ag_2S(s) + 2HNO_3(aq) \rightarrow 2AgNO_3(aq) + H_2S(g)$

(f) $NH_3(g) + HCl(g) \rightarrow NH_4Cl(s)$

(g) $2Na(s) + 2H_2O(l) \rightarrow 2NaOH(aq) + H_2(g)$

(h) $PbSO_4(s) \rightarrow PbO(s) + SO_2(g)$

(i) $Fe(s) + CuCl_2(aq) \rightarrow FeCl_2(aq) + Cu(s)$

(j) $FeO(s) + H_2SO_4(aq) \rightarrow FeSO_4(aq) + H_2O(l)$

Oxidation–Reduction

4.6. In the 1800s, diarsenic trioxide was considered the archetype of a poison. It is produced when arsenic burns in air. This reaction is illustrated by the following unbalanced equation:

$$As(s) + O_2(g) \longrightarrow As_2O_3(s)$$

(a) Balance the equation.

(b) Identify the substance oxidized and the substance reduced.

(c) Identify the oxidizing agent and the reducing agent.

4.7. Hydrogen peroxide is a substance that has great practical importance. Its 3% aqueous solution by volume is frequently used as an antiseptic, whereas its 30% solution is used to bleach textile fibers, waxes, fats, hair, wood, and other materials. This utility of hydrogen peroxide is based in part on the fact that it may act in specific chemical reactions as either an oxidizing agent or a reducing agent, or even as both. Determine the role that hydrogen peroxide plays in the reactions corresponding to the following equations:

(a) $H_2O_2(aq) + Cl_2(g) \rightarrow 2HCl(aq) + O_2(g)$

(b) $H_2O_2(aq) + Na_2SO_3(aq) \rightarrow H_2O(l) + Na_2SO_4(aq)$

(c) $2H_2O_2(aq) \rightarrow 2H_2O(l) + O_2(g)$

Factors Affecting Reaction Rates

4.8. Among the various factors that influence reaction rates, which is involved in each of the following observations:

(a) In the northern United States, an automobile engine starts with difficulty during winter months, but with ease during summer months.

(b) Flour in bulk ignites with difficulty, but dispersed in air, flour ignites explosively.

(c) Hospital patients under oxygen therapy breathe more easily compared to when such therapy is not used.

(d) Wood lying exposed to the atmosphere does not burn, whereas it does burn when exposed to an ignition source.

(e) The roasting time for a 20-lb turkey is $5\frac{1}{2}$ hours at 325°F, but only 4 hours at 450°F.

Combustion Phenomena

4.9. Why is it impossible for water and carbon dioxide to burn in the open atmosphere?

4.10. When placed over a campfire, why does the bottom of a skillet often become black with soot (carbon)?

4.11. When wood and coal have burned, an ash generally remains. What matter constitutes this ash?

4.12. Beeswax is almost pure myricyl palmitate, a relatively complex substance having the chemical formula $C_{47}H_{94}O_2$. Explain why the burning of a candle made of beeswax does not ordinarily leave a residue.

4.13. Piles of coal occasionally ignite by spontaneous combustion. Why is this hazard greater in coal piles containing freshly pulverized coal as compared to those consisting of coal in large chunks?

Chemistry of Fire Extinguishment

4.14. What is the most likely reason that USDOT limits the quantity of Dry Ice that may be offered for transportation on board aircraft?

4.15. Why do free radicals only possess a transient existence?

4.16. Halon 1211 is bromochlorodifluoromethane.
   (a) Draw its Lewis structure.
   (b) Draw the structure of the free radical that forms when Halon 1211 is exposed to the high temperatures accompanying a fire.

4.17. Grease fires occurring in kitchens may usually be extinguished with ordinary table salt (sodium chloride). Describe the mechanism by which table salt is able to extinguish grease fires.

4.18. Ammonium dihydrogen phosphate is often used for flameproofing wood, paper, and textiles. It is also used to coat vegetation when fighting forest fires. Describe the mechanism by which this substance is able to retard the spread of fire.

5

Some Aspects of the USDOT Hazardous Materials Regulations

Certain aspects of regulations promulgated by the U.S. Department of Transportation (USDOT) are so important when studying the chemistry of hazardous materials that this entire chapter is devoted to their review. This particular set of regulations was published pursuant to the *Hazardous Materials Transportation Act*. Implementation of these regulations has a particularly significant impact on emergency-response personnel, since the regulations provide a means for the rapid identification of hazardous materials when they are encountered during transportation mishaps. Certain aspects, like marking and labeling requirements on packaging, may even provide helpful information when hazardous materials are encountered during more routine emergency-response actions. In order to properly handle such incidents, swift and accurate identification of the material involved is always highly advantageous.

While the USDOT regulations* are applicable to those who offer or accept hazardous materials for transportation within the borders of the United States, they have been written so as to conform to generally accepted international standards. Thus, these regulations may potentially aid anyone who encounters hazardous materials at virtually any point worldwide.

*The intent of the USDOT hazardous materials regulations is to protect transportation personnel and equipment from materials with dangerous properties. To further promote safety within the transportation industry, these regulations are periodically revised and amended. Until 1988, they were highly complex in both content and organization. Then proposals were made to simplify them so as to produce consistent modal and regional transportation requirements. To the extent possible, the discussion in this chapter follows these more simplified proposals.

Since the hazardous materials regulations change from time to time, users of the regulations should consult Title 49 of the *Code of Federal Regulations*, Parts 171 through 179, for the latest amendments.

5-1 SHIPPING PAPERS

A *hazardous material* is defined by USDOT as any designated material "capable of posing an unreasonable risk to health, safety and property when transported." Practically speaking, the transportation of over 2700 hazardous materials is regulated by USDOT. These materials are tabulated in a Hazardous Materials Table, a small section of which is illustrated in Table 5-1. Note that the Hazardous Materials Table provides ten columnar entries for each hazardous material. One or more symbols may be entered in column 1, each of which relates to certain transportation aspects. For instance, one symbol is the letter A; when this letter is displayed in column 1, the specified requirements typically apply to hazardous materials offered or intended for transportation by aircraft. The symbols that may be entered in Column 1 are not directly useful to emergency-response personnel.

With few exceptions, USDOT requires each person who offers a hazardous material for transportation to describe it on *shipping papers.* These shipping papers may take a number of forms, such as a shipping order, bill of lading, manifest, or similar document. Shipping papers provide the following information for each hazardous material intended for transportation:

Proper shipping name
Hazard class
Identification number
Packing group
Total quantity by weight, volume, or as otherwise appropriate

This information must be properly entered on shipping papers in a specific manner, like that shown in Fig. 5-1. We shall note some features of each requirement.

Proper Shipping Name

The *proper shipping name* of a hazardous material is the name that appears in roman type in column 2 of the Hazardous Materials Table, not the alternative names sometimes listed in italics or other names that are used to describe the material. A hazardous material is not always described by its chemical name. Sometimes, a technical name is provided; and some entries refer to commercial preparations or mixtures as opposed to pure substances. Thus, a carrier who desires to ship ten drums containing gasoline enters the name of this hazardous material as "Gasoline" on shipping papers accompanying the shipment, as Table 5-1 provides. This material may not be described as "automobile fuel" or any other similar term. On the other hand, when appropriate, the use of one or more italicized names from the table is permitted, when used *in addition to* the proper shipping name.

When a consignment consists of hazardous materials and other materials whose shipment is not regulated by USDOT, the hazardous materials, identified by

TABLE 5-1 SELECTED ENTRIES FROM THE HAZARDOUS MATERIALS TABLE[a]

Symbols (1)	Hazardous materials descriptions and proper shipping names (2)	Hazard class (3)	Identification numbers (4)	Packing group (5)	Labels (6)	Special provisions (7)	Packaging authorizations (§173.***)			Quantity limitations		Vessel stowage requirements		
							Exceptions (8A)	Non-bulk packaging (8B)	Bulk packaging (8C)	Passenger aircraft or railcar (9A)	Cargo aircraft only (9B)	Cargo vessel (10A)	Passenger vessel (10B)	Other stowage Provisions (10C)
	Acetone	3	UN1090	II	FLAMMABLE LIQUID	T8	150	202	242	5 L	60 L	1,3	5	
	Acetonitrile, see Methyl cyanide													
	Ammonia solutions, *density (specific gravity) between 0.880 and 0.957 at 15 degrees C in water, with more than 10% but not more than 35% ammonia.*	8	UN2672	III	CORROSIVE	T14	154	203	241	5 L	60 L	1,2	1,2	40,85
	Arsine	2.3	UN2188	I	POISON GAS, FLAMMABLE GAS.	10	None	192	245	Forbidden	Forbidden	1	5	40,95
	Asbestos, white (*chrysolite, actinolite, anthophyllite, tremolite*)	9	UN2590	III	CLASS 9		155	216	240	200 kg	200 kg	1,2	1,2	
	Battery fluid, acid	8	UN2796	II	CORROSIVE	B2, B15, N1, N6 N26, N34, T9, T27.	154	202	242	1 L	30 L	1,2	1	33
	Bromine *or* Bromine solutions	8	UN1744	I	CORROSIVE, POISON.	10, B12N 1, N11, N34, T18, T41	None	227	249	Forbidden	Forbidden	1		
	Calcium or calcium alloys	4.3	UN1401	II	DANGEROUS WHEN WET.		None	212	241	15 kg	50 kg	1,3	5	
	Calcium oxide	8	UN1910	III	CORROSIVE		154	213	240	25 kg	100 kg	1,2	1,2	
	Caustic soda, (etc.) *see* Sodium hydroxide *etc.*													
	Chlorobenzene	3	UN1134	III	FLAMMABLE LIQUID.	T1	150	203	241	60 L	220 L	1,3	1	
	Corrosive liquids, n.o.s.	8	UN1760	I	CORROSIVE	B4, N1, N11, N26, N34, T42.	None	201	242	0.5 L	2.5 L	1,2	1	40
	Corrosive liquids, poisonous, n.o.s	8	UN2922	I	CORROSIVE, POISON.	N1, N11, N34.	None	201	243	0.5 L	2.5 L	1,2	1	40
	Dichloromethane	6.1	UN1593	III	KEEP AWAY	N36, T13	153	203	241	60 L	220 L	1,2	1,2	25, 34

Ethyl mercaptan	3	UN2363	I	FLAMMABLE LIQUID.		None	202	243	Forbidden	30 L	1,3	5	12, 13 34, 35 40
Ethyl methyl ketone or Methyl ethyl ketone	3	UN1193	II	FLAMMABLE LIQUID.	T8	150	202	242	5 L	60 L	1.3	1	
Flammable liquids, poisonous, n.o.s.	3	UN1992	I	FLAMMABLE LIQUID, POISON.	B38, T42	None	201	243	Forbidden	30 L	1,3	5	40
Flammable solids, n.o.s	4.1	UN1325	II	FLAMMABLE SOLID.		151	212	240	15 kg	50 kg	1,3	1	
D[b] Fuel oil (No. 1, 2, 4, 5, or 6)	3	NA1202	III	None		150	203	241	60 L	220 L	1,3	1,3	
Gasoline	3	UN1203	II	FLAMMABLE LIQUID.	T8	150	202	242	5 L	60 L	1,3	5	12
Hydrazine, anhydrous or Hydrazine aqueous solutions *with more than 64% hydrazine, by weight*	3	UN2029	I	FLAMMABLE LIQUID, POISON, CORROSIVE.	B16, B17, B24, N1, N11, N26, N35, T25.	None	201	243	Forbidden	2.5 L	1	5	40
Hydrofluoric acid, anhydrous, see Hydrogen fluoride, anhydrous.													
Hydrogen fluoride, anhydrous	8	UN1052	I	CORROSIVE, POISON.	B12, T24, T27.	None	163	243	Forbidden	Forbidden	1	5	40
Lime, unslaked, see Calcium oxide *Lye, see* Sodium hydroxide, solutions Liquefied petroleum gas *see* Petroleum gases, liquefied													
Mercury oxide	6.1	UN1641	II	POISON		None	212	242	25 kg	100 kg	1,2	1,2	95
Methyl cyanide	3	UN1648	II	FLAMMABLE LIQUID POISON.	T14	None	202	243	1 L	60 L	1,3	1	40
Methyl cyclohexane	3	UN2296	II	FLAMMABLE LIQUID.	B1, T1	150	202	242	5 L	60 L	1,3	1	
Methylene chloride, see Dichloromethane *Methyl ethyl ketone, see* Ethyl methyl ketone.													

[a] Reproduced from Title 49, *Code of Federal Regulations*, §172.101.
[b] The letter "D" identifies proper shipping names which are appropriate for describing hazardous materials for domestic transportation but which may be inappropriate for international transportation under the provisions of international regulations.

137

TABLE 5-1 SELECTED ENTRIES FROM THE HAZARDOUS MATERIALS TABLE (cont.)

Symbols (1)	Hazardous materials descriptions and proper shipping names (2)	Hazard class (3)	Identification numbers (4)	Packing group (5)	Labels (6)	Special provisions (7)	(8) Packaging authorizations (§173.***)			(9) Quantity limitations		(10) Vessel stowage requirements		
							Exceptions (8A)	Non-bulk packaging (8B)	Bulk packaging (8C)	Passenger aircraft or railcar (9A)	Cargo aircraft only (9B)	Cargo vessel (10A)	Passenger vessel (10B)	Other stowage Provisions (10C)
	Nitric acid, red fuming	8	UN2032	I	CORROSIVE, OXIDIZER, POISON.	B17, B28, B30, 10.	None	158	244	Forbidden	Forbidden	1	5	33, 37, 38, 40, 63
	Nitrogen, compressed	2.2	UN1066		NONFLAMMABLE GAS		306	302	314, 315	75 kg	150 kg	1,3	1,3	85
	Organophosphorus pesticides, solid, toxic, n.o.s.	6.1	UN2783	I	POISON	N77	None	211	242	5 kg	50 kg	1,2	1,2	40, 95
	Petroleum gases, liquefied	2.1	UN1075		FLAMMABLE GAS.		306	304 315	314,	Forbidden	150 kg	1,3	1	40, 85
	Phosgene	2.3	UN1076	I	POISON GAS, CORROSIVE.	10, B7, B45, B46	None	192	245	Forbidden	Forbidden	1	5	40, 95
	Phosphine	2.3	UN2199	I	POISON GAS, FLAMMABLE GAS	B7, 10	None	192	245	Forbidden	Forbidden	1	5	40, 95
	Sodium	4.3	UN1428	II	DANGEROUS WHEN WET.	A19,A20, B22, N2, N16, N26, N34, T15, T29.	None	212	243	Forbidden	50 kg	1	5	
	Sodium hydrate, see Sodium hydroxide													
	Sodium hydroxide, solid	8	UN1823	II	CORROSIVE		154	212	240	15 kg	50 kg	1,2	1,2	
	Sodium hydroxide solution	8	UN1824	II	CORROSIVE	B2, N34, T8.	154	202	242	1 L	30 L	1,2	1,2	
	Stannic chloride, anhydrous	8	UN1827	II	CORROSIVE	B2, T8, T26.	154	202	242	1 L	30 L	1	1	8
	Toluene	3	UN1294	II	FLAMMABLE LIQUID	T1	150	202	242	5 L	60 L	1,3	1	
	1,1,1-Trichloroethane	6.1	UN2831	III	KEEP AWAY FROM FOOD	N36, T7	153	203	241	60 L	220 L	1,2	1,2	34, 40

SHORT FORM — ORIGINAL — NOT NEGOTIABLE

RECEIVED, subject to the classifications and tariffs in effect on the date of the issue of this Bill of Lading.

DATE	PAGE: OF	FREIGHT PREPAID except when box at right is checked	COLLECT ☐	CARRIER	NO.	ROUTE/DEL. CARRIER

SHIPPER'S NO'S

CUSTOMER'S ORDER NO'S

CONSIGNED TO

THE PROPERTY DESCRIBED BELOW, IN APPARENT [illegible]

Received $ ______ to apply in prepayment of the charges on the property described hereon.

Agent or Cashier

Per ______ (The signature here acknowledges only the amount prepaid)

Charges Advanced:

$

Subject to Section 7 of conditions, of applicable Bill of Lading, if this shipment is to be delivered to the consignee without recourse on the consignor, the consignor shall sign the following statement:
The carrier shall not make delivery of this shipment without payment of freight and all other lawful charges.

(Signature of Consignor)

The fibre boxes used for this shipment conform to the specifications set forth in the box maker's certificate thereon, and all other requirements of the Consolidated Freight Classification.

H M	DESCRIPTION OF ARTICLES IF HAZARDOUS MATERIAL - PROPER SHIPPING NAME (CHEMICALS AND ACIDS - RVNX - 50¢/LB.)	EXCEP-TIONS	EXEMPTIONS	** WEIGHT (SUB TO CORR)		CLASS OR RATE	NO PKGS	TYPE PKG.
x	10 drums, Gasoline, 3, UN1203, PG II			4,500	LBS			
x	40 Cylinders, Nitrogen, compressed, 2.2, UN1066			800	LBS			
x	1 Drum, Flammable solids, n.o.s., 4.1, UN1325, PG II			452	LBS			
	4 Boxes, Advertising Materials, Paper, NOI			60	LBS			
	1 Roll, Paper printing, newsprint			690	LBS			
	12 Sets, Carbon paper			22	LBS			
					LBS			
					LBS			
					LBS			
					LBS			
					LBS			

Shipper —

Per ______

This is to certify that the above named materials are properly classified, described, packaged, marked and labeled, and are in proper condition for transportation, according to the applicable regulations of the Department of Transportation.

This shipment is within the limitations prescribed for Passenger/Cargo only Aircraft. (Delete nonapplicable).

Per ______

**If the shipment moves between two ports by a carrier by water, the law requires that the Bill of lading shall state whether it is carrier's or shipper's weight.
† NOTE—Where the rate is dependent on value, shippers are required to state specifically, in writing, the agreed or declared value of the property. The agreed or declared value of the property is to be specifically stated by the shipper to be not exceeding RVNX 50¢/lb.

Agent ______ Per ______

AGENT MUST SIGN ORIGINAL BILL OF LADING AND MUST DETATCH AND RETAIN THE SHIPPING ORDER

Figure 5-1 Shippers are required to provide a description of the hazardous materials offered for transportation on shipping papers in a prescribed manner, including their proper shipping names, hazard classes, identification numbers, packing groups, and total quantity per description. When hazardous materials are involved in transportation mishaps, the nature of the consignment may be readily identified from such papers.

their proper shipping names, are entered *first* on shipping papers. Thus, in Fig. 5-1, gasoline, nitrogen, and an unspecified flammable solid are listed on the shipping paper before the materials that are not subject to the USDOT regulations. Alternatively, a contrasting color or an × may be used to identify the hazardous materials description entries.

When carriers of hazardous materials select the proper shipping name to describe a particular commodity, they select the name from the Hazardous Materials Table that most accurately describes the material to be shipped. In the event that the correct technical name of a material is neither listed nor entirely accurate, selection is then made from certain general descriptions; or carriers may use the letters n.o.s. (meaning not otherwise specified), n.o.i. (meaning not otherwise indexed), or n.o.i.b.n. (meaning not otherwise indexed by name).

On occasion, the letters RQ are used with a hazardous material description on shipping papers. For instance, the proper shipping description for aqueous ammonia is the following: "RQ, Ammonia solutions, 8, UN2672, PG III." RQ refers to the minimum amount that constitutes a *reportable quantity* of a hazardous substance as that term is used in CERCLA (Sec. 1-5). Reportable quantities of hazardous substances are provided in federal transportation and environmental regulations.* Nine of the hazardous materials in Table 5-1 are hazardous substances: ammonia solutions; chlorobenzene; anhydrous hydrogen fluoride; red fuming nitric acid; phosgene; sodium; solid sodium hydroxide; sodium hydroxide solution; and toluene. The reportable quantity is 5,000 lb (2268 kg) for anhydrous hydrogen fluoride and phosgene. The reportable quantity is 1,000 lb (454 kg) for the other hazardous substances listed here.

CERCLA requires carriers to notify the National Response Center[†] as soon as a discharge or other release to the environment is detected of a hazardous substance (for example, during a transportation mishap), whenever an amount equal to or greater than the reportable quantity is released to the environment.

There are a variety of general requirements that carriers follow when properly describing hazardous materials on shipping papers. Some pertinent examples are provided in Table 5-2.

Hazard Classes

Each hazardous material regulated by USDOT is associated with a specific *hazard class,* which meets the defining criterion of a specific hazard. Hazard criteria are examined in Sec. 5-7. Nine hazard classes are symbolized by a number from 1 to 9,

---

*For example, see Title 40, *Code of Federal Regulations,* §302.4, Table 302.4.

[†] The National Response Center may be initially contacted at 800-424-8802 or 202-426-2675. A follow-up report in writing is also required to the Director, Office of Hazardous Materials Regulations, Materials Transportation Bureau, Department of Transportation, Washington, D.C. 20590.

TABLE 5-2 SOME EXAMPLES OF PROPER HAZARDOUS MATERIALS DESCRIPTIONS
ON SHIPPING PAPERS

Example	Appropriate regulation
Corrosive liquids, n.o.s. (Valeric acid), 8, UN1760, PG I	If a hazardous material is described by an n.o.s. entry in the table, the name of substance is entered in parenthesis immediately following the proper shipping name.
Flammable liquids, poisonous, n.o.s. (contains xylene and methanol), 3, UN1992, PG I	The name of a mixture may be most aptly described by use of an appropriate modifier.
Waste flammable solids, n.o.s., 4.1, UN2926, PG I	The proper shipping name for a hazardous waste under RCRA (Sec. 1-5) includes the word "Waste" before the basic description of the hazardous material.
DOT-E***, RQ, Hydrogen fluoride, anhydrous, 8, UN1052, PG I	DOT-E-*** refers to an exemption that USDOT has provided (*** is the appropriate exemption number); RQ means reportable quantity (see text).
RQ, Toluene, 3, UN1294, PG II, Placarded FLAMMABLE	When a hazardous material is shipped by rail tank car, the shipping papers bear the notation "Placarded," followed by the name of the required placard after the basic description.
Petroleum gases, liquefied, 2.1, UN1075, noncorrosive, DOT-113A (Do Not Hump or cut Off Car While in Motion)	When a flammable gas is shipped in a rail tank car, such as a DOT-113 rail tank car, the shipping papers note for each such rail tank car an appropriate distinction, like A, B, or C, in addition to the words "Do Not Hump or Cut Off Car While in Motion." The word "noncorrosive" may be entered on shipping papers describing liquefied petroleum gas to indicate that the hazardous material is chemically compatible with the structural material of the cargo tank.
RQ, Sodium, 4.3, UN1428, PG II (Dangerous When Wet)	The words "Dangerous When Wet" are entered on shipping papers when a material, by chemical interaction with water, is liable to become spontaneously flammable or to give off flammable gases in dangerous quantities.
Stannic chloride, anhydrous, 8, UN1827, PG II, Ltd Qty	The use of "Limited Quantity" or "Ltd Qty" means that this hazardous material may be only transported in limited amounts by either passenger aircraft, railcar, or cargo aircraft.
Fuel oil, 3 (combustible liquid), NA1202, PG III	If the hazardous material is a combustible liquid, the description "combustible liquid" appears in parentheses immediately following the hazard class.
Corrosive liquids, poisonous, n.o.s. (contains o-cresol), 8, UN2922, PG I	When the proper shipping name of a hazardous substance (Sec. 1-5) does not identify the constituent that makes it a hazardous substance, the name of the constituent is entered in association with the basic description.

TABLE 5-2 SOME EXAMPLES OF PROPER HAZARDOUS MATERIALS DESCRIPTIONS ON SHIPPING PAPERS (cont.)

Example	Appropriate regulation
EMPTY: last contained methyl ethyl ketone, 3, UN1193, PG II	The shipment of packaging that contains a residue of a hazardous material identifies the last hazardous material contained therein.
Ethyl mercaptan, 3, UN2363, PG II (Cargo Aircraft Only)	When a hazardous material is to be shipped by air and USDOT prohibits the shipment aboard passenger-carrying aircraft, the words "Cargo Aircraft Only" are entered after the basic description.
Organophosphorus pesticides, solid, toxic, n.o.s., 6.1, UN2783, PG I (contains 3-hydroxy-*N*-methyl-*cis*-crotonamide dimethyl phosphate)	When the name of a poison is not included in the proper shipping name, the name of the compound that causes it to be poisonous is entered along with the basic description.
RQ, Phosgene, 2.3, UN1076, PG I (Poison Inhalation Hazard)	Hazardous materials that are toxic by inhalation are described by the words "Poison Inhalation Hazard," in addition to the proper shipping name.

and one additional hazard class is represented by the letters ORM-D or ORM-E. Certain of the initial nine hazard classes are further divided into divisions, which are symbolized by a number following the number of the hazard class and separated by a period; thus, 1.2 refers to division 1.2 of class 1.

Table 5-3 lists the class numbers, division numbers, and the class or division names assigned to USDOT-regulated hazardous materials. For a given hazardous material, the hazard class is identified from the entries in column 3 of the Hazardous Materials Table. To further aid in describing a hazardous material, the column 3 designation is entered on shipping papers immediately adjacent to the proper shipping name. Thus, a carrier desiring to transport gasoline enters a 3 next to the name Gasoline (see Table 5-1 and Fig. 5-1).

Sometimes, the word "forbidden" appears in column 3 of the Hazardous Materials Table. USDOT prohibits such hazardous materials from being offered or accepted for transportation.

Identification Numbers

The identification number is either of the prefixes UN (United Nations) or NA (North America), followed by a four-digit number. The UN numbers are derived from an international system designed by the United Nations Committee of Experts on Transportation of Dangerous Goods. Since some USDOT-regulated hazardous materials are not included in the UN system, USDOT adopted numbers used in the North American Materials Identification System for those hazardous materials not

TABLE 5-3   HAZARD CLASSES AND DIVISIONS OF HAZARDOUS MATERIALS

Hazard class number	Division number (if any)	Name of hazard class or division
None		Forbidden materials
None		Forbidden explosives
1	1.1	Explosives (with a mass explosion hazard)
1	1.2	Explosives (with a projection hazard)
1	1.3	Explosives (with predominantly a fire hazard)
1	1.4	Explosives (with no significant blast hazard)
1	1.5	Very insensitive explosives; blasting agents
2	2.1	Flammable gas
2	2.2	Nonflammable compressed gas
2	2.3	Poisonous gas
3		Flammable and combustible liquids
4	4.1	Flammable solids
4	4.2	Spontaneously combustible materials
4	4.3	Dangerous when wet materials
5	5.1	Oxidizers
5	5.2	Organic peroxides
6	6.1	Poisonous materials
6	6.1	Irritating materials
6	6.2	Etiologic or infectious substances
7		Radioactive materials
8		Corrosive materials
9		Miscellaneous hazardous materials
None		Other regulated materials: ORM-D and ORM-E

covered by the UN system. These USDOT identification numbers are provided in the Hazardous Materials Table in column 4.

USDOT requires the identification number for a hazardous material to be entered on shipping papers immediately following the hazard class. Thus, a carrier desiring to transport gasoline must enter the number UN1203 next to the hazard class 3 (again see Table 5-1 and Fig. 5-1).

The purpose of the identification number is to rapidly provide information to an emergency-response team under conditions when this information is warranted. When the identification number is known, team members may quickly identify the nature of a hazardous material and respond effectively to the incident in which it is involved. One way by which this may be accomplished is through the use of a guidebook published by USDOT, the *Emergency Response Guidebook*, which references each identification number to a specific response action. We shall review the use of this guidebook in Sec. 5-8.

Packing Groups

Packaging is defined by USDOT to be "the receptacle and any other components or materials necessary for the receptacle to perform its containment function and to ensure compliance with the minimum [USDOT] packing requirements . . . ." These receptacles may be fiber drums, wooden boxes, steel drums, glass bottles, or similar items. The packaging used for the shipment of hazardous materials is required to "be designed, constructed, maintained, filled, its contents so limited, and closed," so that under conditions normally incident to transportation:

1. There will be no release of hazardous materials to the environment;
2. The effectiveness of the packaging will not be significantly reduced; and
3. There will be no mixture of gases or vapors in the package which could, through any credible spontaneous increase of heat or pressure, significantly reduce the effectiveness of the packaging.

USDOT's packaging requirements match the type of shipping container with the degree of hazard of the material to be transported; that is, materials of similar hazard are, in general, packaged in the same manner. Any packaging that is suitable for a material, because of the properties of the material, is authorized for the transportation of that material. More secure and stronger packages are required for containing materials that pose the greatest hazard. USDOT further requires the manufacturers of all containers and packages to fulfill certain performance testing requirements.

The authorized means of packaging hazardous materials are distinguished for bulk and nonbulk shipments. For most hazard classes, USDOT assigns a *packing group* and identifies it in column 5 of the Hazardous Materials Table for each hazardous material intended for transportation.

There are only three packing groups, identified as packing groups I, II, and III. These groups indicate that the degree of hazard is either great, medium, or minor, respectively. Classes 1 and 7 and divisions 2.1 and 2.2 of class 2 do not have packing groups; nor does USDOT assign a packing group to ORM-D materials. Referring to Table 5-1, gasoline is assigned to packing group II. A shipper interested in transporting gasoline would then consult the appropriate section of the USDOT regulations to identify the authorized packaging for transporting gasoline, either in bulk or in nonbulk.

Column 7 of the Hazardous Materials Table refers to special codes that stipulate certain packaging provisions, prohibitions, and exceptions that may apply to specific modes of transport. For instance, the T8 noted for gasoline in Table 5-1, applies only to transportation in IM (intermodal) portable tanks.

Column 8 of the Hazardous Materials Table specifies applicable sections of the regulations for exceptions (column 8A), nonbulk packaging requirements (column 8B), and bulk packaging requirements (column 8C). When "none" appears in column 8A, no packaging exceptions are authorized by USDOT. The sections of column 8 are completed so that §173 precedes the designated numerical entry. For instance, the entry 202 in column 8B associated with the proper shipping name "Gasoline" indicates that §173.202 of Title 49, *Code of Federal Regulations* should be consulted for nonbulk packaging requirements. This information, while essential for gasoline shippers, is of little value for emergency-response personnel; hence, it is not discussed further.

When describing a hazardous material on shipping papers, USDOT requires identification of the packing group, if any, immediately following the USDOT identification number. For such purposes, the words "packing group" may be abbreviated as PG. Thus, when gasoline is to be shipped, the complete hazardous material description on shipping papers reads as follows: "Gasoline, 3, UN1203, PG II."

Total Quantities

USDOT requires the entry on shipping papers of the total quantity of a hazardous material intended for shipment by weight, volume, or as otherwise appropriate. The entry appears either before or after the description of the hazardous material, or both. Abbreviations may be used to express units of measurement. Furthermore, the type of packaging (drums, boxes, and the like) in which the quantity is contained may be entered in any appropriate manner either before or after the basic description. For example, the shipping papers shown in Fig. 5-1 identify a total quantity of 4500 lb of gasoline shipped in 10 drums, 800 lb of nitrogen shipped in 40 cylinders, and 452 lb of a flammable solid shipped in 1 drum, in addition to other nonregulated materials.

Column 9 of the Hazardous Materials Table specifies the maximum net quantities that may be offered for transportation in any one package by passenger-carrying aircraft or rail tank car (column 9A) or by cargo aircraft only (column 9B). For instance, no more than 5 L of gasoline may be shipped by passenger-carrying aircraft or by rail tank car; and no more than 60 L may be transported by cargo-carrying aircraft. When "Forbidden" appears in either column 9A or 9B, the hazardous material may not be offered for transportation by the applicable mode of transport.

5-2 LOCATION OF SHIPPING PAPERS DURING TRANSIT

In a transportation mishap, where may emergency-response personnel expect to locate the shipping papers? USDOT requires shipping papers to be carried on transport vehicles in a specific location. These requirements are briefly summarized next.

Motor Vehicles

The driver of a motor vehicle transporting hazardous materials and each carrier using the vehicle must clearly distinguish the appropriate shipping papers from papers of any other kind. When the driver is at the vehicle's controls, the shipping papers must be within immediate reach and either readily visible to a person entering the driver's compartment *or* in a holder mounted to the inside of the door on the driver's side of the vehicle. Figure 5-2 illustrates how immediately available these papers are. When the driver is not at the vehicle's controls, the shipping papers are either in the holder previously described or on the driver's seat in the vehicle.

Shipping
papers

Figure 5-2   When the driver of a motor vehicle is at the vehicle's controls, shipping papers describing hazardous materials are required by USDOT to be kept within immediate reach. This generally means that the papers are in a holder mounted to the inside of the door on the driver's side of the vehicle.

Rail Tank Cars

Shipping papers describing a hazardous materials consignment for transportation by rail must be in the possession of a member of the train crew transporting the hazardous material. Furthermore, the train crew is required to possess a document that indicates the position in the train of each loaded placarded car containing a hazardous material.

Aircraft

Shipping papers describing a hazardous materials consignment for transportation by aircraft must always be in the possession of the pilot in command.

Watercraft

A carrier, its agents, and any person designated for this purpose to transport hazardous materials aboard a vessel on water must prepare a dangerous cargo manifest, lost or stowage plan. This document is kept in a designated holder on or near the vessel's bridge.

5-3 LOCATION OF HAZARDOUS MATERIALS ON BOARD WATERCRAFT

The manner by which nonbulk quantities of hazardous materials are placed aboard most transport vehicles is not regulated by USDOT. However, when shipment occurs by means of a vessel on water, certain stowage requirements apply. These requirements are coded in column 10 of the Hazardous Materials Table.

Column 10 is divided into three sections. Column 10A applies to cargo-carrying vessels, whereas 10B applies to passenger-carrying vessels. Each column specifies the authorized stowage locations on board vessels by means of a numerical code ranging from 1 to 6. These codes identify the following authorized locations on board vessels:

1 The material may be stowed on deck.

2 The material must be stowed under deck.

3 The material must be stowed under deck away from heat.

1,2 The material may be stowed on deck or under deck.

1,3 The material may be stowed on deck or under deck away from heat.

4 The material may be transported on a passenger vessel in only the quantity specified in column 9A of the table and is subject to the stowage requirements specified for a cargo vessel for the same material.

5 The material is forbidden and may not be offered for transportation or transported by vessel.

6 The material shall be transported in a magazine subject to certain requirements of USDOT.

Hazardous materials are stowed under deck away from heat, if possible, when the code is identified as either 1,2 or 1,3.

Column 10C identifies certain other stowage provisions for specific hazardous materials by means of a numerical code ranging from 1 to 99. Examples of such codes are provided in Table 5-4.

TABLE 5-4 EXAMPLES OF ADDITIONAL USDOT STOWAGE AND SEGREGATION REQUIREMENTS ON BOARD WATERCRAFT[a]

Code	Provision
2	Temperature-controlled material
8	Glass bottles not permitted on passenger vessels
12	Keep cool
13	Keep dry
14	Metal drums only permitted under deck
25	Shade from radiant heat
32	Stow away from combustible materials
39	Stow away from liquid, halogenated hydrocarbons
61	Stow separated from corrosive materials
71	Stow separated from nitric acid
91	Stow separated from flammable solids
98	Stow away from all flammable materials

[a] Title 49 *Code of Federal Regulations*, §176.84.

For gasoline, Table 5-1 notes the entries 1,3 in column 10A, 5 in column 10B, and 12 in column 10C. The first two codes tell us that gasoline may only be shipped on board cargo vessels. The code 12 means "keep cool."

5-4 USDOT LABELS

Shippers are required to affix a warning label directly to the outside surface of some packages or containment devices of hazardous materials intended for shipment. These labels are color coded, usually square, but tilted 90° on point, and may illustrate a hazardous symbolic message. The information conveyed by these labels correlates with the principal hazard of the lading being shipped. The hazard class or division number is designated in the lower corner of most labels. The specific warning labels required by USDOT are illustrated in Sec. 5-7.

A hazardous material label is required to be affixed on the following: a nonbulk package; a portable tank of less than 1000-gal (3785-L) capacity; a DOT-106 or -110 multiunit tank car; and an overpack, freight container, or unit load device of no greater than 640-ft^3 (18.1-m^3) capacity that contains a package for which labels are required.

USDOT requires labels to be affixed on packages of hazardous materials as provided in column 6 of the Hazardous Materials Table. For instance, containers of gasoline are labeled with the FLAMMABLE LIQUID label. A steel drum containing gasoline that has been properly labeled for shipment is illustrated in Fig. 5-3. Occasionally, column 6 directs the use of more than a single label on packaging, depending on the unique properties of the material to be transported. The use of multiple labels on a hazardous material package is required when the material meets the definition of more than one hazard class. When their use is warranted, multiple labels are displayed next to each other.

Figure 5-3 When gasoline is shipped in a 30-gallon steel drum, USDOT requires the drum to be marked with the proper shipping name and USDOT identification number; furthermore, USDOT requires the FLAMMABLE LIQUID label to be affixed to the outside surface of the drum.

Finally, packages containing hazardous materials that are authorized only on cargo aircraft are labeled with a CARGO AIRCRAFT ONLY label, like that illustrated in Fig. 5-4.

5-5 USDOT-REQUIRED MARKINGS

With few exceptions, USDOT requires nonbulk packages of hazardous materials to be legibly marked with certain information. For instance, such packages are marked

Figure 5-4 The CARGO AIRCRAFT ONLY label, required on hazardous materials packaging when USDOT authorizes shipment only on board cargo aircraft (and *not* on board passenger-carrying aircraft). The label is black on an orange background.

with the material's proper shipping name and identification number. When the proper shipping name is described by an n.o.s. entry, the package must be marked with the technical name of the material in parentheses, immediately following or below the proper shipping name. On nonbulk packaging, these markings are required on only one side.

When transporting liquid hazardous materials in nonbulk packages, the following requirements are generally applicable:

1. Closure of the packaging must be upward.
2. Except for liquefied compressed gas cylinders, packages must be marked with package orientation markings on two opposite vertical sides of the package, with arrows pointing in the correct upright direction.

An example of this marking is illustrated in Fig. 5-5.

Figure 5-5 When gasoline is shipped in a 5-gallon metal can inside a cardboard box, USDOT requires orientation markings on two opposite vertical sides of the package with arrows pointing in the correct upright direction.

Nonbulk packages containing division 2.3 materials and poisonous liquids that pose an inhalation hazard are required to be marked "Inhalation Hazard." Each nonbulk plastic packaging used for containing Division 6.1 materials must be permanently marked POISON.

Each nonbulk packaging containing a material classed as either ORM-D or ORM-E is required to be marked on at least one side or end with the appropriate

ORM designation immediately following or below the proper shipping name of the material. The appropriate ORM designation is marked inside a rectangle, as illustrated in Fig. 5-6. The package of an ORM-D material prepared for shipment by air is marked ORM-D-AIR.

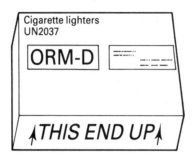

Figure 5-6 The proper marking on a package containing a hazardous material designated as ORM-D. When authorized for shipment only by cargo aircraft, the package is marked ORM-D-AIR.

The packaging for bulk shipments is required to be marked with the identification number of the lading, according to the following:

1. On each side and each end, if the capacity of the container is 1000 gal (3785.4 L) or more.
2. On two opposing sides if the capacity of the container is less than 1000 gal (3785.4 L).

Certain special marking requirements apply to shipments of hazardous materials in portable tanks, cargo tanks, and multiunit tank car tanks. For instance, a cargo tank used for shipping hazardous materials has displayed appropriate identification numbers on white square-on-point placards or orange panels, like that illustrated in Fig. 5-7. Portable tanks, tank cars, and multiunit tank cars containing a hazardous material are also marked with the proper shipping name and identification number. However, an identification number must *never* be displayed directly on a POISON GAS, RADIOACTIVE, EXPLOSIVES, BLASTING AGENT, or DANGEROUS placard.

When hazardous waste (Sec. 1-5) is transported in a container of 110 gal or less, the information illustrated in Fig. 5-8 is marked on the container.

5-6 USDOT PLACARDS

When a hazardous material is shipped either in bulk or nonbulk packagings, a warning placard is generally required to be affixed on the motor vehicle, unit load device, rail tank car, bulk packaging or freight container used during transportation. Each type of placard is specified in relation to the appropriate hazard class in Tables 5-5 and 5-6; the nature of each placard is described in Sec. 5-7. The required placard must be displayed on each side and each end of the freight container or transport vehicle, as appropriate.

Figure 5-7   USDOT requires the identification number of a hazardous material to be displayed on certain transport vehicles (without the UN or NA prefix). It may be displayed either on placards affixed to or on orange panels marked on the vehicle.

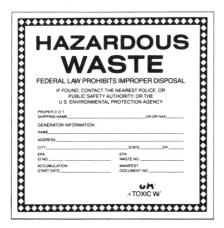

Figure 5-8   The marking required on hazardous waste containers. Before transporting hazardous waste or offering hazardous waste for transportation offsite, the generator of the waste is required to mark this information on each hazardous waste container.

When a hazardous material is shipped whose hazard class is noted in Table 5-5, the appropriate container, device, or transport vehicle is always placarded, irrespective of the amount shipped. Placarding is also required for a hazardous material shipped *in bulk* whose hazard class is noted in Table 5-6. However, for *nonbulk*

TABLE 5-5 USDOT HAZARD CLASSES THAT ALWAYS REQUIRE PLACARDING

Hazard class or division number	Placard name
1.1	EXPLOSIVES 1.1
1.2	EXPLOSIVES 1.2
1.3	EXPLOSIVES 1.3
2.3	POISON GAS
4.3	DANGEROUS WHEN WET
6.1[a]	POISON
7[b]	RADIOACTIVE

[a] Packing group I inhalation hazard only.
[b] Radioactive Yellow-III label only.

TABLE 5-6 HAZARD CLASSES THAT REQUIRE PLACARDING ONLY UNDER CERTAIN CONDITIONS

Hazard class or division number		Placard name
1.4		EXPLOSIVES 1.4
1.5		EXPLOSIVES 1.5
2.1		FLAMMABLE GAS
2.2		NONFLAMMABLE GAS
3	(flammable liquid)	FLAMMABLE
3	(combustible liquid)	COMBUSTIBLE
4.1		FLAMMABLE SOLID
4.2		SPONTANEOUSLY COMBUSTIBLE
5.1		OXIDIZER
5.2		ORGANIC PEROXIDE
6.1	(PG I or II, other than PG I inhalation hazard)	POISON
6.1	(PG III)	None
6.2		None
8		CORROSIVE
9		None
ORM-D		None
ORM-E		None

packagings, the use of placards is not routinely required when the gross weight of all hazardous materials covered only by Table 5-6 is less than 1000 lb (454 kg).

A freight container, unit load device, motor vehicle, or railcar that is used to ship nonbulk packagings with two or more categories of hazardous materials covered by Table 5-6 may display separate placards. Alternatively, instead of using multiple placards, the shipper may affix DANGEROUS placards, like the one illustrated in Fig. 5-9. However, when 5000 lb (2268 kg) or more of one category of hazardous

Figure 5-9 The DANGEROUS placard. USDOT permits the use of the DANGEROUS placard when two or more categories of hazardous materials covered by Table 5-6 are shipped in the same transport vehicle *and* when the quantity of a specific category of hazardous material does not exceed 5,000 lb (2,268 kg). The DANGEROUS placard has red upper and lower triangles, with a white rectangular central area and white outer border; the inscription is black.

material is loaded therein at one loading facility, the placard specified in Table 5-6 for that category must be used.

When the use of multiple placards is warranted because two or more categories of hazardous materials are shipped (because of the gross weight), certain exceptions should be noted, as follows:

1. When both division 1.1 and 1.2 explosives are transported in the same container or transport vehicle, only the EXPLOSIVES 1.1 placard is displayed.
2. When both flammable and combustible liquids are shipped in a cargo tank, a portable tank, or a compartmented tank car, only the FLAMMABLE placard is displayed.
3. When a nonflammable gas is shipped by motor vehicle with a flammable gas, only the FLAMMABLE GAS placard is displayed.
4. When division 1.4 explosives, division 1.5 explosives, or oxidizers are shipped with division 1.1 or 1.2 explosives, only the EXPLOSIVES 1.1 or EXPLOSIVES 1.2 placard is displayed, as appropriate.
5. When an oxidizer is shipped with division 1.5 explosives by motor vehicle or rail tank car, only the EXPLOSIVES 1.5 placard is displayed.

5-7 USDOT CLASSIFICATION OF HAZARDOUS MATERIALS

As noted earlier, nine hazard classes have been designated from 1 through 9, in addition to "other regulated materials." We shall now briefly review the basic features of these classes, as well as the labelling and placarding requirements imposed by USDOT when hazardous materials are offered or accepted for shipment.

Class 1: Explosives

An *explosive* is defined by USDOT as "any chemical compound, mixture or device, the primary or common purpose of which is to function by explosion, *i.e.,* with substantially instantaneous release of gas and heat . . . ." There are five divisions within class 1: divisions 1.1, 1.2, 1.3, 1.4, and 1.5.

Explosives in *division 1.1* are substances and articles that have a mass explosion hazard. *Division 1.2* explosives are substances and articles that have a projection hazard, but not a mass explosion hazard. *Division 1.3* explosives are substances and articles that have a fire hazard and either a minor blast hazard or a minor projection hazard, or both, but not a mass explosion hazard. *Division 1.4* explosives are substances and articles that present no significant hazard of exploding. *Division 1.5* explosives are very insensitive substances, such as blasting agents.

Packages of explosives are labeled with warning labels like those shown in Fig. 5-10(a). Motor vehicles, rail tank cars, and freight containers in which explosives are transported have placards affixed on each side and each end, like those shown in Fig. 5-10(b). The EXPLOSIVES 1.1, EXPLOSIVES 1.2, and EXPLOSIVES 1.3 placards are displayed whenever any quantity of the corresponding substances is shipped.

Class 2: Gases

There are three divisions within this hazard class corresponding to the following hazardous materials: division 2.1 (flammable gas), division 2.2 (nonflammable compressed gas), and division 2.3 (poisonous gas).

A *flammable gas* is any material that is a gas at 68°F (20°C) or less and 1 atm of pressure, and:

1. is ignitable at 1 atm when in a mixture of 13% or less by volume with air; or
2. has a flammable range of 1 atm with air of at least 12% regardless of the lower limit.

A *nonflammable compressed gas* is any material or mixture that exerts in its container a pressure of 40 psia (275.8 kPa) at 70°F (21.1°C), or regardless of the pressure at 70°F (21.1°C), exerts in the container a pressure of 104 psia (717.1 kPa) at 130°F (54.4°C), and does not meet the definition of division 2.1 or 2.3.

A *poisonous gas* means a material that is a gas at 68°F (20°C) or less and 1 atm of pressure and:

1. is known to be so toxic to humans as to pose a hazard to health during transportation, or
2. in the absence of adequate data on human toxicity, is presumed to be toxic to humans, because when tested on laboratory animals, it has an LC_{50} (Sec. 9-4) less than 5000 ppm.

When their use is warranted, USDOT requires that the appropriate warning labels be affixed to the packaging in which the class 2 hazardous material is contained; these labels resemble those shown in Fig. 5-11(a).

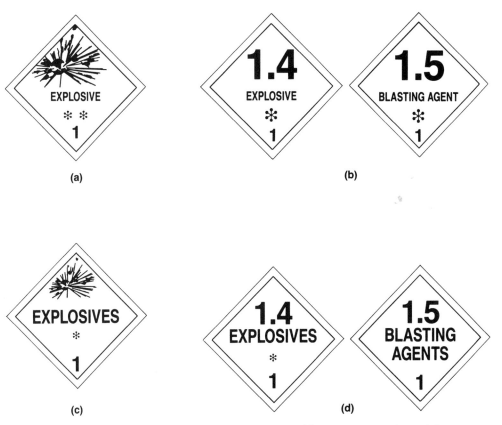

Figure 5-10   As appropriate the label indicated in (a) is affixed to containers of division 1.1, division 1.2, and division 1.3 explosives intended for shipment. The ** is replaced with the appropriate division number and compatibility group. (Compatibility groups are discussed in Sec. 13-3.) As appropriate, the label in (b) is affixed to containers of division 1.4 and division 1.5 explosives intended for shipment; the * is replaced with the appropriate compatibility group. Each label affixed on packages of explosives is orange with black writing, inscription, and inner border. When required, placards like those indicated in (c) and (d) are displayed on the motor vehicle, rail tank car, or freight container used for transporting explosives, as appropriate. These placards are orange with a white outer border and black inscription. The placards in (c) are displayed when division 1.1, division 1.2, and division 1.3 explosives are shipped; and the placards in (d) are displayed when division 1.4 and division 1.5 explosives are shipped, as appropriate. The * is replaced with the appropriate division number.

When placards are required to be displayed on motor vehicles, cargo tanks, and other transport vehicles carrying gases, they resemble those shown in Fig. 5-11(b). POISON GAS placards are required on transport vehicles carrying any amount of a poisonous gas.

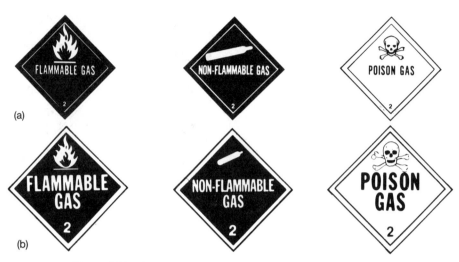

Figure 5-11 As appropriate, the label indicated in (a) is affixed on containers of compressed gases offered for shipment, except on cylinders containing nonpoisonous gases (see text). The FLAMMABLE GAS label is red with either a black or white symbol and inner border. The NONFLAMMABLE GAS label is green with either a black or white inscription and inner border. The POISON GAS label is white with black writing and inner border and a black and white skull and crossbones. When required, placards like those indicated in (b) are displayed on the motor vehicle, rail tank car, or freight container used for transporting compressed gases, as appropriate. The FLAMMABLE GAS placard is red with a white symbol, inscription, and outer border. The NONFLAMMABLE GAS placard is green with a white symbol, inscription, and outer border. The POISON GAS placard is white with a black solid line inner border; the symbol and inscription are also black.

Class 3: Flammable Liquids

A *flammable liquid* is any liquid having a flash point of not more than 141°F (60.5°C), except any mixture having one or more components with a flash point greater than 141°F (60.5°C) or higher, that makes up at least 99%, of the total volume of the mixture.

A *combustible liquid* is either of the following: (1) any liquid that has a flash point above 141°F (60.5°C) and below 200°F (93.3°C), or (2) any material that has a flash point of 200°F (93.3°C) or greater and is transported as a liquid at a temperature at or above its flash point.

USDOT requires a warning label on flammable liquid containers like that shown in Fig. 5-12(a). When placards are required to be displayed on a transport vehicle shipping a flammable or combustible liquid, they resemble those shown in Fig. 5-12(b). The word GASOLINE may be used in place of FLAMMABLE on a placard that is displayed on a cargo tank or a portable tank being used to transport gasoline by highway.

(a)   (b)

Figure 5-12   The label indicated in (a) is affixed to containers of flammable liquids offered for shipment. The FLAMMABLE LIQUID label is red with a white or black border, symbol, and inscription. When required, placards like either of those indicated in (b) are displayed on the motor vehicle, rail tank car, or freight container used for transporting flammable and combustible liquids, as appropriate. Both placards are red with white symbol, inscription, and outer border.

Class 4: Flammable Solids, Spontaneously Combustible Materials, and Dangerous When Wet Materials

There are three divisions within this hazard class corresponding to the following hazardous materials: division 4.1 (flammable solid), division 4.2 (spontaneously combustible material), and division 4.3 (dangerous when wet material).

A *flammable solid* is any solid material, other than an explosive, that, under conditions normally incident to transportation is likely to cause fires through friction or retained heat from manufacturing or processing, or that can be ignited readily, and when ignited burns so vigorously and persistently as to create a serious transportation hazard.

A *spontaneously combustible material* is a material that is likely to heat spontaneously under conditions normally incident to transportation or to heat up in contact with air and then is likely to catch fire. Pyrophoric liquids, like those noted in Chapter 8, are division 4.2 materials.

A *dangerous when wet material* is a material that, by interaction with water, is likely to become spontaneously flammable or to give off a flammable gas in dangerous quantities.

Packages containing a division 4.1, 4.2, or 4.3 hazardous material display the appropriate USDOT label shown in Fig. 5-13(a) corresponding to its lading. When required on transport vehicles, appropriate placards are displayed, like those shown in Fig. 5-13(b). DANGEROUS WHEN WET placards are displayed whenever any quantity of a dangerous when wet material is shipped.

Class 5: Oxidizers and Organic Peroxides

There are two divisions in class 5: division 5.1 (oxidizer) and division 5.2 (organic peroxide). An *oxidizer* is a material such as a chlorate, permanganate, inorganic

Figure 5-13   As appropriate, the label indicated in (a) is affixed to containers of flammable solids, spontaneously combustible materials, and dangerous when wet materials offered for shipment. The FLAMMABLE SOLID label is white with three vertical red stripes equally spaced on each side of a red stripe in the center of the label; the overprinted symbol, inscription, and inner border line are black. The SPONTANEOUSLY COMBUSTIBLE label is red in the lower half and white in the upper half, with a black symbol and printing. The DANGEROUS WHEN WET label is blue with a black or white border line, symbol, and inscription. When required, placards like those indicated in (b) are displayed on the motor vehicle, rail-car, or freight container used for transporting flammable solids, spontaneously combustible materials, and dangerous when wet materials, as appropriate. The FLAMMABLE SOLID placard is white with seven vertical red stripes and a black inner border with a black overwritten inscription and symbol. The SPONTA-NEOUSLY COMBUSTIBLE and DANGEROUS WHEN WET placards resemble the corresponding labels.

peroxide, or a nitrate that yields oxygen readily to stimulate the combustion of organic matter. An *organic peroxide* is an organic compound containing the —(O—O)— structure and which may be considered a derivative of hydrogen peroxide.

Packages containing a division 5.1 or 5.2 hazardous material display the appropriate USDOT label shown in Fig. 5-14(a) corresponding to its lading. When required on transport vehicles, appropriate placards are displayed, like those shown in Fig. 5-14(b).

Class 6: Poisonous Materials and Infectious Materials

There are two divisions in class 6: division 6.1 (poisonous material) and division 6.2 (infectious substance). A *poisonous material* is a material, other than a gas, that is known to be toxic to humans such as to afford a hazard to health during transportation, or which, in the absence of adequate data on human toxicity, is presumed to be

(a)

(b)

Figure 5-14   As appropriate, the label indicated in (a) is affixed to containers of oxidizers and organic peroxides offered for shipment. Each label is yellow with black printing, symbol, and inner border. When required, placards like either of those indicated in (b) are displayed on the motor vehicle, railcar, or freight container used for transporting oxidizers and organic peroxides, as appropriate. Each placard is yellow with a black symbol, inscription, and inner border.

toxic to humans because of data obtained from tests performed on animals. An *infectious material,* or *etiologic agent,* means a viable microorganism or its toxin that causes or may cause human disease.

Packages containing a division 6.1 or 6.2 hazardous material display the appropriate USDOT label shown in Fig. 5-15(a) corresponding to its lading. USDOT does not require placards on transport vehicles used only for the shipment of division 6.2 materials. However, POISON placards resembling that in 5-15(b) are required to be displayed whenever any quantity of a poisonous material is shipped that represents an inhalation hazard.

Class 7: Radioactive Materials

A *radioactive material* is any material or combination of materials that spontaneously emits ionizing radiation having a "specific activity" greater than 0.002 microcuries per gram. The meaning of these terms is related to certain features of nuclear phenomena; they are discussed in Chapter 14.

Packages containing a class 7 hazardous material display the appropriate USDOT label shown in Fig. 5-16(a) corresponding to its lading. These are called the *RADIOACTIVE WHITE-I, RADIOACTIVE YELLOW-II,* and *RADIOACTIVE YELLOW-III* labels, respectively. The RADIOACTIVE WHITE-I label signifies the minimum hazard associated with radioactive materials, while the RADIOACTIVE YELLOW-III signals the maximum hazard. The proper type of label to affix to packages containing radioactive material is decided in part by the specific activity of the lading. We shall discuss this aspect of labeling radioactive materials packages in Chapter 14.

When required on transport vehicles carrying radioactive materials, the placard like that shown in Fig. 5-16(b) is displayed. RADIOACTIVE placards are

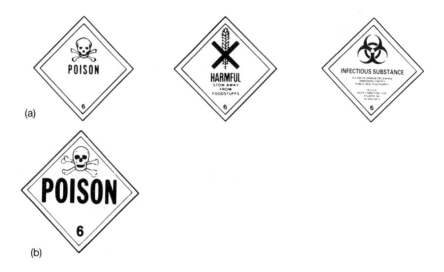

Figure 5-15 As appropriate, the label indicated in (a) is affixed to containers of poisonous materials, irritating materials, and infectious materials when offered for shipment. The POISON and KEEP AWAY FROM FOOD labels are white with black writing, inscription, and inner border. The INFECTIOUS SUBSTANCE label is white with either red or black inscription and border. When required, placards like that indicated in (b) are displayed on the motor vehicle, rail tank car, or freight container used for transporting certain poisonous materials and irritating materials.

required whenever any quantity of a radioactive material is shipped requiring RADIOACTIVE YELLOW-III labels to be displayed on the accompanying packaging.

Class 8: Corrosive Materials

A *corrosive material* is a liquid or solid that causes visible destruction or irreversible alterations in human skin tissue at the site of contact, or a liquid that has a severe corrosion rate on steel or aluminum, in accordance with certain proscribed USDOT testing procedures.

USDOT requires a warning label on containers of a corrosive material like that shown in Fig. 5-17(a). When a placard is required on a transport vehicle carrying a corrosive material, it resembles that shown in Fig. 5-17(b).

Class 9: Miscellaneous Hazardous Materials

A *miscellaneous hazardous material* is any material that presents a hazard during transportation, but that is not included in any other hazard class. Included in this class is any material that has an anesthetic, noxious, or other similar property.

Figure 5-16   As appropriate, the label indicated in (a) is affixed to containers of radioactive materials offered for shipment. The RADIOACTIVE WHITE-I label is white with a black symbol and printing, except for the I, which is red. The RADIOACTIVE YELLOW-II and RADIOACTIVE YELLOW-III labels are yellow in the top half and white in the lower half; each bears a black symbol and printing, except for the II or III, as appropriate, which is red. When required, placards like that indicated in (b) are displayed on the motor vehicle, rail tank car, or freight container used for shipping containers of radioactive materials bearing the RADIOACTIVE YELLOW-III label. The top portion of the RADIOACTIVE placard is yellow with the symbol black; the lower portion is white with a black inscription.

(a)   (b)

Figure 5-17   When required, the label indicated in (a) is affixed to containers of corrosive materials offered for shipment. The CORROSIVE label is white in the top half and black in the bottom half; the printing is white, the inner border is black, and the symbol is black and white. When required, placards like that indicated in (b) are displayed on the motor vehicle, rail tank car, or freight container used for transporting corrosive materials. The center and lower half of the CORROSIVE placard is black except for the printing, which is white; the symbol is black and white on a white background.

When required, packages of miscellaneous materials bear a label like that shown in Fig. 5-18. However, USDOT does not require the use of a specific placard on transport vehicles used to ship packages containing only miscellaneous hazardous materials.

Figure 5-18 The CLASS 9 label is affixed to containers of miscellaneous hazardous materials offered for shipment. The CLASS 9 label is white with seven black vertical stripes in the upper triangle.

Aside from the hazardous materials previously described in nine classes, there is one additional group of hazardous materials corresponding to the designations ORM-D and ORM-E. ORM is an acronym for "other regulated materials." A hazardous material designated as ORM-D is a "material such as a consumer commodity which . . . presents a limited hazard during transportation due to its form, quantity, and packaging." Hair spray packed in an aerosol container is an example of such a consumer commodity. A hazardous material designated as ORM-E is one "that is not included in any other hazard class, but is subject to the USDOT regulations." Certain hazardous wastes as defined by RCRA (Sec. 1-5) are examples of hazardous materials designated as ORM-E.

5-8 THE EMERGENCY RESPONSE GUIDEBOOK

Suppose we know the USDOT identification number for a particular hazardous material, either by reference to the proper description on shipping papers or by observing it displayed on portable tanks, rail tank cars, or multiunit tank cars. How may this number be used advantageously when the hazardous material is involved in a transportation mishap?

Identification numbers are catalogued in a USDOT publication, *The Emergency Response Guidebook*.* In one section of the guidebook, the identification numbers are arranged by increasing number, beginning at 1001; in another section, chemical substances are arranged alphabetically. The guidebook cross-references each identification number to the chemical substance it represents and to one of 55 *guides,* beginning at guide 11 and proceeding to guide 66. Each individual guide consists of a single page on which two major points of information are provided: *potential hazards* and *emergency action*. Thus, when a hazardous materials incident occurs, the guidebook may be used by emergency-response personnel to locate the appropriate guide. The guide directs personnel to certain initial actions which, when implemented, should protect lives, property, and the environment. The *Emergency Response Guidebook* should be readily available in any vehicle that is routinely used during emergency-response actions so that it may be consulted when the need arises.

How is the guidebook used? Suppose we are concerned with a transportation

*Available from the U.S. Department of Transportation, Research and Special Programs Administration, Materials Transportation Bureau, U.S. Department of Transportation, Washington, D.C. 20590.

incident involving an initially unidentified hazardous material that is being shipped by truck. Upon securing the shipping papers during this mishap, personnel note the following hazardous materials description: "Glass bottles, RQ, Phosphoric acid, 8, UN 1805, PG III; 485 lb." Suppose further that other materials, not regulated by USDOT, constitute the rest of the consignment. This description informs personnel that the truck was used to ship bottles of phosphoric acid.

The identification number, UN1805, is first in importance. Referring to the *Emergency Response Guidebook,* the number 1805 leads the reader to guide 60, as illustrated in Fig. 5-19 on page 164. Guide 60 is reproduced in Fig. 5-19 on page 165; it directs personnel to follow certain indicated procedures, depending on whether the acid is still confined or leaking from containers. First-aid directions are also provided by the guide in the event that the acid has contacted skin tissue.

The value of the *Emergency Response Guidebook* is self-evident. When properly used, it may protect the public, as well as the emergency-response personnel who are called upon to handle incidents involving hazardous materials. It is designed to provide guidance primarily during the initial phases of a response action. When bulk quantities are involved in transportation mishaps, or whenever deemed necessary, a telephone call should be placed to CHEMTREC (Sec. 1-7) for further assistance.

REVIEW EXERCISES[a]

Shipping Papers

5.1. Provide the proper shipping name only for the following hazardous materials as required by USDOT:
  (a) Acidic battery fluid   (b) Methylene chloride
  (c) Acetonitrile   (d) Mercuric oxide
  (e) Red fuming nitric acid

5.2. Identify the hazard class of each of the following substances when offered for transportation:
  (a) Asbestos, white   (b) Methyl cyclohexane
  (c) Unslaked lime   (d) Calcium metal
  (e) Bromine   (f) Arsine
  (g) Anhydrous hydrazine

5.3. A carrier is retained to deliver 12,000 lb of chlorobenzene by rail tank car to a given destination.
  (a) Based on USDOT regulations, provide the complete description to be entered on the waybill accompanying this consignment.
  (b) What additional information does the train crew document concerning this consignment?

5.4. A carrier is retained to deliver sixty 55-gal drums of waste acetone by motor vehicle to

---

[a] Solutions to many of the following exercises require the use of Table 5-1.

ID No.	Guide No.	Name of Material	ID No.	Guide No.	Name of Material
1787	60	HYDROGEN IODIDE SOLUTION	1801	60	OCTYL TRICHLOROSILANE
1788	60	HYDROBROMIC ACID	1802	45	PERCHLORIC ACID, not more than 50% acid, by weight
1788	60	HYDROGEN BROMIDE SOLUTION	1803	60	PHENOLSULPHONIC ACID, liquid
1789	60	HYDROCHLORIC ACID SOLUTION	1804	29	PHENYL TRICHLOROSILANE
1789	60	HYDROGEN CHLORIDE SOLUTION	1805	60	PHOSPHORIC ACID
1789	60	MURIATIC ACID	1806	39	PHOSPHORUS PENTA-CHLORIDE
1790	59	ETCHING ACID, liquid	1807	39	PHOSPHORIC ANHYDRIDE
1790	59	FLUORIC ACID	1807	39	PHOSPHORUS PENTOXIDE
1790	59	HYDROFLUORIC ACID SOLUTION	1808	39	PHOSPHORUS TRIBROMIDE
1790	59	HYDROGEN FLUORIDE SOLUTION	1809	39	**CHLORIDE OF PHOSPHORUS ***
			1809	39	**PHOSPHORUS TRICHLORIDE ***
1791	60	HYPOCHLORITE SOLUTION with more than 5% available chlorine	1810	39	**PHOSPHORUS OXYCHLORIDE ***
			1810	39	PHOSPHORYL CHLORIDE
1791	60	POTASSIUM HYPOCHLORITE SOLUTION	1811	60	POTASSIUM BIFLUORIDE
1791	60	SODIUM HYPOCHLORITE SOLUTION	1811	60	POTASSIUM HYDROGEN FLUORIDE
			1812	54	POTASSIUM FLUORIDE
1792	59	IODINE MONOCHLORIDE	1813	60	BATTERY, electric, storage, dry, containing POTASSIUM HYDROXIDE
1793	60	ISOPROPYL ACID PHOSPHATE			
1794	60	LEAD SULFATE, with more than 3% free acid	1813	60	CAUSTIC POTASH, dry, solid
1796	73	ACID MIXTURE, nitrating	1813	60	POTASSIUM HYDROXIDE, dry, solid
1796	73	MIXED ACID	1814	60	CAUSTIC POTASH SOLUTION
1796	73	NITRATING ACID			
1796	73	NITRATING ACID, mixture	1814	60	POTASH LIQUOR
1798	60	NITROHYDROCHLORIC ACID	1814	60	POTASSIUM HYDROXIDE SOLUTION
1798	60	NITROMURIATIC ACID	1815	29	PROPIONYL CHLORIDE
1799	60	NONYL TRICHLOROSILANE	1816	29	PROPYL TRICHLOROSILANE
1800	39	OCTADECYL TRICHLORO-SILANE	1817	39	PYROSULFURYL CHLORIDE

* Look for information next to this **NAME** in the TABLE OF EVACUATION DISTANCES in the back of this book. Use this in addition to the Guide Page if there is NO FIRE.

Figure 5-19 The USDOT identification number for phosphoric acid directs emergency-response personnel to the page of the *Emergency Response Guidebook* shown above, which further directs the reader to Guide No. 60, reproduced on page 165. In the guidebook itself, the names of certain substances are highlighted in bold print. When the name of a substance appears in the guidebook in bold print, the reader should also consult its Table of Evacuation Distances.

GUIDE 60

POTENTIAL HAZARDS

HEALTH HAZARDS

Contact causes burns to skin and eyes.

If inhaled, may be harmful.

Fire may produce irritating or poisonous gases.

Runoff from fire control or dilution water may cause pollution.

FIRE OR EXPLOSION

Some of these materials may burn, but none of them ignites readily.

Flammable/poisonous gases may accumulate in tanks and hopper cars.

Some of these materials may ignite combustibles (wood, paper, oil, etc.).

EMERGENCY ACTION

Keep unnecessary people away; isolate hazard area and deny entry.

Stay upwind; keep out of low areas.

Self-contained breathing apparatus (SCBA) and structural firefighter's protective clothing will provide limited protection.

CALL CHEMTREC AT 1-800-424-9300 FOR EMERGENCY ASSISTANCE. If water pollution occurs, notify the appropriate authorities.

FIRE

Some of these materials may react violently with water.

Small Fires: Dry chemical, CO_2, Halon, water spray or standard foam.

Large Fires: Water spray, fog or standard foam is recommended.

Move container from fire area if you can do it without risk.

Cool containers that are exposed to flames with water from the side until well after fire is out. Stay away from ends of tanks.

SPILL OR LEAK

Do not touch spilled material; stop leak if you can do it without risk.

Small Spills: Take up with sand or other noncombustible absorbent material and place into containers for later disposal.

Small Dry Spills: With clean shovel place material into clean, dry container and cover; move containers from spill area.

Large Spills: Dike far ahead of liquid spill for later disposal.

FIRST AID

Move victim to fresh air; call emergency medical care.

Remove and isolate contaminated clothing and shoes at the site.

In case of contact with material, immediately flush skin or eyes with running water for at least 15 minutes.

Keep victim quiet and maintain normal body temperature.

Figure 5-19 (continued)

an incineration facility. Each drum weighs 362 lb. Based on USDOT regulations, provide the complete description that must be entered on the manifest accompanying this consignment.

5.5. Determine whether USDOT allows or prohibits the shipment of each of the following substances by passenger-carrying or cargo aircraft, both or neither, *and,* when shipment is permitted, identify the maximum quantity that can be shipped in one package.

(a) Acetonitrile
(b) Methyl cyclohexane
(c) Chlorobenzene
(d) Unslaked lime
(e) Arsine
(f) Methylene chloride

5.6. An oil company intends to ship 50,000 gal of fuel oil from Corpus Christi, Texas, to New Orleans, Louisiana, on board a vessel by water.

(a) Specify the complete description of fuel oil to be entered on the shipping papers accompanying this consignment.
(b) Must the oil be stowed in any special fashion?
(c) Where must the shipping papers for this consignment be kept?

Labels, Markings, and Placards

5.7. During a potential arson investigation, an inspector identifies an empty container with an affixed hazardous materials label. Most of the label has been scarred by fire, and its warning information has been largely obliterated. However, the inspector discerns the number 3 in the lower part of this label.

(a) What class of hazardous materials is represented by this number?
(b) What potential value may this knowledge have for the investigator?

5.8. Determine which warning label, if any, is required on packages and containers when the following hazardous materials are offered for transportation by motor vehicle:

(a) Acidic battery fluid
(b) Methylene chloride
(c) Acetonitrile
(d) Mercuric oxide
(e) Methyl cyclohexane
(f) Arsine
(g) Bromine
(h) Anhydrous hydrazine

5.9. A transport company is retained to deliver eighty 55-gal drums of 1,1,1-trichloroethane by motor vehicle to a given destination. Each drum weighs approximately 600 lb.

(a) Based on USDOT regulations, provide the complete description to be entered on the shipping papers accompanying this consignment.
(b) While the driver is at the wheel of the motor vehicle, where are the shipping papers kept?
(c) Which warning label, if any, is affixed to each drum?
(d) Which placards, if any, are affixed to the motor vehicle?

5.10. Determine the total number of required placards on a motor vehicle when the vehicle is used to ship each of the following mixed loads of hazardous materials:

(a) 1000 lb of bromine, 2500 lb of methyl cyclohexane, and 4000 lb of mercuric oxide
(b) 1000 lb of bromine, 2500 lb of methyl cyclohexane, and 5100 lb of barium chlorate

5.11. Describe the nature of markings required by USDOT on steel drums containing waste methyl ethyl ketone, whose flash point is 20°F (−7°C).

5.12. A carrier in Maryland is retained to deliver the following mixed load of hazardous materials to a chemical research facility by motor vehicle:

> Twenty 55-gal drums of 1,1,1-trichloroethane, each weighing 655 lb
>
> Eight 55-gal drums of acetone, each weighing 345 lb
>
> Ten 55-gal drums of methyl cyclohexane, each weighing 405 lb
>
> One box containing fifteen 1-lb bottles of mercuric oxide
>
> Seven 5-gal glass bottles of 1,1,1-trichloroethane, each weighing 66 lb and packaged individually in cardboard boxes

(a) Identify the proper description for these hazardous materials that should be entered on the shipping papers accompanying the consignment.
(b) What warning labels, if any, must be affixed to the packaging?
(c) Which placards, if any, must be displayed on the motor vehicle?
(d) Where does USDOT require the carrier to display the placards?

Emergency-Response Actions

5.13. Arriving at the scene of a train wreck, the members of an emergency-response crew observe an orange panel on the side of a leaking, overturned rail tank car with the letters 1789 in black. Placards and markings on the rail tank car are otherwise obscured by smoke from a nearby fire.
(a) Using Fig. 5-19, what is the name of the hazardous material contained in the rail tank car?
(b) What action with respect to the leak, if any, does the *Emergency Response Guidebook* recommend?

6

Chemistry of Some Common Elements

The following eight elements are especially important in the study of hazardous materials: oxygen, hydrogen, fluorine, chlorine, bromine, phosphorus, sulfur, and carbon. These elements are associated with a special chemical reactivity, or they are regularly used as raw materials by certain manufacturing and process industries. The chemistry of these elements is also particularly important, as it may affect the chemistry of many other substances.

6-1 OXYGEN

Oxygen has properties that make it our most important element. This necessitates that the study of this element be given foremost attention.

Oxygen is the most abundant element on Earth. It makes up approximately 21% by volume of our atmosphere. However, most oxygen on Earth is not found in the free state, but in combination with other elements as chemical compounds. Water and carbon dioxide are common examples of compounds that contain oxygen, but there are countless others, many of which exist in the rocks and minerals that comprise Earth's crust. Oxygen accounts for approximately 50% by mass of all substances in Earth's crust, the atmosphere, oceans, and other bodies of water (Table 3-1).

Oxygen is represented by the chemical formula O_2. At ordinary temperatures, it is a colorless, odorless, and tasteless gas, only slightly more dense than air. At low temperatures and under pressure, however, oxygen exists as a pale blue liquid or solid. Liquid and solid oxygen have the curious feature of being attracted to a magnet. Some other physical properties of oxygen are listed in Table 6-1.

Commercially, oxygen is usually obtained from liquid air. Air liquefies when it is compressed at pressures greater than 545 psia (37.1 atm) while being simultaneously cooled below $-318°F$ ($-190°C$). The industrial process of preparing liquid

TABLE 6-1   PHYSICAL PROPERTIES OF OXYGEN

Freezing point	−361.12°F (−218.4°C)
Boiling point	−297.33°F (−182.96°C)
Heat of fusion	5.95 Btu/lb (3.3 cal/g)
Heat of vaporization	91.70 Btu/lb (50.9 cal/g)
Density of liquid at boiling point	87.94 lb/ft³ (1,410 g/L)
Density of gas at boiling point	0.268 lb/ft³ (4.3 g/L)
Density of gas at room temperature	0.081 lb/ft³ (1.3 g/L)
Vapor density (air = 1)	1.105
Liquid-to-gas expansion ratio	875

air consists of a series of compression, expansion, and cooling operations. Liquid air is essentially a mixture of about 43% nitrogen and 54% oxygen, with less than 3% by mass of argon. When this mixture is allowed to boil at 1 atm, the vapor phase above the liquid becomes enriched in nitrogen. Simultaneously, the liquid phase slowly becomes enriched in oxygen. It is often called *LOX*.

Facilities that use large supplies of oxygen often store it as the cryogenic liquid in tanks like that illustrated in Fig. 6-1. Storage of the liquid rather than the gas constitutes a space saving, since 875 volumes of the gas may be compressed into 1 volume of the liquid at 68°F (20°C). While stored as the liquid, oxygen may be dis-

Figure 6-1   A fixed storage tank of liquid oxygen (LOX) outside a major hospital in Allentown, Pennsylvania. A backup liquid tank is also shown. The liquid oxygen is vaporized for delivery to house lines and points of use. Hospitals and clinics use large amounts of oxygen for certain types of routine patient care, for example, oxygen-enriched therapy. Storing oxygen as the cryogenic liquid is often more efficient and economical, as compared to storing an equal quantity of the compressed gas in cylinders. (Courtesy of Air Products and Chemicals, Inc., Allentown, Pennsylvania)

charged from the tank as either the liquid or gas. However, the discharge of gaseous oxygen is often considered more desirable.

Aerospace industries use liquid oxygen as the oxidizer of liquid fuels in the propellant systems of missiles and rockets. But liquid oxygen is more typically encountered in everyday practice in less exotic industries. Cold-storage and food-processing plants, hospitals, metal-fabrication plants, electrical power plants, and the steel industry use liquid oxygen for various purposes.

Commercially, oxygen may be stored and transported as a compressed gas in steel cylinders, single-unit rail tank cars, tank trucks, and multiunit tanks or as liquid oxygen in special cryogenic vessels. In the former case, its proper shipping name is "oxygen, compressed"; in the latter case, it is "oxygen, refrigerated liquid." US-DOT regulates the transportation of oxygen as a nonflammable gas, but, when required, oxygen containers are labeled NONFLAMMABLE GAS and OXIDIZER. When required, the transport vehicle is placarded NONFLAMMABLE GAS.

Oxygen possesses two chemical properties of which firefighters should be aware. First, as previously noted in Sec. 4-6, oxygen supports combustion; that is, whenever substances burn in air, they usually unite with atmospheric oxygen. Thus oxygen is an oxidizing agent. In everyday practice, it is probably the most commonly encountered oxidizer.

Second, oxygen is capable of supporting the life process. Humans and animals inhale air and assimilate its oxygen through the lungs. The oxygen is carried by the bloodstream to the cells, where it is used to oxidize foodstuffs. The product of this oxidation is carbon dioxide, which is exhaled. The overall phenomenon releases energy, which the body uses to maintain life; it is called *respiration*. Obviously, without sufficient oxygen, we cannot long survive.

Most likely, as the human race evolved, we learned to tolerate an atmosphere containing oxygen at a concentration of 21% by volume. Exposure to other oxygen concentrations may cause adverse health effects. For instance, lung damage may result from inhaling an atmosphere for several days whose oxygen concentration exceeds 60% by volume; and infants exposed to an atmosphere containing 35% to 40% oxygen by volume may suffer permanent visual impairment or blindness due to retrolental fibroplasia.

Oxygen may indirectly cause a hazardous situation for firefighters and others who are forced to breathe an atmosphere reduced in oxygen. When a fire burns in a confined area, the atmosphere soon becomes reduced or even depleted of oxygen. When individuals must breathe this air, they are likely to suffer certain adverse health effects. Some symptoms associated with breathing atmospheres reduced in oxygen below 21% by volume are identified in Table 6-2. In particular, note that death may result within only a few short minutes from breathing air whose oxygen concentration is less than 6% by volume.

It is well to remember that, as it exists in the atmosphere, oxygen has been diluted with other gases. But as a compressed gas, and especially as the cryogenic liquid, oxygen is highly concentrated. In these forms, it is more chemically reactive than atmospheric oxygen. This can be easily verified by noting that materials burn

TABLE 6-2 SIGNS AND SYMPTOMS EXPERIENCED FROM INHALING REDUCED LEVELS OF OXYGEN

Percent oxygen in air	
20 (or above)	Normal
12–15	Muscular coordination for skilled movements is lost.
10–14	Consciousness continues, but judgment is faulty and muscular effort leads to rapid fatigue.
6–8	Collapse occurs rapidly, but quick treatment prevents fatal outcome.
6 (or below)	Death occurs in 6 to 8 minutes.

much more rapidly in an atmosphere of pure oxygen than they do in air. For this reason, valves, regulators, gauges, or fittings associated with containers of compressed oxygen should never be lubricated with oil. This effect of concentration on the rate of combustion is evident when burning materials are observed in contact with liquid oxygen. Hot iron, for example, continues to burn so vigorously when immersed in liquid oxygen that the metal melts from its own heat of combustion, despite the fact that it is surrounded by a liquid colder than $-292°F$ ($-180°C$).

Some substances that ordinarily burn with difficulty are likely to burn vigorously when they encounter liquid oxygen. Fuels, oils, greases, tar, asphalt, paper, textiles, and paint often burn with an increased intensity when they contact liquid oxygen. LOX easily penetrates into wood, concrete, asphalt, and other porous materials. Combustion or explosion of these materials may be triggered by a spark or shock, even after the oxygen appears to have vaporized. Finally, certain finely divided combustible metals are likely to burn at explosive rates when in contact with liquid oxygen. For instance, when LOX is spilled on magnesium shavings, the mixture may appear to detonate.

Cylinders and other containers of compressed oxygen may be encountered during fire incidents. If so, they should be cooled with streams of water to prevent rupturing. USDOT specifies the use of fusible plugs for oxygen cylinders, which are designed to melt at elevated temperatures. But these plugs only decrease the possibility of cylinder rupture during fires; they do not totally eliminate this risk. Thus great care should be taken when dealing with containers of compressed oxygen during firefighting, as well as other compressed gases.

Ozone

Oxygen also exists as an allotropic modification called *ozone*. An *allotrope* is a variation of an element that possesses a set of physical and chemical properties significantly different from the "normal" form of the element. Only a few elements have allotropic forms; oxygen, sulfur, and phosphorus are some of them. Ozone is just another form of oxygen, but it has three instead of the usual two atoms of oxygen per molecule. Thus, its chemical formula is represented as O_3.

At room conditions, ozone is a pale blue gas with a pungent odor. It is more dense than oxygen (vapor density = 1.7) and considerably more soluble in water. Two characteristics make ozone one of the most hazardous materials known: an exceptional chemical reactivity and pronounced toxicity. Ozone is considerably more reactive than oxygen. In fact, ozone is one of the most energetic oxidizing agents known.

Individuals are often familiar with ozone as the gas that forms in low concentration around operating electric motors or following lighting storms. Actually, while lightning produces some ozone, surprisingly little remains after a thunderstorm. The pleasant odor of air following a thunderstorm is due more to well-scrubbed air, not ozone. While it appears to possess a sweet smell in low concentrations, ozone has an irritating, pungent odor at moderate to high concentrations.

Ozone serves us in a very important way. Approximately 10 to 19 miles (16 to 30 km) from Earth's surface is a region of the atmosphere containing a relatively high concentration of ozone, appropriately called the *ozone layer;* it is illustrated in Fig. 6-2. At this altitude, the concentration of oxygen is greater than it is in other parts of the atmosphere. In addition, at this height from Earth's surface, some oxygen molecules are dissociated into their individual atoms (O·). Oxygen atoms and

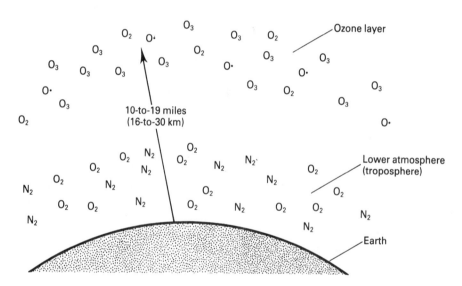

Figure 6-2 The atmosphere near Earth's surface consists predominantly of nitrogen and oxygen, along with lesser amounts of other gases. But at the outer edge of the stratosphere is a region called the *ozone layer* in which oxygen is constantly converted into ozone. This ozone prevents much of the sun's ultraviolet radiation from reaching Earth. However, atmospheric scientists have discovered that destruction of the ozone layer is ongoing. The depletion of ozone causes more harmful ultraviolet radiation to strike Earth. This is a serious cause for concern as ultraviolet radiation causes sunburn and skin cancer; in addition, it has been linked to cataract formation and weakening of the immune system.

oxygen molecules frequently collide in the ozone layer; such collisions often result in the formation of ozone.

$$O_2(g) + O \cdot (g) \longrightarrow O_3(g)$$

The benefit derived from the presence of the ozone layer lies in the capability of ozone to absorb low-energy ultraviolet radiation originating in the sun. By so doing, ozone molecules revert back to "normal" oxygen by the two-step reaction shown by the following equations:

$$O_3(g) \longrightarrow O_2(g) + O \cdot (g)$$

$$O_3(g) + O \cdot (g) \longrightarrow 2O_2(g)$$

Thus, ozone simultaneously undergoes formation and photodecomposition. This results in a steady-state concentration of ozone, which persists as a layer in the outer stratosphere.

No other substance in the atmosphere absorbs ultraviolet radiation to the same extent as ozone. Thus, without the existence of an ozone layer, ultraviolet radiation from the sun would not be filtered, and this could cause serious consequences on our planet. At the worst, plant life would most likely no longer survive; furthermore, ozone depletion could cause climatic changes and thus disturb the balance of aquatic and land ecosystems.

Notwithstanding the potential benefits of ozone in the stratosphere, the presence of ozone in the lower atmosphere can be hazardous. Ozone often occurs as an air pollutant in extremely low concentrations, especially as a component of the photochemical smog commonly observed near major metropolitan areas during warm weather.

Ozone is produced within the lower atmosphere by a two-step mechanism associated with the bombardment of nitrogen dioxide by solar rays. Nitrogen dioxide exists in the polluted atmosphere in conjunction with the oxidation of nitrogen monoxide (Sec. 9-9), a constituent of vehicular exhaust and other atmospheric emissions. Exposed to sunlight, nitrogen dioxide dissociates into nitrogen monoxide and oxygen atoms. Ozone forms when the oxygen atoms react with molecular oxygen. These reactions are summarized by the following equations:

$$NO_2(g) \longrightarrow NO(g) + O \cdot (g)$$

$$O_2(g) + O \cdot (g) \longrightarrow O_3(g)$$

Ozone pollution tends to reach a peak atmospheric concentration in the midafternoon when the intensity of sunlight is usually greatest. When warranted, weather forecasters announce an "ozone alert," which serves to warn the public of excessive ozone pollution.

Inhalation of an ozone-polluted atmosphere often causes people, especially the elderly, to struggle for breath and feel dizzy; they are likely to experience fatigue; and the eyes, nose, throat, and lungs feel irritated. In scientific studies, healthy persons exercising in a controlled environment reported difficulty in breathing after two hours when their breathing atmosphere contained ozone at a concentration of only 0.15 part per million (ppm). Based on the results of these studies, in 1979 USEPA established a standard limiting the concentration of ozone in the atmosphere to 0.12 ppm during any hour of the day, which includes a 0.03 ppm margin of safety. To comply with this standard, some American cities limited the number of motor vehicles on roadways in order to curb the major source of nitrogen dioxide. Nonetheless, many cities were unable to comply with the standard by the statutory deadline of December 31, 1987.

Ozone may also be encountered industrially, but because of its innate instability, ozone is neither stored nor transported from one location to another. Instead, it must be synthesized and used directly. Since ozone is so highly reactive, it is being used for more and more purposes. In contemporary times, ozone is employed primarily as an antiseptic and bactericide, especially at waste water treatment plants, and to bleach oils, fats, textiles, and sugar solutions. When employed in the workplace, OSHA and NIOSH have established the maximum permissible exposure limit to ozone as 0.1 ppm.

6-2 HYDROGEN

Hydrogen is an odorless, colorless, tasteless, and nontoxic substance. It ranks ninth in natural abundance by mass (see Table 3-1). However, only traces of hydrogen exist in the free state in the lower atmosphere, that is, as H_2. Ordinarily, hydrogen exists in combination with other elements in such compounds as water, acids, and fuels.

At ordinary temperature and pressure, hydrogen exists as a gas. But hydrogen liquefies under a pressure of 294 psi (20 atm) as it is cooled to temperatures less than $-390°F$ ($-234.5°C$). Liquid hydrogen is often called LH2. As the liquid, hydrogen is still colorless and odorless. Commercially, hydrogen is available as either of these forms: the compressed gas or cryogenic liquid. Use of the latter form, however, is limited mainly to the aerospace industry.

Hydrogen is produced industrially by either of four common methods. In the first method, steam is passed over red-hot coke or coal, which forms a mixture of gases called *water gas*. The latter consists mainly of carbon monoxide and hydrogen. Steam reacts with carbon in the coke or coal, as illustrated by the following equation:

$$C(s) + H_2O(g) \longrightarrow CO(g) + H_2(g)$$

The hydrogen may be separated from the carbon monoxide, or the carbon monoxide may be converted to carbon dioxide by passing water gas over a catalyst like iron-III) oxide with more steam.

$$CO(g) + H_2O(g) \longrightarrow CO_2(g) + H_2(g)$$

Here again, the hydrogen may be separated from the carbon dioxide.

The second industrial method of producing hydrogen involves the action of steam on natural gas at high temperatures. This reaction is illustrated by the following equation:

$$CH_4(g) + H_2O(g) \longrightarrow 3H_2(g) + CO(g)$$

Water gas is again produced, which is treated by either of the methods previously noted to isolate the hydrogen.

The third method of producing hydrogen, although to a limited extent, involves the use of iron to displace hydrogen from steam, as shown by the following equation:

$$3Fe(s) + 4H_2O(g) \longrightarrow Fe_3O_4(s) + 4H_2(g)$$

Finally, when the cost of electricity is relatively low, hydrogen is often produced by the electrolysis of water to which sodium hydroxide has been added. This electrolytic method also produces oxygen as the following equation illustrates:

$$2H_2O \longrightarrow 2H_2(g) + O_2(g)$$

Hydrogen produced electrolytically is higher in purity than the hydrogen produced by other industrial methods.

Hydrogen is regularly transported as a flammable compressed gas in steel cylinders or in rail or truck tanks. Liquid hydrogen is transported in cryogenic tanks or portable containers by either rail, motor vehicle, or barge. When required by US-DOT, such transport vehicles are placarded FLAMMABLE GAS.

Hydrogen is used by the chemical industry to make ammonia, methanol, and hydrochloric acid, as well as for the process called *hydrogenation*. The latter is a chemical reaction in which hydrogen is combined with certain organic compounds at high temperatures and pressures under the influence of a catalyst. Vegetable oils that have been hydrogenated become suitable for the manufacture of soaps, lubricating agents, and shortening. Hydrogen is used by the petroleum industry for the production of petroleum fuels. It is also used to produce high-purity metals.

Hydrogen is also used for high-temperature cutting and welding. For such purposes, the oxyhydrogen torch shown in Fig. 6-3 is employed. The torch is often

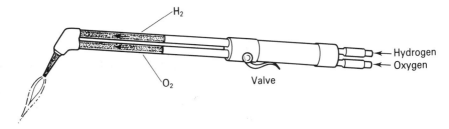

Figure 6-3 The oxyhydrogen torch. Hydrogen and oxygen enter the mixing chamber separately. The mixture discharges from the nozzle at such a rate that the flame burns at the tip of the torch. Using the oxyhydrogen torch, temperatures between 3300°F (1800°C) and 4400°F (2400°C) may be attained.

used, for example, to melt and work platinum, whose melting point is 3191°F (1755°C). Typically, compressed hydrogen and oxygen are stored separately in cylinders and then pressure-fed through tubing into the torch's mixing chamber. The mixture is then discharged into the nozzle, at which point the hydrogen is ignited.

Hydrogen possesses the least density of all substances; its vapor density is only 0.07. The combination of low density and high fuel value makes hydrogen useful as a rocket fuel. Hydrogen actually produces more energy for the least mass of any material. Liquid hydrogen and liquid oxygen were the propellants that powered the second and third stages of the Apollo/Saturn V rocket shown in Fig. 6-4.

3rd stage

2nd stage

1st stage

Figure 6-4 The Apollo 11/Saturn V rocket beginning its ascent into space from the Kennedy Space Center, Cape Carnaveral, Florida. The three stages of this rocket are depicted on the left. The propellants for the second and third stages were hydrogen and liquid oxygen. In these two stages, respectively, 945,000 lb (430,000 kg) and 230,000 lb (105,000 kg) of propellants were used. (Courtesy of the National Aeronautics and Space Administration, Washington, D.C.)

It is the density of hydrogen that gives it *lifting power*. The lifting power of 1 liter of hydrogen at 0°C and 1 atm is equal to the difference in mass of 1 liter of air and 1 liter of hydrogen at these conditions. One liter of air under these conditions has a mass of 1.2930 g, whereas 1 liter of hydrogen has a mass of 0.0899 g. Hence, the lifting power of hydrogen is 1.2930 − 0.0899, or 1.2031 g/L in air. Other physical properties of hydrogen are listed in Table 6-3.

TABLE 6-3 PHYSICAL PROPERTIES OF HYDROGEN

Freezing point	−434.45°F (−259.14°C)
Boiling point	−423.13°F (−252.85°C)
Heat of fusion	25.04 Btu/lb (13.9 cal/g)
Heat of vaporization	191.87 Btu/lb (106.5 cal/g)
Density of liquid at boiling point	4.428 lb/ft³ (71 g/L)
Density of gas at boiling point	0.081 lb/ft³ (1.3 g/L)
Density of gas at room temperature	0.0051 lb/ft³ (0.082 g/L)
Heat of combustion	57.8 kcal/mol
Lower explosive limit	4%
Upper explosive limit	75%
Vapor density (air = 1)	0.0695
Liquid-to-gas expansion ratio	865

The density of hydrogen is also responsible for its rapid rate of diffusion. Hydrogen diffuses in the atmosphere more rapidly than any other substance. In other words, molecules of hydrogen move about with faster velocities than the heavier molecules of other gases at the same temperature. This also causes hydrogen to leak through tiny openings, like pipe joints and valve connections, especially when compressed. Consequently, it is more difficult to store compressed hydrogen than other gases.

Hydrogen is a highly flammable gas. When sparked, a mixture of hydrogen and air ignites when the concentration of hydrogen is from 4% to 75% by volume. This combustion forms water vapor, as illustrated by the following equation:

$$2H_2(g) + O_2(g) \longrightarrow 2H_2O(g)$$

Due to its vapor density, hydrogen burns upward. But it is very difficult to visibly observe this combustion during daylight, since burning hydrogen has an almost nonluminous flame. Most of the heat of combustion evolves as heat, not light. When hydrogen burns in an atmosphere of pure oxygen, the accompanying heat of combustion is 60,000 Btu/lb (28.7 kcal/g). As noted earlier, this is the most energy for the least mass that any material provides during combustion. This amount of heat is frequently sufficient to kindle the fires of other nearby materials that burn.

The only condition under which any attempt should be made to extinguish hydrogen fires is when the flow of hydrogen from its containment vessel can be stopped. Otherwise, as a practical matter, extinguishing hydrogen fires is rarely possible. Often, the best advice is to allow the fire to burn itself out, although precaution should be taken to prevent the fire from spreading to adjacent locations.

Perhaps the best-known disaster involving hydrogen was the explosion of the German dirigible, the *Hindenburg*, shown in Fig. 6-5. Hydrogen held this airship aloft, stored in large gas bags. Diesel fuel was used to run its engines. Germany had intended the *Hindenburg* to inaugurate a new era in fast transatlantic travel. But, while attempting its landing approach on May 6, 1937, at Lakehurst, New Jersey, the airship mysteriously caught fire and exploded. Following this incident, the use of hydrogen was replaced with helium in airships. Although possessing only about 93% as much lifting power as hydrogen, helium is safer to use since it is a nonflammable gas.

Figure 6-5 The German dirigible, the *Hindenburg*, exploding over its landing site at the Naval Air Station in Lakehurst, New Jersey, on May 6, 1937. This was the first and only time the swastika was permitted to be displayed on a German air vessel entering the United States. Flammable hydrogen gas had held the *Hindenburg* aloft from its departure point in Frankfurt, Germany. (Courtesy of Wide World Photos, New York, N.Y.)

Hydrogen is frequently encountered as the product of certain chemical reactions. For instance, hydrogen may be displaced by metals from water. The alkali metals and alkaline earth metals react violently with water. The hydrogen that forms generally bursts spontaneously into flame. We shall examine the water reactivity of these metals in Chapter 8.

Hydrogen is also displaced from acids by metals. The following equations illustrate the reactions of tin and aluminum with hydrochloric acid and sulfuric acid, respectively:

$$Sn(s) + 2HCl(aq) \longrightarrow SnCl_2(aq) + H_2(g)$$

$$2Al(s) + 3H_2SO_4(aq) \longrightarrow Al_2(SO_4)_3(aq) + 3H_2(g)$$

With these metals, hydrogen is released slowly.

The chemical reactivity of metals is often measured by their ability to displace hydrogen from water and acidic solutions under identical conditions. Table 6-4 is an arrangement of the metals by decreasing chemical reactivity. It is called an *activity series*. The metals listed at the top of the table are so chemically reactive that they react with both water and acids. Those below magnesium release hydrogen from steam and acid solutions, but not from liquid water. Those metals below iron in the series do not displace hydrogen from steam, even when the temperature is elevated. Those below lead possess insufficient chemical reactivity and do not release hydrogen from either water or acids.

TABLE 6-4   ACTIVITY SERIES OF THE ELEMENTS

Elements that release hydrogen from water:	Cesium
	Lithium
	Rubidium
	Potassium
	Barium
	Sodium
	Calcium
	Magnesium
Elements that release hydrogen from acids:	Aluminum
	Manganese
	Zinc
	Chromium
	Iron
	Nickel
	Tin
	Lead
Elements that do not release hydrogen from acids:	Bismuth
	Copper
	Mercury
	Silver
	Platinum
	Gold

Hydrogen is also displaced by certain metals from solutions of sodium hydroxide or potassium hydroxide. Aluminum, zinc, and silicon are the most common metals capable of displacing hydrogen from these solutions.

One of the more commonly encountered hazards associated with hydrogen occurs during the charging of storage batteries, as shown in Fig. 6-6. Storage batteries produce electric current by chemical action and operate by means of either of two reversible chemical reactions, commonly referred to as *charging* and *discharging*. A battery is *charged* by sending an outside current through the battery; then, when the battery *discharges,* it produces electricity through chemical reactions involving the substances that formed during the charging process.

Figure 6-6   A battery-charging station, such as that illustrated here, should be located in an open, well-ventilated location to prevent the accumulation of hydrogen. Even a mixture containing only 4% by volume of hydrogen in air may explode when ignited by a spark or flame. (Courtesy of Argonne National Laboratory, Argonne, Illinois)

Let's consider the common lead storage battery, like that found in most automobiles. When the battery is being charged, bubbles of oxygen slowly accumulate near its positive plate. Simultaneously, hydrogen slowly accumulates near the negative plate. Both gases then dissipate into the atmosphere, but the hydrogen tends to rise in an undisturbed environment. When the battery is fully charged, however, the applied current decomposes the water more rapidly into hydrogen and oxygen. Thus, the amount of hydrogen and oxygen released to the atmosphere increases sharply near the completion of the charging process.

When banks of storage batteries are simultaneously charged in confined areas, the concentration of hydrogen in the atmosphere is likely to exceed its lower explosive limit. A flammable mixture in air is achieved when a minimum concentration of only 4% by volume of hydrogen has accumulated. Thus, to prevent the risk of fire and explosion, adequate ventilation must be provided in rooms containing banks of storage batteries.

6-3 FLUORINE

Elemental fluorine possesses a certain distinction as a hazardous material: It is the most chemically active nonmetal. Fluorine forms compounds with every other element, except possibly the lighter noble gases. For this reason, fluorine is never encountered in nature as the free element; but it is fairly widely distributed in certain rocks and minerals, to the extent of 0.027% by mass (see Table 3-1).

The extremely high reactivity of fluorine causes considerable difficulties both in the preparation and handling of this element. Industrially, it must be prepared by passing an electric current through a molten mixture of potassium fluoride and anhydrous (without water) hydrofluoric acid in an especially designed apparatus.

At room conditions, fluorine is a pale yellow gas with a pungent, irritating odor. Its chemical formula is F_2. Some of the important physical properties of this halogen are provided in Table 6-5. Fluorine is ordinarily encountered as a compressed gas, which is transported in cylinders made of an especially designed steel or nickel to which fluorine is relatively noncorrosive. USDOT regulates the transportation of fluorine as a poisonous gas; cylinders of compressed fluorine are labeled POISON GAS and OXIDIZER. When any quantity of fluorine is shipped, the transport vehicle is placarded POISON GAS.

TABLE 6-5 PHYSICAL PROPERTIES OF FLUORINE

Freezing point	−363.37°F (−34.05°C)
Boiling point	−306.31°F (−100.98°C)
Heat of fusion	18 Btu/lb (9.8 cal/g)
Heat of vaporization	71.52 Btu/lb (39.7 cal/g)
Density of liquid at boiling point	93.86 lb/ft³ (1,505 g/L)
Density of gas at boiling point	0.069 lb/ft³ (1.108 g/L)
Density of gas at room temperature	0.097 lb/ft³ (1.56 g/L)
Vapor density (air = 1)	1.312
Liquid-to-gas expansion ratio	965

Industrially, fluorine is used mainly for the preparation of various fluorides, like sulfur hexafluoride, uranium hexafluoride, and fluorocarbons (Sec. 11-7).

Like oxygen, fluorine supports combustion but does not itself burn. In other words, it chemically acts as an oxidizing agent. Ordinary materials, including wood, plastics, and metals, spontaneously burn in an atmosphere of fluorine. Even "fireproof" asbestos burns in fluorine. Fluorine also reacts vigorously with hydrogen-containing compounds, like water. When fluorine contacts water, the two substances chemically react, as illustrated by the following equation:

$$5F_2(g) + 5H_2O(l) \longrightarrow 8HF(aq) + O_2(g) + H_2O_2(aq) + OF_2(g)$$

This is one of the few ways by which water may be oxidized to hydrogen peroxide.

Fires involving fluorine are especially hazardous. Generally, the combustion of materials with fluorine provides a fire having a white-hot intensity. Such heat kindles the combustion of virtually any other nearby material capable of burning.

The ability to oxidize many other substances is fluorine's primary hazard. A secondary hazard is its highly toxic nature. Even small traces of fluorine can be irritating to skin tissue. Fluorine may cause deep, penetrating burns on contact. This corrosive effect may be delayed and progressively worsen with time. When inhaled, the element reacts with moisture in the lungs and respiratory tract; continued exposure causes serious edema and death within minutes. In the workplace, OSHA and NIOSH stipulate only 0.1 part per million (ppm) as the maximum permissible exposure limit to fluorine.

As noted earlier, fluorine cylinders are not equipped with fusible plugs. Hence, when they are encountered during fires or transportation mishaps, fluorine cylinders are likely to rupture if they have been exposed to prolonged intense heat. Due to the exceptional reactivity and poisonous nature of fluorine, attempts are not usually made to extinguish its fires. Instead, fluorine fires are generally permitted to burn until the fuel has been totally exhausted.

6-4 CHLORINE

Chlorine is also a chemically reactive halogen. Due to its reactivity, chlorine is not found in nature in the free state as Cl_2. Instead, it is found naturally in combination with metals like sodium, potassium, and magnesium, to the extent of 0.19% by mass.

At room conditions, chlorine is a yellow-green gas with a characteristic penetrating, irritating odor. It is about 2.5 times heavier than air and highly poisonous. Several other physical properties of chlorine are noted in Table 6-6.

TABLE 6-6 PHYSICAL PROPERTIES OF CHLORINE

Freezing point	−29.29°F (−34.05°C)
Boiling point	−149.76°F (−100.98°C)
Density of gas at 0°C (32°F)	0.2003 lb/ft³ (3.209 g/L)
Density of liquid at 0°C (32°F)	91.67 lb/ft³ (1,468 g/L)
Heat of vaporization at boiling point	123.7 Btu/lb (68.7 cal/g)
Heat capacity between 30° and 80°F	0.113 Btu/lb °F (0.113 cal/g °C)
Vapor pressure	53.155 psia (3.617 atm)
Vapor density (air = 1)	2.486
Liquid-to-gas expansion ratio	457.6

Commercially, chlorine is generally prepared by passing an electric current through molten sodium chloride or through aqueous solutions of either sodium chloride or magnesium chloride. Chlorine is very versatile in its chemical reactivity; thus, it has many industrial uses. The chemical industry requires relatively large quantities of chlorine in the production of a wide range of chlorine-containing compounds. Many such compounds are industrially used as solvents, pesticides, dyes, and bleaching powders; still others are used to produce plastics, fluorocarbons, refrigerants, fire extinguishers, and other commercially important substances. It is also used as a germicide for the purification of water and sewage, a bactericide for the sterilization of water for drinking and for use in swimming pools, and as a bleach for paper pulp and certain textiles.

Elemental chlorine should not be confused with certain chlorine-containing compounds used primarily as chlorinating agents in swimming pools. These latter substances are frequently called "chlorine," but they are not actually elemental chlorine. Instead, they are oxidizing agents that generate chlorine by chemical action for bactericidal purposes within the environment of a swimming pool. These oxidizers are more appropriately discussed in Chapter 10.

Chlorine is available commercially in 100- and 150-lb (45.4- and 68-kg) and 1-ton (908-kg) steel cylinders under a pressure of 84 psi (5.7 atm) at 70°F (21°C) as the liquid, although chlorine is generally discharged from these containers as a gas. USDOT requires the use of seamless cylinders for the 100- and 150-lb cylinders, such as those shown in Fig. 6-7. Chlorine may also be transported by motor vehicle or in single or multiunit rail tank cars. USDOT regulates the transportation of chlorine as a poisonous gas. Steel cylinders and other containers of chlorine are labeled POISON GAS. Vehicles used to transport any quantity of chlorine are placarded POISON GAS.

Rail tank cars and cylinders containing chlorine are equipped with pressure-relief valves (that is, fusible plugs located on the valve just below the valve seat). These valves are designed to melt between 158° and 165°F (70° and 74°C). Thus, when exposed to intense sources of heat, these tanks and containers slowly release their contents to the environment and do not ordinarily rupture.

The 1-ton container used for transporting chlorine is a welded tank having a loaded mass of as much as 3700 lb (1680 kg). Barges are often used to transport chlorine in such ton-containers along inland waterways. As Fig. 6-8 illustrates, these vessels are a cause for concern when they are involved in mishaps, since large quantities of chlorine may be released to the atmosphere from just one ton-container (see also Exercise 6.20).

Rail tank cars and tank trucks that transport chlorine are surrounded by 4 inches of insulation. Thus, if fire impinges on either of these vessels, the heat does not immediately threaten to rupture the container. This provides time during which firefighters may cool the tanks with water to prevent such an accident. In Fig. 6-9, firefighters are being shown the method of properly cooling a chlorine rail tank car during a training exercise.

When cylinders or rail tank cars containing chlorine are involved in transporta-

(b)

(a)

Figure 6-7 Three transport vessels for shipping chlorine in the United States. (a) Steel cylinders used to transport chlorine as the compressed gas are constructed without seams and have capacities ranging from 1 to 150 lb (0.45 to 68 kg). (b) Insulated tank cars used to transport liquid chlorine are fabricated from certain carbon steels. The manner by which they must be constructed is specified by USDOT. These tank cars are nominally of 55 or 90 tons (50,000 kg or 82,000 kg) capacity and are equipped with safety relief devices designed to relieve excess internal pressure. (c) Chlorine barges, also used to transport liquid chlorine, are used primarily on inland waters. Each barge has four independent, cylindrical, uninsulated pressure tanks mounted longitudinally. The most common barge capacities are 600 tons (540,000 kg); each of the four tanks holds 150 tons (140,000 kg). (Courtesy of the Chlorine Institute, Inc., Washington, D.C.)

184

Figure 6-8 Liquid chlorine is often shipped in portable tanks by barge. This barge, carrying four tanks of liquid chlorine, partially submerged after it rammed into a dam along the Ohio River. Each tank contained enough chlorine to threaten the health and safety of the local residents within a 35-mile radius of the incident. Fortunately, in this accident, none of the containers lost its contents. (Courtesy of the U.S. Environmental Protection Agency, Region V, Chicago, Illinois)

Figure 6-9 In this training drill, a chlorine tank was kept cool until the car could be moved away from a nearby brush fire. Note where the firefighters from the Baltimore City Fire Department are directing their fog applicator. It is aimed toward the top housing, which contains the car valves, rather than at the tank itself. All chlorine tank cars are protected by 4 in. of insulation. Hence, the most likely spot at which the tank would rupture is near the valve. (Courtesy of the Chlorine Institute, Inc., Washington, D.C.)

tion mishaps, it is advisable to locate spots from which chlorine may escape from its container. This should only be undertaken by individuals experienced in the use of positive-pressure, self-contained breathing apparatus. *Liquid* chlorine is present in these containers. Hence, liquid chlorine may drip from valves, fittings, or openings.

Liquid chlorine is much more concentrated than the gas. Unconfined at ordinary temperature and pressure conditions, the liquid readily evaporates, producing a substantially larger volume as gas. One volume of liquid chlorine evaporates into 460 volumes of gas. Hence, even when leaks cannot be stopped, an attempt should be made to prevent liquid chlorine from further escaping its container. This may be accomplished by rolling the container so that the opening is located upward. While chlorine gas may continue to escape from the container, liquid chlorine remains confined. Ultimately, any remaining liquid should be safely transferred to another container by qualified personnel.

It is not always a simple matter to locate chlorine leaks, especially when chlorine has been escaping from its container for some time. One method whereby leaks may be easily located is based on a chemical reaction that forms a white cloud of ammonium chloride. A rag soaked with ammonium hydroxide (from household ammonia) and tied to a stick can be passed along the surface of a container to pinpoint the location of leaks. This practice is safe unless the area is heavily concentrated with either chlorine or ammonia. Atmospheres heavily concentrated with chlorine should never be neutralized with ammonia. Attempts to do so have caused explosions, most likely from the formation of the unstable compound nitrogen trichloride.

Chlorine is a nonflammable gas, but it does support combustion. In an atmosphere of chlorine, several metals and nonmetals burn with incandescence, as the following equations illustrate:

$$Cu(s) + Cl_2(g) \longrightarrow CuCl_2(s)$$

$$2As(s) + 3Cl_2(g) \longrightarrow 2AsCl_3(s)$$

$$2Sb(s) + 3Cl_2(g) \longrightarrow 2SbCl_3(s)$$

$$P_4(s) + 6Cl_2(g) \longrightarrow 4PCl_3(l)$$

Chlorine also supports the combustion of many organic compounds.

While chlorine is chemically reactive, toxicity is considered its principal hazard. In the workplace, OSHA and NIOSH stipulate only 1 part per million as the maximum permissible exposure limit to chlorine. Inhalation may cause inflammation of the nose and throat and congestion of lung tissue; prolonged inhalation could result in death. Individuals exposed to chlorine should be removed from the area, and blankets should be supplied to keep them warm. A physician should be contacted immediately.

6-5 BROMINE

Bromine is another reactive halogen. Like fluorine and chlorine, it is never found in nature in its elemental form, Br_2. Instead, bromine is found naturally in compounds, usually in salt deposits (for example, the Great Salt Lake) or the ocean.

At room conditions, bromine is a reddish-brown liquid that readily volatilizes to an intensely irritating vapor. Several other physical properties of bromine are noted in Table 6-7.

TABLE 6-7   PHYSICAL PROPERTIES OF BROMINE

Boiling point	137.80°F (58.78°C)
Freezing point	19.04°F (−7.2°C)
Density of gas at 59°C (138°F)	0.182 lb/ft³ (2.928 g/L)
Heat of vaporization at the boiling point	78.66 Btu/lb (43.7 cal/g)
Vapor pressure	3.385 psia (0.2303 atm)

Most bromine is industrially prepared from seawater, which is processed for its bromine by mixing it with chlorine in an amount equivalent to the bromine content of the water. This brine is then passed into towers, where bromine is blown out by air and absorbed into hot sodium carbonate. The following equations illustrate this production process:

$$2NaBr(aq) + Cl_2(g) \longrightarrow Br_2(aq) + 2NaCl(aq)$$

$$3Na_2CO_3(aq) + Br_2(aq) \longrightarrow 5NaBr(aq) + NaBrO_3(aq) + 3CO_2(g)$$

The final solution is then treated with sulfuric acid to recover the bromine. It is then stored in either glass bottles such as shown in Fig. 6-10 or drums made of nickel or Monel, an alloy that is very resistant to corrosion.

Bromine is used mainly by the chemical industry for the production of fumigants, fire extinguishers (for example Halon 1301), dyes, pharmaceuticals, photographic chemicals, and other substances. Formerly, a great amount of bromine was used to produce ethylene bromide, which was needed for the production of tetraethyl lead. However, regulations promulgated under the Clean Air Act have severely limited the use of this antiknock agent in motor fuels; consequently, the demand for bromine production has sharply decreased.

USDOT regulates the transportation of bromine as a corrosive material. Bromine packages are labeled CORROSIVE and POISON. Bromine may also be transported by means of rail tank cars or portable tanks. When required, the tank car is placarded CORROSIVE.

Bromine is a nonflammable substance, but it does support combustion. In this sense, the element closely resembles chlorine, although bromine is less chemically reactive than chlorine. Bromine reacts with both metals and nonmetals. Aluminum reacts with bromine vigorously, and potassium reacts explosively. However, below 572°F (300°C), bromine is inert to lead, nickel, magnesium, tantalum, iron, zinc, and even metallic sodium.

Since bromine is a resonably good oxidizer, spills of bromine on wood, excel-

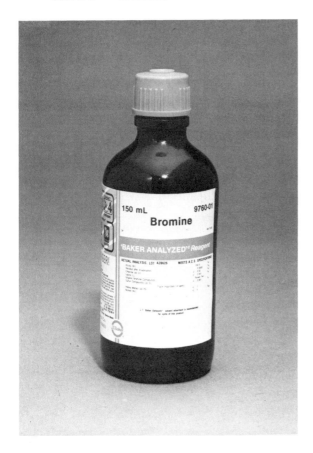

Figure 6-10 When encountered for laboratory use, liquid bromine is often stored in relatively small volumes, such as the 0.3-pint (150-mL) container shown here. This volume is approximately 1 lb. Bulk quantities of bromine may also be encountered in much larger amounts, like 55-ton tank cars. (Courtesy of J. T. Baker, Inc., Phillipsburg, New Jersey)

sior, and sawdust are likely to cause the combustion of these materials. For this reason, these cellulosic materials should never be used to remove bromine spills; nor should they be used to cushion bromine containers during transportation. Spills of bromine should either by absorbed into an inert material or diluted substantially with water prior to further cleanup.

Notwithstanding its ability to oxidize certain substances, the principal hazard of bromine is its corrosive nature. Vapors of bromine cause the eyes to tear; when inhaled, they can produce serious inflammation of the lungs and respiratory tract. The liquid burns the flesh and forms wounds that heal very slowly. Spills of bromine on skin tissue should be immediately flushed with cold water, followed by the application of an aqueous solution containing 5% to 10% by mass of sodium thiosulfate.

In the workplace, OSHA regulates the exposure of employees to bromine. OSHA and NIOSH stipulate only 0.1 ppm as the maximum permissible exposure limit to bromine.

6-6 PHOSPHORUS

Phosphorus is also a chemically reactive element. Hence, it is found in nature only in combination with other elements as chemical compounds, to the extent of 0.12% by mass (see Table 3-1).

Elemental phosphorus has several allotropic modifications, two of which are commercially important, *white phosphorus* and *red phosphorus*. Some physical properties of these allotropes are provided in Table 6-8. The allotropes of phosphorus are used to make special alloys (for example, phosphor bronze), rodenticides, fireworks, and "strike-anywhere" and safety matches (Sec. 10-10).

TABLE 6-8   PHYSICAL PROPERTIES OF PHOSPHORUS

	White phosphorus	Red phosphorus
Melting point	111°F (44°C)	1094°F (590°C)
Boiling point	535°F (280°C)	535°F (280°C)
Density	114 lb/ft³ (1.82 mL)	146 lb/ft³ (2.34 mL)
Autoignition temperature	86°F (30°C)	500°F (260°C)
	(spontaneous in dry air)	

The chemical properties of these allotropes are so strikingly different that we shall discuss them separately.

White Phosphorus

This is a waxy, translucent solid with a density about twice that of water. It consists of tetraatomic molecules and thus has the chemical formula P_4. White phosphorus is the unstable form of this element at room conditions; hence, on standing, it acquires a yellow coloration due to partial conversion to the more stable red allotrope. Sometimes white phosphorus is called *yellow phosphorus*, but these two names refer to the same allotrope.

White phosphorus is prepared from calcium phosphate rock by heating it with sand and coke in an electric furnace. The following equation represents the overall chemistry:

$$2Ca_3(PO_4)_2(s) + 6SiO_2(s) + 10C(s) \longrightarrow 6CaSiO_3(s) + 10CO(g) + P_4(g)$$

The phosphorus vapors are conducted from the furnace and condensed under water.

Exposed to air, white phosphorus fumes and spontaneously ignites with an odor resembling that of burning matches. Vigorous combustion in air produces either of two oxides, tetraphosphorus hexoxide or tetraphosphorus decoxide, as the following equations illustrate:

$$P_4(s) + 3O_2(g) \longrightarrow P_4O_6(s)$$

$$P_4(s) + 5O_2(g) \longrightarrow P_4O_{10}(s)$$

These oxides are white compounds. Hence, when white phosphorus burns in bulk, billows of dense white, choking smoke are produced, as Fig. 6-11 illustrates. To extinguish phosphorus fires, any type of commercial fire extinguisher may be used, but wet sand works most effectively. Precaution should be taken to prevent exposure to the oxides of phosphorus, as they irritate the eyes, throat, and lungs.

Figure 6-11   When elemental phosphorus or certain phosphorus-bearing compounds burn in air, the smoke or fumes accompanying the combustion are white. This coloration is caused by the oxides of phosphorus, which are white compounds. The train noted in this incident derailed while hauling a consignment of diphosphorus pentasulfide, a water-reactive substance. (Courtesy of the U.S. Environmental Protection Agency, Region V, Chicago, Illinois)

To prevent contact with atmospheric oxygen, white phosphorus is stored under water as the solid or under a blanket of an inert gas like nitrogen as the liquid. White phosphorus is often said to possess a garliclike odor, but this is very likely due to the presence of phosphine. Phosphine is a poisonous gas (Sec. 8-7) that slowly forms when white phosphorus is stored under water.

From Table 6-8, we note that the ignition temperature of white phosphorus is so low that even body heat is sufficient to raise it above its ignition temperature. Consequently, white phosphorus should never be handled without gloves. Burns produced from phosphorus are extremely painful and slow to heal. In the past, when burning pieces of phosphorus stuck to skin, application of a concentrated copper(II) sulfate solution to the affected area was recommended. White phosphorus is chemically reactive with copper(II) ions, reducing them to free copper. Thus, this reaction made it easier to identify the location of small pieces of phosphorus by the brown coloration of copper. However, today we know that copper(II) sulfate is toxic in its own right and should be permitted to touch skin only briefly as an aqueous solution of only 0.5% to 1% by mass. Then soap and water should be used repeatedly, while pieces of phosphorus are simultaneously removed.

White phosphorus is quite volatile at room conditions and is highly poisonous. The lethal dose is only about 0.1 g for the average adult. OSHA and NIOSH stipulate 0.1 mg/m³ as the maximum permissible exposure limit to this element. Repeated inhalation causes an affliction known as *phossy jaw* (phosphorus necrosis), the actual rotting or disintegration of the jawbone.

The proper shipping name of white phosphorus depends on the condition in which it is transported. Any of the following names is correct, as appropriate: "RQ, phosphorus, white dry"; "RQ, phosphorus, yellow dry"; "RQ, phosphorus, under water"; "RQ, phosphorus, in solution"; and "RQ, phosphorus white, molten." USDOT regulates the transportation of white phosphorus as a spontaneously combustible material, and containers of white phosphorus are labeled SPONTANEOUSLY COMBUSTIBLE and POISON. When offered for shipment in nonbulk quantities, solid white phosphorus is placed in water inside approved containers.

As shown in Fig. 6-12, white phosphorus may also be transported in bulk by rail tank car or tank truck as either the solid or liquid. The element is either immersed in water or blanketed with a chemically inert gas. USDOT requires each rail tank car to be marked PHOSPHORUS. Motor vehicles, rail tank cars, and freight containers carrying bulk quantities of white phosphorus are placarded SPONTANEOUSLY COMBUSTIBLE.

When rail tank cars have been unloaded of a consignment of white phosphorus, USDOT requires the tanks to be filled to their entire capacity with an inert gas or to their entire capacity and dome to not more than 50% its capacity with water having a temperature not exceeding 140°F (60°C). Then the tank cars are placarded with a warning placard reading "SPONTANEOUSLY COMBUSTIBLE/EMPTY."

Red Phosphorus

This allotropic form of phosphorus is a dark red solid, believed to consist of molecules containing a number of P_4 units. Its chemical formula is usually given as P, or as P_x. Red phosphorus is made by heating white phosphorus at 482°F (250°C) in an iron container from which air has been excluded.

Red phosphorus has chemical properties that are profoundly different from those noted for white phosphorus. Pure red phosphorus is not poisonous, nor does it burn spontaneously in air. It does burn in air, however, when exposed to an ignition source, forming a mixture of the oxides of phosphorus.

While red phosphorus is less reactive than white phosphorus, it still reacts chemically. Furthermore, red phosphorus is a moderately unstable element and oxidizes slowly in air. This reaction is exothermic and is accelerated by an increase in temperature. Consequently, the necessary conditions exist for spontaneous combustion, but this generally occurs only when red phosphorus is stored in bulk quantities.

Water has a tendency to react with hot red phosphorus, forming toxic phosphine. For this reason, water is not recommended as a fire extinguisher on fires of red phosphorus. Instead, such fires should be extinguished with foam or dry chemicals.

USDOT regulates the transportation of red phosphorus as a flammable solid. Its proper shipping name is "Phosphorus, amorphous." Containers of red phosphorus are labeled FLAMMABLE SOLID. When required, motor vehicles and freight containers carrying packages of red phosphorus are placarded FLAMMABLE SOLID.

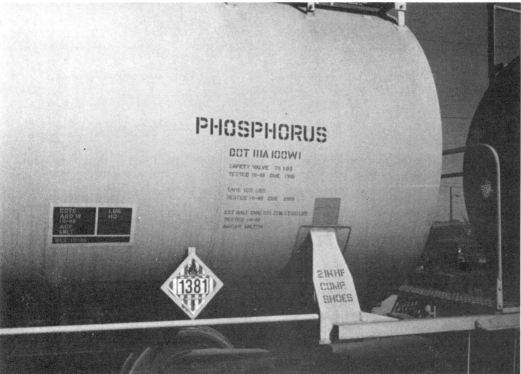

Figure 6-12 A tank car used for transporting elemental phosphorus by rail. The enlargement illustrates some USDOT-required markings and placards. By rail, phosphorus is generally transported in the liquid state of matter. USDOT requires it to be covered with water or blanketed with an inert gas. USDOT formerly required FLAMMABLE SOLID placards to be displayed as noted here; however, recently proposed regulations require SPONTANEOUSLY COMBUSTIBLE placards to be displayed (see text). (Courtesy of FMC Corporation, Phosphorus Chemicals Division, Pocatello, Idaho)

6-7 SULFUR

Nonfire

Fire

Elemental sulfur has been known since before recorded history. It occurs naturally as the element, particularly near the Gulf of Mexico and in Japan and Italy, as well as in the combined state in minerals and ores too numerous to mention. It accounts for 0.06% by mass of all the elements found in Earth's crust, atmosphere, and oceans (see Table 3-1).

Sulfur is one of the world's most important raw materials. There is hardly a chemical industry that does not use either elemental sulfur or one of its compounds during some point of manufacture or production. In particular sulfur is employed in the manufacture of sulfuric acid, which is by far the most widely used of all inorganic compounds. About four-fifths of all sulfur is used in sulfuric acid production. Sulfur is used either directly or indirectly in making matches; sulfur is also used to produce vulcanized rubber, fertilizers, dyes and chemicals, drugs and pharmaceuticals, explosives, pesticides, and other products. Figure 6-13 illustrates some of these uses of elemental sulfur.

Figure 6-13 Elemental sulfur is a constituent of many industrial and domestic products, like gunpowder, matches, insecticides, fertilizers, and vulcanized rubber, some of which are noted here. When the sulfur in these products burns, sulfur dioxide is formed.

Elemental sulfur is likely to be encountered in any one of several ways. Solid chunks of naturally occurring sulfur, frequently called *brimstone,* are one form. The pure solid state of sulfur occurs in either of two allotropic forms, *orthorhombic* and *monoclinic.* A finely divided powder of solid sulfur, called *flowers of sulfur,* is a commercially important agricultural product. Finally, sulfur is also likely to be encountered in the molten, or liquid, state of matter.

However, these forms are not uniquely associated with distinctively different hazardous properties. Hence, for our purposes, sulfur may be regarded as a single hazardous material. Some of its physical properties as the solid and liquid are noted in Tables 6-9 and 6-10, respectively.

TABLE 6-9 PHYSICAL PROPERTIES OF SOLID SULFUR

Melting point	230–246°F (110–119°C)
Autoignition temperature	
Dispersed in air	374°F (190°C)
Undispersed	428°F (220°C)
Density	129.2 lb/ft^3 (2.07 g/mL)
Heat capacity	0.16 Btu/lb °F (0.16 cal/g °C)
Heat of combustion	3982.2 Btu/lb (2,210.4 cal/g)
Explosive limits of dust in air	0.025–1.4 oz/ft^3
Solubility in water	Insoluble

TABLE 6-10 PHYSICAL PROPERTIES OF LIQUID SULFUR

Boiling point	832°F (444°C)
Ignition temperature in air	478°–502°F (248°–261°C)
Flash point	335°–370°F (168°–188°C)
Heat capacity at 290°F (143°C)	0.25 Btu/lb °F (0.25 cal/g °C)
Vapor pressure at 284°F (140°C)	0.002 psia (0.0001447 atm)
Density at 280°F (138°C)	111.7 lb/ft^3 (1.79 g/mL)

Orthorhombic sulfur is the stable form of solid sulfur at room conditions. It is a pale yellow, crystalline solid. When held at a temperature between 204.8°F (96°C) and 235.2°F (112.8°C), orthorhombic sulfur changes to monoclinic sulfur, which exists as masses of long transparent needles. Since monoclinic sulfur is not the stable allotrope of sulfur, it slowly changes back to the orthorhombic form. Either form exists as molecular aggregates of eight sulfur atoms; hence, the chemical formula of sulfur is S_8.

At room conditions, solid sulfur is a combustible material. As the finely divided solid, sulfur is prone to undergo spontaneous combustion. Flowers of sulfur form explosive mixtures with air, which readily ignite from their own static electricity. The flash point of sulfur is only 405°F (207°C). Furthermore, elemental sulfur readily melts under fire conditions and may thereby flow to adjacent areas. Hence, sulfur is often regarded as a hazardous material in the fire service not only because it burns, but also since it may kindle the ignition of other materials.

Sulfur is often stored and transported as a molten liquid. This form of the material possesses several potentially hazardous features. Molten sulfur burns skin tissue on contact. It also possesses a toxic hazard due to the presence of limited quantities of hydrogen sulfide. Brimstone generally contains sulfurous organic compounds as impurities in the concentration range of 0.3 to 0.8% by mass. These compounds are called *mercaptans*. They are immediately identifiable by their odor which resembles the vile, penetrating smell of expired skunk. At the temperatures required to melt brimstone, mercaptans slowly decompose, forming the poisonous and flammable gas, hydrogen sulfide. This fact has important implications when storing or transporting molten sulfur. For instance, molten sulfur storage tanks should be appropriately vented to slowly release the hydrogen sulfide and prevent the build-up of internal pressure.

Solid sulfur is also transported and stored in burlap bags, wooden barrels, fiberboard boxes, and other packages. When fire is encountered in these containers, water may be used to effectively extinguish the fire. Whenever possible, however, the burning containers should be segregated from other containers to prevent the spread of fire. Water should be discharged on sulfur fires as a fog as opposed to streams, since the use of fog avoids generating steam under a layer of sulfur. Finally, if molten sulfur ignites, or if solid sulfur melts during a fire, an attempt should be made to limit the material's tendency to flow to adjacent locations. This may be accomplished by constructing a nonflammable barrier or trench.

USDOT only regulates the transportation of molten sulfur. Containers are labelled FLAMMABLE SOLID, and when necessary, their transport vehicles are placarded FLAMMABLE SOLID and are marked MOLTEN SULFUR.

Sulfur burns in air, forming sulfur dioxide, a poisonous gas having a suffocating, choking odor. This combustion is denoted by the following equation:

$$S_8(s) + 8O_2(g) \longrightarrow 8SO_2(g)$$

The formation of sulfur dioxide during sulfur fires generally poses a serious hazard. The gas is relatively dense, and it diffuses slowly through calm air. Thus, breathing becomes difficult near the immediate scene of sulfur fires. Since sulfur dioxide is toxic, proper firefighting strategy necessitates use of positive-pressure, self-contained breathing apparatus.

6-8 CARBON

Elemental carbon occurs naturally as either of two allotropic forms, *graphite* and *diamond,* or as one of several amorphous (that is, noncrystalline) forms including *coal, coke,* and *charcoal.* In each of these forms, elemental carbon is represented by its chemical symbol, C. Carbon ranks fourteenth in abundance when expressed as percentage by mass in Earth's crust, the atmosphere, oceans, and other bodies of water (Table 3-1). Yet carbon ranks much higher in terms of importance, since it is found in all living things in the form of compounds like proteins and carbohydrates.

Graphite and Diamond

Graphite and diamond possess diversely different physical properties. Graphite is always black and greasy, while diamond may be cut and polished to a crystalline transparent luster. Graphite is one of the softest known elements, while diamond is the hardest natural substance. Graphite is 2.25 times as dense as water, but diamond is 3.5 times denser than graphite. At moderate pressures, graphite is the stable allotropic form of carbon. Only a slow rate of transformation prevents diamond from being converted to graphite at moderate temperatures.

Graphite and diamond are fairly inert to chemical attack, although they burn at elevated temperatures and react with oxidizing agents. On complete combustion in air, they form carbon dioxide. Graphite is often considered a fire risk, but this hazard actually refers more to the amorphous forms of carbon. The transportation of graphite is not specifically regulated by USDOT.

Graphite possesses an anomalous property when compared to other nonmetals. It is a good conductor of heat. This feature, coupled with the fact that graphite melts at 6300°F (3500°C), is sometimes industrially useful. For instance, graphite can be molded into crucibles used for melting steel and other metals with high melting points. Graphite is also used as a component in several commercially available fire extinguishers called *dry powders,* not to be confused with the *dry chemical* extinguishers (Sec. 4-9). Dry powders are often used to extinguish class D fires.

Amorphous Forms of Carbon

Almost without exception, the amorphous forms of carbon are believed to have originated from plants. These plants lived millions of years ago, during the carboniferous age. At this time, giant plants and trees flourished more than they do today. As these plants died and decayed, they became buried deep below Earth's surface. Here extreme pressure and temperature converted them into coal.

The extent to which the coal-forming process has occurred determines the *rank* or class of coal; some ranks are shown in Fig. 6-14. Each rank differs from the others by its age. Peat is the youngest rank of coal and anthracite the oldest. Each also possesses a different fixed carbon content, which gives each rank a different range of heating values. Minerals are often present in all ranks of coal, especially minerals containing sulfur.

Many organic compounds are locked within the structure of coal. This is most easily demonstrated by heating coal in the absence of air in an apparatus like that shown in Fig. 6-15. Certain volatile gases evolve as the coal is heated, including methane and ammonia; this mixture of gases is called *coal gas.* Simultaneously, a relatively viscous liquid condenses near room temperature. The latter material is called *coal tar* and consists of a mixture of many organic compounds, like benzene, toluene, phenol, naphthalene, anthracene, and others.

Coal tar may be subjected to *distillation,* a process that separates groups of the components of coal tar from groups of others. The coal tar is heated, causing a

Figure 6-14 Some ranks or classes of coal: peat, lignite, three varieties of bituminous coal, and anthracite. All ranks of coal are thought to have originated from the decay of plants under extreme temperatures and pressures deep below the surface of Earth. Each rank differs from the others by the time necessary to achieve its formation in nature. (Courtesy of the National Coal Association, Washington, D.C.)

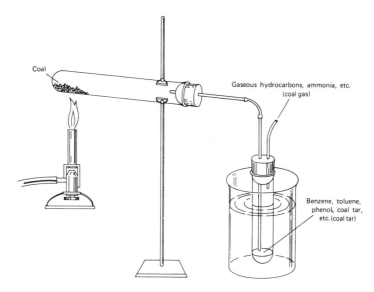

Coal

Gaseous hydrocarbons, ammonia, etc.
(coal gas)

Benzene, toluene,
phenol, coal tar,
etc. (coal tar)

Figure 6-15 Heating coal in the absence of air results in the production of *coal gas* and *coal tar*, both of which contain components that are either flammable or combustible. When coal burns in the presence of air, it is this combination of components that burns.

number of the components to vaporize at their boiling points; these vapors are then condensed to liquids. The substances derived from this process are often called *coal tar distillates*. The latter term is loosely used, but light and heavy fractions of coal tar are sometimes differentiated. The light coal distillate possesses the lower flash point and thus burns more readily than the heavy coal distillate. USDOT regulates the transportation of coal tar distillates as flammable liquids.

A sufficient activation energy must generally be provided to vaporize the volatile organic components of coal from its complex structure. Only freshly mined coal and coal that has been recently pulverized are likely to burn spontaneously, since near their surfaces these flammable and combustible materials are often present in concentrations above the lower explosive limit.

The solid residue remaining after coal (usually bituminous coal) has been heated in the absence of air is called *coke*, a sample of which is shown in Fig. 6-16. Coke is mainly elemental carbon mixed with minerals. Most coke produced in the United States is used for reducing iron ore in blast furnaces. This process, called *smelting*, is illustrated by the following equation:

$$C(s) + FeO(s) \longrightarrow Fe(s) + CO(g)$$

Coke may also be used to reduce ores of arsenic, tin, copper, zinc, and other elements. Hot coke, or coke having retained heat from its use in manufacturing processes, is prone to undergo spontaneous combustion. USDOT forbids its transportation.

When wood is heated in the absence of air, it similarly decomposes. A mixture

Figure 6-16 A sample of coke, the carbonaceous residue that remains when bituminous coal is heated in the absence of air. (Courtesy of the National Coal Association, Washington, D.C.)

of gases evolves consisting of two fractions: a noncondensable fraction and a condensable fraction (near room temperature). The noncondensable fraction is composed mostly of hydrogen, methane, carbon monoxide, and carbon dioxide. The condensable fraction is a yellow-to-red liquid consisting of a mixture of acetic acid, methyl alcohol, acetone, and other materials. The mixture of substances is called *pyroligneous acid.*

Materials other than coal and wood may also be decomposed in the absence of air, like animals bones, nut shells, corn cobs, peach pits, and similar items. The solid residue remaining after heating these materials as described is called *charcoal.* USDOT regulates the transportation of charcoal as a spontaneously combustible material.

Fires in ground charcoal or charcoal screenings are properly handled by locating, removing, and isolating the burning packages. Water may be used to effectively extinguish charcoal fires, but only when applied at a relatively high flow rate. The water absorbs heat. However, if water is applied at an insufficient flow rate, the hot charcoal may react with the water. This reaction forms hydrogen, which can serve to reignite the charcoal.

$$C(s) + H_2O(g) \longrightarrow H_2(g) + CO(g)$$

Elemental carbon exhibits an important physical property known as *adsorption.* This property refers to the surface retention of certain solid, liquid, or gaseous molecules and is not to be confused with *absorption,* which involves the penetration of one substance into the bulk of another one. All forms of elemental carbon exhibit the ability to adsorb substances to their surfaces, but charcoal is the most popular of the carbon-containing adsorbing agents. Charcoal retains the skeletal or cellular structure of the material from which it was made and is therefore highly porous. This porosity gives charcoal a very large surface area per unit mass. For this reason, charcoal is an excellent adsorbant.

Charcoal is used in gas mask filters and canisters, gas stacks, and cigarette filter tips. In these items, the charcoal filters or reduces the concentration of undesirable toxic gases by adsorption. Carbon is preferential in its adsorbing feature. While some substances are adsorbed to its surface, other substances are not. Because of this feature, charcoal is often used industrially to purify products intended for commercial use. For instance, charcoal may be used to purify raw sugar. The impure sugar, which is brown from the presence of impurities, is dissolved in water, and the sugar solution is mixed with charcoal. The colored matter in raw sugar, but not the sugar itself, adsorbs to the surface of charcoal. After the charcoal is filtered, the sugar solution is permitted to evaporate. The white residue is *refined sugar*.

REVIEW EXERCISES

Oxygen and Ozone

6.1. Many interior areas within large buildings are relatively confined. During a fire, why is the atmospheric oxygen often reduced in such areas?

6.2. Oxygen therapy involves providing certain sick individuals, such as persons suffering from respiratory diseases, with an atmosphere of enriched oxygen to breathe. Why is such an atmosphere likely to be more beneficial to such individuals than ordinary air?

6.3. Why should hoses and regulators from oxygen cylinders never be interchanged with similar equipment intended for use with other gases?

6.4. Why should piping never be painted when it is to be used to transfer liquid oxygen between two storage containers?

6.5. What is the most likely reason why USDOT does not regulate the transportation of ozone as a hazardous material?

6.6. How is it possible for Earth to be surrounded by an ozone layer if ozone itself is an unstable substance?

Hydrogen

6.7. Why is the concentration of hydrogen greater in the upper stratosphere than in the lower atmosphere?

6.8. Why has the use of hydrogen for the inflation of manned balloons been largely superseded by the use of helium?

6.9. Why is hydrogen more difficult to store as a compressed gas than most other gaseous substances?

6.10. Why is a leak of escaping hydrogen from piping generally more difficult to locate than leaks of most other flammable gases?

6.11. Write a balanced chemical equation for the displacement of hydrogen by the following metals from either water or the indicated acid:

(a) Sodium from cold water
(b) Calcium from cold water
(c) Magnesium turnings with hot water
(d) Aluminum with aqueous hydrochloric acid
(e) Zinc with aqueous sulfuric acid

6.12. When banks of storage batteries are kept in a windowless basement room having minimum ventilation, in what part of the room is a fire hazard most likely to exist immediately after the batteries have been charged?

Fluorine

6.13. What is the most likely reason USDOT limits the quantity of fluorine per metal cylinder to only 6 lb at room temperature and 400 psig?

6.14. A mixture of hydrogen and fluorine gases explodes spontaneously, even in the dark, producing hydrogen fluoride.
(a) Identify the oxidizing agent and reducing agent.
(b) Write the chemical equation that illustrates this chemical reaction.

6.15. Why should piping intended for transfer of fluorine be absolutely free of moisture, grease, and pipe dope?

6.16. Why does the *Emergency Response Guide* recommend the use of a water spray to reduce the concentration of leaking vapors of poisonous gases, like fluorine, but emphatically recommends that water not be discharged on the leak area itself?

Chlorine

6.17. A mixture of hydrogen and chlorine reacts violently when exposed to sunlight, forming hydrogen chloride.
(a) Write the balanced chemical equation for this reaction.
(b) Identify the oxidizing agent and reducing agent.
(c) What role does sunlight play?

6.18. Why do safety engineers generally recommend that cylinders of chlorine and ammonia be stored in nonadjacent locations?

6.19. Considering maximum safety, where should liquid chlorine be stored prior to use as a bactericide in an indoor public swimming pool: indoors or outdoors? Why?

6.20. Suppose a barge inadvertently rams into a bridge, rupturing a 1-ton transport container of liquid chlorine. What immediate action should be taken if this incident occurs near a heavily populated residential area? (*Hint:* Use data in Table 6-6 to determine the number of cubic feet of chlorine gas contained in 1 ton of liquid chlorine.)

6.21. A chemical gas supplier intends to transport ten steel cylinders of chlorine by motor vehicle, each containing 150 lb of chlorine. The shipping papers accompanying this consignment bear the following information: 10; Cyl., RQ, Chlorine, 2.3, UN1017, PG IB; 1500 lb.
(a) What hazard information should the supplier mark on each cylinder?
(b) How should the supplier label each cylinder?
(c) Which placard(s), if any, should the motor carrier affix to the transport vehicle?

Bromine

6.22. A bottle containing 5 lb (2.27 kg) of bromine is inadvertently dropped on a concrete floor. Is it a relatively safe practice to soak the liquid into sawdust prior to disposal? Why?

6.23. Which of the following metals may be used as the material for caps on glass bottles containing bromine: aluminum, lead, or nickel?

Phosphorus

6.24. What precaution always needs to be taken when storing the following:
  (a) White phosphorus
  (b) Red phosphorus

6.25. Which of the following may be used to extinguish white phosphorus fires: water, carbon dioxide, nitrogen, Halon 1301, sand?

6.26. Why does white phosphorus turn yellow?

6.27. During World War I, white phosphorus was used to produce smoke screens. Describe how this may be accomplished.

Sulfur

6.28. Identify the choking gas that forms when stacks of rubber tires burn.

6.29. What precaution needs to be taken when sulfur is ground into a fine powder?

6.30. Sulfur reacts with powdered zinc explosively, forming zinc sulfide.
  (a) Write the chemical equation illustrating this reaction.
  (b) Identify the oxidizing agent and reducing agent.

6.31. Why is the use of positive-pressure, self-contained breathing apparatus recommended when combating sulfur fires?

Carbon

6.32. Write the chemical equation that illustrates the complete combustion of diamond.

6.33. How is it known that coal originated from plants?

6.34. Explain the following observation: Charcoal has a specific gravity of 1.5, but it floats on water.

6.35. Why do farmers often char the ends of wooden fence posts prior to burying an end in earth?

6.36. What is the function of the charcoal in gas masks used for industrial and military purposes?

7

Chemistry of Some Corrosive Materials

In a general sense, corrosion refers to a chemical process in which metals and minerals are converted into undesirable by-products. This process is often caused by the action of atmospheric oxygen, in which case the resulting corrosion by-products are metallic oxides. For instance, a common example of corrosion is the rusting of items made of iron. When iron rusts, the element combines with the oxygen to form either of the iron oxides, iron(II) oxide or iron(III) oxide. These latter compounds are produced as soft scales that peel from the iron, leaving the underlying metallic surface exposed to the atmosphere. The metal gradually wears away and destroys the integrity of the item. In this case, which is the more common situation, corrosion is detrimental.

However, corrosion may be a beneficial process, too. This is true when aluminum, chromium, copper, nickel, and certain types of steel corrode. In the case of aluminum, corrosion by atmospheric oxygen results in the formation of aluminum oxide. This substance strongly adheres to the metal as a film, which effectively prevents further degradation of the aluminum. Metals protected in this manner are said to be *passive,* or corrosion-resistant.

Corrosion is not limited to the action of atmospheric oxygen on materials. Corrosion is caused by any substance that spontaneously eats into or destroys either metals or minerals, like many acids and bases (Sec. 7-1). Furthermore, corrosion is not limited to the action of such substances on just metals and minerals alone. Corrosion also refers to the destructive action such substances have on certain metallic compounds, as well as on skin tissue.

It is well to note here the following definition of a *corrosive material* as used by USDOT:

". . . a liquid or solid that causes visible destruction or irreversible alterations in human skin tissue at the site of contact, or a liquid that has a severe corrosion rate on steel or aluminum in accordance with the following criteria:

Figure 7-1 A typical tank truck used for shipping bulk quantities of corrosive materials. USDOT requires each person offering such consignments to provide the motor carrier with the required placards (generally CORROSIVE placards) and the identification number corresponding to the nature of the consignment (either on placards or orange panels), which are displayed on each side and each end of the transport vehicle. (Courtesy of Rogers Cartage Company, Oak Lawn, Illinois)

(1) A material is considered to be destructive or to cause irreversible alteration in human skin tissue if, when tested on the intact skin of the albino rabbit [by a specified biological technique], the structure of the tissue at the site of contact is destroyed or changed irreversibly after an exposure period of 4 hours or less.

(2) A liquid is considered to have a severe corrosion rate if its corrosion rate exceeds 6.25 mm (0.250 in.) per year on a specified type of steel or aluminum at a test temperature of 55°C (131°F)." [USDOT specifies a type of steel and aluminum, as well as the test procedure.]

These aspects relating to corrosion are illustrated on the USDOT CORROSIVE labels and placards, which serve to emphasize the potentially destructive nature of corrosive materials when they inadvertently contact skin and metals.

USDOT regulates the transportation of corrosive materials. Packages containing corrosive materials are labeled CORROSIVE. When required, the vehicles that ship these hazardous materials are placarded CORROSIVE. Bulk quantities of corrosive materials are often transported in tank trucks resembling the type shown in Fig. 7-1.

7-1 CHEMICAL NATURE OF ACIDS AND BASES

Several theories have been proposed to account for the properties of acids and bases, but we shall only need to note the theory proposed in 1887 by the Swedish chemist, Svante Arrhenius, modernized to reflect current scientific thought.

Arrhenius proposed that an *acid* is a substance that produces hydrogen ions (H^+) when dissolved in water. Hydrogen ions are hydrogen atoms, but stripped of their electrons. A single hydrogen ion is nothing more than the nucleus of a hydrogen atom.

Today, chemists know that free hydrogen ions cannot exist alone in aqueous solution due to their high charge density. Instead, they rapidly become solvated; that is, they loosely bond to water molecules. These solvated hydrogen ions are very complex species, which we can represent as $(H_2O)_x H^+$, where x is a number. We collectively represent these various aggregates of solvated hydrogen ions by the notation H^+(aq).

Based on this concept, we can readily represent the ionization in water of typical acids. For instance, the ionization in water of hydrochloric acid (HCl) may be represented as follows:

$$HCl(aq) \longrightarrow H^+(aq) + Cl^-(aq)$$

Arrhenius further proposed that a *base* is a substance that produces hydroxide ions (OH^-) when dissolved in water. These ions also become solvated, which we represent as OH^-(aq). For instance, sodium hydroxide produces hydroxide ions when dissolved in water, as the following equation illustrates:

$$NaOH(aq) \longrightarrow Na^+(aq) + OH^-(aq)$$

Certain water-insoluble metallic hydroxides are not ordinarily regarded as bases, since they do not produce a significant concentration of hydroxide ions in water. Examples of water-insoluble metallic hydroxides are iron(II) and iron(III) hydroxides, aluminum hydroxide, and nickel hydroxide.

The ability of acids and bases to act as corrosive materials depends in part on their relative strengths when dissolved in water. The relative strengths of acids and bases refer to the tendency that these individual substances have to form hydrated hydrogen ions and hydroxide ions, respectively, when dissolved in water. Those acids and bases that yield a high concentration in water of hydrogen and hydroxide ions are called *strong acids* and *strong bases,* respectively. For instance, hydrochloric acid is an example of a *strong acid,* because it almost completely ionizes into solvated hydrogen ions and chloride ions when dissolved in water. Furthermore, sodium hydroxide is an example of a *strong base,* since it almost completely ionizes into solvated sodium and hydroxide ions when dissolved in water.

On the other hand, some substances tend to retain the identity of their unit formulas when dissolved in water more than they tend to dissociate into ions. Such substances yield relatively low concentrations of hydrogen or hydroxide ions in water and are called *weak acids* and *weak bases,* respectively. Acetic acid is an example of a weak acid; when dissolved in water, it mainly exists as molecules of acetic acid, although some smaller number of solvated hydrogen and acetate ions also exist. Ammonium hydroxide is an example of a weak base; when dissolved in water, it exists primarily as units of ammonium hydroxide and does not significantly dissociate into ammonium and hydroxide ions.

TABLE 7-1 RELATIVE STRENGTHS OF ACIDS IN WATER[a]

Perchloric acid	$HClO_4$
Sulfuric acid	H_2SO_4
Hydrochloric acid	HCl
Nitric acid	HNO_3
Phosphoric acid	H_3PO_4
Nitrous acid	HNO_2
Hydrofluoric acid	HF
Acetic acid	CH_3COOH
Carbonic acid	H_2CO_3
Hydrocyanic acid	HCN
Boric acid	H_3BO_3

[a] Listed in descending order.

Table 7-1 classifies the common acids and bases according to their chemical nature as strong and weak acids and bases, respectively.

The degree of corrosiveness of acids and bases may depend on the amounts of these substances that are contained in a given volume of water. Occasionally, such amounts are qualitatively described by referring to the substances as either *concentrated* or *diluted*. Concentrated may mean that the given substance contains no water at all. But generally, when an acid or base is described as concentrated, it actually contains some amount of water. When water is strictly absent from a material, it is said to be *anhydrous*.

For example, concentrated hydrochloric acid is a commercially available substance that contains roughly 37% hydrogen chloride by mass in water. When an arbitrary amount of water has been added to this commercial material, the resulting solution is said to have been diluted. Concentrated corrosive materials are hazardous for the reasons previously noted. On the other hand, diluted corrosive materials may or may not be hazardous, depending on the amount of water present. Corrosive materials that have been highly diluted with water may exhibit a very low degree of corrosivity.

The degree to which an acid corrodes is also dependent on whether or not the acid is an oxidizing agent. An oxidizing acid corrodes metals by oxidizing them to their corresponding positive ions. Nitric acid is an example of an oxidizing acid, while hydrochloric acid is an example of a nonoxidizing acid. We shall note whether individual acids exhibit oxidizing properties as we examine their unique properties, beginning in Sec. 7-5.

Finally, we should note that all acids may be divided into two main categories: inorganic or mineral acids and organic acids. *Mineral acids* consist of molecules having atoms of hydrogen, an identifying nonmetal (typically chlorine, sulfur, or phosphorus), and maybe oxygen. *Organic acids* consist of molecules having atoms of carbon, hydrogen, and oxygen only, and their molecular structures always contain the following characteristic group of atoms, called the *carboxyl group:*

$$-C\!\!\underset{\textstyle OH}{\overset{\textstyle O}{\diagup\!\!\!\!\diagdown}}$$

Both the mineral and organic acids corrode metals and body tissue. However, the organic acids possess another hazard: They burn at elevated temperatures. Nevertheless, none of the organic acids is generally considered a major fire risk.

7-2 CORROSIVITY OF ACIDS AND BASES

Acids are associated with certain common properties. They taste sour, cause the colors of indicator dyes to change to other characteristic colors, and react with bases to form salts and water. On the other hand, bases taste bitter, feel slippery, cause the colors of indicator dyes to take on characteristic colors, and react with acids to form salts and water. Litmus is probably the most common indicator used to differentiate acids from bases. Derived from certain lichens, a solution of litmus is allowed to impregnate strips of paper, which are then dried. When moistened with an aqueous acid, litmus paper turns red; when moistened with a basic solution, it turns blue.

When acids and bases chemically interact, they neutralize each other. An example of neutralization is represented by the following equation:

$$HCl(aq) + NaOH(aq) \longrightarrow NaCl(aq) + H_2O(l)$$

The compound other than water that forms from the neutralization of acids and bases is called a *salt*. A salt may also be regarded as the compound that results when the hydrogen in an acid is replaced by a metallic ion. NaCl is the chemical formula of sodium chloride, which is ordinary *table salt*.

Acids

We shall consider only the manner by which acids act as corroding agents when they chemically react with metals, metallic oxides, and metallic carbonates. At room conditions, these latter substances are almost always encountered as relatively hard, dense, water-insoluble solids. For this reason, they are often chosen as structural materials for buildings, statues, and the like. But when these materials react with acids, they are often chemically converted into water-soluble compounds. This chemical reaction forms the basis for the corrosion and ultimate deterioration of structures made from them.

Let's consider limestone, which is a mixture of certain metallic oxides and carbonates like calcium carbonate. When acid is allowed to contact limestone on buildings, statuary, and other structures, water-soluble compounds are produced, principally calcium compounds. In our modern world, acids are frequently present as pollutants in the atmosphere, having originated as simple nonmetallic oxides associated with the burning of coal (for example, sulfur dioxide) or from the operation of motor vehicles (for example, nitrogen dioxide). As rainwater strikes such limestone structures, water-soluble compounds are slowly removed, thus exposing new limestone to the polluted atmosphere. Given adequate time, complete deterioration of the structure by corrosion is inevitable.

Reactions of acids and metals.   Acids react with all common metals, except copper, silver, gold, and mercury. These are simple displacement reactions (see Sec. 4-3) that produce hydrogen and a compound of the metal, as illustrated by the following equations:

$$Mg(s) + 2HCl(aq) \longrightarrow MgCl_2(aq) + H_2(g)$$

$$2Al(s) + 3H_2SO_4(aq) \longrightarrow Al_2(SO_4)_3(aq) + 3H_2(g)$$

Because of this property of acids, it is logical to expect that the use of metal barrels and drums as acid containers would be sharply curtailed. However, some metals, like steel, may be made passive to chemical attack by certain acids. Consequently, drums made from certain types of steel may be used to contain some acids. Nevertheless, acids always present an indirect fire and explosion hazard due to their potential for generating hydrogen in metal shipping containers, piping, or equipment used for their processing or storage.

Reaction of acids and metallic oxides.   An acid reacts with a metallic oxide to form a compound of the metal and water. Examples of this type of reaction are illustrated by the following equations:

$$FeO(s) + 2HCl(aq) \longrightarrow FeCl_2(aq) + H_2O(l)$$

$$Al_2O_3(s) + 6HNO_3(aq) \longrightarrow 2Al(NO_3)_3(aq) + 3H_2O(l)$$

Frequently, acids are beneficially used to remove such metallic oxides and other impurities from the surface of metals. When used in this fashion, the acid is commonly called *pickle liquor,* and the process is called *pickling.* Large quantities of pickle liquor are used annually by the steel industry to remove rust from the surface of steel in conjunction with the manufacture of wire, rod, nuts and bolts, and other steel products. The common acids used in pickle liquor are sulfuric acid, hydrochloric acid, or phosphoric acid.

Reaction of acids with metallic carbonates.   Acids react with metallic carbonates to produce carbon dioxide, water, and a compound of the metal. Examples of this type of reaction are illustrated by the following equations:

$$CaCO_3(s) + 2HCl(aq) \longrightarrow CaCl_2(aq) + CO_2(g) + H_2O(l)$$

$$ZnCO_3(s) + H_2SO_4(aq) \longrightarrow ZnSO_4(aq) + CO_2(g) + H_2O(l)$$

As noted earlier, the reaction of acidic atmospheric pollutants with metallic carbonates in limestone causes the corrosion of structures made from limestone.

Reaction of Acids on Skin Tissue

Each acid reacts with skin tissue by a unique mechanism, but the result is often independent of the acid. The corrosive effect which an acid has on skin tissue varies with the acid's concentration. Contact with a diluted acid may only redden the area of exposure, while contact with a concentrated acid for the same duration often blisters it. Such swelling of skin tissue caused by exposure to a concentrated acid is illustrated in Fig. 7-2. We say that the skin is *burned,* since the appearance of the skin often resembles a first-, second-, or third-degree thermal burn.

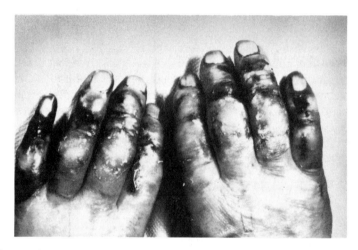

Figure 7-2 An acid burn on the hands. Such damage to the skin from corrosive materials may be irreversible, leaving ugly scars after wounds have healed. To correct the physical appearance of damaged skin, reconstructive cosmetic surgery is often necessary. (Courtesy of Dr. Sophie M. Worobec, M.D., University of Illinois, College of Medicine, and Dr. Paul Lazar, M.D., Chicago, Illinois)

The wounds developed from exposure to acids may form scars that heal and leave no outward evidence of dermatological damage. But acids may also damage skin deeply, especially when the exposure has been prolonged. Tissue damage is also likely to be irreversible. In such incidents plastic surgery is required to correct the surficial appearance of the skin.

Bases

Bases may also act as corrosive materials, but the manner by which they corrode is more narrowly limited.

Reaction of bases and metals.   Other than their destructive action on skin tissue, bases act as corrosive materials in only one other way: The concentrated solutions of strong bases react with certain metals to produce hydrogen and a compound of the metal.

Aluminum, zinc, and lead are the only three common metals that react in this fashion. The following equations illustrate the chemical reactions of these metals with sodium hydroxide:

$$2Al(s) + 6NaOH(aq) \longrightarrow 2Na_3AlO_3(aq) + 3H_2(g)$$

$$Zn(s) + 2NaOH(aq) \longrightarrow K_2ZnO_2(aq) + H_2(g)$$

$$Pb(s) + 2NaOH(aq) \longrightarrow Na_2PbO_2(aq) + H_2(g)$$

Reaction of bases on skin tissue. The corrosive effect that a base has on skin tissue varies with the concentration of the base. Contact of the skin with a diluted base usually only reddens the area of exposure. Thus, the corrosive action of a base is similar to that caused by exposure of the skin to a diluted acid. However, the corrosive effect that concentrated bases have on skin tissue is decidedly different, especially following prolonged exposure. Bases tend to emulsify skin at the site of contact; that is, bases cause the appearance of skin to change in consistency, resembling a thick liquid. The tissue also becomes very sticky. Continued exposure to such concentrated bases causes deep, slow-healing wounds. As in the case of exposure to concentrated acids, the damage to the skin tissue may be irreversible at the site of contact.

7-3 THE pH SCALE

The pH is a number that describes the acidity or alkalinity of an aqueous solution. The range of the pH scale is from 0 to 14. Aqueous solutions having a pH from 0 to 7 are *acidic,* while aqueous solutions having a pH from 7 to 14 are *basic* (or *caustic* or *alkaline*). The pH of pure water is 7, since water is neither acidic nor basic. Table 7-2 lists the pH of some commonly encountered solutions and mixtures.

In a chemical laboratory, the pH is frequently determined by means of an electrometric measurement using an apparatus called a *pH meter*. A pH meter is a voltage-measuring device attached to a pair of electrodes, like the one shown in Fig. 7-3. When the tips of the electrodes have been immersed in a solution, the pH of the solution can be readily established simply by reading the scale.

The definition of the pH is the following: the negative power to which 10 must be raised to equal the hydrogen ion concentration. In algebra, we learn that if $10^x = y$, then x is the logarithm of y to the base 10. Hence, if $[H^+] = 10^{-pH}$, then the pH is expressed as follows:

$$pH = -\log [H^+]$$

It is important to note that the pH scale is an exponential one. Thus, vastly different concentrations in the hydrogen ion may exist in given solutions than their pH values seemingly reflect.

TABLE 7-2 THE pH VALUES OF SOME COMMONLY ENCOUNTERED SOLUTIONS
AND MIXTURES

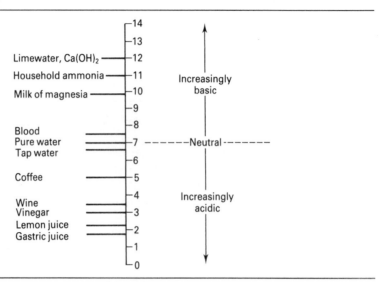

It is appropriate to note that corrosivity is one of four characteristics used to characterize a hazardous waste under RCRA (Sec. 1-5). A hazardous waste exhibits the characteristic of corrosivity if it possesses either of the following properties:

1. It is aqueous and has a pH less than or equal to 2 or greater than or equal to 12.5.
2. It is a liquid and corrodes steel at a rate greater than 0.250 in. (6.35 mm) per year at a test temperature of 130°F (55°C) using a specified test method.

Hazardous wastes that possess the characteristic of corrosivity are assigned an EPA hazardous waste number of D002.

7-4 RESPONDING TO EMERGENCY INCIDENTS INVOLVING CORROSIVE MATERIALS

Safe response to emergencies involving corrosive materials often requires the use of acid-resistant apparel, such as the acid-entry suit with clear face shield shown in Fig. 7-4. Such a suit is designed to protect the wearer from inadvertently splashing corrosive materials on the skin and minimize or eliminate the hazards associated with handling incidents involving corrosive materials. In all such incidents, members of emergency-response teams should at least wear protective eyeglasses.

When corrosive materials are encountered during emergencies, either of two response actions should normally be followed, depending on the circumstances in

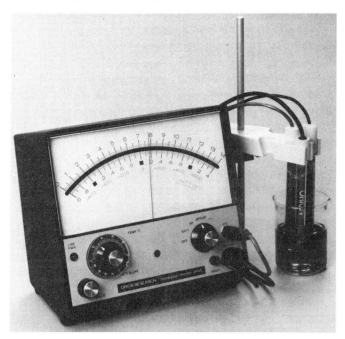

Figure 7-3 A pH meter used for measuring the pH of aqueous solutions. Electrodes are immersed into the solution (at the right) whose pH is desired. The scale graduated from 0 to 14 is the pH scale; the pH of this solution is 7.8. The other scales shown on the meter are used by chemists for other electrometric measurements. (Courtesy of Orion Research, Inc., Boston, Massachusetts)

which the materials are involved. The first action that is often useful is dilution of the corrosive material with streams of water. This is usually an adequate response action when relatively small quantities of either an acid or base are at issue. For instance, suppose a gallon container of concentrated liquid sulfuric acid has spilled on a laboratory floor and has flowed toward a drain that leads to a waste-water treatment plant. In this example, dilution of the acid spill with copious amounts of water lessens the corrosive nature of the acid by reducing the concentration of hydrogen ions in a given volume.

The second potentially useful response action for situations involving corrosive materials relates to the phenomenon of neutralization. As we noted in Sec. 7-2, acids and bases react with each other, neutralizing their respective counterparts. Acids may also be effectively neutralized through the use of metallic carbonates and bicarbonates. For instance, caustic spills can be neutralized with a weak acid, like acetic acid. Response teams may thus eliminate the corrosive nature of either an acid or base by neutralizing the material with the opposite substance. All emergency-response teams should have ready access to such materials.

Suppose that a rail tank car transporting 10,000 gal of sulfuric acid is involved in a transportation incident and leaks its contents on the underlying soil. In this in-

Figure 7-4 This fully encapsulating suit is worn when responding to incidents involving the spillage of acids and other corrosive materials. The suit is made of a material that is resistant to permeation by corrosive materials and thus prevents bodily contact with them. (Courtesy of Mine Safety Appliances Co., Pittsburgh, Pennsylvania)

stance, diluting the acid with water may be impractical, since a relatively large quantity of water would be needed to effectively reduce the acid's corrosive nature. Instead, a dike should ordinarily be dug to confine the material to a relatively small area; then the acid can be neutralized by the addition of lime, sodium carbonate (soda ash) or sodium bicarbonate (baking soda). Lime is the common name of calcium hydroxide, which neutralizes sulfuric acid as the following equation illustrates:

$$H_2SO_4(aq) + Ca(OH)_2(s) \longrightarrow CaSO_4(s) + 2H_2O(l)$$

Finally, the residue should be removed to a state-approved chemical waste disposal facility.

Emergency-response forces are often called on to assist in situations where individuals have been exposed to corrosive materials. It is well to recall that corrosive materials may irreversibly alter tissue at the site of contact. The most potentially hazardous situation is one in which corrosive materials have been inadvertently splashed in the eyes. In such instances, it is vital to immediately flush the eyes with a gentle stream of running water for at least 15 minutes. If the afflicted individual wears contact lenses, the eyes should first be irrigated for several minutes. Then the lenses should be removed and the eyes again irrigated. Thereafter, an ophthalmologist should be promptly contacted.

Sometimes emergency teams must assist an individual who has unintentionally swallowed a corrosive material. In such situations, the afflicted individual should *not* be induced to vomit. The stomach wall is relatively tough; it normally withstands the presence of gastric juices having a pH of less than 2. Vomiting should not normally be induced, since the corrosive vomitus may contact sensitive bronchial tissues. When swallowed, corrosive materials should be neutralized in the stomach. If the material is an acid, the individual should consume milk of magnesia. (Milk of magnesia is the common name for an aqueous suspension of magnesium hydroxide.) If the material is a base, the individual should consume vinegar or citrus fruit juices. In both instances, enough neutralizing agent (at least equal to the amount swallowed) should be consumed to assure that neutralization of the corrosive material has been complete.

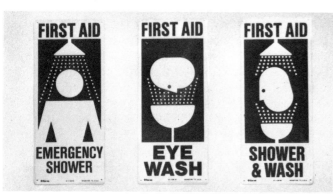

Figure 7-5   When corrosive materials are splashed in the eyes or when they otherwise come in contact with skin tissue, they must be removed immediately to protect the skin against permanent damage by flushing the afflicted area with water. For this reason, eye-wash stations and showers are necessarily located in science laboratories and in other areas where corrosive materials are regularly stored and used. (Courtesy of Lab Safety Supply Company, Janesville, Wisconsin)

In the workplace, when the eyes or body of any person may be exposed to injurious corrosive materials, OSHA requires suitable facilities to be available for quick drenching or flushing of the eyes and body within the work area for immediate emergency use. An eye-wash station and shower that fulfill this requirement are illustrated in Fig. 7-5.

7-5 SULFURIC ACID

So many different industrial processes use sulfuric acid that the sulfuric acid market is considered a better guide to general business conditions than even the steel industry. In fact, it has been said that the economic status of a nation may be measured by its consumption of sulfuric acid. This is due to the fact that sulfuric acid is used to manufacture so many other chemical substances. Even other acids are often produced through the use of sulfuric acid.

Pure *sulfuric acid* is a colorless, oily liquid with a density about twice that of water. Its chemical formula is H_2SO_4. Some of its important physical properties are noted in Table 7-3. Commercially, sulfuric acid is encountered as a clear to brownish-colored liquid, depending on the degree of purity that is required by the consumer.

Sulfuric acid was first obtained by collecting the distillate on heating green vitriol, a hydrated form of ferrous sulfate; the liquid was called *oil of vitriol*. In modern times, however, sulfuric acid is prepared by first burning sulfur to produce sulfur dioxide, and then oxidizing sulfur dioxide further in the presence of a catalyst to sulfur trioxide. Sulfur trioxide is said to be the *anhydride* of sulfuric acid. An anhydride is a chemical compound derived from an acid by elimination of a molecule of water.

Sulfur trioxide unites with water to form sulfuric acid, as illustrated by the following equation:

$$H_2O(l) + SO_3(g) \longrightarrow H_2SO_4(l)$$

However, sulfur trioxide does not readily unite with pure water. Hence, when sulfuric acid is manufactured, sulfur trioxide is sometimes absorbed in 97% sulfuric acid by mass, in which it readily dissolves. Liquid solutions of sulfur trioxide and water boil at 640°F (338°C) at 1 atm, and contain 98.3% sulfuric acid by mass. This material is the *concentrated sulfuric acid* of commerce.

TABLE 7-3 PHYSICAL PROPERTIES OF CONCENTRATED SULFURIC ACID

Concentration of H_2SO_4 in water	98.33%
Specific gravity	1.84
Boiling point	640°F (338°C)
Freezing point	50°F (10°C)
Solubility in water	Infinitely soluble

An excess of sulfur trioxide may also be dissolved in concentrated sulfur
acid. The resulting material is called *oleum,* or *fuming sulfuric acid.* Its chemical
formula is sometimes given as $H_2S_2O_7$. However, this is the chemical formula of
disulfuric acid, which is only one of several acids present in oleum. Hence, it is generally felt that $xH_2SO_4 \cdot ySO_3$ is a more correct representation of this substance.
Commercially, oleum is available in concentrations from 10% to 70% sulfur trioxide
by mass. Oleum whose concentration is 65% sulfur trioxide by mass is expressed as
the formula $4H_2SO_4 \cdot 9SO_3$.

Some hazardous features of sulfuric acid are illustrated in Fig. 7-6. The first
hazard is associated with the heat liberated when sulfuric acid is diluted. When the
concentrated acid is dissolved in water, considerable heat is evolved, about 84,000 J
(20 kcal or 79 Btu) per mole. Since this dilution reaction is so exothermic, extreme
caution must be exercised when aqueous solutions of sulfuric acid solutions are prepared. When preparing *any* aqueous acid solution, the acid should always be carefully poured into the water with stirring. When the reverse process is performed, localized boiling and violent spattering may occur.

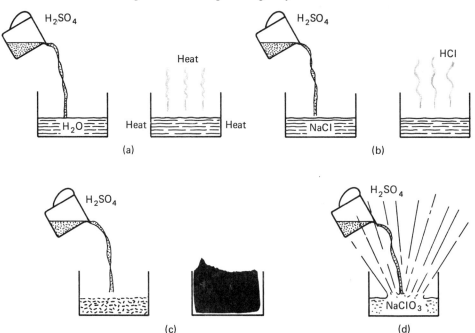

Figure 7-6 Some potentially hazardous features of concentrated sulfuric acid. In (a), the
mixture of concentrated sulfuric acid with water causes the evolution of sufficient heat to
trigger the self-ignition of some materials. In (b), concentrated sulfuric acid acts as a reactant
in a double-replacement reaction with sodium chloride; in this instance, toxic vapors of
hydrogen chloride are produced. In (c), concentrated sulfuric acid dehydrates sugar; that is, it
removes the elements of water from sugar, leaving a residue of carbon. In (d), concentrated
sulfuric acid reacts with sodium chlorate, again extracting the elements of water; these two
substances are chemically incompatible, exploding on contact.

Another hazardous feature of sulfuric acid illustrated in Fig. 7-6 concerns the ability of sulfuric acid to extract water from certain materials. Sulfuric acid has such a tremendous affinity for water that it even abstracts the elements of water from some compounds. Several representative dehydration reactions involving concentrated sulfuric acid are noted in Table 7-4. The first two reactions are illustrative of the action concentrated sulfuric acid has on simple organic compounds: The water is so strongly extracted that only carbon remains. For this reason, sulfuric acid completely destroys textiles and paper. This is also the chemical action that concentrated sulfuric acid has on body tissue.

Sometimes the dehydration reactions of sulfuric acid occur explosively. For instance, the third reaction cited in Table 7-4 illustrates the explosive nature of a mixture of concentrated sulfuric acid and concentrated perchloric acid. Since a mixture of these two acids is potentially explosive, the acids should never be stored side by side. Special note should also be given to reactions between concentrated sulfuric acid and oxidizing agents like metallic permanganates, dichromates, and chlorates. These reactions appear to proceed stepwise, first forming an unstable acid, which subsequently dehydrates to the corresponding anhydride, sometimes explosively.

The third hazardous feature of sulfuric acid is associated with its potential for

TABLE 7-4 REPRESENTATIVE DEHYDRATION REACTIONS OF CONCENTRATED SULFURIC ACID

Substance	Equation	Comments
Sugar ($C_{12}H_{22}O_{11}$)	$C_{12}H_{22}O_{11}(s) \rightarrow 12C(s) + 11H_2O(g)$	Reaction runs smoothly, but is very exothermic
Cellulose (wood products)	$(C_6H_{10}O_5)_x(s) \rightarrow 6xC(s) + 5xH_2O(g)$	Reaction runs smoothly, except for finely divided materials; very exothermic
Concentrated perchloric acid	$2HClO_4(l) \rightarrow Cl_2O_7(g) + H_2O(g)$	Violent explosion occurs
Formic acid	$HCOOH(l) \rightarrow H_2O(g) + CO(g)$	Reaction runs smoothly, but CO is toxic
Oxalic acid	$H_2C_2O_4(s) \rightarrow H_2O(g) + CO(g) + CO_2(g)$	Reaction runs smoothly, but CO is toxic
Ethyl alcohol (C_2H_5OH)	$C_2H_5OH(l) \rightarrow C_2H_4(g) + H_2O(g)(T = 150°C)$	Reaction runs smoothly, but product is flammable
	$2C_2H_5OH(l) \rightarrow C_2H_5-O-C_2H_5(g) + H_2O(g)$ ($T = 125°C$)	Reaction runs smoothly, but product is flammable
$KMnO_4$	$2KMnO_4(s) + H_2SO_4(l) \rightarrow K_2SO_4(l) + 2HMnO_4(s)$ $2HMnO_4(s) \rightarrow Mn_2O_7(s) + H_2O(g)$ $2Mn_2O_7(s) \rightarrow 4MnO_2(s) + 3O_2(g)$	Violent explosion occurs
$K_2Cr_2O_7$[a]	$K_2Cr_2O_7(s) + H_2SO_4(l) \rightarrow H_2Cr_2O_7(l) + K_2SO_4(s)$ $H_2Cr_2O_7(l) \rightarrow 2CrO_3(s) + H_2O(g)$	Reaction runs smoothly, but is very exothermic
$NaClO_3$	$NaClO_3(s) + H_2SO_4(l) \rightarrow NaHSO_4(s) + HClO_3(l)$ $3HClO_3(l) \rightarrow HClO_4 + 2ClO_2 + H_2O$	Violent explosion; ClO_2 is unstable and toxic

[a] A substance having the chemical formula $H_2Cr_2O_7$ has actually never been identified.

reacting with so many other chemical substances. Table 7-5 illustrates several of these reactions and their potentially hazardous consequences. Industrially, many reactions involving sulfuric acid are controlled and result in the production of substances that are beneficial to society. However, it is well to note that sulfuric acid is chemically incompatible with a large group of substances. When chemical reactions involving sulfuric acid occur in an uncontrolled fashion, they are likely to form products that are toxic, flammable, or explosive.

One unique type of chemical reaction should be specifically noted for concentrated sulfuric acid: its oxidizing potential. When concentrated and hot, sulfuric acid is a strong oxidizer. For instance it rapidly oxidizes copper, carbon, and lead, as the following equations illustrate:

$$Cu(s) + 2H_2SO_4(conc) \longrightarrow CuSO_4(aq) + SO_2(g) + 2H_2O(l)$$

$$C(s) + 2H_2SO_4(conc) \longrightarrow CO_2(g) + SO_2(g) + 2H_2O(l)$$

$$Pb(s) + 3H_2SO_4(conc) \longrightarrow Pb(HSO_4)_2(s) + SO_2(g) + 2H_2O(l)$$

When diluted with water, sulfuric acid loses much of its oxidizing capability.

USDOT regulates the transportation of sulfuric acid, its spent solutions, and fuming sulfuric acid as corrosive materials. These regulations distinguish between concentrations that are less than and greater than 62.5% sulfuric acid by mass. Packages of sulfuric acid are labeled CORROSIVE in either situation. But packages of fuming sulfuric acid are labeled CORROSIVE and POISON. Certain concentrations of sulfuric acid may be contained for transportation in metal drums made of a spe-

TABLE 7-5 HAZARDOUS REACTIONS INVOLVING CONCENTRATED SULFURIC ACID

Reactant	Equation	Hazardous feature
NaBr	$2NaBr(s) + 2H_2SO_4(l) \rightarrow Br_2(g) + SO_2(g) + Na_2SO_4(aq) + 2H_2O(l)$	Reaction runs relatively smoothly, but SO_2 and Br_2 are toxic
NaI	$2NaI(s) + 2H_2SO_4(l) \rightarrow I_2(s) + SO_2(g) + Na_2SO_4(aq) + 2H_2O(l)$	Reaction runs relatively smoothly, but SO_2 and vapors of I_2 are toxic
NaCN	$2NaCN(s) + 3H_2SO_4(l) + 2H_2O(l) \rightarrow$ $(NH_4)_2SO_4(aq) + 2NaHSO_4(aq) + 2CO(g)$	Reaction runs relatively smoothly but CO is toxic
NaSCN	$NaSCN(s) + 2H_2SO_4(l) + H_2O(l) \rightarrow$ $COS(g) + NaHSO_4(aq) + NH_4HSO_4(aq)$	Violent explosion; COS (carbonyl sulfide) is extremely toxic and flammable
HI	$8HI(aq) + H_2SO_4(l) \rightarrow H_2S(g) + 4I_2(g) + 4H_2O(l)$	Reaction runs relatively smoothly, but H_2S is toxic

cial steel that has been passivated. Sulfuric acid may also be transported by rail or motor vehicle in tank cars made either of a special steel or lined with materials impervious to the lading. When required, transport vehicles shipping sulfuric acid and fuming sulfuric acid are placarded CORROSIVE.

In the workplace, OSHA regulates the exposure of employees to sulfuric acid. OSHA stipulates a permissible exposure limit to sulfuric acid of 1 mg/m^3.

7-6 NITRIC ACID

Next to sulfuric acid, nitric acid is industrially the most important acid. It is primarily used in the manufacture of ammonium nitrate fertilizers, explosives, and nitrated organic compounds. For instance, nitric acid is required for the production of nitroglycerin (Sec. 13-5) and trinitrotoluene (TNT) (Sec. 13-8). Nitric acid is also used in the synthetic fiber industry for the production of rayon cloth.

Pure *nitric acid* is a colorless liquid, having a density almost 1.5 times that of water. When commonly encountered, however, it is often yellow to red-brown in color, depending on the concentration of dissolved nitrogen dioxide. Its chemical formula is HNO_3. Nitrogen dioxide arises from the slow decomposition of nitric acid, a reaction catalyzed by sunlight, as the following equation illustrates:

$$4HNO_3(l) \longrightarrow 4NO_2(g) + 2H_2O(l) + O_2(g)$$

A mixture of nitric acid and water boils at 187°F (86°C) when the concentration of nitric acid is 68.2% by mass. This is the *concentrated nitric acid* of commerce. Several physical properties of concentrated nitric acid are noted in Table 7-6.

Other grades of nitric acid, however, are commercially available, even anhydrous nitric acid. The commercial acid containing 97.5% nitric acid, less than 2% water, and less than 0.5% nitrogen oxides by mass is called *white fuming nitric acid*. The acid containing more than 86% nitric acid, less than 5% water, and from approximately 6% to 15% nitrogen oxides by mass is called *red fuming nitric acid*.

In modern times, most nitric acid is industrially manufactured from ammonia by means of a stepwise series of reactions. Gaseous ammonia is first mixed with about ten times its volume of air and then exposed to platinum gauze, which acts catalytically. Nitric monoxide (NO) first forms, as the following equation illustrates:

$$4NH_3(g) + 5O_2(g) \longrightarrow 4NO(g) + 6H_2O(g)$$

Additional air is permitted to enter the reaction system, which oxidizes the nitric ox-

TABLE 7-6   PHYSICAL PROPERTIES OF CONCENTRATED NITRIC ACID

Concentration of HNO_3 in water	68%–70%
Specific gravity	1.50
Boiling point	187°F (86°C)
Freezing point	−44°F (−42°C)
Solubility in water	Infinitely soluble

ide to nitrogen dioxide:

$$2NO(g) + O_2(g) \longrightarrow 2NO_2(g)$$

The nitrogen dioxide is then reacted with water:

$$3NO_2(g) + H_2O(l) \longrightarrow 2HNO_3(l) + NO(g)$$

The nitrogen monoxide may be recycled through the system.

Nitric acid acts as a corrosive material in several distinct ways, some of which are illustrated in Fig. 7-7. Nitric acid chemically attacks certain metals. This chemical action is caused by the strong oxidizing nature of nitric acid. The metals are oxidized to their positive ions, while the nitric acid is reduced to any of the following, depending on the concentration of the nitric acid: nitrogen, nitric monoxide, nitric dioxide, dinitrogen monoxide, or the ammonium ion. The more reactive metals, like zinc, actually form all these nitrogenous products under the proper conditions. Reactions illustrating the oxidation of zinc by nitric acid are represented by the following equations:

$$5Zn(s) + 12HNO_3(aq) \longrightarrow 5Zn(NO_3)_2(aq) + 6H_2O(l) + N_2(g)$$

$$3Zn(s) + 8HNO_3(aq) \longrightarrow 3Zn(NO_3)_2(aq) + 4H_2O(l) + 2NO(g)$$

$$Zn(s) + 4HNO_3(conc) \longrightarrow Zn(NO_3)_2(aq) + 2H_2O(l) + 2NO_2(g)$$

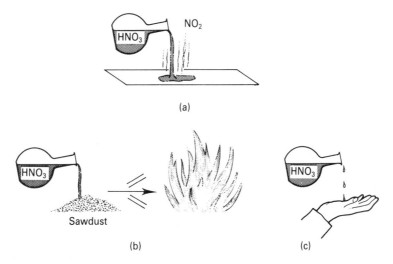

Figure 7-7 Some potentially hazardous features of concentrated nitric acid. In (a), the hot acid acts as a corrosive material; nitric acid corrodes certain metals and nonmetals alike, simultaneously producing an oxide of nitrogen. When nitric acid corrodes these substances, it chemically reacts as a strong oxidizing agent. This is further illustrated in (b), which shows concentrated nitric acid causing the ignition of sawdust. In (c), concentrated nitric acid acts as a corrosive material on skin tissue, leaving ugly, yellow scars.

$$4Zn(s) + 10HNO_3(aq) \longrightarrow 4Zn(NO_3)_2(aq) + 5H_2O(l) + N_2O(g)$$

$$4Zn(s) + 10HNO_3(aq) \longrightarrow 4Zn(NO_3)_2(aq) + 3H_2O(l) + NH_4NO_3(aq)$$

In general, however, only one of these products is formed from the oxidation of a given metal. Nitrogen dioxide typically forms when *concentrated* nitric acid corrodes metal, while nitrogen monoxide typically forms when *diluted* nitric acid corrodes metal.

Finely divided metals often react explosively with concentrated nitric acid. Yet, on the other hand, aluminum, iron, chromium, and cobalt are not attacked at all by concentrated nitric acid. Corrosion of aluminum by nitric acid does not occur at room temperature if the concentration of nitric acid is greater than 80% by mass, and corrosion of iron does not occur if the concentration is greater than 70% by mass. Chromium resists chemical attack by nitric acid in all proportions, and stainless steel resists attack by all concentrations of nitric acid except those greater than 97% by mass. In each instance, corrosion of the metal is prevented by the formation of a protective coating of the associated metallic oxide.

Hot nitric acid corrodes nonmetals, like carbon and sulfur. The following equations illustrate these chemical reactions:

$$C(s) + 4HNO_3(conc) \longrightarrow CO_2(g) + 4NO_2(g) + 2H_2O(l)$$

$$S_8(s) + 48HNO_3(conc) \longrightarrow 8H_2SO_4(l) + 48NO_2(g) + 16H_2O(l)$$

The corrosive effect that nitric acid has on metals and nonmetals is only one of its hazardous features. Nitric acid also oxidizes many organic compounds, sometimes explosively, and causes their subsequent ignition. Turpentine, acetic acid, acetone, ethyl alcohol, nitrobenzene, and aniline are among the substances that explode and burn when mixed with hot, concentrated nitric acid.

Nitric acid also causes the spontaneous ignition of wood, excelsior, and other cellulosic materials. The potential for spontaneous ignition is particularly good when these materials have been finely divided. Consequently, sawdust and similar cellulosic materials should never be used to cushion bottles of nitric acid intended for transportation.

Finally, nitric acid corrodes body tissue by reacting with complex proteins that make up the structure of tissues, converting them to a yellow substance called *xanthoproteic acid*. Spills of nitric acid on skin tissue thus result in ugly, yellow burns.

USDOT regulates the transportation of nitric acid and red fuming nitric acid as corrosive materials. These regulations distinguish between concentrations that are less than and greater than 70% nitric acid by mass. Packages of nitric acid are labeled CORROSIVE in either situation. But packages of red fuming nitric acid are labeled CORROSIVE, OXIDIZER, and POISON. When required, the transport vehicles shipping nitric acid and red fuming nitric acid are placarded CORROSIVE.

USDOT requires that nitric acid be contained in specified steel or aluminum drums. It also specifies the use of special aluminum alloy tank cars and tank trucks for transportation of nitric acid by rail freight, highway, or water. Each rail tank car is marked NITRIC ACID.

USDOT also recognizes a related hazardous material known as *nitrating acid*. This is a mixture of approximately 61% sulfuric acid and 36% nitric acid by mass. Nitrating acid is often used industrially in the manufacture of explosives and plastics. Like both acids of which it is composed, nitrating acid is a corrosive material. USDOT regulates the transportation of nitrating acid and spent nitrating acid. The appropriate regulations distinguish between concentrations that are less than or greater than 50% nitric acid by mass. In the former case, packages of nitrating acid are labeled CORROSIVE; in the latter case, they are labeled CORROSIVE and OXIDIZER. When required, the transport vehicles shipping nitrating acid are placarded CORROSIVE.

In the workplace, OSHA regulates the exposure of employees to nitric acid. OSHA stipulates a permissible exposure limit of 2 ppm.

7-7 HYDROCHLORIC ACID

Hydrochloric acid is one of our most industrially important acids. The steel industry uses it as pickle liquor, especially prior to galvanizing, tinning, and enameling. The food industry uses it as a processing agent for manufacturing products like corn syrup. The petroleum industry uses it to activate petroleum wells. The chemical industry uses it for the production of dozens of important compounds containing chlorine, including synthetic rubber. Hydrochloric acid is also likely to be found in the home as a cleaning agent in various products, like toilet bowl cleaners. It is also used to maintain the proper pH of water in swimming pools.

Pure *hydrochloric acid* is a colorless, fuming, pungent liquid. It is composed of hydrogen chloride dissolved in water. The chemical formula of both hydrochloric acid and hydrogen chloride is HCl. The *concentrated* acid of commerce generally contains from 36% to 38% hydrogen chloride by mass. Some of its physical properties are noted in Table 7-7. Hydrochloric acid is also commercially available as diluted aqueous solutions containing either 28%, 31%, or 35% hydrogen chloride by mass. Its industrial grade, called *muriatic acid,* is slightly yellow due to the presence of dissolved compounds of iron.

Hydrochloric acid is industrially prepared by dissolving hydrogen chloride in water. Some hydrogen chloride is prepared by the direct combination of hydrogen and chlorine, as the following equation illustrates:

$$H_2(g) + Cl_2(g) \longrightarrow 2HCl(g)$$

TABLE 7-7 PHYSICAL PROPERTIES OF CONCENTRATED
HYDROCHLORIC ACID

Concentration of HCl in water	36%–38%
Specific gravity	1.20
Boiling point	−121°F (−85°C)
Freezing point	−175°F (−115°C)
Solubility in water	85 g/100 g of water

However, significant quantities of hydrogen chloride are also produced by the chlorination of organic compounds. Hydrogen chloride may also be produced by reacting sodium chloride and sulfuric acid, as noted by the following equation:

$$NaCl(s) + H_2SO_4(l) \longrightarrow NaHSO_4(s) + HCl(g)$$

When a container of hydrochloric acid is opened, hydrogen chloride is readily detectable as a pungent vapor. Hydrochloric acid is hazardous as a corrosive material, but its vapor is also hazardous as a poisonous material. The gas is one-fifth heavier than air. Inhalation of hydrogen chloride may lead to complete degeneration of cells in the respiratory tract and may even destroy the lining of the tract. In the workplace, OSHA and NIOSH recommend a ceiling limit of only 5 parts per million (ppm) of hydrogen chloride. Symptoms arising from the inhalation of various concentrations of hydrogen chloride are noted in Table 7-8.

Both the concentrated and dilute forms of hydrochloric acid are strong acids, but hydrochloric acid is not an oxidizing acid. Hence, it chemically corrodes by means of the reactions previously noted in Sec. 7-2. On the other hand, hydrochloric acid is chemically incompatible with oxidizing agents, like metallic permanganates, dichromates, and chlorates. Often the reactions between concentrated hydrochloric acid and oxidizing agents result in the formation of chlorine, as the following equations illustrate:

$$2KMnO_4(s) + 16HCl(conc) \longrightarrow$$
$$2MnCl_2(aq) + 2KCl(aq) + 8H_2O(l) + 5Cl_2(g)$$

$$K_2Cr_2O_7(s) + 14HCl(conc) \longrightarrow$$
$$2KCl(aq) + 2CrCl_3(aq) + 7H_2O(l) + 3Cl_2(g)$$

$$KClO_3(s) + 6HCl(conc) \longrightarrow KCl(aq) + 3H_2O(l) + 3Cl_2(g)$$

USDOT regulates the transportation of anhydrous hydrogen chloride and hydrochloric acid as a poison gas and corrosive material, respectively. When metal drums and tank cars are used to transport hydrochloric acid, they are lined with rubber or some other acid-resistant material. Anhydrous hydrogen chloride cylinders and other containers are labeled POISON GAS and CORROSIVE; transport vehicles

TABLE 7-8 INHALATION EFFECTS OF HYDROGEN CHLORIDE ON HUMANS

Hydrogen chloride Concentration in air (ppm)	Symptoms
1–5	Limit of odor
5–10	Mild irritation of mucous membranes
35	Irritation of throat on short exposure
50–100	Barely tolerable
1000	Danger of lung edema after short exposure

shipping them are placarded POISON. Hydrochloric acid packages are labeled COR-ROSIVE; when required, transport vehicles transporting hydrochloric acid are plac-arded CORROSIVE.

In the workplace, OSHA regulates the exposure of employees to hydrogen chloride. OSHA stipulates a permissible exposure limit of 5 ppm.

7-8 PERCHLORIC ACID

Another industrially important mineral acid is perchloric acid. Its chemical formula is $HClO_4$. Perchloric acid possesses the unique feature of being the strongest mineral acid. It is primarily used by chemical and electroplating industries.

Perchloric acid is industrially prepared by distilling a mixture of potassium perchlorate and sulfuric acid. The chemical reaction associated with its production is illustrated by the following equation:

$$2KClO_4(s) + H_2SO_4(aq) \longrightarrow 2HClO_4(l) + K_2SO_4(aq)$$

Distillation of the reaction mixture under reduced pressure yields a colorless liquid having a composition by mass of 72.4% perchloric acid and water, which boils at 397°F (203°C). This is the *concentrated* perchloric acid of commerce. Some of its important physical properties are noted in Table 7-9.

Aqueous solutions of perchloric acid containing less that 85% perchloric acid by mass are completely stable and generally incapable of oxidizing most substances at ambient temperature. Nonetheless, concentrated perchloric acid is a powerful oxidizing agent; when hot, concentrated perchloric acid is capable of oxidizing many substances, especially certain organic compounds.

Care should always be exercised to avoid mixing concentrated perchloric acid with cellulosic materials like paper, wood, and cotton, as the acid could cause them to burn; for instance, concentrated perchloric acid absorbed into sawdust readily causes the sawdust to burst into flame. For this reason, sawdust should never be used to cushion bottles of concentrated perchloric acid that are intended for shipment.

Cellulosic materials that have been soaked with concentrated perchloric acid should be regarded as a fire and explosion hazard. When the acid has soaked into a wooden floor, as during a transportation mishap, great care should be taken to saturate the area thoroughly with water, followed by treatment with sodium bicarbonate.

TABLE 7-9 PHYSICAL PROPERTIES OF
CONCENTRATED PERCHLORIC ACID

Concentration of $HClO_4$ in water	72.4%
Specific grravity	1.70
Boiling point	397°F (203°C)
Freezing point	0°F (−18°C)
Solubility in water	Very soluble

Unless it has been properly treated to destroy the acid, wood soaked with perchloric acid may be very prone to ignite when it subsequently dries.

While the diluted and concentrated forms of perchloric acid are stable, *anhydrous* perchloric acid is extremely unstable. While anhydrous perchloric acid is not routinely available commercially, special care should be exercised to avoid inadvertently dehydrating the acid during its use or storage. For instance, since concentrated sulfuric acid dehydrates perchloric acid (see Table 7-4), bottles of sulfuric acid and perchloric acid should never be stored side-by-side.

USDOT regulates the transportation of two specific concentrations: perchloric acid exceeding 50% but not exceeding 72% strength by mass and perchloric acid whose strength does not exceed 50% acid. In the latter case, USDOT regulates perchloric acid as a corrosive material; in the former case, as an oxidizer. Packages of solutions having either concentration are labeled CORROSIVE and OXIDIZER. Vehicles used to ship perchloric acid in a concentration that does not exceed 50% are placarded CORROSIVE, whereas vehicles used to ship perchloric acid solutions from 50% to 72% are placarded OXIDIZER. USDOT forbids the transportation of perchloric acid whose concentration exceeds 72% by mass.

7-9 HYDROFLUORIC ACID

Hydrofluoric acid has always been an industrially important acid, but its importance has increased immeasurably with the advent of the computer industry. In the past, the primary uses of hydrofluoric acid have been to polish, etch, and frost glass, to pickle brass, copper, and certain steel alloys, and to serve as either the catalyst or fluorinating agent in certain chemical reactions. In modern times, hydrofluoric acid is also being used either directly or as a component of a "mixed acid" with sulfuric acid in the manufacture of computer chips. It is also used by petroleum companies to convert petroleum-derived gasoline into high-octane gasoline blending stock.

Hydrofluoric acid consists of hydrogen fluoride dissolved in water. The chemical formula of both hydrofluoric acid and hydrogen fluoride is HF. The industrial preparation often involves the chemical reaction between sulfuric acid and calcium fluoride, found in the ores *fluorspar* and *fluorite*, as represented by the following equation:

$$CaF_2(s) + H_2SO_4(conc) \longrightarrow CaSO_4(s) + 2HF(aq)$$

When a solution of hydrofluoric acid is boiled, a composition is attained consisting of 43.2% hydrogen fluoride by mass, which boils at 68°F (20°C). This is the usual *concentrated* hydrofluoric acid of commerce. Some of its important physical properties are noted in Table 7-10. Various other grades of hydrofluoric acid are available, however, even anhydrous hydrofluoric acid (minimum 99% hydrogen fluoride by mass).

Hydrofluoric acid is a colorless, fuming liquid. Its most distinguishing chemical property is the ability to react with silicon dioxide (sand) and glass to form sili-

TABLE 7-10 PHYSICAL PROPERTIES OF CONCENTRATED HYDROFLUORIC ACID

Concentration of HF in water	Concentrated solutions usually contain 48% to 60% HF, but commercial strengths up to 100% HF are available.
Specific gravity	1.0
Boiling point	68°F (20°C)
Freezing point	−117°F (−83°C)
Solubility in water	Infinitely soluble

con tetrafluoride. Calcium silicate, a component of glass, reacts with hydrofluoric acid as the following equation illustrates:

$$CaSiO_3(s) + 6HF(conc) \longrightarrow CaF_2(s) + SiF_4(g) + 3H_2O(l)$$

Hydrofluoric acid is the only common acid that corrodes glass.

Chemically, hydrofluoric acid is a weak, nonoxidizing acid. As a corrosive material, it attacks metals as previously noted in Sec. 7-2, but less vigorously than hydrochloric acid. Nevertheless, concentrated hydrofluoric acid and hydrogen fluoride are strongly corrosive to body tissue, producing severe burns that heal very slowly. These burns are generally more painful than those caused by other acids. Extremely severe burns may result from mild exposure to hydrofluoric acid. Externally, the skin whitens, but the tissues beneath the skin may also be destroyed. Unattended, the destruction may even spread to the bones. When dilute solutions of hydrofluoric acid contact body tissues, they may not cause pain or visibly alter the skin at the site of contact until hours after exposure. By this time, the acid may have penetrated the tissue several millimeters, which is likely to ultimately cause skin ulcers.

Skin that has been exposed to hydrogen fluoride or hydrofluoric acid should be thoroughly cleansed with cold water until the whitening of the tissue disappears. Then an ointment consisting of 3 oz of magnesium oxide, 4 oz of heavy mineral oil, and 11 oz of white petroleum jelly should be applied.* This dressing is intended for use only on the skin, *not the eyes*. When the eyes have been exposed to hydrogen fluoride or hydrofluoric acid, they should be irrigated immediately with water for at least 15 minutes. Medical assistance should be sought.

USDOT regulates the transportation of three specific concentrations of hydrofluoric acid as corrosive materials: anhydrous hydrofluoric acid; hydrofluoric acid, not more than 60%; and hydrofluoric acid, more than 60%. Packages of solutions having either concentration are labeled CORROSIVE and POISON. Vehicles used to ship them are placarded CORROSIVE.

Since hydrofluoric acid chemically attacks glass, it obviously cannot be containerized in glass vessels for storage or transportation. Formerly, the interior of glass bottles was coated with wax, with which hydrofluoric acid is chemically compatible, and these waxed bottles were used to containerize hydrofluoric acid. But to-

---

*H. C. Hodge and F. A. Smith, *Fluorine Chemistry*, Vol. IV J. H. Simons, ed. (New York: Academic Press, 1965), p. 36.

day hydrofluoric acid is generally transported in passivated metal drums or in rail tank cars or tank trucks like those shown in Fig. 7-8.

In the workplace, OSHA regulates the exposure of employees to anhydrous hydrofluoric acid. OSHA stipulates a permissible exposure limit of 3 ppm to the anhydrous acid; NIOSH stipulates a limit of only 2.5 ppm.

Figure 7-8 Hydrofluoric acid is commercially available as its anhydrous liquid and as aqueous solutions. Rail tank cars, tank trucks, and other vehicles used to transport the forms of hydrofluoric acid are placarded CORROSIVE in accordance with USDOT regulatory requirements. (Courtesy of E. I. du Pont de Nemours & Company, LaPorte, Texas)

7-10 PHOSPHORIC ACID

Phosphorus forms at least eight acids, but phosphoric acid is the only one in common industrial use. It is primarily used for pickling steel and for manufacturing a number of commercially important phosphates. For instance, a mixture of calcium dihydrogen phosphate and calcium sulfate is called *superphosphate,* an important synthetic fertilizer. Many detergents contain metallic phosphates, and ammonium dihydrogen phosphate is a dry chemical fire extinguisher.

Phosphoric acid is a clear, odorless, sparkling liquid or transparent solid, depending on its concentration and temperature. Its chemical formula is H_3PO_4. Various grades of phosphoric acid are commercially available. When an aqueous solution of phosphoric acid is allowed to boil at atmospheric pressure, a syrupy solution containing about 85% phosphoric acid by mass is obtained. This is the concentrated phosphoric acid of commerce. Some of its physical properties are noted in Table 7-11. A food grade of phosphoric acid is also commercially available, which is used in some soft drinks and other food products.

Phosphoric acid is manufactured from phosphate rock, which contains calcium phosphate. In one production method, phosphate rock is reacted with sulfuric acid. This chemical reaction is illustrated by the following equation:

$$Ca_3(PO_4)_2(s) + 3H_2SO_4(aq) \longrightarrow 3CaSO_4(s) + 2H_3PO_4(aq)$$

In an alternative method, elemental phosphorus is produced from phosphate rock (see Sec. 6-6), the phosphorus is burned to form tetraphosphorus decoxide, and the oxide is then reacted with water. This reaction is illustrated by the following equation:

$$P_4O_{10}(s) + 6H_2O(l) \longrightarrow 4H_3PO_4(aq)$$

Tetraphosphorus decoxide is thus the anhydride of phosphoric acid. In fact, *phosphoric anhydride* is a name by which this oxide is more commonly known. This substance has such a tremendous affinity for water that it is often used industrially as a drying agent. The reaction with water is violent, evolving considerable heat. While phosphoric anhydride is itself nonflammable, the heat evolved by its reaction with water is sufficient to ignite nearby flammable and combustible materials. For this reason, phosphoric anhydride is often considered a dangerous fire risk. Exposure of

TABLE 7-11 PHYSICAL PROPERTIES OF
CONCENTRATED PHOSPHORIC ACID

Concentration of H_3PO_4 in water	85% (generally)
Specific gravity	1.69
Boiling point	500°F (260°C)
Freezing point	108°F (42°C)
Solubility in water	Very soluble

the skin, mucous membranes, and eyes to phosphoric anhydride is also a major concern, since phosphoric acid locally forms, which subsequently burns the tissue.

Phosphoric acid is a moderately strong, nonoxidizing acid. It acts as a corrosive material in the manner previously noted in Sec. 7-2 for most corrosive materials.

USDOT regulates the transportation of tetraphosphorus decoxide and phosphoric acid as corrosive materials. In the former case, the proper shipping name is *phosphorus pentoxide*. Packages of both materials are labeled CORROSIVE; when required, their transport vehicles are placarded CORROSIVE.

In the workplace, OSHA regulates the exposure of employees to phosphoric acid. OSHA stipulates a permissible exposure limit of 1 mg/m^3.

7-11 CHLOROSULFONIC ACID

Chlorosulfonic acid is an important acid used primarily by the chemical industry for production of certain detergents, dyes, explosives, and other products. Its preferred name is *chlorosulfuric acid,* but in the United States, it is still known mainly as *chlorosulfonic acid.*

Chlorosulfonic acid is a colorless to light-yellow, fuming liquid. Its chemical formula is $ClSO_3H$. It is normally prepared by combining anhydrous sulfur trioxide and hydrogen chloride, as the following equation illustrates:

$$SO_3(g) + HCl(g) \longrightarrow ClSO_3H(l)$$

As the 704 symbol indicates, chlorosulfonic acid is a water-reactive material. Hence, it is commercially available only as the anhydrous liquid. Some physical properties of this acid are noted in Table 7-12.

The reaction of chlorosulfonic acid with water is violent, forming hydrochloric and sulfuric acids. This reaction is illustrated by the following equation:

$$ClSO_3H(l) + H_2O(l) \longrightarrow HCl(aq) + H_2SO_4(aq)$$

Vapors generally escape and fume from openings in containers of chlorosulfonic acid. These fumes are the acid products that form when chlorosulfonic acid reacts with atmospheric moisture. They are also particularly irritating to the eyes and sensitive mucous membranes.

Chlorosulfonic acid is also a strong oxidizing acid. The oxidation–reduction

TABLE 7-12   PHYSICAL PROPERTIES OF CONCENTRATED CHLOROSULFONIC ACID

Specific gravity	1.79
Boiling point	306°F (152°C)
Freezing point	−112°F (−80°C)
Solubility in water	Decomposes in water

reactions in which this substance participates are similar to those previously noted for sulfuric acid.

Finally, chlorosulfonic acid decomposes easily at elevated temperatures into sulfuryl chloride (SO_2Cl_2) and sulfuric acid. Sulfuryl chloride also decomposes into sulfur dioxide and chlorine. The overall chemical reaction may be illustrated as follows:

$$2ClSO_3H(l) \longrightarrow H_2SO_4(l) + SO_2(g) + Cl_2(g)$$

USDOT regulates the transportation of chlorosulfonic acid as a corrosive material. Packages of chlorosulfonic acid are labeled CORROSIVE; when required, their transport vehicles are placarded CORROSIVE.

7-12 ACETIC ACID

Acetic acid is the substance responsible for the sour taste and sharp odor of vinegar, which contains from 3% to 6% acetic acid by volume. It is also a constituent of the pyroligneous acid obtained from the destructive distillation of wood (see Sec. 6-8). It is used mainly by the chemical industry for the synthesis of substances like acetone, dyes, plastics, and pharmaceuticals.

Acetic acid is generally the most commonly encountered of the organic acids. Its chemical formula is CH_3COOH, represented by the following Lewis structure (Sec. 6-8):

$$CH_3C \overset{\displaystyle O}{\underset{\displaystyle OH}{\diagup \!\!\!\! \diagdown}}$$

Some of its physical properties are noted in Table 7-13.

Industrially, acetic acid is manufactured by a number of methods. However, the most common method involves the chemical union of acetylene and water vapor forming acetaldehyde, followed by the gas-phase catalyzed oxidation of acetaldehyde to acetic acid. These chemical reactions are illustrated by the following equa-

TABLE 7-13   PHYSICAL PROPERTIES OF CONCENTRATED ACETIC ACID

Concentration of CH_3COOH in water	99%–100%
Specific gravity	1.05
Boiling point	244°F (118°C)
Freezing point	61°F (17°C)
Solubility in water	Infinitely soluble
Flash point	109°F (43°C)
Autoignition temperature	800°F (426°C)
Lower explosive limit	4%
Upper explosive limit	16%

tions:

$$C_2H_2(g) + H_2(g) \longrightarrow CH_3CHO(g)$$
$$\text{acetaldehyde}$$
$$2CH_3CHO(g) + O_2(g) \longrightarrow 2CH_3COOH(g)$$

When aqueous solutions of acetic acid are subjected to low temperatures, liquid and solid phases normally result. The liquid phase contains impurities and is discarded, but the solid phase generally contains greater than 99% acetic acid by mass. The latter is called *glacial acetic acid*. When purified in this fashion, it is the concentrated acetic acid of commerce.

At room temperature, concentrated acetic acid is a colorless, pungent liquid. Its vapors are choking and suffocating and may easily damage the lungs and associated bronchial passageways. Severe painful burns may also result from exposure of other skin tissues to concentrated acetic acid and its vapors, particularly the eyes.

Acetic acid is a weak, nonoxidizing acid. However, it is chemically incompatible with all common oxidizing agents. These oxidation–reduction reactions are not generally explosive, but they do evolve considerable heat. Hence, acetic acid should never be stored in the near vicinity of oxidizing agents. Concentrated acetic acid is also a flammable material. Fires involving acetic acid may be readily extinguished by diluting the acid with water; aqueous solutions of acetic acid are generally nonflammable.

USDOT regulates the transportation of acetic acid as a corrosive material. Packages of acetic acid are labeled CORROSIVE; when required, their transport vehicles are placarded CORROSIVE.

In the workplace, OSHA regulates the exposure of employees to glacial acetic acid. OSHA stipulates a permissible exposure limit of 10 ppm.

7-13 SODIUM HYDROXIDE

The most commercially important of the alkaline corrosive materials is *sodium hydroxide,* known commercially as either *lye* or *caustic soda.* It finds countless applications throughout industry, including the purification of petroleum products, reclaiming of rubber, and processing of textiles and paper. The chemical industry uses large quantities of sodium hydroxide to manufacture soap, rayon, and cellophane. In the home, it is used to clean clogged plumbing pipes.

At room temperature, sodium hydroxide is a white solid, commercially available as flakes, pellets, sticks, and granulated or as a concentrated aqueous solution. Its chemical formula is NaOH. Sodium hydroxide is manufactured primarily by passing an electric current through solutions of sodium chloride (brine). This chemical reaction is represented by the following equation:

$$2NaCl(aq) + 2H_2O(l) \xrightarrow{\;e^-\;} 2NaOH(aq) + H_2(g) + Cl_2(g)$$

TABLE 7-14 PHYSICAL PROPERTIES OF SODIUM HYDROXIDE
AND POTASSIUM HYDROXIDE

	NaOH	KOH
Specific gravity	2.13	2.04
Boiling point	2534°F (1390°C)	2408°F (1320°C)
Melting point	599°F (315°C)	680°F (360°C)
Solubility in water	42 g/100 g of H_2O	107 g/100 g of H_2O

Sodium hydroxide is a strong base. Its physical properties are noted in Table 7-14. Its concentrated solutions corrode aluminum, zinc, and lead, as noted in Sec. 7-2, as well as body tissues. In fact, it is the ability of sodium hydroxide to react with water-insoluble animal products (greases and fats) that often makes it effective at clearing clogged plumbing.

Concentrated solutions of sodium hydroxide also corrode glass, especially on prolonged contact, forming sodium silicates. For this reason, sodium hydroxide solutions are not typically stored or transported in glass containers.

Considerable heat, about 42,000 J (10 kcal or 40 Btu) per mole, is evolved when sodium hydroxide dissolves in water. This is sufficient heat to initiate some fires. Hence, during storage, sodium hydroxide should be segregated from flammable and combustible materials.

USDOT regulates the transportation of solid and solution forms of sodium hydroxide as corrosive materials. Packages of these hazardous materials are labeled CORROSIVE; when required, vehicles used for their transportation are placarded CORROSIVE.

In the workplace, OSHA regulates the exposure of employees to sodium hydroxide. OSHA stipulates a permissible exposure limit of 2 mg/m^3.

7-14 POTASSIUM HYDROXIDE

Another alkaline corrosive material of commercial importance is *potassium hydroxide*. It is also known by the common names *caustic potash* and *potash lye*. It is used mainly by the chemical industry for the production of compounds used in fertilizers, photography, and pharmaceutical preparations. It is also used to make soft soaps and liquid soaps. Potassium hydroxide is the electrolyte in alkaline storage batteries.

At room temperature, potassium hydroxide is a white solid. Its chemical formula is KOH. It is prepared by passing an electric current through aqueous solutions of potassium chloride. This chemical reaction is illustrated by the following equation:

$$2KCl(aq) + 2H_2O(l) \xrightarrow{e^-} 2KOH(aq) + Cl_2(g) + H_2(g)$$

It is commercially available as flakes, pellets, sticks, and granulated and as a concentrated aqueous solution. The physical properties of potassium hydroxide are noted in Table 7-14.

Sodium and potassium hydroxides chemically react almost identically. Potassium hydroxide is actually regarded as a stronger base than sodium hydroxide, but the difference is of little consequence here.

USDOT regulates the transportation of the solid and solution forms of potassium hydroxide as corrosive materials. Packages of these hazardous materials are labeled CORROSIVE; when required, vehicles used for their transportation are placarded CORROSIVE.

REVIEW EXERCISES

General Features of Corrosive Materials

7.1. Which of the following methods of storing a piece of steel wool are likely to extend its effectively useful lifetime, and which methods are likely to maximize the rate of its corrosion?
  (a) Immerse the piece of steel wool in water prior to each use.
  (b) Moisten the piece of steel wool prior to each use, but do not totally immerse it in water.
  (c) When not using the piece of steel wool, store it in a container of liquid vegetable oil.
  (d) When not using the piece of steel wool, keep it dry but exposed to the atmosphere.
  (e) When not using the piece of steel wool, keep it dry and sealed in a bottle from the atmosphere.

7.2. Some metals that have been treated with an oxidizing agent become passive to corrosion. Explain how this chemical treatment reduces the extent to which such metals corrode.

7.3. What features of the USDOT definition of a corrosive material are depicted on labels and placards associated with the consignment of corrosive materials for transportation?

General Nature of Acids and Bases

7.4. Illustrate by means of chemical equations how Arrhenius's theory interprets the acidic or basic nature of aqueous solutions that result when the following compounds are dissolved in water: (a) nitric acid; (b) sulfuric acid; (c) potassium hydroxide; (d) ammonium hydroxide.

7.5. A firefighter argues that the degree of corrosiveness of a given acid is *always* proportional to its concentration of solvated hydrogen ions. Why is this not always true?

7.6. Give the chemical formula of the following acids and bases:
  (a) Sulfurous acid
  (c) Hypochlorous acid
  (e) Phosphoric acid
  (g) Magnesium hydroxide
  (i) Ammonium hydroxide
  (b) Nitric acid
  (d) Hydrochloric acid
  (f) Acetic acid
  (h) Sulfuric acid

7.7. Perchloric acid is an example of a "strong" acid. Using a chemical equation, describe what this means.

Corrosivity of Acids and Bases

7.8. The exterior structural material of Notre Dame Cathedral in Paris is limestone. The surface of the limestone now shows signs of powdering and blistering. What chemical process has caused this damage?

7.9. Acids generally corrode metals. Why does USDOT authorize the transportation of some acids, like sulfuric acid, when containerized in steel drums?

7.10. Acid spills may be neutralized with sodium bicarbonate (baking soda). Write the chemical equation for this neutralization reaction.

7.11. Which of the following four alternatives is preferable when cleaning up 5 gal of an unidentified concentrated acid from a concrete floor?
  (a) Soak up the residue with sawdust or excelsior, sweep it into a drum for disposal, and flush the area thoroughly with water.
  (b) Neutralize the acid with soda ash, collect the residue for disposal, and flush the area thoroughly with water.
  (c) Flush the area thoroughly with water without first neutralizing the spill.
  (d) Use a mop to collect the residue in a bucket for ultimate disposal.

7.12. Acid vendors often recommend venting a steel drum containing an acid on initial receipt, and on weekly intervals thereafter, by slowly removing a bung.
  (a) What gas is vented from the drum?
  (b) Why should sources of ignition be absent during this exercise?
  (c) Why is venting such a drum a good safety practice?

7.13. Write the chemical equation that illustrates the following phenomena:
  (a) Tin reacts with aqueous hydrochloric acid.
  (b) Iron(III) oxide "dissolves" in aqueous sulfuric acid.
  (c) Sodium carbonate reacts with aqueous acetic acid.
  (d) Zinc carbonate reacts with aqueous perchloric acid.

The pH Scale

7.14. Chemists often measure the concentration of a solution in terms of the *molarity*. The molarity (M) of a solution means the number of moles of a substance (called the solute) contained in 1 liter of solution. For instance, when 1 mole of a substance is contained in one liter of solution, that solution is said to possess a concentration of $1M$. Suppose the concentration of a solution of hydrochloric acid is $0.001M$. What is its pH?

7.15. Pure water has a pH of 7.0. However, rainwater is usually slightly acidic, even in unpolluted environments. What causes the acidity of rainwater?

Sulfuric Acid

7.16. Describe the mode of action by which concentrated sulfuric acid corrodes skin tissue.

7.17. Explain why pouring concentrated sulfuric acid on sawdust is likely to cause the sawdust to burn, while pouring dilute sulfuric acid on sawdust is unlikely to cause it to burn.

7.18. Why do safety officers often recommend that oxidizing agents be stored in different areas from those used for storage of concentrated sulfuric acid?

Nitric Acid

7.19. Acids generally corrode metals. Why does USDOT authorize the transportation of some acids, like nitric acid, when containerized in stainless steel or aluminum drums?

7.20. What is the most likely reason why USDOT prohibits the use of certain cushioning materials between glass bottles of concentrated nitric acid during transportation, such as hay, excelsior, and ground cork?

7.21. How may skin tissue be identified when it has been scarred by nitric acid?

Hydrochloric Acid

7.22. Chlorine pellets, an oxidizing agent, and muriatic acid are chemical substances used in home swimming pools. Why is it an unsafe practice for homeowners to store these materials in the same area?

7.23. Why are vapors of hydrogen chloride corrosive when inhaled or allowed to contact the eyes?

7.24. Brickmasons sometimes use a dilute solution of hydrochloric acid to remove excess mortar from the surface of bricks. What chemical action is responsible for mortar removal by this acid? [*Hint:* Mortar contains compounds of calcium.]

7.25. Anhydrous hydrogen chloride may be transported in USDOT-approved cylinders. What marking, label, and placard are required when shipping a consignment of anhydrous hydrogen chloride in cylinders?

Perchloric Acid

7.26. What is the most likely reason that USDOT prohibits the use of certain cushioning materials between glass bottles of concentrated perchloric acid during transportation, such as hay, excelsior, and ground cork?

7.27. What is the most likely reason that USDOT prohibits the transportation of perchloric acid whose concentration exceeds 72% by mass?

7.28. Is there danger of forming anhydrous perchloric acid when boiling the concentrated acid under normal atmospheric conditions?

Hydrofluoric Acid

7.29. When storing hydrofluoric acid in glass bottles, why is it essential to first coat their interior surfaces with wax?

7.30. Why do sealed glass bottles containing hydrofluoric acid often burst?

7.31. When working with hydrofluoric acid, what general precaution should be taken to pre-vent it from inflicting biological harm?

Phosphoric Acid

7.32. Why is it an unsafe practice to store phosphoric anhydride in the near vicinity of flammable and combustible materials?

7.33. Phosphoric acid is the only common mineral acid that is added to processed foods. What property does it possess that results in its potential use in foods without simulta-neously causing the consumer to experience adverse health effects?

Chlorosulfonic Acid

7.34. What is the most desirable method to clean up a spill of 1 gallon of chlorosulfonic acid?

Acetic Acid

7.35. Thirteen-gallon glass carboys containing glacial acetic acid are stored in a warehouse with other corrosive materials. During a fire in the warehouse, the temperature of the ambient environment is heated to 1000°F (540°C).
(a) Assuming the carboys rupture, is acetic acid likely to burn at this temperature?
(b) How should fires involving acetic acid be fought?

Sodium and Potassium Hydroxides

7.36. Why do the directions on a container of lye inform the user to avoid aluminum con-tainers when dissolving the contents in water?

7.37. Some commercial oven cleaners for intended use in home ovens contain sodium hy-droxide in an aerosol. Why is sodium hydroxide an effective oven cleaner?

7.38. When concentrated potassium hydroxide solutions are stored in glass bottles, they ap-pear milky after some time. Why is this so?

8

Chemistry of Some Water-Reactive Materials

Water may react or interact with hazardous materials in a number of potentially dangerous ways. Hence, when water is used to extinguish chemical fires, some degree of precaution is always warranted. Water may react with a substance to form a product that is either flammable, explosive, toxic, or corrosive. The process whereby water causes the decomposition of a substance is called *hydrolysis*. It is generally represented by an equation like the following:

$$A + H_2O(l) \longrightarrow C + D$$

Here, A represents a reactant and C and D represent the products of the reaction; water is also a reactant.

Not all hydrolysis reactions result in the formation of products that have potentially dangerous properties. But, if either C or D is a gas or vapor that achieves a concentration within its flammable or explosive range or any substance that causes toxic or corrosive effects, the contact of A with water must generally be avoided. For instance, water chemically reacts with certain metals, generating hydrogen, which catches fire and even initiates combustion of the metal. Such metals constitute the fuels of class D fires. Some are so chemically reactive that they are spontaneously combustible.

Water may cause or aggravate a potentially hazardous situation in other ways. For instance, some substances absorb atmospheric water vapor. Such substances are said to be *hygroscopic*. Two examples of hygroscopic substances are sodium hydroxide and concentrated sulfuric acid. On the positive side, hygroscopic substances may be used commercially as drying agents or desiccants; but on the negative side, hygroscopic substances may absorb so much water that they overflow unattended containers. Concentrated sulfuric acid, for instance, actually overflows its container when left to stand in humid air.

Certain substances are capable of spontaneously igniting when exposed to air; they are said to be *pyrophoric*. USDOT defines a *pyrophoric liquid* as any liquid that ignites spontaneously in dry or moist air at or below 130.1°F (54.5°C).

USEPA characterizes RCRA hazardous wastes (Sec. 1-5) by means of a property called *reactivity*. A chemical waste may exhibit the characteristic of reactivity in several ways, including the following

1. It reacts violently with water.
2. It forms potentially explosive mixtures with water.
3. When mixed with water, it generates toxic gases, vapors, or fumes in a quantity sufficient to present a danger to human health or the environment.

Wastes exhibiting the characteristic of reactivity possess the USEPA hazardous waste number D003.

8-1 ALKALI METALS

The alkali metals are lithium, sodium, potassium, cesium, rubidium and francium. Cesium, rubidium, and francium have little or no commercial value and thus are unlikely ever to be encountered. On the other hand, lithium, sodium, and potassium are such industrially important metals that any could be encountered in a transportation mishap, chemical plant, or elsewhere.

The alkali metals are the most chemically reactive of all metals. This special chemical reactivity gives rise to their potential hazards. As a group of elements, these metals are commonly considered dangerous fire risks, since most of them may self-ignite and all are water reactive. For this latter reason, the alkali metals are ordinarily transported and stored under mineral oil. USDOT regulates their transportation as dangerous-when-wet materials. Packages of the alkali metals are labeled DANGEROUS WHEN WET; vehicles shipping such packaged consignments are placarded DANGEROUS WHEN WET.

Some physical properties of the alkali metals are compared in Table 8-1.

Lithium

Lithium is a soft, silvery metal. It is the least dense solid element, a property that causes metallic lithium to float on petroleum products. Lithium also possesses an extraordinarily high specific heat, which would ordinarily make it an excellent coolant. However, the cost and corrosiveness of liquid lithium cause this application to be relatively impractical. Nevertheless, lithium is valuable commercially in the production of porcelain, ceramics, castings, fungicides, bleaching agents, pharmaceuticals, and greases. It is also used to degas copper alloys and is itself a component of a lightweight magnesium alloy.

Like all alkali metals, lithium is a water-reactive element. It replaces the hydrogen in water, as the following equation illustrates:

TABLE 8-1 PHYSICAL PROPERTIES OF THE ALKALI METALS

	Lithium	Sodium	Potassium
Density at 68°F (20°C)	33.3 lb/ft³ (0.534 g/mL)	60.6 lb/ft³ (0.972 g/mL)	51.1 lb/ft³ (0.819 g/mL)
Melting point	354°F (179°C)	207.5°F (97.5°C)	147°F (63.7°C)
Boiling point	2403°F (1317°C)	1621°F (883°C)	1400°F (760°C)
Heat of fusion	186 Btu/lb (103.2 cal/g)	49.0 Btu/lb (27.2 cal/g)	26.3 Btu/lb (14.6 cal/g)
Heat of vaporization	8430 Btu/lb (4680 cal/g)	1809 Btu/lb (1005 cal/g)	893 Btu/lb (496 cal/g)
Heat capacity	0.90 Btu/lb °F (0.90 cal/g °C)	0.30 Btu/lb °F (0.30 cal/g °C)	0.19 Btu/lb °F (0.19 cal/g °C)

$$2Li(s) + 2H_2O(l) \longrightarrow 2LiOH(aq) + H_2(g)$$

This chemical reaction occurs more slowly, however, than the water reactions of other alkali metals. Lithium is the only alkali metal whose melting point is greater than the boiling point of water. Consequently, bulk lithium remains as the solid state of matter during the course of its reaction with water. The other alkali metals generally melt, which causes a larger surface area of the metals to be exposed for further reaction with water.

The reaction between lithium and water also differs from the reactions of the other alkali metals with water in that the evolved hydrogen does not usually self-ignite. This is due to the fact that the reaction occurs slowly enough that the accompanying heat of reaction dissipates.

Solid *bulk* lithium does not ordinarily ignite spontaneously at room conditions, even when the metal is exposed to an atmosphere of pure oxygen. In absolutely dry air, lithium does not burn spontaneously. Even molten lithium oxidizes so slowly that it may be poured in air without losing its bright luster. However, at temperatures exceeding 400°F (200°C), lithium glows in an atmosphere of oxygen with a characteristic crimson color, forming lithium oxide. This oxidation is illustrated by the following equation:

$$4Li(s) + O_2(g) \longrightarrow 2Li_2O(s)$$

By contrast with the other alkali metals, lithium also may combine directly with atmospheric nitrogen, forming lithium nitride. This reaction is illustrated by the following equation:

$$6Li(s) + N_2(g) \longrightarrow 2Li_3N(s)$$

Thus, lithium fires may not be extinguished by blanketing them with nitrogen.

Sodium

Sodium is the sixth most abundant element by mass in the earth's crust and oceans (see Table 3-1). It is also a soft, silvery bright metal. Sodium is the most common alkali metal. In bulk, metallic sodium is available commercially as blocks or bricks, like those shown in Fig. 8-1.

Figure 8-1 Metallic sodium is often encountered commercially in the shape of bricks, like those shown here. Individual sodium bricks should only be handled with dry gloves made of loose-fitting canvas or similar material, previously dusted with dry soda ash to remove any adhering moisture. Such bricks are ordinarily immersed in a neutral oil within 55-gallon steel drums. These containers should be stored in a dry, fireproof building or room used exclusively for sodium storage. (Courtesy of E. I. du Pont de Nemours & Company, Wilmington, Delaware)

In the past, most metallic sodium was used industrially in conjunction with the production of tetraethyl and tetramethyl lead. However, this use has been sharply curtailed, since these components of motor vehicle fuels are being phased out of use (Sec. 8-3). Today, most metallic sodium is used in connection with the production of titanium metal and as a catalyst for synthetic rubber production. Sodium metal is also used as a reducing agent in some organic synthetic procedures and is usually the raw material used for synthesizing highly reactive sodium compounds like sodium peroxide and sodium hydride.

Metallic sodium is also used in two modified forms: sodium amalgam and with potassium as an alloy. An *amalgam* is a special alloy of a metal with mercury. Reactions of sodium proceed less vigorously when using sodium amalgam as compared to metallic sodium.

Metallic sodium ignites spontaneously in air at room temperature with a characteristic yellow flame, forming sodium oxide. This oxidation is illustrated by the following equation:

$$4Na(s) + O_2(g) \longrightarrow 2Na_2O(s)$$

The surface of the metal becomes coated with sodium oxide and, in a moist atmosphere, sodium hydroxide and sodium carbonate. In fact, the oxidation reaction is most likely initiated by water, since sodium is not oxidized in absolutely dry air. By contrast with metallic lithium, sodium does not unite with atmospheric nitrogen.

Sodium burns in an atmosphere of pure oxygen to form a mixture of sodium oxide and sodium peroxide. The formation of sodium peroxide is illustrated by the following equation:

$$2Na(s) + O_2(g) \longrightarrow \underset{\text{sodium peroxide}}{Na_2O_2(s)}$$

Sodium peroxide is an oxidizing agent (Sec. 8-6).

When large pieces of sodium react with water, the reaction evolves much heat, causing the metallic sodium to melt, spatter, and spontaneously ignite. With small pieces, spontaneous ignition may not occur, especially if the pieces are so small that they can float about on the water surface and if the heat of reaction is absorbed by the water. However, in ordinary circumstances, the evolved hydrogen bursts into flame. The production of hydrogen is denoted by the following equation:

$$2Na(s) + 2H_2O(l) \longrightarrow 2NaOH(s) + H_2(g)$$

Potassium

Potassium in the seventh most abundant element by mass in the earth's crust and oceans (see Table 3-1). It is also a soft, silvery metal. Most potassium used commercially is for the production of a sodium–potassium alloy called NAK (pronounced "nack"), a heat exchange fluid.

Potassium exhibits the chemistry of the alkali metals, but is much more chemically reactive than either lithium or sodium. On exposure to air at room temperature, metallic potassium burns with a characteristic purple flame, principally forming potassium oxide. This reaction is illustrated by the following equation:

$$4K(s) + O_2(g) \longrightarrow 2K_2O(s)$$

Potassium burns in an atmosphere of pure oxygen, forming a mixture of potassium peroxide and potassium superoxide. Superoxides, more properly called *hyperoxides,* are metallic compounds containing the O_2^- ion. They are very reactive oxidizing agents. The chemical equation illustrating the formation of potassium superoxide is the following:

$$K(s) + O_2(g) \longrightarrow KO_2(s)$$
$$\text{potassium superoxide}$$

Superoxides are hydrolyzed readily, forming oxygen and hydrogen peroxide, as illustrated by the following equation:

$$2KO_2(s) + 2H_2O(l) \longrightarrow 2KOH(aq) + O_2(g) + H_2O_2(aq)$$

This reaction occurs with such ease that it may be used as a source of oxygen in self-contained breathing apparatus (Sec. 9-15). Only a trace of potassium superoxide remains when potassium burns in air, since it reacts so rapidly with atmospheric moisture. By contrast, however, the superoxides of lithium and sodium are extremely difficult to prepare.

As noted earlier, metallic potassium possesses an exceptional chemical reactivity. Chemists have conjectured that this reactivity is due to minute amounts of potassium superoxide, which often accompany potassium metal. Even this minute quantity is an extremely hazardous material, since the substance is such a potent oxidizing agent. Supporting this conjecture is the fact that potassium detonates on contact with bromine, while lithium and sodium react with bromine at room temperature without incident. An explosion also occurs when potassium contacts kerosene, while lithium and sodium may be safely stored under kerosene.

The reaction between potassium and water is also relatively violent, again most likely due to the presence of small amounts of potassium superoxide. The reaction is illustrated by the following equation:

$$2K(s) + 2H_2O(l) \longrightarrow 2KOH(s) + H_2(g)$$

The evolved hydrogen almost always bursts into flame.

Fighting Alkali Metal Fires

Since the alkali metals react with water to form flammable hydrogen, water will not extinguish alkali metal fires. Carbon dioxide and the Halon agents are equally ineffective extinguishers of alkali metal fires. These substances react furiously with the alkali metals, producing clouds of carbon, as the following equations illustrate:

$$4Na(s) + CO_2(g) \longrightarrow 2Na_2O(s) + C(s)$$

$$4Na(s) + CCl_4(l) \longrightarrow 4NaCl(s) + C(s)$$

In the second equation, carbon tetrachloride (Halon 104) is used to represent a typical Halon agent (see footnote, p. 125).

Since these common fire extinguishers are useless, it is apparent that a special extinguisher must be employed on alkali metal fires. To extinguish sodium or potassium fires, the use of dry powder fire extinguishers (p. 128), like that illustrated in Fig. 8-2, is normally recommended. Dry powder contains graphite, a solid. Hence, it may be sprinkled on small fires or shoveled on large fires. Graphite smothers the fire and conducts heat from the immediate fire zone.

When dry powder is used to extinguish an alkali metal fire, some caution needs to be exercised when disposing of the remaining residue. A significant quantity of unburned metal may remain under the extinguisher; or small bits and pieces of metal may remain dispersed within the bulk of the extinguisher. Either situation is potentially dangerous: Exposure of segments of this residue to atmospheric moisture is likely to cause the fire to rekindle.

Even when all the metal has been consumed by chemical action, a hazard may still exist. At the high temperatures accompanying alkali metal fires, metallic carbides often form. For instance, sodium and graphite unite to form sodium carbide, as the following equation represents:

$$2Na(s) + 2C(s) \longrightarrow Na_2C_2(s)$$

Such metallic carbides remain as components of the fire residue. They pose a fire risk, since metallic carbides are water-reactive solids; their reaction with water forms flammable acetylene (Sec. 8-7). Hence, even if all the alkali metal has burned, these metallic carbides may react with atmospheric moisture. Acetylene readily catches fire, which may cause the remaining alkali metal to reignite. To as-

Figure 8-2 Lith-X dry powder, a graphite-based fire extinguishing agent. In general, the use of Lith-X is recommended for extinguishing class D fires. (Courtesy of Ansul Fire Protection, Marinette, Wisconsin)

sure that an unwanted fire does not rekindle, small quantities of the fire residue must be reacted with water in a controlled fashion and in a location removed from flammable and combustible materials.

Lithium fires are unique as class D fires, and additional care must be taken to properly extinguish them. The ideal fire extinguisher for use on lithium fires is a dry chemical containing lithium chloride. This dry chemical smothers lithium fires without subsequent formation of any other chemical substance.

Dry powder containing graphite may also be used to extinguish lithium fires. *Lith-X,* shown in Fig. 8-3, is the trademark for a graphite-based dry chemical suitable for use on lithium fires. However, some lithium carbide remains in the residue following the extinguishing of a lithium fire. As previously noted, this residue must be disposed of cautiously to ensure that a fire does not rekindle.

Dry chemicals containing sodium salts are *not* recommended for use on lithium fires, since they chemically react with lithium. For instance, when dry chemicals containing either sodium carbonate or sodium chloride are used on lithium fires, metallic sodium is liberated. The sodium is equally as hazardous as lithium. This formation of metallic sodium is represented by the following equations:

$$Na_2CO_3(s) + Li(s) \longrightarrow Li_2O(s) + CO_2(g) + 2Na(s)$$

$$NaCl(s) + Li(s) \longrightarrow LiCl(s) + Na(s)$$

Obviously, these dry chemicals do not effectively extinguish lithium fires.

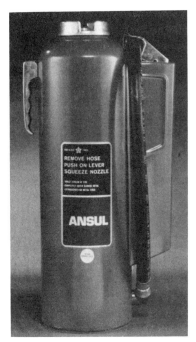 

Figure 8-3   A commercially available portable fire extinguisher for class D fires (left) and an illustration of its use (right). (Courtesy of Ansul Fire Protection, Marinette, Wisconsin)

Finally, it should be noted that the atmosphere of these class D fires may be extremely caustic due to the formation of the corresponding metallic oxide, hydroxide, or carbonate. Particulates of such compounds are likely to be present in the smoke accompanying alkali metal fires. Their inhalation may cause adverse health effects ranging from minor irritation and congestion of the nose, throat, or bronchi to severe lung injury.

8-2 MAGNESIUM, ZIRCONIUM, TITANIUM, ALUMINUM, AND ZINC

Magnesium, zirconium, titanium, aluminum, and zinc possess a hazardous feature in common: When they are in the form of a dust or powder, these metals may spontaneously explode. Consequently, the dust or powder of these metals presents a serious explosion and fire risk.

The purity of these metals is important as it relates to the risk of fire or explosion. If the surface of the metal has been coated by its oxide (which is relatively unreactive), the explosion and fire risk decidedly diminishes. Other factors also affect the potential explosion and fire risk associated with these metals, such as particle size, distribution and dispersion of the particles, moisture content, ignition temperature of the metal, and the amount of adsorbed gases, particularly oxygen.

TABLE 8-2 PHYSICAL PROPERTIES OF SEVERAL WATER-REACTIVE METALS

	Magnesium	Zirconium	Titanium
Density at 68°F (20°C)	108.57 lb/ft³ (1.74 g/mL)	405.58 lb/ft³ (6.5 g/mL)	280.79 lb/ft³ (4.5 g/mL)
Melting point	1204°F (651°C)	3362°F (1850°C)	3146°F (1800°C)
Boiling point	2025°F (1107°C)	5252°F (3578°C)	5432°F (3262°C)
Heat of fusion	160.2 Btu/lb (88.9 cal/g)	108.1 Btu/lb (60 cal/g)	188.1 Btu/lb (104.4 cal/g °C)
Heat capacity	0.25 Btu/lb °F (0.25 cal/g °C)	0.068 Btu/lb °F (0.068 cal/g °C)	0.10 Btu/lb °F (0.10 cal/g °C)
Ignition temperature of dust of metal (with an ignition source)	968°F (520°C)	70°F (21°C)	896°F (480°C)

	Aluminum	Zinc	
Density at 68°F (20°C)	168.47 lb/ft³ (2.7 g/mL)	461.74 lb/ft³ (7.4 g/mL)	
Melting point	1220°F (660°C)	787°F (419°C)	
Boiling point	3733°F (2056°C)	1661°F (905°C)	
Heat of fusion	170.4 Btu/lb (94.5 cal/g)	44.0 Btu/lb (24.4 cal/g)	
Heat capacity	0.21 Btu/lb °F (0.21 cal/g °C)	—	
Ignition temperature of dust of metal (with an ignition source)	1193°F (645°C)	—	

Each of these five pyrophoric metals is also water reactive, particularly when the metal is very hot, as when it is burning. The metals cause the release of flammable hydrogen from water at varying rates. The rate of hydrogen production depends on factors like the purity of the metal and its state of subdivision. These chemical reactions of the metals and water proceed exothermically, and the hydrogen that forms often spontaneously bursts into flame. In certain instances it is likely that the spontaneous ignition of these metallic dusts is first triggered by the initial combustion of hydrogen, generated by the action of atmospheric moisture with the metal.

Some physical properties of these metals are noted in Table 8-2.

Magnesium

Magnesium is the eighth most abundant element by mass in the earth's crust and oceans (see Table 3-1). It is an exceptionally lightweight metal and is therefore often employed for the construction of aircraft, racing cars, transportable machinery, engine parts, automobile frames and bumpers, and other items for which the mass of the object is an important factor. Because of the popularity of magnesium as a structural metal, it is commercially available in all sizes from powders to large ingots.

A dull film of magnesium oxide on the surface of magnesium often makes the metal appear unreactive. But magnesium is actually a very reactive metal when exposed to an ignition source. It burns readily in air with a brilliant, blinding flame, as Fig. 8-4 illustrates. When magnesium burns in air, only about 75% by mass combines with oxygen to form magnesium oxide. The remaining 25% combines with atmospheric nitrogen to form magnesium nitride. These chemical reactions are repre-

Figure 8-4 Bumpers, chassis, wheel frames, and other automotive parts are often made of a magnesium alloy. When they become engulfed in a fire, such as the car-dealership structural fire illustrated here, the magnesium itself burns, producing an intensely brilliant, hot flame.

sented by the following equations:

$$2Mg(s) + O_2(g) \longrightarrow 2MgO(s)$$

$$3Mg(s) + N_2(g) \longrightarrow Mg_3N_2(s)$$

Burning magnesium generates brilliant white flames and intense heat. This combustion is put to use commercially in camera photoflash bulbs to provide a source of instantaneous ultraviolet light. Firefighters should be wary of magnesium fires, since the brilliance of these flames may damage the retina of the eyes. Firefighters should also avoid breathing the smoke evolved during magnesium fires. This smoke contains particulates of magnesium oxide, which react with water in the bronchial passageways to form magnesium hydroxide. This latter substance is very caustic and causes considerable discomfort and localized lung injury.

The burning of magnesium metal consitutes a class D fire. The recommended fire extinguishers on magnesium fires are either dry sand, Met-L-X, or G-1 graphite. Water is not generally recommended as a fire extinguisher on magnesium fires. While magnesium reacts slowly with water at room temperature, when either the magnesium or water is hot, a chemical reaction proceeds vigorously to produce hydrogen, as the following equation illustrates:

$$Mg(s) + 2H_2O(l) \longrightarrow Mg(OH)_2(s) + H_2(g)$$

Thus, to be effective as a fire extinguisher on magnesium fires, water must be applied in volumes that deluge the fire and cool the metal. When water is applied too slowly on magnesium fires, hydrogen is produced, which burns and rekindles the fire. This means that relatively large volumes of water must be applied within a short time on magnesium fires. Since this is often impossible, water is not generally recommended as a fire extinguisher on magnesium fires.

Hot magnesium also reacts with carbon dioxide, as the following equation illustrates:

$$2Mg(s) + CO_2(g) \longrightarrow 2MgO(s) + C(s)$$

Since this reaction is exothermic, heat is not removed when carbon dioxide is used on magnesium fires, and combustion is sustained. The use of Halon agents as fire extinguishers is similarly ineffective on magnesium fires for the same reason. For instance, Halon 104 reacts to form magnesium chloride and carbon dioxide, as the following equation illustrates:

$$2MgO(s) + CCl_4(g) \longrightarrow 2MgCl_2(s) + CO_2(g)$$

USDOT regulates the transportation of magnesium as a flammable solid when more than 50% of the metal exists as pellets, turnings, or ribbons. Packages of these forms of the metal are labeled FLAMMABLE SOLID and, when required, vehicles transporting them are placarded FLAMMABLE SOLID. USDOT regulates the transportation of magnesium powder as a dangerous-when-wet material. Packages are labeled DANGEROUS WHEN WET and SPONTANEOUSLY COMBUSTIBLE.

Vehicles shipping packages of magnesium powder are placarded DANGEROUS WHEN WET. On the other hand, USDOT forbids transportation of magnesium *dross*, a scum that forms on the surface of the molten metal.

Zirconium

Powder or sponge

Zirconium is the twentieth most abundant element by mass in the earth's crust and oceans (see Table 3-1). It has become a very useful metal in contemporary times, but its main use is limited almost entirely to the nuclear and steel industries. In the nuclear industry, zirconium is used in reactor cores and to clad uranium fuel rods. In the steel industry, zirconium is used as a *deoxidizer*, that is, as a substance that removes oxygen from molten steel. Some zirconium is also used commercially in camera photoflash bulbs.

Finely divided zirconium (as a dust or powder) is one of the most pyrophoric metals known. This is due to the fact that zirconium dust possesses an autoignition temperature of only 70°F (21°C). Zirconium dust has been known to spontaneously ignite with explosive force at room temperature from heat generated by the mere friction of its component particles on one another. Hence, zirconium dust poses a very serious fire and explosion risk.

When zirconium burns in air, the fire provides an extremely brilliant white flame. A mixture of zirconium oxide and zirconium nitride form as the following equations illustrate:

$$Zr(s) + O_2(g) \longrightarrow ZrO_2(s)$$

$$2Zr(s) + N_2(g) \longrightarrow 2ZrN(s)$$

Zirconium does not react with water at room temperature. Hence, zirconium may be stored under water (as well as other liquids) to reduce its susceptibility to spontaneous combustion. Hot, burning zirconium, however, does react with water, displacing hydrogen. This chemical reaction is illustrated by the following equation:

$$Zr(s) + 2H_2O(l) \longrightarrow ZrO_2(s) + 2H_2(g)$$

Fires involving the burning of zirconium metal are class D fires. Hence, it is generally recommended that dry chemical soda ash, G-1 graphite, or Met-L-X be used as fire extinguishers on zirconium fires. Water may only be used as a fire extinguisher when applied in amounts sufficient to cool the metal. Furthermore, carbon dioxide and Halon agents react chemically with hot zirconium, which makes the use of either of these fire extinguishers ineffective on zirconium fires.

USDOT regulates the transportation of zirconium metal in the following six forms:

1. Zirconium, dry [coiled wire, finished metal sheets, strip (thinner than 254 micrometers, but not thinner than 18 micrometers)]
2. Zirconium, dry (finished sheets, strip or coil wire)

3. Zirconium powder, dry [(a) mechanically produced, particle size between 3 and 53 micrometers); (b) chemically produced, particle size between 10 and 840 micrometers]

4. Zirconium powder, wetted [with not less than 25% water (visible excess of water must be present): (a) mechanically produced, particle size less than 53 micrometers; (b) chemically produced, particle size less than 840 micrometers]

5. Zirconium scrap

6. Zirconium suspended in a liquid

Transportation of the first and fourth forms is regulated as a flammable solid. Packages are labeled FLAMMABLE SOLID; when required, their transport vehicles are placarded FLAMMABLE SOLID. Transportation of the second, third and fifth forms is regulated as a spontaneously combustible material. Packages are labeled SPONTANEOUSLY COMBUSTIBLE; when required, their transport vehicles are placarded SPONTANEOUSLY COMBUSTIBLE. Finally, transportation of the sixth form is regulated as a FLAMMABLE LIQUID. Packages are labeled FLAMMABLE LIQUID; when required, their transport vehicles are placarded FLAMMABLE LIQUID.

Titanium

Titanium is the tenth most abundant element by mass in the earth's crust and oceans (see Table 3-1). Like magnesium, it possesses a low density relative to most other metals. Furthermore, it is as strong as steel, but 45% lighter in mass. Titanium thereby has the potential of being a promising structural material. In fact, the combination of lightness and strength is the reason titanium is used for the manufacture of aircraft parts, jet engines, and missiles. However, production of metallic titanium is very costly, which limits its common use. Titanium is also resistant to corrosion by seawater and thus is often used in ships and underwater machinery.

Titanium is chemically similar to zirconium. Finely divided titanium is a pyrophoric metal and is considered a dangerous fire and explosion risk. Even the bulk metal autoignites at only 1300°F (700°C). When it burns in air, hot titanium forms a mixture of its oxide and nitride, as the following equations illustrate:

$$Ti(s) + O_2(g) \longrightarrow TiO_2(s)$$
$$2Ti(s) + N_2(g) \longrightarrow 2TiN(s)$$

Titanium fires are class D fires and should be extinguished with dry chemical soda ash, Met-L-X, or G-1 graphite. The use of water, carbon dioxide, and Halon agents is ineffective on titanium fires for reasons similar to those indicated earlier for the ineffectiveness of these extinguishers on zirconium fires.

USDOT regulates the transportation of the following forms of titanium metal:

1. Titanium powder, dry [(a) mechanically produced, particle size between 3 and

53 micrometers; (b) chemically produced, particle size between 10 and 840 micrometers]

2. Titanium powder, wetted [with not less than 25% water (a visible excess of water must be present): (a) mechanically produced, particle size less than 53 micrometers; (b) chemically produced, particle size less than 840 micrometers]

3. Titanium sponge granules or titanium sponge powders

Transportation of the first form is regulated as a spontaneously combustible material. Packages are labeled SPONTANEOUSLY COMBUSTIBLE; when required, their transport vehicles are placarded SPONTANEOUSLY COMBUSTIBLE. Transportation of the second and third forms is regulated as a flammable solid. Packages are labeled FLAMMABLE SOLID; when required, their transport vehicles are placarded FLAMMABLE SOLID.

Aluminum

Aluminum is undoubtedly more familiar to nonscientists than any other metal discussed in this section. It is the third most abundant element in the earth's crust and oceans (see Table 3-1). Aluminum is lighter in mass than either titanium or zirconium, but not as lightweight as magnesium. Aluminum may be rolled into very thin sheets, a property that makes it useful for a variety of domestic and industrial purposes. For instance, numerous structural components of buildings are made of aluminum: siding, eaves, screens, and window and door frames. However, aluminum melts at only 660°C (1220°F). Hence, unprotected aluminum used in buildings and other structures is likely to collapse during many fires.

Pure aluminum is one of the most chemically reactive metals. However, aluminum oxide readily forms on the surface of aluminum and protects the metal from further chemical attack. When we encounter aluminum, its surface has usually been oxidized. It is this oxide coating that often causes individuals to regard aluminum as a relatively nonhazardous metal. However, the coating is not totally impervious to chemical attack. Seawater, for instance, rapidly corrodes aluminum. Furthermore, heavy deposits of the oxide on aluminum electrical wiring may cause a fire hazard. The oxide restricts the flow of electrical current, which may cause the wire to become overheated and cause fire.

The chemical affinity of metallic aluminum for oxygen may be demonstrated by the reaction between powdered aluminum and iron(III) oxide. Some features of a laboratory demonstration of this reaction are illustrated in Fig. 8-5. A mixture by mass of 27% powdered aluminum and 73% iron(III) oxide is called *thermite*. When enclosed in a metal cylinder, it potentially serves as an incendiary bomb. When thermite is heated to the combustion temperature of aluminum, a chemical reaction occurs, called the *thermite reaction,* as illustrated by the following equation:

$$2Al(s) + Fe_2O_3(s) \longrightarrow 2Fe(l) + Al_2O_3(s)$$

The thermite reaction evolves 184 kcal (770,000 J or 730 Btu) and produces temper-

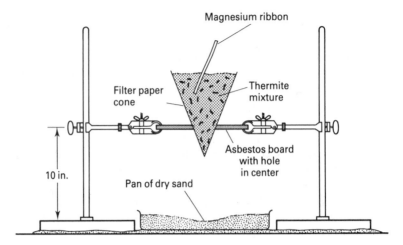

Figure 8-5 This laboratory demonstration should only be conducted by skilled individuals as it constitutes a dangerous fire risk. The thermite mixture [powdered aluminum and iron(III) oxide] is contained in a paper cone over a pan of dry sand, which protects the table top against possible damage. Ignition of the magnesium ribbon serves to initiate the decomposition of iron(III) oxide. The resulting chemical reaction makes a striking spectacle as molten iron drips from the cone. Since sparks are likely to jettison several feet, a portable fire extinguisher should be readily available.

atures of approximately 2200°C (3992°F). Iron melts at 1535°C (2795°F). Hence, the iron produced by this chemical reaction is a white-hot liquid, which may readily spread fire to adjacent locations.

Pure aluminum metal as the dust or powder is pyrophoric and represents another serious fire and explosion risk. When hot aluminum burns in air, it forms a mixture of its oxide and nitride, as the following equations illustrate:

$$4Al(s) + 3O_2(g) \longrightarrow 2Al_2O_3(s)$$
$$2Al(s) + N_2(g) \longrightarrow 2AlN(s)$$

Aluminum fires should be extinguished with dry chemical soda ash, Met-L-X, or G-1 graphite. The use of water, carbon dioxide, or the Halon agents as fire extinguishers is not recommended on aluminum fires.

USDOT regulates the transportation of two forms of aluminum powder: (1) aluminum powder, coated (not less than 20% aluminum powder, particle size less than 250 micrometers), and (2) aluminum powder, uncoated. Transportation of the coated form is regulated as a flammable solid. Packages of this form of the metal are labeled FLAMMABLE SOLID; when required, their transport vehicles are placarded FLAMMABLE SOLID. The transportation of the second form is regulated as a dangerous-when-wet material. Packages are labeled DANGEROUS WHEN WET, and their transport vehicles are placarded DANGEROUS WHEN WET.

Zinc

Zinc is rarer than all the other metals discussed in this section, but it is generally more commonly known, with the possible exception of magnesium and aluminum. In fact, zinc has been known since times before Christ. This early use of zinc is associated with the ease by which zinc may be extracted from its naturally occurring ores, which is untrue of zirconium and titanium.

Zinc metal is commonly used for a number of purposes. It is used with copper to form the common alloy *brass*. It is also used to coat iron and protect it from corrosion; zinc-coated iron is said to have been *galvanized*. It is also used in dry-cell batteries and structural materials. Zinc dust is used as a commercial paint pigment.

Hot zinc powder or dust, sometimes called zinc ashes, is a pyrophoric metal. Thus, it represents a fire and explosion risk. Hot zinc powder is most likely encountered during its manufacture. When zinc burns in air, it appears to form only zinc oxide and not zinc nitride:

$$2Zn(s) + O_2(g) \longrightarrow 2ZnO(s)$$

Spontaneous ignition of zinc dust is catalyzed by the presence of moisture. Powdered zinc reacts slowly with water to produce hydrogen, as the following equation illustrates:

$$Zn(s) + 2H_2O(l) \longrightarrow Zn(OH)_2(s) + H_2(g)$$

Hence, the catalysis is most likely associated with the presence of a flammable mixture of hydrogen and air.

Zinc fires are class D fires. They should be extinguished with either dry chemical soda ash, Met-L-X, or G-1 graphite. The use of water, carbon dioxide, and Halon agents is not recommended. Firefighters should take proper precaution to avoid inhaling the smoke generated from zinc fires, as zinc oxide and all other compounds of zinc are poisonous.

USDOT regulates the transportation of metallic zinc in the following two forms: zinc ashes and zinc powder. Both forms are regulated as dangerous-when-wet materials. Packages of zinc ashes are labeled DANGEROUS WHEN WET, but packages of zinc powder are labeled DANGEROUS WHEN WET and SPONTANEOUSLY COMBUSTIBLE. In both instances, their transport vehicles are placarded DANGEROUS WHEN WET.

8-3 ORGANOMETALLIC COMPOUNDS

An industrially important group of compounds is comprised of metal atoms bonded directly to carbon atoms; these substances are collectively called *organometallic compounds*. They are frequently used as polymerization catalysts (Sec. 12-1) and to synthesize certain organic compounds whose preparation is otherwise difficult or impossible. Approximately 50 organometallic compounds are available commercially,

TABLE 8-3 PROPERTIES OF SEVERAL ORGANOMETALLIC COMPOUNDS

Chemical name	Formula	Properties
Trimethylaluminum	$Al(CH_3)_3$	Colorless liquid; decomposes in water and detonates; freezes at 59.7°F (15.4°C); inflames spontaneously in air; vapor decomposes above 300°F (150°C).
Dimethylcadmium	$Cd(CH_3)_2$	Colorless liquid having a musty odor; explosively decomposes above 212°F (100°C); freezes at 24.4°F (−4.2°C); pyrophoric; decomposes in water.
Tetramethyltin	$Sn(CH_3)_4$	Colorless, highly volatile liquid; insoluble in water; stable to moisture and air; freezes at −63.4°F (−53°C); thermally stable to 750°F (400°C).
Tri(isobutyl)aluminum	$Al(C_4H_9)_3$	Colorless liquid; pyrophoric; freezes at 33.8°F (1°C); thermally unstable above 165°F (75°C); water reactive.

containing from one to ten carbon atoms per molecule. Some properties of four of them are listed in Table 8-3.

Diethylzinc

The first organometallic compound ever prepared was diethylzinc. Its chemical formula is $(C_2H_5)_2Zn$. Diethylzinc is a foul-smelling, colorless liquid that immediately inflames on exposure to air. It was prepared by reacting ethyl iodide with zinc filings. Ethyl zinc iodide (C_2H_5ZnI) first forms, which when heated strongly, decomposes and forms diethylzinc. The following equations illustrate these reactions:

$$C_2H_5I(l) + Zn(s) \longrightarrow C_2H_5ZnI(l)$$

$$2C_2H_5ZnI(l) \longrightarrow (C_2H_5)_2Zn(l) + ZnI_2(s)$$

Diethylzinc is used to synthesize certain organic compounds, especially other organometallic compounds, and to catalyze the polymerization of ethenes (Sec. 12-4). It is typical of several organometallic compounds in that it is simultaneously pyrophoric and water reactive. Diethylzinc burns spontaneously in air, forming zinc oxide, carbon dioxide, and water, and it reacts violently with water, forming ethane. These chemical reactions are noted by the following equations:

$$(C_2H_5)_2Zn(l) + 7O_2(g) \longrightarrow ZnO(s) + 4CO_2(g) + 5H_2O(g)$$

$$(C_2H_5)_2Zn(l) + 2H_2O(l) \longrightarrow Zn(OH)_2(s) + C_2H_6(g)$$

Since many organometallic compounds are pyrophoric and water reactive, they may not be handled in the presence of air or moisture without rapid decomposition.

USDOT regulates the transportation of diethylzinc as a spontaneously combustible material. When shipped, USDOT requires that packages of this substance be labeled SPONTANEOUSLY COMBUSTIBLE. When required, the transport vehicles shipping these packages are placarded SPONTANEOUSLY COMBUSTIBLE.

Tetraethyl lead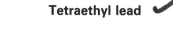

Perhaps the most well-known organometallic compound is tetraethyl lead. Its chemical formula is $(C_2H_5)_4Pb$. Tetraethyl lead is a colorless, poisonous, oily liquid having a pleasant odor. It may be prepared by reacting ethyl chloride with a sodium–lead alloy, as the following equation notes:

$$2C_2H_5Cl(g) + 4NaPb(s) \longrightarrow (C_2H_5)_4Pb(l) + 4NaCl(s) + 3Pb(s)$$

Tetraethyl lead was formerly popular as an antiknock agent in fuels for motor vehicles (see Sec. 9-11). Addition of only a thimblefull of this substance to each gallon of gasoline prevents audible pings, which could otherwise be detected in the vehicle's combustion chambers as the fuel burned. However, tetraethyl lead converts to lead(II) oxide as it burns. The following equation summarizes this combustion reaction:

$$2Pb(C_2H_5)_4(l) + 27O_2(g) \longrightarrow 2PbO(s) + 16CO_2(g) + 20H_2O(l)$$

The lead(II) oxide is a constituent of the exhaust gases expelled from the combustion chamber to the atmosphere. The release of lead in exhaust gases for decades has seriously polluted the environment. For this reason, USEPA was forced to require the abandonment of tetraethyl lead in vehicular fuels.

Tetraethyl lead is unlike most organometallic compounds. While it is combustible, it is not pyrophoric; nor is it water reactive. Its major hazard is the poisonous nature of its liquid or vapor when inhaled, ingested, or absorbed through the skin. Tetraethyl lead fires may be extinguished with water or carbon dioxide, but the toxic nature of the accompanying smoke requires the use of positive–pressure, self-contained breathing apparatus. Furthermore, when tetraethyl lead leaks from its container, every effort should be made to prevent it from entering soil, water, or sewer systems, as the substance may cause untold public health and environmental problems.

USDOT regulates the transportation of tetraethyl lead as a poisonous material. Packages of tetraethyl lead are labeled POISON, and transport vehicles shipping tetraethyl lead are placarded POISON.

OSHA regulates the exposure of workers to tetraethyl lead. OSHA stipulates a permissible exposure limit of 0.075 mg/m^3.

Aluminum Alkyl Compounds

An industrially important group of organometallic compounds consists of substances in which an aluminum atom is covalently bonded to a carbon atom. These substances are called *aluminum alkyl compounds*. Examples of aluminum alkyl compounds are trimethylaluminum and tri(isobutyl)aluminum, whose properties are listed in Table 8-3. They are used primarily as polymerization catalysts.

All known aluminum alkyl compounds are pyrophoric, violently water reactive, and highly toxic. Their fires evolve intense heat. To capture this heat value, some aluminum alkyl compounds are used in certain special fuels. For instance, triethylaluminum thickened with polyisobutylene is a pyrophoric fuel used in some military rockets.

USDOT does not regulate the transportation of specific aluminum alkyl compounds; but the transportation of this class of compounds is regulated. The proper shipping name is aluminum alkyls or aluminum alkyl halides, as appropriate. USDOT regulates them as spontaneously combustible materials. Packages are labeled SPONTANEOUSLY COMBUSTIBLE; when required, their transport vehicles are placarded SPONTANEOUSLY COMBUSTIBLE.

Fighting Fires Involving Organometallic Compounds

As we noted earlier, tetraethyl lead is the only common organometallic compound that is not pyrophoric and water reactive. Hence, with this single exception, the fighting of fires involving these substances must be performed in a manner that totally avoids the use of water and contact of the substance with the atmosphere.

Fires involving bulk quantities of an organometallic compound are usually extinguished by absorbing it in dry sand, diatomaceous earth, or G-1 graphite. To be totally effective, several inches of the absorbent should be built up over the surface of the material. Thereafter, small quantities of the absorbed compound may be safely moved to other locations and treated with water to destroy its chemical reactivity.

Fires involving small quantities of an organometallic compound may be extinguished with carbon dioxide. However, a blanket of carbon dioxide may not be permanently maintained over any substance. Once the carbon dioxide dissipates into the atmosphere, the remaining organometallic compound is again exposed to air and may burst into flame.

The intense heat that accompanies many fires of organometallic compounds requires that firefighters wear special apparel. One type consists of a totally contained aluminized suit like that illustrated in Fig. 8-6. These suits are designed to reduce the amount of heat that contacts the body by reflecting radiation. Nevertheless, these suits have limited applicability. No one can remain in the near vicinity of an organometallic fire for very long. Since aluminum alkyl compounds and tetraethyl lead are highly toxic, individuals must generally use positive-pressure, self-contained breathing apparatus when responding to incidents involving these substances.

8-4 METALLIC HYDRIDES

The most common commercially available metallic hydrides are those that are composed of hydrogen, either an alkali metal or aluminum, and sometimes boron. These substances are often used industrially as reducing agents. Unless heated to their melt-

Figure 8-6 Water-reactive organo-metallic compounds may be highly pyrophoric. Consequently, when responding to fire incidents in which they are involved, special protective clothing must be worn. This specialized aluminum suit reflects heat and provides protection against potential bodily contact with the reactive substance. (Courtesy of Ethyl Corporation, Baton Rouge, Louisiana)

ing points, they are relatively stable compounds. But they also possess several hazardous features in common. In particular, they react with water, including atmospheric moisture, forming flammable hydrogen. They also react violently with oxidizing agents. For these reasons, the common metallic hydrides represent a serious fire risk.

The simple alkali metal hydrides form by the direct combination of the metal with hydrogen between 570° and 1300°F (300° and 700°C). Lithium hydride and sodium hydride are white solids whose chemical formulas are LiH and NaH, respectively. This water reactivity of these two metallic hydrides is represented by the following equations:

$$LiH(s) + H_2O(l) \longrightarrow LiOH(s) + H_2(g)$$

$$NaH(s) + H_2O(l) \longrightarrow NaOH(s) + H_2(g)$$

Other commercially available alkali metal hydrides are lithium and sodium borohydride, more properly named lithium and sodium tetrahydridoborate, respectively. These substances are prepared by chemically reacting the respective alkali metal hydride with trimethyl borate. They are white to gray solids.

Each of these alkali metal borohydrides is water reactive. For instance, sodium borohydride reacts with water to form hydrogen and sodium triborate, as noted by the following equation:

$$3NaBH_4(s) + 6H_2O(l) \longrightarrow Na_3B_3O_6(s) + 12H_2(g)$$

A commercial advantage of this water reaction is that it occurs more slowly when compared to the rates of reaction between water and other metallic hydrides.

There are at least three metallic hydrides of aluminum: aluminum hydride, lithium aluminum hydride, and aluminum borohydride. Aluminum hydride, whose chemical formula is AlH_3, is unstable and easily oxidized. Hence, it has only limited commercial usefulness. However, lithium aluminum hydride and aluminum borohydride are often used as reducing agents in organic syntheses.

Lithium aluminum hydride is more properly named *lithium tetrahydridoaluminate*. It is prepared by chemically reacting lithium hydride with anhydrous aluminum chloride. The nature of this reaction is illustrated in the next equation:

$$4LiH(s) + AlCl_3(s) \longrightarrow LiAlH_4(s) + 3LiCl(s)$$

Although pure lithium hydride is a white solid, it slowly turns gray on standing. The solid forms available for laboratory and commercial use are a powder and tablets. Ethereal solutions are also encountered commercially. Notwithstanding the hazardous nature of the metallic hydride, ethers are highly flammable liquids (Sec. 11-8); thus, these ethereal solutions constitute a dangerous fire and explosion risk.

Lithium aluminum hydride is sensitive to heat, moisture, and oxygen. Great care must be taken when grinding it, as static sparks can cause lithium aluminum hydride to spontaneously ignite. When the dust of lithium aluminum hydride is mixed with air, explosions often occur. In moist atmospheres, the compound ignites with ease. To assure its safe use, lithium aluminum hydride is often introduced into chemical reactors in a fashion whereby it does not encounter air.

Lithium aluminum hydride is a water-reactive substance. When lithium aluminum hydride reacts with water, hydrogen is generated, as the following equation illustrates:

$$LiAlH_4(s) + 4H_2O(l) \longrightarrow Al(OH)_3(s) + LiOH(s) + 4H_2(g)$$

The hydrogen first burns with a quiet flame, but such intense heat is generated that the metallic residue glows with incandescence.

Aluminum borohydride, more properly named *aluminum tetrahydridoborate,* is prepared by reacting sodium borohydride with anhydrous aluminum chloride as the following equation notes:

$$3NaBH_4(s) + AlCl_3(s) \longrightarrow Al(BH_4)_3(l) + 3NaCl(s)$$

It is a volatile liquid which ignites spontaneously in air. It also reacts violently with water, generating hydrogen as noted by the following equation:

$$Al(BH_4)_3(l) + 3H_2O(l) \longrightarrow Al(OH)_3(s) + 3H_3BO_3(s) + 3H_2(g)$$

Figure 8-7   A container of lithium aluminum hydride with warning labels in German and English (left) and the enlarged English label (right). The major properties of this substance are provided on this label, thus warning those who handle and use lithium aluminum hydride of its potential hazards. The umbrella warning is used by the vendor to imply "Keep Away from Water." (Courtesy of Henley Chemicals, Inc., Montvale, New Jersey; Chemetall, Frankfurt a.M., Federal Republic of Germany)

The fires of these metallic hydrides should be extinguished by using either lime or limestone or a suitable powder composition based on sodium chloride. Metallic hydrides react violently with water, carbon dioxide, nitrogen, and the Halon agents; thus these fire extinguishing agents are not effective on metallic hydride fires.

USDOT specifically regulates the transportation of lithium and sodium hydride and of lithium and sodium borohydride as dangerous-when-wet materials. Packages of these substances are labeled DANGEROUS WHEN WET like the container in Fig. 8-7; and the transport vehicles shipping any quantity of them are placarded DANGEROUS WHEN WET.

OSHA regulates the exposure of employees to lithium hydride. OSHA stipulates a permissible exposure limit of 0.025 mg/m^3.

USDOT also regulates the transportation of aluminum hydride, lithium aluminum hydride, and the ethereal solution of the latter substance as dangerous-when-wet materials. Packages of lithium borohydride and lithium aluminum hydride are labeled DANGEROUS WHEN WET and their transport vehicles are placarded DANGEROUS WHEN WET. Packages of the ethereal solution of lithium aluminum hydride are labeled DANGEROUS WHEN WET and FLAMMABLE LIQUID; their transport vehicles are placarded DANGEROUS WHEN WET. Aluminum borohydride is regulated as a spontaneously combustible material. Packages of aluminum borohydride are labeled SPONTANEOUSLY COMBUSTIBLE and their transport vehicles are placarded SPONTANEOUSLY COMBUSTIBLE.

8-5 BORANES

Compounds of hydrogen and one or more nonmetals are called *molecular hydrides*. They exist as molecular units. A common molecular hydride is water, whose molecules are composed of hydrogen and oxygen atoms. Other examples of some simple molecular hydrides are noted in Table 8-4.

The molecular hydrides of boron are water-reactive compounds; they are

TABLE 8-4 EXAMPLES OF SOME COMMON MOLECULAR HYDRIDES

Substance	Formula	Hazardous properties
Hydrogen sulfide	H_2S	Highly toxic by inhalation; highly flammable and dangerous fire risk
Ammonia	NH_3	Sharp, intensely irritating odor; highly toxic and irritating; moderate fire risk
Phosphine	PH_3	Highly toxic by inhalation; spontaneously flammable
Methane	CH_4	Severe fire and explosion hazard
Arsine	AsH_3	Highly toxic; spontaneously flammable
Hydrogen chloride	HCl	Highly toxic by inhalation; strongly irritating
Hydrogen fluoride	HF	Highly toxic by inhalation; strongly irritating

called *boranes*. At least 14 boranes are known, but most are too unstable at room temperature to have commercial applicability. In fact, all boranes decompose to boron and hydrogen above 570°F (300°C). These substances are represented by the general chemical formula B_mH_n, where m and n are integers. As a rule, they burn to form boron(III) oxide, react readily with water, and are highly toxic.

The simplest of the boranes (BH_3) is unstable at atmospheric pressure and becomes *diborane* (B_2H_6), sometimes called *boroethane*. Diborane is a flammable gas having a vapor density of 0.96. A concentration of only 0.8% by volume of diborane in air is spontaneously combustible, burning with a characteristic green flame and evolving 527 kcal/mol (2090 kJ/mol or 2210 Btu/mol). Thus, the intensity of heat from diborane fires is similar to the intensity from liquid natural petroleum (LNP) fires. The combustion reaction is represented by the following equation:

$$B_2H_6(g) + 3O_2(g) \longrightarrow B_2O_3(s) + 3H_2O(g)$$

This extremely high heat of combustion coupled with its low molecular weight makes diborane potentially useful as a rocket fuel.

Diborane violently reacts even with atmospheric moisture. The reaction between diborane and water forms hydrogen, as the following equation notes:

$$B_2H_6(g) + 6H_2O(l) \longrightarrow 3H_3BO_3(s) + 6H_2(g)$$

Water has only limited success as a fire extinguisher on borane fires. For instance, it may be used to cool containers. Carbon dioxide and nitrogen are unreactive with diborane. Hence, either of these substances is an effective fire extinguisher on diborane fires, particularly when the fires are small. However, the use of either carbon dioxide or nitrogen on burning diborane is at best only temporarily effective. Since diborane is lighter in mass than air, the natural movement of the gas assists in dissipating both carbon dioxide and nitrogen. Nevertheless, the use of these fire extinguishers may be successful long enough to allow an individual to stop the flow of gas from its container, as Fig. 8-8 illustrates. When it is impossible to stop the flow, it is generally best to allow the gas to burn itself out.

Halon agents should be avoided as fire extinguishers on diborane fires, since diborane and Halon agents are chemically reactive. For instance, diborane reacts with carbon tetrachloride (Halon 104), forming hydrogen, as the following equation notes:

$$2B_2H_6(g) + 3CCl_4(g) \longrightarrow 4BCl_3(l) + 3C(s) + H_2(g)$$

Diborane is toxic. In the workplace, OSHA and NIOSH recommend only 0.1 ppm as the permissible exposure limit to diborane.

USDOT regulates the transportation of diborane as a poisonous gas. When diborane is transported, its cylinders and other packages are labeled POISON GAS and FLAMMABLE GAS. Transport vehicles shipping diborane are placarded POISON GAS.

The most stable borane is *decaborane*. Its chemical formula is $B_{10}H_{14}$. It is a highly toxic, white solid that easily vaporizes at room conditions. The vapor of de-

Figure 8-8 Pyrophoric materials ignite spontaneously when exposed to air. Such behavior is typical of diborane, phosphine (Sec. 8-7), and certain organometallic compounds. Hence, if their fires are to be effectively extinguished, it is essential to stop the flow of the gas or vapor from its container. Otherwise, once the extinguisher has dissipated, fire again erupts. (Courtesy of Ethyl Corporation, Baton Rouge, Louisiana)

caborane has a penetrating, pungent odor and flashes at 176°F (80°C). This combustion is represented by the following equation:

$$B_{10}H_{14}(g) + 11O_2(g) \longrightarrow 7H_2O(l) + 5B_2O_3(s)$$

Decaborane is also water reactive, although its rate of hydrolysis is slower than the corresponding reaction involving diborane. The reaction between decaborane and water is represented by the following equation:

$$B_{10}H_{14}(s) + 30H_2O(l) \longrightarrow 10H_3BO_3(s) + 22H_2(g)$$

Fires involving decaborane are best extinguished with carbon dioxide, but flooding volumes of water may also be used on small fires. The use of Halon agents is ineffective, since these substances chemically react with decaborane and help to sustain combustion. Firefighters should be wary of the toxic nature of decaborane. OSHA and NIOSH stipulate only 0.05 ppm as the permissible exposure limit.

USDOT regulates the transportation of decaborane as a flammable solid. Containers of decaborane are labeled FLAMMABLE SOLID and POISON, and their transport vehicles are placarded FLAMMABLE SOLID.

8-6 METALLIC PEROXIDES

Compounds comprised of metal and peroxide ions [O_2^{2-} or $(-O-O-)^{2-}$] are called *metallic peroxides*. They are primarily hazardous as oxidizers, but metallic peroxides are also water-reactive compounds.

Industrially important metallic peroxides are those of the alkali and alkaline earth metals, especially sodium peroxide and barium peroxide. The peroxides of sodium and barium form when these metals burn in an atmosphere of pure oxygen. They are white solids used mainly as oxidizing agents or to initiate polymerization (Sec. 12-1).

When heated, sodium and barium peroxide decompose to form oxygen, as the following equations note:

$$2Na_2O_2(s) \longrightarrow 2Na_2O(s) + O_2(g)$$

$$2BaO_2(s) \longrightarrow 2BaO(s) + O_2(g)$$

Such reactions are actually typical of all oxidizers: They decompose on heating to form oxygen, which supports combustion. These metallic peroxides may also react as oxidizing agents in chemical reactions, as Fig. 8-9 illustrates. The ability of metallic peroxides to react as oxidizers is often considered their primary hazard.

When metallic peroxides react with water, they form either hydrogen peroxide or oxygen. Hydrogen peroxide is typically the product of the reaction between metallic peroxides and cold water, whereas oxygen is typically the product of the reaction between metallic peroxides and hot water. This is illustrated with sodium peroxide by the following equations:

$$Na_2O_2(s) + H_2O(l) \longrightarrow H_2O_2(aq) + NaOH(aq)$$

$$2Na_2O_2(s) + 2H_2O(l) \longrightarrow 4NaOH(aq) + O_2(g)$$

Thus, containers of metallic peroxides should be protected against absorption of moisture.

Metallic peroxides do not burn, but they may be involved in fires. Water must be used cautiously when fighting such fires. A stream of water aimed at large amounts of metallic peroxides may cause the peroxide to react explosively. Since the atmosphere near such fires is enriched in oxygen, the ongoing fire is likely to intensify and other fires may be more readily initiated.

USDOT regulates the transportation of several metallic peroxides as oxidizers, including the peroxides of lithium, sodium, potassium, magnesium, calcium, and

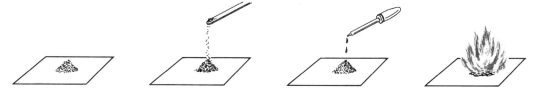

Figure 8-9 The oxidizing power of metallic peroxides is illustrated by this experiment. Arrange 5 g of powdered aluminum into the shape of a cone 0.5-in. high on a sheet of nonflammable material and sprinkle 0.5 g of sodium peroxide loosely over its surface. Use an eye dropper to add one or two drops of water, as illustrated. The chemical reaction that occurs involves the ignition of the metallic aluminum. Caution should be exercised on disposing of the residue, as the hot aluminum glows for some time.

barium. Packages of these metallic peroxides are labeled OXIDIZER; packages of barium peroxide are also labeled POISON. When required, their transport vehicles are placarded OXIDIZER.

8-7 METALLIC CARBIDES AND PHOSPHIDES

Carbon ions existing as C_2^{2-}, C^{4-}, or C_3^{4-} are called *carbides;* the compounds containing metallic and carbide ions are called *metallic carbides.* A number of metallic carbides are known, but only calcium carbide has significant industrial value.

Calcium carbide is manufactured by heating a mixture of coke and lime in an electric furnace at high temperatures. This reaction is noted by the following equation:

$$CaO(s) + 3C(s) \longrightarrow CaC_2(s) + CO(g) \quad (T > 2000°C)$$

Calcium carbide is used industrially as a major source of acetylene and for the production of calcium cyanamide, a fertilizer.

When calcium carbide reacts with water, acetylene is formed, as the following equation notes:

$$CaC_2(s) + 2H_2O(l) \longrightarrow Ca(OH)_2(s) + C_2H_2(g)$$

Thus, the principal recommendation to be made regarding the storage and handling of calcium carbide is to keep it dry. Calcium carbide that has been exposed to moisture poses a flammable and explosive hazard.

Industrial-grade calcium carbide normally contains calcium phosphide as an impurity. Calcium phosphide is an example of a *metallic phosphide,* that is, a compound containing metallic and phosphide (P^{3-}) ions. Metallic phosphides are also water-reactive compounds. When calcium phosphide reacts with water, phosphine forms, as the following equation denotes:

$$Ca_3P_2(s) + 6H_2O(l) \longrightarrow 3Ca(OH)_2(s) + 2PH_3(g)$$

Phosphine is a toxic, flammable gas (see Table 8-4). It possesses the characteristic putrefying odor of a mixture of garlic and rotten fish. Acetylene prepared from calcium carbide generally possesses a trace of this odor due to the presence of a low concentration of phosphine.

While neither metallic carbides nor metallic phosphides burn, they may be involved in fires. The use of water as a fire extinguisher is certain to intensify any fire involving either of these hazardous materials. Furthermore, carbon dioxide and the Halon agents are usually ineffective extinguishers of fires involving metallic carbides or metallic phosphides, especially when the fires have been caused by the addition of water to either of these substances. This is due to the fact that the flammable gas is generated from the solid mass of either the metallic carbide or phosphide, which dissipates the blanket of carbon dioxide or Halon agent. As acetylene or phosphine contacts atmospheric oxygen, it again bursts into flame.

USDOT regulates the transportation of calcium carbide and calcium phosphide as dangerous when wet materials. Packages of both substances are labeled DANGEROUS WHEN WET, and their transport vehicles are placarded DANGEROUS WHEN WET.

8-8 SOME WATER-REACTIVE METALLIC AND NONMETALLIC CHLORIDES

Several compounds containing chlorine and either metals or nonmetals are water-reactive substances. These compounds are hazardous materials because they react with water, often violently, to form hydrogen chloride. Hydrogen chloride is a toxic, irritating gas, and its aqueous solution, hydrochloric acid, is corrosive (Sec. 7-7). When the water-reactive metallic and nonmetallic chlorides react with water, large amounts of heat are typically evolved. If the heat is absorbed by flammable or combustible materials, it may serve as a source of ignition.

We shall briefly note some features of several of these compounds.

Anhydrous Aluminum Chloride

This is a white to yellow solid. Its chemical formula is $AlCl_3$. Anhydrous aluminum chloride is often used as the catalyst for production of certain organic compounds, including aluminum alkyl compounds. It is also used as the raw material for preparation of certain metallic hydrides (Sec. 8-4).

Anhydrous aluminum chloride reacts violently with water, forming hydrochloric acid, as the following equation depicts:

$$2AlCl_3(s) + 3H_2O(l) \longrightarrow Al_2O_3(s) + 6HCl(aq)$$

USDOT regulates the transportation of anhydrous aluminum chloride as a corrosive material. Packages containing anhydrous aluminum chloride are labeled CORROSIVE; when required, their transport vehicles are placarded CORROSIVE.

Anhydrous Antimony Pentachloride

The chemical formula of antimony pentachloride is $SbCl_5$. Anhydrous antimony pentachloride is a water-reactive compound. It is a reddish-yellow, oily liquid; it does not have widespread use.

When anhydrous antimony chloride reacts with water, antimony(V) oxide and hydrochloric acid form, as the following equation notes:

$$2SbCl_5(l) + 5H_2O(l) \longrightarrow Sb_2O_5(s) + 10HCl(aq)$$

Antimony pentachloride is also hygroscopic, and the resulting aqueous solution is corrosive. In the workplace, OSHA and NIOSH recommend 0.5 mg/m^3 as the permissible exposure limit to antimony and its compounds.

USDOT regulates the transportation of anhydrous antimony pentachloride as a corrosive material. Packages containing anhydrous antimony chloride are labeled CORROSIVE; when required, their transport vehicles are placarded CORROSIVE.

Boron Trichloride

Boron trichloride is a colorless, fuming liquid that boils at 65°F (18°C). Its chemical formula is BCl_3. Boron trichloride is frequently used as the raw material for synthesizing boron compounds and as a catalyst in certain organic synthetic procedures. It is decomposed by water, forming boric acid and hydrochloric acid.

$$BCl_3(l) + 3H_2O(l) \longrightarrow H_3BO_3(aq) + 3HCl(aq)$$

USDOT regulates the transportation of boron trichloride as a poisonous gas. Cylinders and other packages of boron trichloride are labeled POISON GAS and CORROSIVE, and their transport vehicles are placarded POISON GAS.

Phosphorus Oxychloride

This substance is also called *phosphoryl chloride*. It is a colorless, fuming liquid with a pungent odor. Its chemical formula is $POCl_3$. Phosphorus oxychloride is used as a chlorinating agent, catalyst, and fire-retarding agent. It is decomposed by water forming phosphoric acid and hydrochloric acid.

$$POCl_3(l) + 3H_2O(l) \longrightarrow H_3PO_4(aq) + 3HCl(aq)$$

USDOT regulates the transportation of phosphorus oxychloride as a corrosive material. Packages containing phosphorus oxychloride are labeled CORROSIVE and POISON; their transport vehicles are placarded POISON.

Phosphorus Pentachloride

This substance is a yellow to green solid. Its chemical formula is PCl_5. Phosphorus pentachloride is used as a chlorinating and dehydrating agent. It is decomposed by water forming phosphorus oxychloride and hydrochloric acid, as the following equation depicts:

$$PCl_5(s) + H_2O(l) \longrightarrow POCl_3(l) + 2HCl(aq)$$

Phosphorus oxychloride also reacts with water as noted earlier, so the overall water reactivity of this substance may be expressed as follows:

$$PCl_5(s) + 4H_2O(l) \longrightarrow H_3PO_4(aq) + 5HCl(aq)$$

These combined reactions evolve considerable heat.

OSHA regulates the exposure of workers to phosphorus pentachloride. The permissible exposure limit is $1 \ mg/m^3$.

USDOT regulates the transportation of phosphorus pentachloride as a corrosive material. Packages containing phosphorus pentachloride are labeled CORRO-SIVE; when required, their transport vehicles are placarded CORROSIVE.

Phosphorus Trichloride

This substance is a colorless, fuming liquid. Its chemical formula is PCl_3. Phosphorus trichloride is used as a chlorinating agent and catalyst and to make phosphorus oxychloride, gasoline additives, and dyes. It is decomposed by water forming phosphorus acid and hydrochloric acid:

$$PCl_3(l) + 3H_2O(l) \longrightarrow H_3PO_3(aq) + HCl(aq)$$

USDOT regulates the transportation of phosphorus trichloride as a corrosive material. Packages containing phosphorus trichloride are labeled CORROSIVE and POISON; their transport vehicles are placarded POISON.

OSHA regulates the exposure of workers to phosphorus trichloride. The permissible exposure limit is 0.5 ppm.

Silicon Tetrachloride and Organosilicon Compounds

Silicon tetrachloride is a colorless, fuming liquid with a suffocating odor. Its chemical formula is $SiCl_4$. Silicon tetrachloride is often used as the raw material from which liquid or semisolid silicon-containing polymers, called *silicones,* are produced. Silicones are used in a number of greases, oils, waxes, and polishes. Silicon tetrachloride is also used extensively in the semiconductor manufacturing industry.

When not confined, silicon tetrachloride readily vaporizes. Furthermore, water causes it to decompose, forming silicic acid and hydrochloric acid, as the following equation notes:

$$SiCl_4(l) + 4H_2O(l) \longrightarrow H_4SiO_4(aq) + 4HCl(aq)$$

Responding to emergencies involving bulk quantities of silicon tetrachloride may represent a serious problem, as Fig. 8-10 illustrates. The formation of hydrogen chloride requires that emergency-response personnel use positive-pressure, self-contained breathing apparatus when incidents arise involving silicon tetrachloride.

USDOT regulates the transportation of silicon tetrachloride as a corrosive material. Packages containing silicon tetrachloride are labeled CORROSIVE, and their transport vehicles are placarded CORROSIVE.

In addition to silicon tetrachloride, a number of silanes are used industrially for various purposes. *Silanes* are compounds whose molecules consist of silicon atoms bonded covalently to hydrogen atoms. Other compounds, called *organosilanes,* have molecules whose silicon atoms are bonded to carbon atoms. Chlorinated derivatives of organosilanes are also industrially useful substances. Examples of chlorinated organosilanes are methyltrichlorosilane, diethyldichlorosilane and allyl trichlorosilane. These substances are usually flammable liquids, highly toxic, and water-reactive. USDOT regulates the transportation of organosilanes and their chlorinated

Figure 8-10 Technicians wearing acid-resistant asbestos suits and self-contained breathing apparatus move toward a leaking tank valve in an effort to stop a leak of silicon tetrachloride from its storage tank. This material is water reactive, forming toxic hydrogen chloride. Such fumes drove thousands of nearby residents from their homes. (Courtesy of the *Chicago Sun-Times*, Jack Lenahan, photographer, Chicago, Illinois)

derivatives. For instance, the transportation of methyltrichlorosilane is regulated as a flammable liquid. Packages of this hazardous material are labeled FLAMMABLE LIQUID, POISON, and CORROSIVE, and their transport vehicles are placarded POISON.

Anhydrous Stannic Tetrachloride

Stannic tetrachloride is also known as tin(IV) chloride. Its chemical formula is $SnCl_4$. Stannic tetrachloride is a colorless, fuming liquid used for the manufacture of blueprint and other sensitized papers. It reacts with water, forming tin(IV) oxide and hydrochloric acid, as noted by the following equation:

$$SnCl_4(l) + 2H_2O(l) \longrightarrow SnO_2(s) + 4HCl(aq)$$

USDOT regulates the transportation of anhydrous stannic tetrachloride as a corrosive material. Packages containing this substance are labeled CORROSIVE; when required, their transport vehicles are placarded CORROSIVE.

OSHA regulates the exposure of workers to inorganic tin compounds, like anhydrous stannic tetrachloride. The permissible exposure limit is 2 mg/m^3.

Sulfuryl Chloride

This substance is also known as *sulfonyl chloride*. It is a colorless liquid with a pungent odor whose chemical formula is SO_2Cl_2. Sulfuryl chloride is mainly used as a chlorinating agent and dehydrating agent. Sulfuryl chloride is rapidly decomposed by water, forming sulfuric acid and hydrochloric acid:

$$SO_2Cl_2(l) + 2H_2O(l) \longrightarrow H_2SO_4(aq) + 2HCl(aq)$$

USDOT regulates the transportation of sulfuryl chloride as a corrosive material. Packages containing this substance are labeled CORROSIVE; when required, their transport vehicles are placarded CORROSIVE.

Anhydrous Titanium Tetrachloride

This is a colorless, volatile liquid with a pungent odor. Its chemical formula is $TiCl_4$. A mixture of anhydrous titanium tetrachloride and an aluminum alkyl compound is industrially used to catalyze the polymerization of certain compounds; it is called a *Ziegler catalyst*. In warfare, anhydrous titanium tetrachloride has been used to establish smoke screens. The latter use is associated with its water reactivity. Titanium(IV) oxide and hydrochloric acid are formed, as the following equation notes:

$$TiCl_4(l) + 2H_2O(l) \longrightarrow TiO_2(s) + 4HCl(aq)$$

USDOT regulates the transportation of anhydrous titanium tetrachloride as a corrosive material. Packages of this hazardous material are labeled CORROSIVE and POISON, and their transport vehicles are placarded POISON.

Thionyl Chloride

This is a red to yellow liquid with a pungent odor. Its chemical formula is $SOCl_2$. Thionyl chloride is used mainly in both inorganic and organic chemical synthetic procedures. Thionyl chloride reacts vigorously with water, forming sulfurous acid and hydrochloric acid, as the following equation notes:

$$SOCl_2(l) + 2H_2O(l) \longrightarrow H_2SO_3(aq) + 2HCl(aq)$$

USDOT regulates the transportation of thionyl chloride as a corrosive material. Packages containing this substance are labeled CORROSIVE; when required, their transport vehicles are placarded CORROSIVE.

8-9 SOME WATER-REACTIVE ORGANIC COMPOUNDS

Several organic compounds react violently with water to form toxic or corrosive products. We shall note here only two such compounds, which react with water to form acetic acid.

Acetic Anhydride

This is a colorless, fuming liquid used principally for the manufacture of aspirin. Its chemical formula is $(CH_3CO_2)O$. When acetic anhydride reacts with water, it forms acetic acid as the sole product, as the following equation notes:

$$(CH_3CO_2)_2O(l) + H_2O(l) \longrightarrow 2CH_3COOH(aq)$$

Breathing may become very difficult in confined areas where large quantities of acetic anhydride have been exposed to the atmosphere. This is caused by the presence of highly irritating vapors of acetic acid.

USDOT regulates the transportation of acetic anhydride as a corrosive material. Packages containing this substance are labeled CORROSIVE; when required, their transport vehicles are placarded CORROSIVE.

OSHA regulates the exposure of workers to acetic anhydride. The permissible exposure limit is 5 ppm.

Acetyl Chloride

This is the colorless, fuming liquid having a pungent odor. Its chemical formula is CH_3COCl. Acetyl chloride is used principally in organic chemical synthetic procedures. On reacting with water, acetic acid and hydrochloric acid are formed, as the following equation notes:

$$CH_3COCl(l) + H_2O(l) \longrightarrow CH_3COOH(aq) + HCl(aq)$$

The flash point of acetyl chloride is only 40°F (4.4°C). Hence, acetyl chloride represents a serious fire risk.

USDOT regulates the transportation of acetyl chloride as a flammable liquid. Packages containing acetyl chloride are labeled FLAMMABLE LIQUID and CORROSIVE. When required, their transport vehicles are placarded FLAMMABLE.

REVIEW EXERCISES

Alkali Metals

8.1. Compare the chemical behavior of metallic lithium with that of sodium and potassium (a) in air, and (b) with water.

8.2. Nitrogen is occasionally used to blanket molten sodium, which prevents its oxidation. Why may nitrogen not be similarly used to blanket molten lithium?

8.3. Which metal, pound for pound, is capable of absorbing more heat within the same range of temperature, lithium or sodium? (*Hint:* Determine the amount of heat in Btu independently absorbed by 100 lb of each metal when it is heated from 68° to 400°F, and compare these values.)

8.4. Why is metallic potassium generally considered more hazardous than either metallic sodium or lithium?

8.5. Briefly compare the chemical behavior of metallic lithium, sodium, and potassium in air at room temperature, and provide relevant equations that illustrate the reactions.

8.6. What precaution needs to be taken when using graphite-based fire extinguishers on alkali metal fires?

8.7. How may the fires of metallic lithium, sodium, and potassium be visually distinguished?

Magnesium, Zirconium, Titanium, Aluminum, and Zinc

8.8. While water is not ordinarily recommended for use as a fire extinguisher on magnesium fires, water can sometimes effectively extinguish them. What precaution needs to be taken by firefighters when using water to extinguish magnesium fires to assure that the fires will be extinguished?

8.9. Following the crash of a race car, the vehicle becomes engulfed in flames, consisting mainly of burning gasoline. Firefighters arriving at the scene of the accident attempt to extinguish the fire by directing a stream of water directly on a metal part of the race car. Instead of extinguishing the fire, however, the water causes it to intensify. What chemistry is associated with this observation?

8.10. Why does powdered zirconium pose an especially serious fire and explosion risk, even more so than most other powdered pyrophoric metals?

8.11. Why is zirconium sponge more prone to spontaneous ignition than most other forms of zirconium?

8.12. When comparing several dispersions of equal quantities of metal dust in air, why is the resulting explosive pressure greatest for the metallic dispersion whose particles are the smallest in size?

8.13. Magnesium and aluminum are very reactive metals. Why is it that neither metal *appears* very reactive when normally encountered?

8.14. An engineer proposes to extinguish a magnesium fire in an industrial furnace by completely shutting off its source of oxygen. Why may the fire continue to burn under such conditions?

8.15. Why must static electricity be totally eliminated in areas near polishing machines that generate aluminum dust?

8.16. Why do fire inspectors usually recommend replacing aluminum electrical wiring that has corroded?

8.17. Zinc dust is often used when synthesizing certain organic chemicals. How should it be stored to reduce its fire and explosion hazard?

8.18. Why does the smoke from class D fires often cause more severe damage to the lungs and associated passageways than the smoke of class A, B, or C fires?

Organometallic Compounds

8.19. What two chemical properties are common to *most* organometallic compounds?

8.20. In what ways does tetraethyl lead chemically differ from most other organometallic compounds?

8.21. Why is the use of positive-pressure, self-contained breathing apparatus recommended when combatting fires of tetraethyl lead?

8.22. Tri(isobutyl)aluminum is a colorless liquid that ignites spontaneously in air, reacts violently with water and many other chemical compounds, and is highly toxic. Based only on this limited information, assign appropriate numbers to each quardrant of the NFPA 704 symbol for tri(isobutyl)aluminum.

8.23. A chemical plant intends to store 125 gallons of liquid triethylaluminum in a 250-gallon stainless steel storage tank.
   (a) What precaution needs to be taken when transferring this hazardous material from its transportation vessel to the storage tank?
   (b) Which gas should be used to fill the void space in the storage tank?
   (c) What fire extingishers should be available for potential use at the plant for combating fires of triethylaluminum?
   (d) Is standard protective clothing adequate for safeguarding those who must combat fires of triethylaluminum?

Metallic Hydrides

8.24. Why do fire inspectors discourage the storage of metallic hydrides containers under automatic water sprinklers?

8.25. Due to the presence of atmospheric moisture, a 5-lb container of sodium hydride self-ignites in a chemistry laboratory. Which of the following actions is most desirable from the viewpoint of protecting life, property, and the environment?
   (a) Toss the container through an open, nearby window, if this involves minimal risk, and allow the material to burn.
   (b) Apply water to the fire.
   (c) Apply dry chemical to the fire.

Boranes

8.26. A chemistry laboratory stores 25 steel cylinders containing different flammable gases, including diborane, in an area isolated from other chemical substances and from the general public. From a safety viewpoint, why should the cylinders of diborane be segregated from the cylinders of other flammable gases?

Metallic Peroxides

8.27. USDOT requires sodium and potassium peroxides to be contained for transportation in either wooden or fiberglass boxes with inside, airtight metal cans. Why does USDOT most likely specify this unique packaging requirement?

8.28. Firefighters have successfully isolated a mixture of unspecified quantities of sodium peroxide and white phosphorous from other substances involved in a chemical fire. Even when isolated, however, the mixture continues to burn. Which is likely to most effectively extinguish the fire of this mixture water, carbon dioxide, or Halon 1301?

Metallic Carbides and Phosphides

8.29. For transportation purposes, USDOT requires calcium carbide to be packaged in either watertight metal drums with rolled, folded top and bottom seams and with welded side seams or in watertight, sift-proof, bulk metal containers. Why does USDOT most likely specify this unique packaging requirement?

8.30. Zinc phosphide is intensely poisonous to mammals and birds and has been used as a rodenticide to control rats, mice, and gophers. Why do manufacturers of zinc phosphide recommend that the rodenticide be stored under completely dry conditions?

Water-Reactive Metallic and Nonmetallic Chlorides

8.31. Containers of water-reactive metallic and nonmetallic chlorides are potentially dangerous, since considerable pressure may build up in them. What substance is responsible for the build up of internal pressure in these containers?

8.32. Titanium tetrachloride is used with ammonia in skywriting. When released to the air, titanium tetrachloride reacts with atmospheric moisture to form hydrogen chloride, which then reacts with ammonia to form clouds of white ammonium chloride. Write the equations illustrating these chemical reactions.

8.33. A worker at a chemical plant was inadvertently sprayed with titanium tetrachloride over his entire body. The worker responded to this incident by jumping into a vat of water. The titanium tetrachloride reacted with the water, releasing so much heat at once that the water boiled, causing the individual to suffer severe burns before he could leave the vat. What should this person have done to remove the chemical substance and reduce the likelihood of chemical burns?

8.34. What is the most likely reason that USDOT forbids the transportation of phosphorous trichloride, phosphorus pentachloride, and phosphorus oxychloride on passenger-carrying aircraft and railcar?

8.35. Write the chemical equations that illustrate the water reactivity of the following substances:
   (a) Anhydrous aluminum chloride   (b) Phosphorus oxychloride
   (c) Anhydrous stannic tetrachloride   (d) Thionyl chloride
   (e) Silicon tetrachloride

Water-Reactive Organic Compounds

8.36. Arriving at the scene of a truck accident, a fire chief secures the shipping papers describing the consignment. These papers provide the following information: 60; Drums, RQ, Acetic anhydride, 8, UN1715, PG II; 24,000 lb. On the nearby soil are hundreds of gallons of a fuming, pungent liquid. Further investigation reveals that at least half of the consignment at issue has spilled on the soil or is still leaking from drums.
   (a) What information is immediately conveyed to the fire chief by the number 8?
   (b) Will this liquid ignite if exposed to an ignition source?
   (c) What immediate orders should be provided to the firefighters responding to this accident?
   (d) What action is the carrier required to take under federal law?

9

Chemistry of Some Toxic Materials

Pharmacology is the study of the nature, properties, and effects that chemical substances have on biological systems. Hence, in its entirety, pharmacology embraces the study of *all* substances known to affect living processes, whether beneficially or detrimentally. For academic purposes, pharmacology is normally divided into several branches. One such branch relates to the specific study of adverse health effects caused by poisonous or toxic substances; it is called *toxicology*. Poisons or toxins are generally regarded as those substances that cause or contribute to illness or death when administered to an organism in good health in relatively small amounts. The basic principles of toxicology comprise an important segment of the study of hazardous materials.

In connection with routine work-related activities, firefighters and other emergency-response forces are regularly exposed to certain poisonous substances, like carbon monoxide. Consequently, acquiring a familiarity with the field of toxicology is appropriate for emergency-response personnel, as the subject matter directly affects their occupational health and safety. We shall be mainly concerned with those aspects of toxicology that relate to the development of irreversible health effects or those that seriously impair the normal functioning of individuals.

Along the latter line, a group of gases is often encountered in connection with the occurrence of fires; they are collectively known as the *fire gases*. They are the following gaseous substances: carbon monoxide, carbon dioxide, hydrogen cyanide, hydrogen sulfide, sulfur dioxide, the oxides of nitrogen, and ammonia. When inhaled, these gases may cause a range of adverse health effects, including death. Since inhalation of these gases may be fatal and because exposure to them most directly impacts the health and safety of firefighters, particular emphasis will be devoted in this chapter to the fire gases.

Firefighters are not the only individuals who are likely to be exposed to toxic substances. Workers in certain industries are also routinely exposed to them. When their use is employed in the workplace, employee exposure to toxic substances is regulated by OSHA. For instance, employers must assure that an employee's exposure to certain air contaminants does not exceed published ceiling values or eight-hour time-weighted average values (see Exercise 9.1).

As first noted in Chapter 5, the transportation of poisonous gases, poisonous materials, irritating materials, and etiologic agents is regulated by USDOT. Any of these materials is likely to be encountered when responding to transportation mishaps.

Finally, USEPA characterizes RCRA hazardous wastes (Sec. 1-5) by means of a property called *EP toxicity* (extraction-procedure toxicity). This property relates only to the following toxic metals and pesticides: arsenic, barium, cadmium, chromium, lead, mercury, selenium, silver, endrin, lindane, methoxychlor, toxaphene, 2, 4-D (2, 4-dichlorophenoxyacetic acid), and 2, 4, 5-TP Silvex [2-(2, 4, 5-trichlorophenoxy) propionic acid]. A hazardous waste exhibits the characteristic of EP toxicity if a sample of the waste is found to contain any of these contaminants at a concentration equal to or greater than the respective values noted in Table 9-1. Hazardous wastes that exhibit the characteristic of EP toxicity are assigned an EPA hazardous waste number from D004 to D017, as tabulated.

USEPA also characterizes RCRA hazardous wastes by means of their potential reactivity. In the introductory paragraphs prior to Sec. 8-1, we noted several ways by which hazardous wastes exhibit this characteristic. The regulations published pursuant to RCRA describe other ways by which reactivity is exhibited. For instance,

TABLE 9-1   CONCENTRATION OF CONTAMINANTS FOR CHARACTERISTIC OF EP TOXICITY

EPA hazardous waste number	Contaminant	Concentration (mg/L)
D004	Arsenic	5.0
D005	Barium	100.0
D006	Cadmium	1.0
D007	Chromium	5.0
D008	Lead	5.0
D009	Mercury	0.2
D010	Selenium	1.0
D011	Silver	5.0
D012	Endrin	0.02
D013	Lindane	0.4
D014	Methoxychlor	10.0
D015	Toxaphene	0.5
D016	2,4-D[a]	10.0
D017	2,4,5-TP Silvex[b]	1.0

[a] 2,4-dichlorophenoxyacetic acid.

[b] 2-(2,4,5-trichlorophenoxy)propionic acid.

cyanide- and sulfide-bearing wastes exhibit reactivity when, exposed to pH conditions between 2 and 12.5, they generate toxic gases, vapors, or fumes in a quantity sufficient to present a danger to human health or the environment. Under these conditions, cyanide- and sulfide-bearing wastes generate the toxic gases hydrogen cyanide and hydrogen sulfide, respectively.

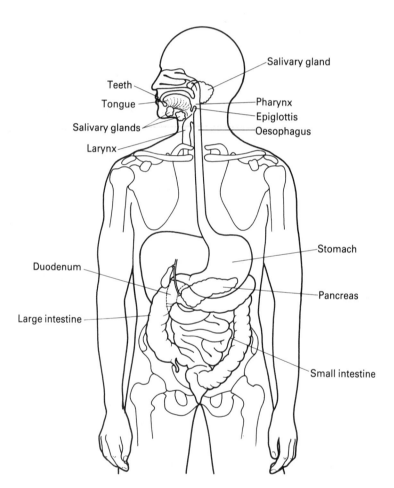

Figure 9-1 The human digestion system, illustrating some of its major components. A subtance taken orally passes through the mouth, into the esophagus, and then into the stomach. The substance may undergo chemical alteration to some degree, since partial digestion occurs in the mouth and stomach. It may even pass through the stomach wall directly into the bloodstream. But, generally, it is not until the substance or its degradation products have passed into the small intestine that absorption into the bloodstream occurs.

9-1 HOW TOXIC SUBSTANCES ENTER THE BODY

There are many routes by which a substance may enter the body, but only three are of concern to us here: ingestion, skin absorption, and inhalation.

Ingestion refers to the swallowing of a substance through the mouth and into the stomach, and generally followed by its entrance into the small intestine; the appropriate physiological aspects are noted in Fig. 9-1. Ingested toxic substances either directly damage the tissues of the mouth or digestive tract at the site of contact, are metabolized to toxic or nontoxic substances, or are absorbed through the intestinal walls to enter the blood system. In the latter case, the circulatory system then disseminates the substances to various body organs and tissues.

Absorption through the skin is a second route by which a substance may enter the body. Skin, the cross section of which is illustrated in Fig. 9-2, is the largest single organ of the body. It often serves as an organ of defense by preventing the direct entry of a substance into the body or its parts, except for the eyes and lungs. But the skin may also act as a portal of entry for some chemical substances. In these instances, the skin acts like a permeable membrane. Generally, liquids absorb most easily, since their concentration is high at a particular point of contact; capillary action aids in their penetration. Some toxic substances, like hydrofluoric acid, absorb through the skin and damage nearby bone matter. But, more commonly, substances that absorb through the skin further assimilate into the blood system and adversely affect a specific site of action or the body as a whole.

Inhalation is a third route by which substances may enter the body. Figure 9-3 illustrates the parts of the human respiratory system through which an inhaled substance travels en route to the lungs. The internal surface of our lungs is so rich in blood vessels that they actually cover an average surface area of approximately 90

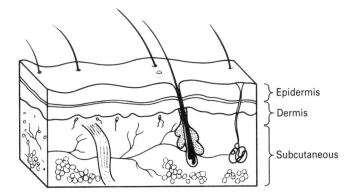

Figure 9-2   The cross section of human skin showng some of its principal components. The outer layer of the skin is the epidermis, a thin surface membrane of dead cells. For a substance to absorb through the skin, it must first penetrate through the epidermis. If this occurs, the substance enters the dermis, a collection of cells that acts as a porous diffusion medium. Here the substance may be further absorbed into the bloodstream.

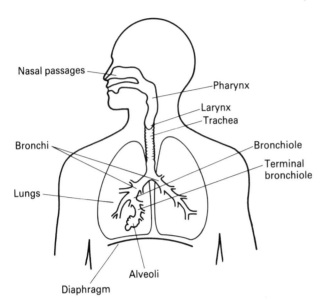

Nasal passages

Pharynx

Larynx
Trachea

Bronchi

Bronchiole

Terminal
bronchiole

Lungs

Alveoli

Diaphragm

Figure 9-3 The human respiratory system, illustrating its major parts. Inhaled vapors and gases, including air, first pass either through the nose or mouth and then through the pharynx, where they enter the trachea (commonly called the windpipe) at the larnyx. Next, inhaled substances enter either of two bronchi, each of which leads to a lung. The individual divisions of each bronchus are called bronchioles.

yd^2 (75 m^2). Thus, a substance inhaled into the lungs is readily absorbed into the bloodstream. This means that the body's response action from inhalation is often relatively swift.

9-2 CLASSIFICATION OF TOXIC EFFECTS

When a substance is toxic, it may cause local damage at the site of contact or at a site of action (that is, at certain tissues or organs), or it may adversely affect the entire organism in some way. These various ways by which a substance may cause biological damage are often used as a broad basis for the classification of toxic effects. In particular, these adverse effects are often called *local* and *systemic* effects. Damage caused by a substance at the site of its contact is a *local* effect; dermatitis caused by strong cleaning agents is an example of a local effect. An adverse health effect that is experienced at a site of action is a *systemic* effect. For instance, when mercury absorbs through the skin, it may damage the kidneys or affect the central nervous system; these are examples of systemic effects.

Toxic effects are also classified according to the time needed for a substance to cause an injury or disease. The following types are commonly noted: acute, short term, chronic, and latent. A toxic substance may produce only one or more than one of these effects.

An *acute* effect is generally regarded as a severe injury caused by a single, relatively short exposure to a substance. For instance, inhalation of hydrogen chloride for only a few seconds may cause lung damage; this is an acute injury. On the other hand, an individual may also be exposed to a substance repeatedly over a relatively short term, say consecutive daily exposures over a few days or weeks, before an adverse health effect is noted. Such an injury is called a *short-term* effect. An example of a short-term effect is illustrated by exposure of skin tissue to certain harsh cleaning agents. While a single handling of such cleaners may not seemingly affect skin tissue, repeated handling may cause dermatitis; this is a short-term injury. Some substances having a relatively low acute toxicity may have significant potential for producing harmful effects by repeated exposures.

A *chronic* effect is an injury or disease that manifests itself after a relatively long period of time has elapsed since initial exposure to the substance causing the ailment. For example, liver angiosarcoma may develop in an individual years after inhalation of vinyl chloride. This development of cancer is an example of a chronic effect.

Finally, a *latent* effect is an injury or disease that remains undeveloped until an incubation period has elapsed. For example, benzene (Sec. 11-3) induces aplastic anemia in humans, but the affliction may not become noticeable until ten years after the initial exposure to benzene.

A toxic substance that enters the body by inhalation may act as either an asphyxiant or irritant. An *asphyxiant* is a chemical substance that causes a loss of consciousness as the result of an insufficient amount of oxygen in the blood. Simple asphyxiants, like nitrogen and hydrogen, are immediately harmful only to the extent that they displace oxygen. Unattended, however, asphyxiated individuals may die from suffocation. Some asphyxiants exercise a detrimental chemical action in the blood [for example, carbon monoxide (Sec. 9-6)], from which death may come rapidly.

A mixture of inhalation asphyxiants with oxygen or air may have an anesthetizing effect on the body. Anesthetized individuals may be either conscious or unconscious, but they experience a partial or total loss of the sense of pain. Such substances are called *anesthetics,* especially when they are intentionally administered for performing surgery. Nitrous oxide and diethyl ether are examples of anesthetics.

An *irritant* is a chemical substance that injures the tissues of the respiratory system and lungs, thereby causing inflammation of the respiratory passages. For instance, a relatively low concentration of hydrogen chloride acts as an irritant. The mechanism of such irritation is generally corrosive; that is, the substance exhibits corrosive action on skin tissue at the site of contact.

9-3 FACTORS AFFECTING TOXICITY

The human body is a very complex and delicately balanced system. Each cell in the body, the basic unit of life, takes in nutrients, resists biological attackers, reproduces, and produces the substances that the host organism needs to survive. Many chemical reactions occur within cells that are responsible for life itself. But when toxins are absorbed across the cellular membrane, they may upset this delicate chemistry. In turn, these cells may malfunction, causing the organism to experience impaired health, disease, or death.

Fortunately, the body possesses natural mechanisms for protecting itself against the adverse action of foreign substances. One such mechanism is provided by the body's immunological system, which is often capable of defending individual cells against toxic invaders. Another mechanism is provided by the normal action of specific organs that possess the capability of converting harmful substances into harmless ones or into substances that are more rapidly excreted than the original toxin. As the result of such biochemistry, foreign substances may be modified in chemical structure, temporarily stored in specific organs, and/or directly eliminated.

Certain organs play particularly important roles in ridding the body of foreign substances. For instance, the liver is capable of detoxifying many foreign substances by biochemical action (for example, by converting them to less harmful substances), and the kidneys often excrete them from the body. Metabolic and excretory processes often work in unison in this manner to keep our bodies free of unwanted substances.

Notwithstanding these facts, however, the body is not always capable of protecting itself against invasion by foreign matter. Whether a substance adversely affects the body depends on several factors, the most important of which are discussed next.

Quantity of Substance

Poisons cause adverse health effects in relatively small amounts, typically 3 g (1/8 oz) or less. Toxicologists refer to the amount of the substance at issue as the *dose*. Some substances are so poisonous that extremely minute doses cause death. For instance, the bacterium *Clostridium botulinum,* the cause of botulism, is a single cell that can release a toxin so potent that four hundred-thousandths of an ounce is enough to kill 1 million laboratory guinea pigs. Fortunately, there are relatively few substances that exhibit this unusually high degree of toxicity.

Nearly all foreign substances adversely affect the body at some dose. We are mainly concerned here only with those substances that cause disease or death when taken in relatively small amounts.

The effect that a substance has on the body often changes with its dose. For example, many adults have experienced the effect that alcoholic beverages can have on the body as a function of dose. When the concentration of alcohol in the bloodstream is around 0.8% by volume, we may feel slightly subdued, relaxed, perhaps even elevated in spirit. But if the concentration reaches 2% to 3%, we usually become intoxicated; at 5%, death can occur.

Length of Exposure

A factor closely related to the dose is the *length of exposure.* Consider the mixture of toxic substances found in tobacco smoke; these combined toxins are the main cause of throat and lung cancer. Yet most of us know people who smoke cigarettes, but have not become afflicted with cancer. On the other hand, those individuals who have contracted throat or lung cancer from smoking have usually smoked habitually for years. This illustrates that some adverse health effects are likely to be manifested only after many repeated doses, each dose lengthening the total period of the body's exposure to the given toxic substance.

The length of exposure is often an important factor when considering the adverse health effects caused by exposure to toxic substances in the work environment. For instance, coal miners often become afflicted with black lung only after having breathed coal dust and its constituents repeatedly for many years.

The length of exposure can also be a factor affecting toxicity in a comparatively beneficial way. Sometimes the body acclimates somehow to repeated exposure to certain toxic substances; that is, while adverse health effects are noted at the time of an initial exposure, such effects are not always repeated upon subsequent exposures to the same dose of the same substance.

Rate and Extent to Which the Substance Is Absorbed into the Bloodstream

Most adverse health effects are experienced only after sufficient time has elapsed for the substance to absorb into the bloodstream. There are seven ways by which this may occur: (1) intravenous, (2) inhalation, (3) intraperitoneal, (4) intramuscular, (5) subcutaneous, (6) oral, and (7) cutaneous. The fastest of these means for entry of liquid poisons into the bloodstream is intravenous, that is, injection directly into a vein; for gaseous poisons, the fastest route of entry is inhalation.

Some time usually elapses before substances taken orally are assimilated into the bloodstream. This is a fortunate circumstance in the case of poisons, since it provides the opportunity to administer an antidote, to induce vomiting, or to take some other action appropriate to saving a life. The time elapse relates with the fact that the absorption of ingested substances into the bloodstream generally occurs in the small intestine, the organ that connects the stomach to the large intestine (see Fig. 9-1). One notable exception is ethyl alcohol, which, when ingested, passes directly through the stomach wall into the bloodstream. This explains why we notice the intoxicating effect of alcohol soon after consumption.

Physical and Chemical Nature of the Substance

This factor is closely related to the previous one. Consider the compounds of chromium, all of which are toxic. Several chromium compounds manifest this toxicity more noticeably, however, because they enter the body more easily than others do. Thus, the fumes or fine powder of metallic chromium may be easily inhaled, causing the formation of ulcerated lesions in the mucous membranes. On the other hand, the systemic effects caused from consumption of chromium compounds in polluted water may seemingly go unnoticed for weeks or even longer.

The water solubility of substances also affects the toxicity of ingested substances. Generally, this is true because substances having a low solubility in water are not significantly absorbed through the gastrointestinal tract. Thus, chromium(III) fluoride, which is very insoluble in water, is far less toxic than chromium(III) sulfate, which is soluble to the extent of approximately 120 g per 100 g of water at room temperature.

Finally, different oxidation states may have a bearing on a substance's toxicity. Compounds containing hexavalent chromium, for instance, are carcinogenic (that is, cancer causing); compounds containing trivalent chromium, while nonetheless toxic, are not known carcinogens.

Age and Health of Afflicted Individual

The degree to which a toxic substance affects individuals may depend on their age and general state of health. In general, youngsters and the elderly are much more susceptible and less tolerant to toxins than the middle-aged. Similarly, individuals already weakened from disease are likely to be more susceptible to poisons than healthy people. This individual sensitivity is typical of individuals who suffer from heart and respiratory ailments; they cannot tolerate moderate concentrations of air pollutants, like ozone, while healthy people are often simultaneously oblivious to their presence.

Cigarette smokers comprise a special group of people possessing individual sensitivity to certain toxins. This is best evidenced by the observation that smokers are more likely than nonsmokers to become afflicted with asbestosis (Sec. 9-12), a progressive, irreversible lung disease associated with exposure to asbestos fibers.

9-4 MEASUREMENT OF TOXICITY

In general toxicological studies, animals are exposed to a given dose of a substance and monitored to establish a defined effect. Then the results of such studies are extrapolated to humans. This method is generally credible for establishing acute or short-term effects.

However, chronic toxicological studies on animals are not always relevant when applied to humans, since the physiology of humans is not always like that of animals. This fact is particularly important as it affects studies on potentially carcinogenic substances. Toxicologists regard a substance as a *suspect human carcinogen* when laboratory studies on animals establish that the substance induces cancer in more than one species or sex in animals, or when they show that the substance increases the incidence of site-specific malignant tumors in a single species or sex, and when there is a statistically significant dose–response relationship in more than one exposed group. When it is established coincidentally that a substance induces cancer in humans, the substance is called a *human carcinogen*.

The following four means are commonly employed to quantify acute toxicity:

1. *Lethal dose, 50% kill (LD_{50})*. The LD_{50} is the amount of a material that, when administered to laboratory animals, kills half of them. It is expressed in milli-

grams of the substance administered per body weight of the animal expressed in kilograms (mg/kg). When extrapolated to humans, the lethal dose for an average person who weighs w kilograms is $LD_{50} \times w$.

2. *Lethal concentration, 50% kill (LC_{50}).* The LC_{50} is the concentration of a material, normally expressed as parts per million (ppm) by volume, that, when administered to laboratory animals, kills half of them during the period of exposure.

3. *Threshold limit value (TLV).* The TLV is the upper limit of a toxin concentration to which an average healthy person may be repeatedly exposed on an all-day, everyday basis without suffering adverse health effects. For gaseous substances in air, the TLV is usually expressed as parts per million (ppm); for fumes or mists in air, it is expressed in milligrams per cubic meter (mg/m^3). Values of the TLV are set by the American Conference of Governmental Industrial Hygienists (ACGIH).*

Many substances possess an odor. The concentration in air at which a substance may be first detected by its odor is called the *odor threshold*. When some substances are first detected by odor, their concentration in an atmosphere may have already exceeded their threshold limit values. For example, the odor threshold of ammonia is 46.8 ppm, but its threshold limit value is 25 ppm. This means that, if the odor of ammonia can be detected in an environment, inhalation of ammonia at the given dose already may pose a risk to one's health.

In the workplace, employees often measure a similar factor called the *time-weighted average threshold limit value* (TWA-TLV). This is the average concentration to which most workers can be exposed during a 40-hour week or 8-hour day without developing adverse health effects.

4. *Immediately dangerous to life and health (IDLH) level.* An IDLH level represents a maximum concentration from which one could escape within 30 minutes without experiencing any escape-impairing symptoms or any irreversible adverse health effect. IDLH levels are published for many substances by OSHA and NIOSH.[†] In practice, when the concentration of a toxic substance in a given area is known, IDLH levels may be used for determining whether self-contained breathing apparatus is needed when entering the area. If the concentration exceeds the IDLH level, positive-demand, self-contained breathing apparatus should be used.

---

*TLVs are published in *Threshold Limit Values for Chemical Substances and Physical Agents in the Workroom Environment with Intended Changes,* American Conference of Governmental Industrial Hygienists, 6500 Glenway Avenue, Building D-5, Cincinnati, Ohio 45211.

†IDHL levels are provided in *NIOSH/OSHA Pocket Guide to Chemical Hazards,* U.S. Department of Health and Human Services/U.S. Department of Labor Publication No. 78–210, available from the Superintendent of Documents, U.S. Government Printing Office, Washington, D.C. 20402.

9-5 SMOKE

Put simply, smoke is the gray to black matter that is generally observed in connection with the incomplete combustion of the material comprising class A and class B fires. In such ordinary fires, smoke consists of a suspension of finely divided carbon particulates in air. These particulates originate as microscopic particles; but they readily conglomerate, thereby becoming visible.

Smoke usually occurs in unison with the products of incomplete combustion, as Fig. 9-4 illustrates. Thus, the appearance of smoke may forewarn the observer of the presence of carbon monoxide in the nearby atmosphere. Together, smoke and combustion products often represent a greater hazard to life and a more serious hindrance to firefighting efforts than the fire itself. Smoke poses such a hazard in four general ways.

1. Smoke obscures vision.
2. When smoke is inhaled, the larger particulates are filtered in the nasal passageway, but the smaller ones may be drawn into the bronchi and lungs. These particulates coat the surfaces of the bronchial passages and lungs, thereby interfering with normal respiration and often causing adverse health effects, ranging from nausea to death.
3. The carbon particulates comprising smoke are adsorbants (Sec. 6-8). Since toxic gases almost always accompany smoke formation and fires, the carbon particulates resemble tiny vehicles transporting toxic gases into the lungs.
4. The smoke from some class A and class B fires contains complex polynuclear aromatic hydrocarbons (Sec. 11-3) like benzo(a)pyrene, a suspect human carcinogen. Inhalation of such smoke may thus induce lung cancer.

Since the dawn of the Industrial Revolution, our air has increasingly become contaminated with the particulates emitted from the smokestacks of various factories and other industrial plants. Particulates are only part of such atmospheric emissions; typically, carbon monoxide and other pollutants are emitted as well. Under the auspices of the Clean Air Act, USEPA now regulates the amounts of such pollutants that may be emitted to the air. With respect to particulates, the primary standard requires that a facility may emit an annual concentration of no more than 75 μg/m^3 and a daily concentration of no more than 260 μg/m^3. The secondary standard requires that no more than 60 μg/m^3 may be emitted annually and no more than 150 μg/m^3 during a 24-hour period.

9-6 COMMON OXIDES OF CARBON

Numerous fires involve the combustion of wood, plastics, heating fuels, and other materials containing carbon in their molecular structures. When such materials burn, the constituent carbon unites with oxygen to form either carbon monoxide or carbon

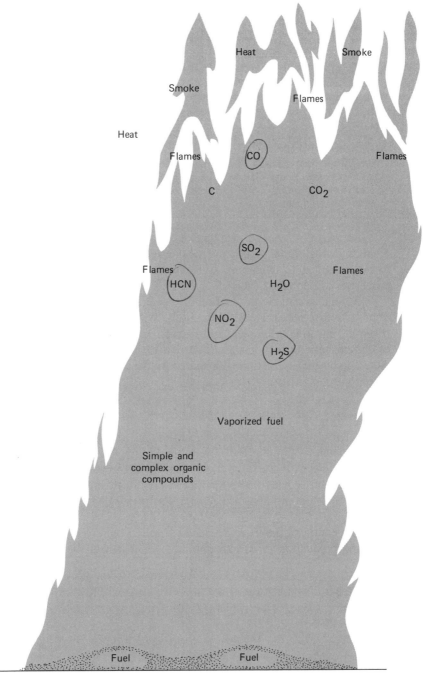

Figure 9-4 Smoke is typically produced during the incomplete combustion of carbon-bearing fuels, along with carbon monoxide and water vapor. Depending on the nature of the fuel, however, the atmosphere surrounding a fire may also contain vapors of the fuel, its decomposition products, and one or more fire gases. When inhaled, these substances are frequently more hazardous to firefighters and fire victims than the accompanying flames or heat.

TABLE 9-2 PHYSICAL PROPERTIES OF CARBON MONOXIDE

Boiling point	$-312.90°F$ ($-191.6°C$)
Heat of vaporization	92.6 Btu/lb (51.4 cal/g)
Liquid density at $-313°F$ ($-191.6°C$)	49.58 lb/ft³ (795 g/L)
Gas density at $-313°F$ ($-191.6°C$)	0.268 lb/ft³ (4.3 g/L)
Gas density at $68°F$ ($20°C$)	0.078 lb/ft³ (1.25 g/L)
Vapor density (air = 1)	0.97
Heat of combustion	67.64 kcal/mol
Autoignition temperature	$1128°F$ ($609°C$)
Lower explosive limit	12.5%
Upper explosive limit	74%
Liquid-to-gas expansion ratio	700

dioxide. These are the common oxides of carbon. Both are colorless, odorless gases whose chemical formulas are CO and CO_2, respectively. Some of their physical properties are noted in Tables 9-2 and 9-3, respectively.

To understand the production of the common oxides of carbon during combustion, it is first necessary to differentiate between the two parallel processes illustrated in Fig. 9-5: incomplete and complete combustion. The *incomplete combustion* of carbon-bearing fuels produces carbon monoxide and soot. Consider the incomplete combustion of methane illustrated by the following equations:

$$2CH_4(g) + 3O_2(g) \longrightarrow 2CO(g) + 4H_2O(g)$$

$$CH_4(g) + O_2(g) \longrightarrow C(s) + 2H_2O(g)$$

Incomplete combustion occurs when the supply of air or oxygen, or access to either air or oxygen, is limited. Put differently, incomplete combustion occurs in fuel-rich fires. This means that carbon monoxide and/or soot are likely combustion products when a carbon-bearing fuel has not been previously mixed with air or oxygen. Thus, these combustion products are generally associated with fires that originate unintentionally.

By contrast, the *complete combustion* of carbon-bearing fuels is associated with the production of carbon dioxide. For instance, the complete combustion of

TABLE 9-3 PHYSICAL PROPERTIES OF CARBON DIOXIDE

Freezing point (at 5.1 atm)	$-69.8°F$ ($-56.55°C$)
Sublimation point (at 1 atm)	$-109.3°F$ ($-78.5°C$)
Heat of fusion	85.68 Btu/lb (47.5 cal/g)
Heat of sublimation	65.30 Btu/lb (36.2 cal/g)
Density of solid (at 1 atm)	97.34 lb/ft³ (1.56 g/mL)
Density of gas at $-109.3°F$ ($-78.5°C$)	0.175 lb/ft³ (2.8 g/L)
Density of gas at $68°F$ ($20°C$)	0.124 lb/ft³ (1.98 g/L)
Vapor density (air = 1)	1.529
Solid-to-gas expansion ratio	790

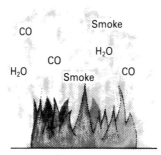 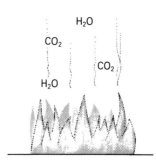

Figure 9-5 The incomplete combustion of carbon-bearing fuels, shown on the left, results in the production of carbon monoxide, carbon particulates (smoke), and water vapor. This situation is typical of a fuel-rich or oxygen-starved fire. On the right, the complete combustion of carbon-bearing fuels results in production of carbon dioxide and water vapor. This combustion is typical of a fuel that had been previously mixed with air or oxygen.

methane is illustrated by the following equation:

$$CH_4(g) + 2O_2(g) \longrightarrow CO_2(g) + 2H_2O(g)$$

Complete combustion occurs when air or oxygen is available in plentiful supply during the burning process. Thus, a mixture of 1 mol of methane and 2 mol of oxygen is more likely to produce carbon dioxide when ignited than carbon monoxide.

Carbon Monoxide

Even more than from exposure to a fire itself, carbon monoxide often causes the death of fire victims. Figure 9-6 demonstrates that individuals who are trapped during a fire may die from exposure to fire gases, like carbon monoxide, before the fire ever reaches them. For those fortunate to survive, carbon monoxide can still inflict other health problems, depending on the amount of carbon monoxide that has been inhaled; some acute effects are demonstrated in Fig. 9-7.

Carbon monoxide may also be encountered in nonfire situations. For instance, we routinely breathe atmospheric carbon monoxide. The atmosphere is polluted with carbon monoxide, since it is almost always a component of the exhaust of automobiles. Carbon monoxide forms when vehicular fuels burn, rather than carbon dioxide, since the supply of oxygen during the combustion of gasoline is usually limited. An automobile's internal combustion engine emits an average of 2.9 lb of carbon monoxide per gallon of gasoline (0.35 kg/L). The amount of carbon monoxide released to the atmosphere by the burning of gasoline in internal combustion engines alone is estimated to be 193 million tons per year. This would ordinarily mean that the atmospheric concentration of carbon monoxide is continuously increasing. Fortunately, however, the carbon monoxide that is released to the atmosphere slowly oxidizes to carbon dioxide.

$$2CO(g) + O_2(g) \longrightarrow 2CO_2(g)$$

Figure 9-6   A victim of a mattress fire. Note the outline of the body on the floor in the photograph at the right, as well as the lack of observable soot under the body. This suggests that death occurred prior to the onset of a flaming combustion and was probably caused by inhalation of a noxious or poisonous atmosphere, such as one containing carbon monoxide. (Courtesy of I. N. Einhorn, Flammability Research Center, University of Utah, Salt Lake City, Utah)

Scientists currently feel that several microorganisms in soil are capable of catalytically controlling this oxidation so that the level of carbon monoxide in the atmosphere remains at the relatively constant value of 0.1 ppm.

Carbon monoxide is also released to the atmosphere by various manufacturing and process industries. Such emissions are regulated by USEPA under the Clean Air Act. In this instance, the primary and secondary standards are the same: The concentration is limited to 40 mg/m^3 (35 ppm) during a 1-hour period and 10 mg/m^3 during an 8-hour period.

Carbon monoxide is a poisonous substance. But how does it adversely affect the body? To answer this question, let's examine the physiology of respiration. When we inhale air, a supply of oxygen is absorbed through the blood vessels in the lungs into the bloodstream. The oxygen is then carried in the bloodstream by a complex component of the blood called *hemoglobin*. The molecular structure of hemoglobin is noted in Fig. 9-8, but due to its complexity, we shall represent it as *Hb*. When hemoglobin picks up oxygen, it does so by forming a compound called *oxyhemoglobin,* represented here as O_2Hb. The production of oxyhemoglobin is then represented by the following equation:

$$Hb(aq) + O_2(g) \longrightarrow O_2Hb(aq)$$

Oxyhemoglobin releases the oxygen at different tissues and organs throughout the body by the reverse reaction.

$$O_2Hb(aq) \longrightarrow Hb(aq) + O_2(g)$$

Having accomplished its function, the hemoglobin returns to the lungs to secure a new supply of oxygen.

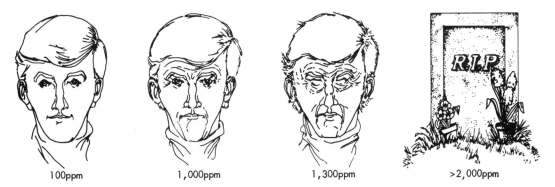

100ppm   1,000ppm   1,300ppm   >2,000ppm

Figure 9-7 Carbon monoxide is a toxic combustion product to which all firefighters are exposed during routine work-related activities. An adult may tolerate carbon monoxide at a concentration in air of 100 ppm without noticeably suffering adverse health effects. However, a 1-hour exposure to an atmosphere containing a concentration of carbon monoxide of 1000 ppm generally causes the individual to experience a mild headache; a reddish coloration of the skin simultaneously develops. A 1-hour exposure to an atmosphere containing 1300 ppm of carbon monoxide usually causes the skin to turn cherry red; the accompanying headache becomes throbbing. A 1-hour exposure to an atmosphere containing 2000 ppm of carbon monoxide is likely to either cause the individual's death or damage to the respiratory and nervous systems of an individual who survives.

When inhaled, carbon monoxide interrupts the respiratory process by interfering with the transport of oxygen. The carbon monoxide readily absorbs into the bloodstream, where it forms a substance called *carboxyhemoglobin* (COHb). This process is represented by the following equation:

$$Hb(aq) + CO(g) \longrightarrow COHb(aq)$$

Carbon monoxide interferes with respiration due to the fact that the chemical affinity for the formation of carboxyhemoglobin is about 300 times greater than the affinity for the formation of oxyhemoglobin. In fact, even oxyhemoglobin readily gives up

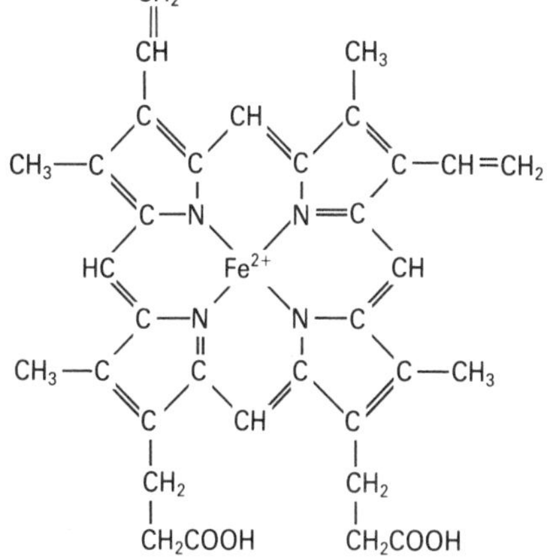

Figure 9-8 The molecular structure of hemoglobin, a complex component of blood whose normal function is to transport oxygen from the lungs to the various tissues and organs of the body.

its oxygen to absorb carbon monoxide, as noted by the following equation:

$$O_2Hb(aq) + CO(g) \longrightarrow COHb(aq) + O_2(g)$$

Carboxyhemoglobin is relatively stable. Hence, when carbon monoxide is inhaled, a quantity of hemoglobin is tied up in carboxyhemoglobin and is unable to perform its normal bodily function. Then the transport of oxygen is hindered and the host suffers from *anoxia,* the lack of oxygen in the blood.

The first noticeable symptom of carbon monoxide poisoning is generally a mild headache. The skin, especially under the fingernails, then takes on a cherry-red coloration. As the concentration of carbon monoxide increases in the bloodstream, the headache may become throbbing and other adverse health effects may become apparent. The TLV for carbon monoxide in air is only 100 ppm. Thus, the human body noticeably tolerates very little of this substance. Other adverse health effects caused by exposure to carbon monoxide are noted in Table 9-4 as a function of the percentage of carboxyhemoglobin present in the bloodstream.

Individuals who have lost consciousness due to the inhalation of too much carbon monoxide should be moved to an open space, where artificial respiration may be applied and oxygen administered. Such individuals should be taken immediately to a medical facility or trauma center capable of providing hyperbaric oxygen therapeutic treatment, like that illustrated in Fig. 9-9.

One of the more serious ways by which we may be exposed to carbon monoxide, although somewhat inadvertently, is through cigarette smoking. Each cigarette is a small cylinder of tightly packed tobacco; thus, the burning of tobacco products is an example of incomplete combustion. This results in the production of carbon monoxide as well as other toxic substances. When inhaled, cigarette smoke results

TABLE 9-4 SIGNS AND SYMPTOMS RESULTING FROM INHALATION OF VARIOUS
CONCENTRATIONS OF CARBOXYHEMOGLOBIN

Percent carboxyhemoglobin	Signs and symptoms
0–10	No signs or symptoms
10–20	Tightness across forehead, possible slight headache, dilation of the cutaneous blood vessels
20–30	Headache and throbbing in the temples
30–40	Severe headache, weakness, dizziness, dimness of vision, nausea, vomiting, and collapse
40–50	Same as above, greater possibility of collapse; cerebral anemia, and increased pulse and respiratory rates
50–60	Cerebral anemia, increased respiratory and pulse rates, coma, and intermittent convulsions
60–70	Coma, intermittent convulsions, depressed heart action and respiratory rate, and possible death
70–80	Weak pulse and slow respiration, leading to death within hours
80–90	Death in less than 1 hour
90–100	Death within a few minutes

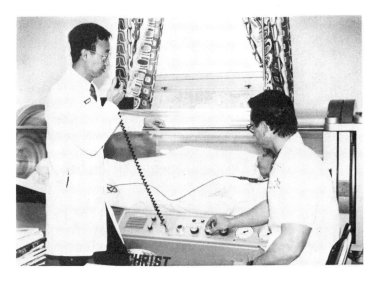

Figure 9-9   A therapist and physician attending to a patient in a hyperbaric oxygen therapy chamber. Hyperbaric oxygen therapy involves administering 100% oxygen at a pressure of up to three times the normal atmospheric pressure for 60 to 90 minutes. The utilization of hyperbaric oxygen therapy at 2 atm accelerates the removal of carbon monoxide from the bloodstream. (Courtesy of St. James Hospital, Medical Center, Chicago Heights, Illinois)

in an increased concentration of carboxyhemoglobin in the bloodstream. The average concentration of carboxyhemoglobin in the blood of heavy smokers is 6.2% by volume.

Aside from its primary hazard as a poisonous gas, carbon monoxide is also a flammable gas. When ignited, it forms carbon dioxide, producing a pale blue flame.

Commercially, carbon monoxide is available as a compressed gas and cryogenic liquid. In the gaseous and liquid states of matter, USDOT regulates the transportation of carbon monoxide as a poisonous gas. Cylinders and other containers of carbon monoxide are labeled POISON GAS and FLAMMABLE GAS; their transport vehicles are placarded POISON GAS.

In the workplace, OSHA stipulates a permissible exposure limit to carbon monoxide at 50 ppm by volume, while NIOSH stipulates only 35 ppm.

Finally, carbon monoxide may be encountered in combined form with metals in hazardous materials called *metallic carbonyls*. An example is nickel tetracarbonyl, whose chemical formula is $Ni(CO)_4$; it is a flammable and poisonous liquid. USDOT regulates the transportation of nickel tetracarbonyl as a poisonous material. Containers of this substance are labeled POISON and FLAMMABLE LIQUID; their transport vehicles are placarded POISON.

Carbon Dioxide

In contrast to carbon monoxide, the body is considerably more tolerant of the presence of carbon dioxide. As normal respiration proceeds, carbon dioxide is formed as a product of oxidation when nutrients in foodstuffs are metabolized. The carbon dioxide remains dissolved in the blood, from which it is thereafter released and exhaled from the lungs. As a by-product of respiration, some concentration of carbon dioxide is always present in the bloodstream at any time.

Because of its role in respiration, the inhalation of relatively small concentrations of carbon dioxide tends to stimulate the rate and depth of normal respiration. This fact has an important impact on firefighters. Since inhalation of only a small amount of carbon dioxide may increase the rate of breathing, the routine duties involved in firefighting may become even more difficult to accomplish than ordinarily.

This is not meant to imply, however, that the body is totally tolerant of carbon dioxide. The TLV for carbon dioxide in air is 5000 ppm (0.5%). But when higher concentrations of this gas are inhaled, the carbon dioxide may cause adverse health effects. Inhalation of an atmosphere containing 3% to 7% percent carbon dioxide by volume may cause asphyxiation. Inhalation of a concentration in excess of 9% by volume may be tolerable for only a few minutes; prolonged inhalation of this concentration may be fatal. At these elevated concentrations, carbon dioxide adversely affects the central nervous system. Individuals who have become asphyxiated due to inhalation of too much carbon dioxide should be promptly removed to an open space and administered oxygen.

Usually large supplies of carbon dioxide are transported and stored for use as a refrigerant and a fire extinguisher (Sec. 4-9) or for the preparation of carbonated beverages. While carbon dioxide is well known as a compressed gas, it may also be encountered commercially as either a cryogenic fluid or solid. Solid carbon dioxide is commonly known as Dry Ice.

In the gaseous and liquid states of matter, USDOT regulates the transportation of carbon dioxide as a nonflammable gas. When required, containers of carbon dioxide are labeled NONFLAMMABLE GAS; their transport vehicles are placarded NONFLAMMABLE GAS. USDOT regulates the transportation of Dry Ice as a miscellaneous hazardous material. Its proper shipping name is "carbon dioxide, solid (Dry Ice)." When required by USDOT, containers of Dry Ice are labeled CLASS 9. Their transport vehicles are not placarded.

In the workplace, OSHA stipulates a permissible exposure limit to carbon dioxide at 5000 ppm by volume.

9-7 HYDROGEN CYANIDE

At room temperature, hydrogen cyanide is a colorless gas; its chemical formula is HCN. The important physical properties of hydrogen cyanide are noted in Table 9-5.

Minute quantities of hydrogen cyanide are found naturally in bitter almonds and in the kernel of the seeds of peaches, apricots, and plums. The odor of hydrogen cyanide is often said to be that of bitter almonds, but some people are incapable of detecting this odor.

TABLE 9-5 PHYSICAL PROPERTIES OF HYDROGEN CYANIDE

Freezing point	6.8°F (−14°C)
Boiling point	79°F (26°C)
Density at 68°F (20°C)	0.075 lb/ft³ (1.20 g/L)
Vapor density	0.93
Lower explosive limit	6%
Upper explosive limit	41%
Autoignition temperature	1000°F (538°C)
Flash point of liquid	0°F (−18°C)

Hydrogen cyanide gas is miscible with water in all proportions; the resulting aqueous solution is called *hydrocyanic acid* or *prussic acid*. Hydrogen cyanide may be encountered industrially as this acid. However, anhydrous hydrogen cyanide is a commercially available liquid. It is an unstable substance, but stabilizers like hydrogen chloride are routinely added to the commercial-grade material. Today, it is in this latter form that hydrogen cyanide is more generally encountered. It is used to fumigate ships, warehouses, and greenhouses.

Hydrogen cyanide is also used extensively for the production of certain plastics. For example, polyacrylonitrile is a plastic consisting of molecules having cyanide groups (—CN). It is in connection with such plastics that hydrogen cyanide is most likely to be encountered as a fire gas. When these plastics are strongly heated, they thermally decompose, forming hydrogen cyanide. When textiles made of silk and wool are exposed to an intense source of heat, they may also generate hydrogen cyanide.

Hydrogen cyanide burns with a purple flame when its concentration in air is between 5.6% and 40% by volume. Thus, when produced during fires, it is likely to be consumed by combustion.

The adverse health effects caused by inhalation of hydrogen cyanide are noted in Table 9-6. The fact that inhalation of this gas may be lethal should be apparent: It once was used in gas chambers to end the lives of criminals who had been sentenced to receive the death penalty. Hydrogen cyanide may also be absorbed through the

TABLE 9-6 TOXICOLOGICAL PROPERTIES OF HYDROGEN CYANIDE

HCN concentration (ppm)	Symptoms
0.2–5.0	Threshold of odor
10	Threshold limit value
18–36	Slight symptoms (headache) after several hours
45–54	Tolerated for $\frac{1}{2}$ to 1 hour without difficulty
100	Death within 1 hour
110–135	Fatal in $\frac{1}{2}$ to 1 hour
181	Fatal after 10 minutes
280	Immediately fatal

skin. The TLV for skin absorption is only 10 ppm. Thus, when a person must spend time in an atmosphere containing hydrogen cyanide (as during rescue operations), the ability of hydrogen cyanide to absorb through the skin should be considered. The use of a rubberized suit and positive-demand, self-contained breathing apparatus is essential.

Like carbon monoxide, hydrogen cyanide interferes with the transportation of oxygen in the bloodstream. Hydrogen cyanide inhibits the action of the enzyme, *cyctochrome oxidase,* which is needed for cellular respiration and energy production.

Many chemical compounds are known in which metallic ions are chemically combined with cyanide ions. Examples of such compounds are sodium cyanide, potassium cyanide, copper cyanide, and zinc cyanide. These substances are normally employed for electroplating certain metals, an operation common to many industries. Like hydrogen cyanide, metallic cyanides are poisonous by skin absorption; this is illustrated in Fig. 9-10. They are also incompatible with acids, forming hydrogen cyanide, as the following illustrative equation indicates:

$$NaCN(s) + HCl(aq) \longrightarrow NaCl(aq) + HCN(g)$$

For this reason, metallic cyanides and acids should always be stored in different locations.

Individuals exposed to excessive amounts of hydrogen cyanide should be provided medical assistance as quickly as possible.

USDOT regulates the transportation of hydrogen cyanide and various metallic cyanides as poisonous materials. Containers of anhydrous, stabilized hydrogen cyanide are labeled POISON and FLAMMABLE LIQUID; when absorbed into an inert material, such containers are labeled POISON. Their transport vehicles are placarded POISON. Containers of metallic cyanides are labeled POISON, and their transport vehicles are placarded POISON.

In the workplace OSHA stipulates a permissible exposure limit to hydrogen cyanide at 10 ppm by volume; NIOSH stipulates only 5 mg/m^3.

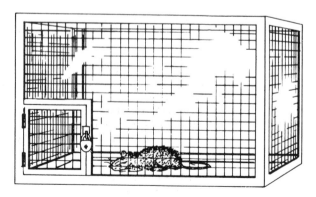

Figure 9-10 Metallic cyanides may absorb through the skin and exert adverse health effects. Here, a mouse is forced to walk on finely ground crystals of sodium cyanide, but is simultaneously prevented from orally ingesting the substance. After the elapse of approximately 1/2 hour, the mouse succumbs to cyanide poisoning.

9-8 SULFUR-BEARING FIRE GASES

Sulfur-containing compounds are common constituents of coal, natural gas, crude oil, wool, hair, animal hides, and several natural and synthetic polymers, including vulcanized rubber. When these materials are exposed to intense heat or when they burn, they form either of the fire gases containing sulfur: hydrogen sulfide or sulfur dioxide.

Hydrogen Sulfide

Hydrogen sulfide is the sulfur analog of water; thus, its chemical formula is H_2S. Some of its physical properties are noted in Table 9-7. It is a colorless gas possessing the disagreeable odor associated with rotten eggs. It is occasionally encountered naturally as the result of the anerobic decay of organic waste. Sewage and swamp water, for example, typically contain dissolved hydrogen sulfide. We sometimes hear that such materials "smell like sulfur." But elemental sulfur is an odorless solid; what is actually meant is that such materials smell like hydrogen sulfide. Some hydrogen sulfide is almost always present in our atmosphere. This is evidenced by the fact that unprotected silverware exposed to air becomes tarnished, a chemical phenomenon caused by the reaction of silver with atmospheric hydrogen sulfide.

The TLV for hydrogen sulfide is only 10 ppm. Continued inhalation in an atmosphere containing hydrogen sulfide causes dizziness and the onset of a headache. One deep breath of pure hydrogen sulfide is fatal; breathing a concentration of 600 ppm by volume is fatal within 30 minutes. Since it possesses such a disagreeable odor, most people are initially aware of its presence. However, hydrogen sulfide also deadens the sense of smell; it is said to cause *olfactory fatigue*. Thus, individuals who remain in an atmosphere containing hydrogen sulfide become oblivious to its presence and may inhale dangerous amounts unknowingly.

Hydrogen sulfide is available industrially, mainly as a liquid, in containers like that shown in Fig. 9-11. It is primarily used in the chemical industry to produce other sulfur-containing compounds, but hydrogen sulfide is also used in the metallurgical industry. USDOT regulates the transportation of hydrogen sulfide as a poi-

TABLE 9-7 PHYSICAL PROPERTIES OF HYDROGEN SULFIDE

Boiling point	$-76°F$ ($-60°C$)
Freezing point	$-117°F$ ($-83°C$)
Density 68°F (20°C)	0.0961 lb/ft^3 (1.539 g/L)
Vapor density (air = 1)	1.2
Autoignition temperature	500°F (260°C)
Lower explosive limit	4.3%
Upper explosive limit	46%
Heat of fusion	0.568 kcal/mol
Heat of vaporization	4.463 kcal/mol

Figure 9-11 Two specially equipped vehicles used to ship bulk quantities of hydrogen sulfide as a liquefied, compressed gas: a 5000-gal (19,000-L) tank truck and a 22,000-gal (83,000-L) rail tank car. These transport vehicles can hold approximately 14 tons (13,000 kg) and 64 tons (58,000 kg) of hydrogen sulfide, respectively. USDOT requires the shipper to placard each vehicle with FLAMMABLE GAS and POISON on both sides and both ends and to display the USDOT identification number 1053. (Courtesy of Montana Sulphur & Chemical Company, Billings, Montana)

sonous gas. Containers are labeled POISON GAS and FLAMMABLE GAS, and their transport vehicles are similarly placarded.

In the workplace, OSHA regulates the exposure of employees to hydrogen sulfide. OSHA stipulates a permissible exposure limit of 50 ppm by volume of hydrogen sulfide for no more than 10 minutes, but NIOSH allows only 10 ppm.

Sulfur Dioxide

Sulfur dioxide is a colorless gas possessing the sharp, pungent odor of burning tires. Its chemical formula is SO_2. Some important physical properties of this substance are noted in Table 9-8.

Sulfur dioxide is a product of combustion that results from the burning of sulfur-containing materials. When hydrogen sulfide burns, for instance, sulfur dioxide forms:

$$2H_2S(g) + 3O_2(g) \longrightarrow 2H_2O(g) + 2SO_2(g)$$

Sulfur-containing substances are often present in coal and other fossil fuels. Consequently, sulfur dioxide results from burning such materials. Near major industrialized areas, it is often encountered as an air pollutant. In the past, our nation's power plants that burned coal emitted almost two-thirds of the sulfur dioxide released into the air.

Emitted to the atmosphere, sulfur dioxide slowly oxidizes to sulfur trioxide, which dissolves in atmosphere moisture, forming sulfuric acid. These phenomena are represented by the following equations:

$$2SO_2(g) + O_2(g) \longrightarrow 2SO_3(g)$$

$$SO_3(g) + H_2O(g) \longrightarrow H_2SO_4(aq)$$

A mixture of sulfur dioxide and sulfur trioxide is sometimes represented as SO_x.

Two environmental problems have developed in highly industrialized regions of the world, where the atmospheric sulfur dioxide concentration has been relatively high: acid rain and sulfurous smog. As the name implies, *acid rain* is precipitation contaminated with dissolved acids like sulfuric acid. Acid rain has posed a threat to the environment by causing certain lakes to become void of aquatic life. The second problem, *sulfurous smog*, is the haze that develops in the atmosphere when molecules of sulfuric acid accumulate, growing in size as droplets until they become sufficiently large to serve as light scatterers.

Under the Clean Air Act, USEPA regulates the amount of the sulfur oxides that facilities may emit to the atmosphere. The primary standard limits the concentration to 80 μg/m^3 (0.03 ppm) during a 24-hour period and to 365 μg/m^3 (0.14 ppm) during a 3-hour period. The secondary standard limits the annual concentration to 1300 μg/m^3 (0.5 ppm). Due to enforcement of the Clean Air Act, great

TABLE 9-8 PHYSICAL PROPERTIES OF SULFUR DIOXIDE

Boiling point	14°F (-10°C)
Freezing point	-105°F (-76°C)
Density at 68°F (20°C)	0.183 lb/ft^3 (2.93 g/L)
Vapor density (air = 1)	2.3
Heat of fusion	1.77 kcal/mol
Heat of vaporization	5.96 kcal/mol

advances have been made in reducing sulfurous atmospheric emissions at modern coal-burning facilities, like that shown in Fig. 9-12.

The threshold limit value for sulfur dioxide is only 5 ppm; at increased concentrations, sulfur dioxide immediately affects the respiratory system. Coughing, chest pains, shortness of breath, and constriction of the airways are immediate symptoms of exposure to this gas. At concentrations exceeding 10 ppm by volume, sulfur dioxide also acts as an eye irritant. In concentrations exceeding 500 ppm by volume, sulfur dioxide may cause death instantly.

Sulfur dioxide is used commercially as a bleaching agent in the pulp and paper industry, in refining sugar, and in processing dried fruit. Some sulfur dioxide is still

Figure 9-12   A modern coal-fired power plant at Colstrip, Montana, which the U.S. Department of Energy used as part of a study to characterize atmospheric emissions from coal-burning facilities. Employees periodically used a helicopter equipped with appropriate instrumentation to analyze the stack plumes for objectionable gaseous constituents. Through such studies, engineering procedures have been devised that effectively reduce the amount of sulfur dioxide emitted to the atmosphere from the burning of coal. (Courtesy of the U.S. Department of Energy, Washington, D.C.)

used as a refrigerant, but generally other compounds have superseded this use. Most commonly, it is available in the liquid state of matter. USDOT regulates its transportation as a poisonous gas. Containers are labeled POISON GAS, and their transport vehicles are placarded POISON GAS.

OSHA stipulates a maximum permissible exposure to sulfur dioxide at 5 ppm by volume, but NIOSH stipulates only 0.5 ppm.

9-9 OXIDES OF NITROGEN

There are six oxides of nitrogen: nitrous oxide or dinitrogen monoxide (N_2O), nitric oxide (NO), dinitrogen trioxide (N_2O_3), nitrogen dioxide (NO_2), dinitrogen tetroxide (N_2O_4), and dinitrogen pentoxide (N_2O_5). The physical properties of several representative oxides of nitrogen are noted in Table 9-9.

Nitric oxide, nitrogen dioxide, and nitrogen tetroxide are fire gases. One or more of them is generated when certain nitrogenous organic compounds burn, like polyurethane (Sec. 12-6). This substance is commercially available as a rigid and flexible plastic; it is used in certain types of home insulation, as well as in bedding and furniture cushioning. Nitric oxide is the product of incomplete combustion, whereas a mixture of nitrogen dioxide and nitrogen tetroxide is the product of complete combustion.

The nitrogen oxides are sometimes collectively symbolized by the formula NO_x. Under the Clean Air Act, USEPA regulates the amount of nitrogen oxides that commercial and industrial facilities may emit to the atmosphere. The primary and secondary standards are the same: The annual concentration of nitrogen dioxide may not exceed 100 $\mu g/m^3$ (0.05 ppm).

Of the six oxides of nitrogen, only dinitrogen trioxide and dinitrogen pentoxide are industrially unimportant. We shall briefly note some features of the other four nitrogen oxides.

Nitrous Oxide

This is a sweet-smelling, nonirritating, colorless gas. It is produced by the thermal decomposition of ammonium nitrate, as the following equation notes:

$$NH_4NO_3(s) \longrightarrow N_2O(g) + 2H_2O(g)$$

TABLE 9-9 PHYSICAL PROPERTIES OF SEVERAL OXIDES OF NITROGEN

	Nitrous oxide	Nitric oxide	Nitrogen dioxide
Boiling point	−127.3°F (−88.49°C)	−243°F (−153°C)	68°F (20°C)
Freezing point	−152.3°F (−102.4°C)	−263°F (−164°C)	12°F (−11°C)
Density	0.124 lb/ft³ (1.978 g/L)	0.084 lb/ft³ (1.34 g/L)	0.093 lb/ft³ (1.49 g/L)
Vapor density (air = 1)	1.53	1.04	1.16

Nitrous oxide is mainly used as an anesthetic by dentists, who refer to the substance as *laughing gas,* since dental patients are apt to laugh hysterically when recovering consciousness. A mixture of approximately 80% by volume of nitrous oxide with oxygen is often administered to prevent anoxia.

Nitrous oxide is a nonflammable gas, but it is an excellent oxidizing agent. Consequently, while it does not burn, nitrous oxide supports the combustion of numerous materials. For instance, sulfur, phosphorus, and carbon burn in an atmosphere of nitrous oxide almost as vigorously as in oxygen. With hydrogen or ammonia, explosive mixtures are produced. Fires supported by nitrous oxide are effectively extinguished by water.

Nitrous oxide is available commercially as a compressed gas and a cryogenic fluid. USDOT regulates its transportation as a nonflammable gas. Containers of nitrous oxide are labeled NONFLAMMABLE GAS and OXIDIZER, and their transport vehicles are placarded NONFLAMMABLE GAS.

Nitric Oxide

This is a colorless gas with a sharp, sweet odor; it is important commercially in connection with the production of nitric acid. Nitric oxide is a flammable gas, forming nitrogen dioxide, as the following equation illustrates:

$$2NO(g) + O_2(g) \longrightarrow 2NO_2(g)$$

It is also an excellent oxidizing agent. In fact, nitric oxide supports combustion almost as well as does oxygen, particularly when the burning material is heated prior to ignition. Phosphorus and magnesium, for instance, burn vigorously in an atmosphere of nitric oxide. Water is generally effective at extinguishing such fires.

Nitric oxide is toxic; its TLV is only 25 ppm, a concentration easily exceeded in fires involving nitrogenous materials. When inhaled, this substance poisons in a manner reminiscent of the mechanism of carbon monoxide poisoning. The nitric oxide chemically combines with hemoglobin to form a substance called *metheglobin* (NOHb), as noted by the following equation:

$$Hb(aq) + NO(g) \longrightarrow NOHb(aq)$$

The production of metheglobin thus reduces the ability of hemoglobin to transport oxygen to bodily tissues and organs. This affliction is called *methemoglobinemia.*

Other than in connection with its nature as a fire gas, nitric oxide is commonly found in the atmosphere as an air pollutant. The origin of this gas in the atmosphere is associated with the union of atmospheric nitrogen and oxygen. Both of these gases are plentiful in the atmosphere, but they normally do not unite to a considerable extent since the appropriate energy of activation is absent.

Nonetheless, in the cylinders of motor vehicles, where a mixture of gasoline vapor and the air ignites, the temperature greatly exceeds that of the ambient air. At this elevated temperature, sufficient energy becomes available such that nitrogen and oxygen chemically combine. This chemical reaction is represented by the following equation:

$$N_2(g) + O_2(g) \longrightarrow 2NO(g)$$

Most of the nitric oxide normally present in the atmosphere originates in this fashion. It occurs in the relatively constant concentration of 0.2 ppm by volume.

Nitric oxide is available commercially as a compressed gas. USDOT regulates its transportation as a poisonous gas. Containers are labeled POISON GAS, and their transport vehicles are placarded POISON GAS.

In the workplace, OSHA regulates the exposure of employees to nitric oxide; the maximum recommended permissible exposure limit is 25 ppm by volume.

Nitrogen Dioxide and Dinitrogen Tetroxide

Nitrogen dioxide is a dark red to brown gas possessing a pungent, acrid odor. At room temperature, this gas has often dimerized (that is, combined with itself, mole for mole), forming dinitrogen tetroxide. The dimerization of nitrogen dioxide is indicated by the following equation:

$$2NO_2(g) \longrightarrow N_2O_4(g)$$

Thus, these two gases are generally found together, although not necessarily in equal concentrations; either is likely to be encountered when nitrogenous materials burn in air. Nitrogen dioxide has been identified as a component of cigarette smoke in an average concentration of 300 ppm by volume.

Aside from being a fire gas, nitrogen dioxide is also frequently encountered as an air pollutant, where it originates from the slow oxidation of nitric oxide.

$$2NO(g) + O_2(g) \longrightarrow 2NO_2(g)$$

Both nitrogen dioxide and dinitrogen tetroxide are powerful oxidizing agents. They support combustion more readily than either nitrous oxide or nitric oxide. It is for this reason that these gases have been used as the oxidizing agents in rockets. Water effectively extinguishes most fires supported by them.

Nitrogen dioxide and dinitrogen tetroxide are toxic substances that may cause methemoglobinemia. The threshold limit value for nitrogen dioxide is only 5 ppm. Inhalation initially causes respiratory distress, such as shortness of breath; inflammable of the lungs is common. However, exposure to nitrogen dioxide may also be lethal. By adversely affecting the pulmonary tissues, this gas causes hemorrhaging in the lungs; death may occur within days following the initial exposure.

Nitrogen dioxide and dinitrogen tetroxide are commercially available as a liquid in which these substances are in chemical equilibrium. USDOT regulates its transportation as a poisonous gas. Containers are labeled POISON GAS and OXIDIZER, and their transport vehicles are placarded POISON GAS.

In the workplace, OSHA regulates employee exposure to these gases. The permissible exposure limit to either gas is only 5 ppm by volume.

TABLE 9-10 PHYSICAL PROPERTIES OF AMMONIA

Boiling point	$-28°F$ ($-33°C$)
Freezing point	$-108°F$ ($-78°C$)
Density	0.048 lb/ft³ (0.771 g/L)
Vapor density (air = 1)	0.596
Autoignition temperature	1204°F (651°C)
Lower explosive limit	16%
Upper explosive limit	25%

9-10 AMMONIA

gas

Ammonia is a colorless gas with a penetrating, suffocating odor; its chemical formula is NH₃. Some of its important physical properties are noted in Table 9-10. Ammonia has been known as a unique chemical substance since the days of alchemy, when it was prepared by heating the hoofs and horns of animals or by heating coal in the absence of air (Sec. 6-8). In the early 1900s, ammonia was still primarily obtained as a by-product of the heating of coal. Today, however, ammonia is largely produced at large manufacturing plants, like that shown in Fig. 9-13, directly from hydrogen and atmospheric nitrogen. At elevated temperatures and pressures, these gases unite, as the following equation illustrates:

$$N_2(g) + 3H_2(g) \longrightarrow 2NH_3(g)$$

As a compressed gas, ammonia is commercially available in steel cylinders and ton containers.

Ammonia is another fire gas. Firefighters may encounter it when materials made from animal products are exposed to an intense source of heat, such as leather items or carpeting made from wool. But fire personnel are undoubtedly more likely to encounter the vapors of ammonia in situations where it is used commercially, or in transportation mishaps. The greatest commercial consumption of ammonia is associated with the manufacture of fertilizers and in connection with its direct application to fields as a fertilizer.

liquefied

Ammonia is easily liquefied by applying low pressure to the confined gas. This form is called *anhydrous liquid ammonia,* which is often stored in 20,000-lb tanks. It is also shipped to various destinations and transferred by pipeline, usually to consuming agricultural areas. Farmers can dispense the anhydrous ammonia directly to soil or irrigation waters. Typically, the ammonia is discharged to soil from a tractor saddle tank or nurse tank mounted behind the tillage tool through a distribution pod, as illustrated in Fig. 9-14. Other than its popular use in agriculture, anhydrous liquid ammonia may be encountered as a refrigerant in large refrigeration installations.

As Table 9-10 illustrates, the vapor density of ammonia is 0.59; hence, ammonia is lighter than air and quickly disperses into the atmosphere when released from its container. Dispersal is even more rapid under windy climatic conditions. Although colorless, when first released to the atmosphere from its liquid storage tank,

Figure 9-13 The major supply of ammonia is produced by the direct synthesis of hydrogen and atmospheric nitrogen under high temperature and pressure. This Phillips Petroleum plant in Beatrice, Nebraska, can produce 575 tons of ammonia daily. The large commercial demand for ammonia is linked with its utilization as a fertilizer for crops, either directly or indirectly, in such products as ammonium nitrate and ammonium phosphate. (Courtesy of the American Petroleum Institute, Washington, D.C.)

ammonia is generally visible as a white fog caused by condensed atmospheric moisture. This feature may permit detection of an ammonia leak.

When large volumes of liquid ammonia are released to the atmosphere at once, as during a transportation accident, unusually large amounts of ammonia gas can concentrate in the immediate area. It is critical in such situations to minimize exposure to this substance. Ammonia acts as an alkali on human skin (Sec. 7-2). Its effect can range from mild irritation to tissue destruction, depending on the length of exposure. The eyes and lungs are particularly susceptible to the caustic action of ammonia. Eye contact causes an immediately noticeable irritation, which, if left unattended, could result in the loss of sight. Ammonia also causes extreme irritation of the bronchial tissues when inhaled; continued inhalation destroys respiratory tis-

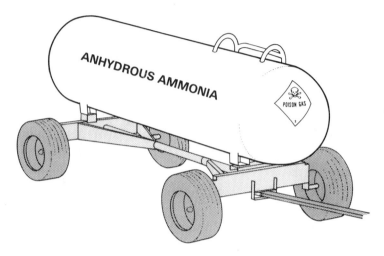

Figure 9-14 Anhydrous ammonia is extremely popular as a fertilizer. The typical "ammonia wagon" used to fertilize soils holds from 100 to 1450 gal (380 to 5500 L) of ammonia. Although confined as its vapor, ammonia is applied to the soil in the liquid state through the use of a converter, which may be mounted on any tillage tool. (Courtesy of DMI, Inc., Goodfield, Illinois)

sue, which causes respiratory and pulmonary diseases. Elevated blood ammonia concentrations may cause death by suffocation. The human response to some concentrations of ammonia is noted in Table 9-11.

Water is capable of absorbing large volumes of ammonia. At room conditions, 1 volume of water absorbs 1176 volumes of ammonia. This is an important factor to recall during emergencies involving ammonia. Water can be applied directly to skin tissue in order to remove any ammonia that has dissolved in surficial body fluids. Water can also be used to effectively disperse the ammonia vapors. The latter is best accomplished by establishing a water curtain downwind from the point where ammonia has been released to the atmosphere.

Ammonia is a flammable gas, but its flammable range is relatively narrow,

TABLE 9-11 INHALATION EFFECTS OF AMMONIA ON HUMANS

Ammonia concentration in air (ppm)	Symptoms
5–10	Detectable limit by odor
50	No chronic effects
150–200	General discomfort; eye tearing; irritation and discomfort of exposed skin; irritation of mucous membranes
400–700	Pronounced irritation and discomfort to the eyes, ears, nose, and throat
2000	Barely tolerable for more than a few moments; serious blistering of the skin; danger of lung edema, asphyxia, and death within minutes following exposure

only from 16% to 25% by volume. Its lower explosive limit is also relatively high. In combination, these factors reduce the likelihood of ammonia fires. In oxygen, ammonia burns with a weak yellow flame to form nitrogen and water, as the following equation notes:

$$4NH_3(g) + 3O_2(g) \longrightarrow 2N_2(g) + 6H_2O(g)$$

USDOT regulates the transportation of anhydrous, liquefied ammonia as a poisonous gas. Its containers are labeled POISON GAS, and their transport vehicles are placarded POISON GAS.

In the workplace, OSHA regulates the design, construction, location, installation, and operation of anhydrous ammonia systems, including refrigerated ammonia storage systems. OSHA also regulates the exposure of employees to ammonia. The permissible exposure limit is 50 ppm.

Many individuals first encounter ammonia as the fumes that escape from *aqueous ammonia*, a water-solution of ammonia often used for cleaning windows. The proper chemical name of aqueous ammonia is *ammonium hydroxide;* its chemical formula is NH_4OH. Commercially available ammonium hydroxide is a liquid containing 27% to 30% ammonia by mass, although the household variety contains approximately 5%. The concentrated solutions are used industrially to produce fertilizers, like ammonium nitrate, ammonium phosphate, and ammonium sulfate.

USDOT regulates the transportation of ammonium hydroxide as either a corrosive material or nonflammable gas. An ammonium hydroxide solution with more than 10% but not more than 35% ammonia by mass is regulated as a corrosive material; with more than 35% ammonia, it is regulated as a nonflammable gas. Containers of ammonium hydroxide are labeled CORROSIVE or NONFLAMMABLE GAS, as appropriate; when required, their transport vehicles are also placarded CORROSIVE or NONFLAMMABLE GAS, as appropriate.

9-11 TOXIC HEAVY METALS

For the purpose of this discussion, a *heavy* metal refers to any metal whose atomic weight is greater than approximately 50. When absorbed into the body, certain heavy metals are toxic. Many are encountered in bulk or solid form and appear quite harmless; yet the fumes, fine powders, or compounds of these metals may be very toxic. The following list notes the metals and their compounds that are considered toxic pollutants by USEPA under the Federal Clean Water Act: antimony, arsenic, beryllium, cadmium, chromium, copper, lead, mercury, nickel, selenium, silver, thallium, and zinc.

In the past, it was generally believed that consumption of water containing low concentrations of heavy metals was not detrimental to an individual's well-being. However, we now know that consumption of certain heavy metals in polluted water (that is, as their dissolved compounds) may cause adverse health effects, even when the concentration of such metals is in the lower parts per million and, in certain instances, even parts per billion.

The mechanism of poisoning from heavy metal exposure is typically independent of the heavy metal. The ions of heavy metals have an extraordinary chemical affinity for sulfur. Hence, when absorbed into the bloodstream, they seek out and combine with the sulfur present in certain of the body's cellular fluids. In particular, heavy metals tend to alter enzymatic and protein action in the body.

Since there are so many toxic heavy metals, this discussion is necessarily limited to the following overview using arsenic, lead, and mercury as representative examples.

Arsenic

By at least the Middle Ages, ingestion of the nearly odorless and tasteless arsenic trioxide was known to cause death. This method of afflicting death on others was used in more than one literary homicide, the most popular of which is Joseph Kesselring's *Arsenic and Old Lace*. Arsenic, like several other heavy metals, tends to accumulate in the body. Thus, ingestion of a small dose of arsenic may seemingly exert no adverse effect at all, while ingestion of multiple small doses could cause death.

Arsenic-containing compounds were once used as components of several inorganic pesticides, like lead(II) arsenate. In the 1940s, such pesticides were used to control insects and rodents. Poisoning of humans from overexposure to such compounds was encountered in areas where fruit and vegetable growing was the major means of livelihood. Arsenic poisoning was also associated with eating unwashed fruit that bore residues of such arsenic compounds. However, encountering arsenic in pesticides is rarely a problem any longer, since such substances have been largely replaced by organic pesticides (Sec. 9-13).

Certain arsenic compounds are employed as pigments in paints; for instance, the pigment *Paris green* is used in some paints. It consists of a mixture of copper(II) arsenite, sodium arsenate, and lead(II) arsenate. It has also been used to a lesser degree as an insecticide; in particular, it has helped to reduce the population of Colorado potato beetles.

Today, arsenic and its compounds are primarily employed in other ways. For instance, arsenic is used as an alloying additive, especially with lead in gunshot pellets and battery grids. In the electronics field, it is used as a doping agent during production of germanium and silicon solid-state semiconductors and light-emitting diodes. Individuals who work in such manufacturing and process industries with arsenic and its compounds represent the group most likely to be exposed to such substances.

For an average adult, the lethal dose for arsenic trioxide is approximately 0.1 g. However, in lesser amounts, arsenic-containing compounds cause other health problems, like mottling of the skin, skin lesions, nervous disorders, and severe, irreversible liver damage. Much fear surrounds its cancer-causing potential: Arsenic is a human carcinogen, causing skin tumors when ingested and lung tumors when inhaled.

The most common cause of chronic arsenic poisoning in the industrial environ-

```
          DANGER

    INORGANIC ARSENIC

      CANCER HAZARD

 AUTHORIZED PERSONNEL ONLY

   NO SMOKING OR EATING

    RESPIRATOR REQUIRED
```

```
            DANGER
  CONTAINS INORGANIC ARSENIC
       CANCER HAZARD
    HARMFUL IF INHALED OR
         SWALLOWED
   USE ONLY WITH ADEQUATE
        VENTILATION
```

Figure 9-15 OSHA requires employers to post the warning sign on the left in areas where worker exposure to airborne inorganic arsenic exceeds a concentration of 10 μg/m^3, averaged over any 8-hour period without the use of respirators. OSHA requires employers to apply the precautionary label on the right to all shipping and storage containers of inorganic arsenic and to all products containing inorganic arsenic.

ment is from inhalation of inorganic arsenic compounds. For this reason, OSHA regulates the airborne exposure of arsenic-bearing inorganic compounds at 5 μg/m^3 as an average concentration over an 8-hour period. To assure that workers are not exposed to higher arsenic concentrations, OSHA requires workers to use air-purifying respirators and to wear protective clothing in areas where airborne arsenic compounds are known to exist. Employers are also required to post a sign indicating areas where airborne arsenic is likely to be present, and they must affix a label on shipping and storage containers of products containing arsenic. Examples of the sign and label are noted in Fig. 9-15.

USDOT regulates the transporation of most arsenic compounds as poisonous materials. Packages containing arsenic compounds are either labeled POISON or KEEP AWAY FROM FOOD; when required, their transport vehicles are placarded POISON.

Lead

Lead-containing compounds are also highly toxic and much more prevalent than arsenic in today's society. Lead pollution is considered by experts to be the chief environmental problem that faces the modern world.

Instances relating to poisoning from lead were recorded as early as the Middle Ages, but, in fact, lead-poisoning had occurred centuries earlier, albeit unknowingly. The related evidence for such cases was identified only in contemporary times. The Romans used lead water pipes and lead storage vessels for storing food products, including wine. When water or wine remained in contact with lead for extended times, lead dissolved in low concentrations. The subsequent consumption of these liquids caused some Romans to die or to suffer from the other adverse health effects of lead poisoning.

Consumption of liquids stored in lead water pipes, as well as in certain lead-containing ceramics and pewter, is still a source of lead poisoning today. The major source of the lead dissolved in modern sources of drinking water comes from the corrosive action of water on the materials formerly used in residential plumbing. The amount of lead that leaches from pipes and soldered pipe joints containing lead depends primarily on the corrosiveness of the water, the time during which the water and plumbing were in contact, and the age and condition of the plumbing. Lead leaches more readily from new solder than from old solder, since a protective covering of water-insoluble lead oxide forms on its surface.

Another common source of lead is associated with the use of certain paints. Many paints contain pigments that are compounds of lead, like lead carbonate (or *white lead*), lead tetroxide (or *red lead*), and lead chromate (or *chrome yellow*). Lead poisoning is associated with these paints primarily in ghetto areas, where hungry children resort to eating old, flaking paint and plaster. This problem is primarily manifested in homes built prior to 1950. A piece of paint or plaster just the size of that indicated below and containing 1% lead by weight, eaten by a child daily for three months may cause lead poisoning:

Notwithstanding these particular sources of lead, the primary source is associated with our prevailing mode of individual transportation: the automobile. Over the past several decades, our atmosphere has slowly become polluted with lead oxide resulting from the combustion of compounds like tetraethyl lead (Sec. 8-3), a former component of many automobile fuels. The problem is particularly prevalent in areas where the density of motor vehicles is high. Tiny particulates of lead oxide are emitted as components of vehicular exhaust, which are subsequently either inhaled or consumed, directly or indirectly. Although the ordinary amount of ingested lead oxide is minutely small, the concentration may cause health problems in certain individuals. For instance, some scientists specifically attribute the brain damage and learning disabilities observed in many inner-city children to routine inhalation of vehicular exhaust containing lead oxide.

Such adverse effects caused by exposure to lead prompted USEPA to establish lead emission standards for new motor vehicles. In practice, this began with the requirement that automakers install emission control devices on new cars. Most automobiles built after 1975 have been equipped with emission control devices that feature catalytic converters. These devices require the use of unleaded gasoline, but they were primarily designed to reduce the amount of carbon monoxide and hydrocarbon emissions. Nonetheless, USEPA took further action by requiring a phase out in the use of leaded gasoline in automobiles. The maximum amount of lead now allowed in unleaded gasoline is 0.1 g/gal.

It was the combination of adverse health effects caused by overexposure to lead that prompted such action. These adverse effects may be conveniently grouped

```
WARNING

LEAD WORK AREA

POISON

NO SMOKING OR EATING
```

Figure 9-16   OSHA requires this warning sign to be posted in work areas where exposure to airborne lead exceeds a concentration greater than 50 $\mu g/m^3$ averaged over an 8-hour period.

into two classes: acute and chronic effects. When taken in a sufficient dose, the ingestion of lead-containing compounds may kill an individual in a matter of days. A condition affecting the brain called *acute encephalopathy* may arise that quickly develops into seizures, coma, and death from cardiorespiratory arrest. One symptom commonly associated with acute lead poisoning is a bluish line on the gums; other symptoms are colic pains, pallor, weakness, constipation, and a paralysis in the forearms and hands. Convulsions are also common, especially in young children who have been overexposed to lead.

Chronic overexposure to lead-containing compounds primarily affects the human blood-forming, nervous, and kidney systems, but it may also harm reproductive, endocrine, hepatic (that is, affecting the liver), cardiovascular, immunologic, and gastrointestinal processes. Exposure to high concentrations of lead may have severe and sometimes fatal consequences, such as brain disease, colic, palsy, and anemia. Damage to the central nervous system in general, and to the brain in particular, is one of the most severe consequences of chronic lead poisoning.

USDOT regulates the transportation of most lead compounds as poisonous materials. Packages containing lead compounds are either labeled POISON or KEEP AWAY FROM FOOD; when required, their transport vehicles are placarded POISON.

In the workplace, OSHA regulates the exposure of workers to lead and inorganic lead compounds; the permissible exposure limit is 50 $\mu g/m^3$. To assure that workers are not exposed to higher lead concentrations, OSHA requires workers to use air-purifying respirators and to wear protective clothing in areas where airborne lead compounds are known to exist. Employers are required to post signs, like that shown in Fig. 9-16, indicating areas where airborne lead is likely to be present.

Mercury

Mercury poisoning has also been known since ancient times, although it does not share the same degree of notoriety as either arsenic or lead poisoning. In the modern world, mercury and mercury-containing compounds are primarily used industrially. In the chemical industry, mercury compounds are often used as catalysts; for instance, mercuric sulfate is used as a catalyst to produce such compounds as acetaldehyde, vinyl chloride, and vinyl acetate. Mercury compounds are also used for other purposes. Mercuric chloride, for instance, known commercially as *corrosive sublimate,* is used to disinfect dishes, bedpans, and other common hospital utensils.

The adverse health effects caused by overexposure to mercury compounds were first signaled in modern times in connection with their existence as chemical pollutants in various rivers and streams. Such mercury compounds are retained by fish that live in these waters. When the fish are consumed by higher animals, the mercury is transmitted in the food chain, ultimately to humans.

The worst incident associated with the consumption of mercury-contaminated fish occurred in 1953 in Minamata, Japan. Mercury compounds had been disposed of with other industrial wastes into Minamata Bay. The disposal was traced to a nearby chemical plant that used mercury compounds as catalysts in conjunction with the production of plastics. The wastes had been discharged to the bay as methyl mercury (CH_3—Hg^+) salts. Fish and shellfish in the bay concentrated the mercury at levels as high as 20 ppm by weight. Nearly 900 people who subsequently consumed these mercury-contaminated fish were afflicted with mercury poisoning. This particular affliction has become known as the _Minamata disease_.

Certain aquatic microorganisms are capable of converting mercury compounds, regardless of their chemical specificity, into compounds in which the mercury becomes _methylated,_ that is, chemically associated with methyl groups. For instance, methogenic bacteria in the mud bottom of lakes convert mercuric chloride to dimethyl mercury as the following equation notes:

$$HgCl_2(aq) + 2CH_4(g) \longrightarrow (CH_3)_2Hg(aq) + 2HCl(aq)$$

Methylated mercury compounds are readily absorbed into animal tissues. This fact has a particularly significant impact on public health and the environment. It means that mercury compounds may pose a health risk to individuals who have been overexposed to either organic or inorganic mercury-containing compounds in polluted water.

Individuals who are afflicted with mercury poisoning may suffer from mild gastritis or may experience severe pain with vomiting. In addition, they are likely to experience any of the following: ataxic gait, convulsions, numbness in the mouth and limbs, constriction in the visual field, or difficulty in speaking. The central nervous system may also be adversely affected.

Elemental mercury is also poisonous. Since it is liquid at room temperature, it possesses an appreciable vapor pressure compared to most other metals. The liquid is also absorbed readily through the skin. The threshold limit value of elemental mercury is only 0.1 ppm by skin absorption. The adverse health effects caused from overexposure to elemental mercury are the same as those caused from mercury compounds.

USDOT regulates the transportation of most mercury compounds as poisonous materials. Packages containing mercury compounds are either labeled POISON or KEEP AWAY FROM FOOD; when required, their transport vehicles are placarded POISON.

In the workplace OSHA regulates the exposure of workers to elemental mercury. OSHA stipulates a permissible exposure limit of 0.1 mg/m³; but NIOSH stipulates only 0.05 mg/m³.

9-12 ASBESTOS

Asbestos is the broad mineralogical term applied to numerous fibrous silicates composed of silicon, oxygen, hydrogen, and metallic ions like sodium, magnesium, calcium, and iron. At least six forms of asbestos occur naturally; a form that crystallizes as asbestos fibers is shown in Fig. 9-17.

Asbestos has a very high melting point, is nonflammable, and is an excellent heat insulator. Because of its fibrous nature, it can be spun into threads and woven into fireproof fabric. This combination of properties was once considered desirable, as noted by the many ways in which asbestos was formerly used. When mixed with magnesium oxide, asbestos was useful for fireproofing, insulating, soundproofing, and decorative purposes. Numerous products containing asbestos became commercially popular: asbestos pipe covering, flooring products, paper products, antifriction materials (like brake lining and clutch facing), roofing materials, and coating and batching compounds. Asbestos was *the* material ideally suitable for all these purposes.

Figure 9-17 Cummingtonite (top left) and cummingtonite–grunerite asbestos (bottom left); serpentenite chrysotile veins (right). In nature, asbestos-bearing minerals are usually found as rocks. However, under certain geological conditions, they crystallize as asbestos fibers. Cummingtonite–grunerite asbestos, produced only in the Transvaal province of South Africa, is the mineral that may crystallize as the form of asbestos fibers known as *amosite,* from the acronym for the company, Asbestos Mines of South Africa. (Courtesy of the U.S. Department of the Interior, Bureau of Mines, Washington, D.C.)

Aside from these beneficial aspects, however, we know today that under special conditions asbestos can inflict serious health problems, including cancer. Unless asbestos has been completely sealed into a product, as in asbestos floor tiles, it may break apart into a dust of tiny fibers, much smaller and more buoyant than ordinary dust. Crumbly, powdery asbestos is said to be *friable*. An example of crumbling asbestos that had previously been coated on heating pipes is noted in Fig. 9-18. It is this friable asbestos that poses a severe hazard to individuals who inhale or swallow its fibers.

The fibers of asbestos are so small that they may float almost indefinitely in air. Their size is related to the size of several other common particles in Fig. 9-19. When inhaled, these fibers tend to deposit deep within the respiratory tract, where they may cause the scarring of lung tissue, a noncancerous affliction called *asbestosis*. But sometimes these fibers are rejected by the lungs, thereby moving up into the throat where they are swallowed. In this fashion, asbestos may cause cancer of the esophagus, stomach, intestines, and rectum. Asbestos may also cause both lung can-

Figure 9-18   Asbestos used as insulation around this piping poses a hazardous condition as it crumbles and flakes. The asbestos separates into a dust of extremely tiny fibers too small to be detected visibly. When these fibers are inhaled or swallowed, they may cause asbestosis, lung cancer, and mesothelioma. In the workplace, when airborne concentrations of asbestos fibers exceed exposure limits established by OSHA, this sign is posted so that employees may take necessary protective steps before entering the area. (Courtesy of the U.S. Environmental Protection Agency, Washington, D.C.)

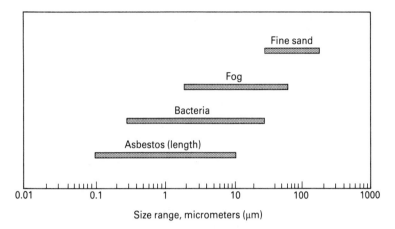

Figure 9-19 The typical size of asbestos fibers, from 0.1 to 10 μm, relative to some other materials. This size is not generally visible to the human eye. (*Source:* U.S. Environmental Protection Agency, "Asbestos-Containing Materials in School Buildings: A Guidance Document," Office of Toxic Substances, Washington, D.C., EPA 450/2-78-014, March 1978)

cer and *mesothelioma,* a cancer of the membranes that line the chest and abdomen. Mesothelioma is characteristic of asbestos overexposure, since it almost never occurs in individuals who have not been exposed to asbestos; it is always fatal.

Certain substances cause a more serious adverse health problem when exposed to the body together with a second substance; this phenomenon is called *synergism.* Asbestos is an example of a substance that exhibits a synergistic effect when inhaled by cigarette smokers. A smoker exposed to asbestos fibers has up to 50 times the chance of developing lung cancer compared to a nonexposed nonsmoker.

Asbestos was the first pollutant to be regulated by OSHA in the workplace. OSHA limits worker exposure to asbestos as the following regulation indicates: "No employee shall be exposed at any time to airborne concentrations of asbestos fibers in excess of 10 fibers, longer than 5 micrometers, per cubic centimeter of air . . . ." The U.S. Food and Drug Administration assures that foods, drugs, and cosmetics are not contaminated with asbestos. The Consumer Products Safety Commission regulates asbestos in consumer products; for instance, it has banned the use of asbestos in consumer clothing, hair dryers, and other products.

USEPA regulates air and water contamination by asbestos and bans the use of asbestos when it may crumble in pipe and boiler coverings. Furthermore, it prohibited virtually all uses of sprayed asbestos materials. To safeguard the health of school children and others who work in schools, USEPA launched a school asbestos program aimed at identifying buildings that contain asbestos materials, inspection of buildings to see if any fibers are being released into the air, and removal or repair of any damaged asbestos material. Finally, USEPA regulates the manner by which asbestos debris is disposed.

USDOT regulates the transportation of asbestos as a miscellaneous hazardous material. When shipped, containers of asbestos are labeled CLASS 9; their transport vehicle is not placarded.

9-13 ORGANIC PESTICIDES

Pesticides are chemical substances used to destroy insects, fungi, rodents, or plants. Most pesticides in use today by either the agricultural community or homeowners are organic compounds. Pesticides are hazardous materials. While designed to destroy pests, they may also adversely affect the health of humans.

Thousands of organic pesticides are known. USDOT regulates their transportation, generally as poisonous materials; OSHA regulates the exposure of workers to pesticides, the permissible exposure limit varying with the specific nature of the pesticide; and USEPA regulates certain aspects of their sale under the Federal Insecticide, Fungicide, and Rodenticide Act [(FIFRA), (Sec. 1-5)].

Before a pesticide may be marketed, it must be registered with USEPA. Pesticides that were already marketed on the enactment date of FIFRA were provided a blanket registration by USEPA; some of these pesticides have been critically evaluated thereafter with respect to the potential impact they were likely to pose on public health and the environment. As the result of government evaluation, the registration of a number of pesticides was canceled or their *registered uses* were suspended. Some of these pesticides are identified in Table 9-12. USEPA also classifies pesticides into general use or restricted use categories based on whether the pesticide is likely to adversely affect the environment. Table 9-13 lists pesticides whose use is restricted; that is, they may only be used by a certified pesticide applicator.

All pesticide containers must be appropriately labeled with the following information "prominently placed with such conspicuousness . . . and in such terms as to

TABLE 9-12   CANCELED OR RESTRICTED-USE PESTICIDES

Aldrin	Kepone
BHC (nongamma)[a]	Mirex
Chlordane	Saffrole
Dibromochloropropane	Sodium cyanide
DDD[b]	Strobane
DDT[c]	Strychnine
Dieldrin	2,4,5-TP/Silvex
Endrin[d]	Thallium sulfate
Heptachlor	Toxaphene[d]

[a] Benzene hexachloride.

[b] Dichlorodiphenyldichloroethane.

[c] Dichlorodiphenyltrichloroethane.

[d] Not all uses have been suspended.

TABLE 9-13 RESTRICTED-USE PESTICIDES

Acrolein	Disulfoton	Mevinphos
Acrylonitrile	Endrin	Monocrotophos
Aldicarb	EPN[a]	Nicotine (alkaloid)
Allyl alcohol	Ethoprop	Paraquat[b]
Aluminum phosphide	Ethyl parathion	Phorate
Azinophos methyl	Fenamiphos	Phosacetim
Calcium cyanide	Fensulfothion	Phosphamidon
Carbofuran	Fluoroacetamide/1081	Picloram
Chlorfenvinphos	Fonofos	Sodium cyanide
Chloropicrin	Hydrocyanic acid	Sodium fluoroacetate
Clonitralid	Methamidophos	Strychnine
Cycloheximide	Methidathion	Sulfotepp
Demeton	Methomyl	Tepp
Dicrotophos	Methyl bromide	Zinc phosphide
Dioxathion	Methyl parathion	

[a] O-ethyl O-para-nitrophenyl phenylphosphonothioate.

[b] Paraquat dichloride and paraquat *bis*-methyl sulfate.

render it likely to be read and understood by the ordinarily individual": the name, brand, or trademark; directions for use that are "necessary for effecting the purpose for which the product is intended" and adequate, if complied with, "to protect health and the environment"; an ingredient statement, giving the name and percentage of each active ingredient and the percentage of all inert ingredients; a statement of the use classification under which it is registered; the name and address of the producer; and a designated warning or caution statement. An example of proper pesticide labeling is provided in Fig. 9-20.

Encountering Pesticides during an Emergency Response Action

In the United States, the National Agricultural Chemicals Association (NACA) has formed a network of safety teams designed to minimize the risk of injury when pesticides are involved in an accident or fire. The Pesticide Safety Team Network was established by NACA to assist in furnishing personnel, equipment, and expertise for the prompt and efficient cleanup and decontamination of pesticides involved in accidents or fires. The Pesticide Safety Team Network may be accessed through CHEMTREC (Sec. 1-7).

The participants are assigned a specific area of the United States for which they act as area coordinators. It is the responsibility of area coordinators to receive reports of any accident involving a pesticide that occurs in their assigned areas and to act in one of several ways to assure that the potential hazard to the public has been reduced or eliminated.

DDVP

TECHNICAL GRADE
ORGANOPHOSPHORUS INSECTICIDE

FOR MANUFACTURING PURPOSES ONLY

ACTIVE INGREDIENTS:

2,2-Dichlorovinyl dimethyl phosphate 93.0%

Related compounds 7.0%

TOTAL 100.0%

KEEP OUT OF REACH OF CHILDREN

POISON

DANGER   **PELIGRO**

"PRECAUCION AL USUARIO: Si usted no lee ingles, no use este producto hasta que la etiqueta haya sido explicado ampliamente."

STATEMENTS OF PRACTICAL TREATMENT

IF SWALLOWED — Call a physician or Poison Control Center immediately. Induce vomiting by giving victim 1 or 2 glasses of water and touching back of throat with finger. Never induce vomiting or give anything by mouth to an unconscious or convulsing person.

IF INHALED — Remove victim to fresh air immediately. Apply artificial respiration if indicated. Get medical attention.

IF IN EYES — Flush eyes with plenty of water for at least 10 minutes. Get medical attention immediately.

IF ON SKIN — Remove contaminated clothing immediately and wash effected area with soap and water. Call a physician immediately. Wash clothing before re-use.

NOTE TO PHYSICIAN — WARNING SYMPTOMS: Symptoms include weakness, headache, tightness in chest, blurred vision, non-reactive pin-point pupil, salivation, sweating, nausea, vomiting, diarrhea, and abdominal cramps.
TREATMENT: Atropine is the specific therapeutic antagonist of choice against parasympathetic nervous stimulation. If there are signs of parasympathetic stimulation, atropine sulfate should be injected at 10-minute intervals, in doses of 1 to 2 milligrams, until complete atropinization has occurred.
Pralidoxime chloride (2-PAM chloride) may also be used as an effective antidote in addition to and while maintaining full atropinization. In adults, an initial dose of 1 gram of 2-PAM should be injected, preferably as an infusion in 250cc of saline over a 15 to 30 minute period. If this is not practical, 2-PAM may be administered slowly by intravenous injection as a 5 percent solution in water over not less than two minutes. After about an hour, a second dose of 1 gram of 2-PAM will be indicated if muscle weakness has not been relieved. For infants and children the dose of 2-PAM is 0.25 grams.
Morphine is an improper treatment.
Clear chest by postural drainage. Oxygen administration may be necessary. Observe patient continuously for 48 hours. Repeated exposure to cholinesterase inhibitors may, without warning, cause prolonged susceptibility to very small doses of any cholinesterase inhibitor. Allow no further exposure until cholinesterase regeneration has been attained as determined by blood test.

SEE BACK/SIDE PANEL FOR ADDITIONAL PRECAUTIONARY STATEMENTS

EPA Reg. No. 5481-96-AA   EPA Est. No. 5481-CA-1

NET CONTENTS 600 POUNDS

AMVAC CHEMICAL CORPORATION

4100 E. WASHINGTON BLVD., LOS ANGELES, CALIF. 90023

PRECAUTIONARY STATEMENTS

HAZARDS TO HUMANS AND DOMESTIC ANIMALS

DANGER: POISONOUS IF SWALLOWED, INHALED OR ABSORBED THROUGH SKIN AND EYES. RAPIDLY ABSORBED THROUGH SKIN. REPEATED INHALATION OR SKIN CONTACT MAY, WITHOUT SYMPTOMS, PROGRESSIVELY INCREASE SUSCEPTIBILITY TO DICHLORVOS (DDVP) POISONING.

Do not swallow or get in eyes, on skin or on clothing. Do not breathe vapor. Do not contaminate food or feed products. Keep away from heat or open flame. Wear clean natural rubber gloves and waterproof protective clothing, and goggles or face shield. Replace gloves frequently and destroy used gloves. Wear a pesticide respirator jointly approved by the Mine Safety and Health Administration and by the National Institute for Occupational Safety and Health under the provisions of 30 CFR Part 11. Wash thoroughly with soap and water after handling and before eating or smoking.

ENVIRONMENTAL HAZARDS

This product is toxic to fish, birds, and other wildlife. Do not discharge effluent containing this product directly into lakes, streams, ponds, estuaries, oceans or public waters unless this product is specifically identified and addressed in an NPDES permit. Do not discharge effluent containing this product to sewer systems without previously notifying the sewage treatment plant authority. For guidance, contact your State Water Board or Regional Office of the Environmental Protection Agency.

DIRECTIONS FOR USE

It is a violation of Federal law to use this product in a manner inconsistent with its labeling.

STORAGE & DISPOSAL

Do not contaminate water, food or feed by storage or disposal...

STORAGE — Store product in original container in a cool, dry, locked place out of reach of children.

PESTICIDE DISPOSAL — Pesticide wastes are acutely hazardous. Improper disposal of excess pesticide, spray mixture or rinsate is a violation of Federal law. If these wastes cannot be disposed of by use according to label instructions, contact your State Pesticide or Environmental Control Agency or the Hazardous Waste representative at the nearest EPA Regional Office for guidance.

CONTAINER DISPOSAL — Triple rinse (or equivalent). Then offer for recycling or reconditioning, or puncture and dispose of in a sanitary landfill, or incineration, or, if allowed by State and local authorities, by burning. If burned, stay out of smoke.

CHEMICAL AND PHYSICAL PROPERTIES: Refer to Technical Bulletin (copies available upon request).

Formulators are responsible for providing data to support their registrations.

DRUM HANDLING: Handle carefully to prevent damage and leakage. Open drums only when set on end and under a ventilated hood. Unscrew bungs slowly to release internal pressure.

NOTICE: This product conforms to its chemical description and is reasonably fit for the purpose stated on the label, when used in accordance with directions under normal conditions of use. Manufacturer is not responsible for the use of this product contrary to the label instructions, or under abnormal conditions or under conditions not reasonably foreseeable to the manufacturer and/or seller, and buyer assumes the risk of any such use.

DICHLORVOS POISON B
RQ DICHLORVOS NA 2783
12/84

Figure 9-20 Label for containers of DDVP. Information on labels should always be read before opening pesticide containers. If a pesticide has been ingested or inhaled, information on the label may be vital for saving lives. Minimally, common and chemical names of the pesticide should be provided to the attending physician. This information is provided in the active ingredient statement. (Courtesy of Amvac Chemical Corporation, Los Angeles, California)

Immediately following receipt of an emergency message, the coordinator communicates with the manufacturer or producer of the pesticide at issue. Together they agree on a procedure to be invoked in the specific instance. The person reporting the incident to the coordinator is then contacted and advised of immediate steps to be taken. If essential, special teams are dispatched to the area under consideration.

Attempts to extinguish fires involving pesticides should only be undertaken when using positive-demand, self-contained breathing apparatus. The use of full protective clothing, including rubber boots, gloves, hats, and coats, is always warranted when fighting fires involving pesticides, since some pesticides can enter the body by skin absorption. Furthermore, inhalation of the smoke or fumes resulting from burning pesticides may cause brain damage or death.

Pesticide poisoning may be identified by any of the following symptoms: blurred vision, headache, vomiting, salivation, and lack of contraction of the pupil of the eye upon exposure to strong light. Some of these symptoms may be mistakenly associated with the symptoms normally observed in connection with heat exhaustion and smoke inhalation.

When fighting pesticide fires, firefighters should observe the following basic principles:

1. *Avoid any direct exposure to smoke or fumes that evolve from pesticide fires.* Thus, such fires should be attacked from the upwind side or at right angles.

2. *Use fog instead of direct streams of water.* Fog is less likely to raise the toxic dust of a pesticide.

3. *Keep runoff water to a minimum.* Ditch banks should be dug to channel runoff water into a temporary reservoir and prevent its entrance into local sewers or waterways. This contaminated water resulting from fighting a pesticide fire should subsequently be disposed of in accordance with applicable regulatory standards.

9-14 CHEMICAL CATEGORIES OF ORGANIC PESTICIDES

Based on their molecular structures, organic pesticides are classified into several groups, the most important of which are the following: organochlorine pesticides, organophosphorus pesticides, carbamate pesticides, and urea pesticides. It is appropriate to discuss each class independently.

Organochlorine Pesticides

solutions

These pesticides are chlorine derivatives of complex hydrocarbons; that is, at least one hydrogen atom in the hydrocarbon molecule has been chemically replaced by a chlorine atom. For example, consider the insecticide known as *Aldrin,* whose molecular formula is $C_{12}H_8Cl_6$. This formula signifies that six hydrogen atoms have been chemically substituted with chlorine atoms in each molecule of a hydrocarbon having the formula $C_{12}H_{14}$. A molecule of Aldrin has the following Lewis structure (Sec. 3-10):

$$
\begin{array}{c}
\text{H} \qquad\quad \text{Cl} \\
\mid \qquad\ \text{H} \quad \mid \\
\text{C} \qquad\ \text{C} \\
\diagup \quad \diagdown \quad \mid \quad \diagdown \\
\text{H--C} \quad\ \mid \quad\ \text{C} \quad\ \mid \quad\ \text{C--Cl} \\
\parallel \quad\ \text{CH}_2 \mid \quad \text{CCl}_2 \parallel \\
\text{H--C} \quad\ \mid \quad\ \text{C} \quad\ \mid \quad\ \text{C--Cl} \\
\diagdown \quad\ \mid \quad \diagup \quad\ \mid \quad \diagup \\
\text{C} \qquad \text{C} \\
\mid \qquad\ \text{H} \quad \mid \\
\text{H} \qquad\quad \text{Cl}
\end{array}
$$

The mode of action by which organochlorine pesticides affect biological activity is not entirely known. It *is* known that these pesticides somehow destroy the delicate balance of sodium and potassium within nerve cells, thereby preventing them from conducting nerve impulses normally. Precisely how this occurs, on the other hand, is not understood.

The fatty tissues of vertebrates are capable of retaining organochlorine pesticides; such animals are said to *bioaccumulate* these substances. Consequently, when birds consume insects or plants dusted or sprayed with an organochlorine pesticide, some fraction of this pesticide is passed along to other animals in the food chain. The organochlorine pesticides are also very stable substances, which causes them to possess a relatively high degree of environmental persistance. This means that they survive for long periods in animal tissues, as well as in soil and aquatic environments, during which they may adversely impact health and the environment.

Perhaps the best known organochlorine pesticide is the substance known by the acronym DDT. The United States and its allies used DDT extensively during World War II to control typhoid fever, malaria, and typhus, all of which were transmitted by insects. However, these beneficial aspects were outweighed by the fact that DDT also entered the food chain and caused untold environmental and health damage worldwide. For this reason, production and sale of this pesticide were canceled by USEPA.

Organophosphorus Pesticides

These pesticides are derivatives of phosphoric acid, H_3PO_4.

$$
\begin{array}{c}
\text{O} \\
\parallel \\
\text{HO--P--OH} \\
\mid \\
\text{OH}
\end{array}
$$

When examined at the molecular level, each molecule of an organophosphorus pesticide is generally observed to possess at least one sulfur atom. Some also possess at least one nitrogen atom.

An example of an organophosphorus pesticide is *Parathion* whose molecular formula is $(C_2H_5)_2PSOC_6H_4NO_2$. The distribution of atoms in *Parathion* is illustrated by the following structural formula:

$$C_2H_5-O \underset{C_2H_5-O}{\overset{\overset{\displaystyle S}{\|}}{>}} P-O-C \underset{C=C}{\overset{C-C}{<}} C-NO_2$$

Organophosphorus compounds are toxic. This adverse effect is caused by the ability of organophosphorus compounds to inactivate or inhibit the normal action of an enzyme called *acetylcholinesterase* (ACHE). This enzyme routinely performs the duty of destroying the substance *acetylcholine,* which is stored in small cavities within certain cells of the nervous systems of insects and vertebrates called *neurons.* The function of acetylcholine is associated with the transmission of nerve impulses from one neuron to another. Once acetylcholine is secreted by nervous activity, it is hydrolyzed to breakdown products; ACHE catalyzes this reaction. Once hydrolyzed, the acetylcholine is destroyed and cannot function further by diffusing to, or having any other effect on, areas other than those in which it was released. But when organophosphorus compounds absorb into the bloodstream, they rapidly move throughout the body and ultimately interact with the cells of the nervous system. Here, they react with ACHE, destroying the chemical nature of this important substance. In turn this prevents the proper transmission of nerve impulses. Instead, the voluntary muscles rapidly twist in insects, which ultimately causes paralysis.

Carbamate Pesticides

These pesticides are derivatives of the organic acid *carbamic acid,* whose chemical formula is

$$H_2N-C \overset{\displaystyle \nearrow O}{\underset{\displaystyle \searrow OH}{}}$$

A typical carbamate pesticide is *Carbyl,* an insecticide. Its structural formula is

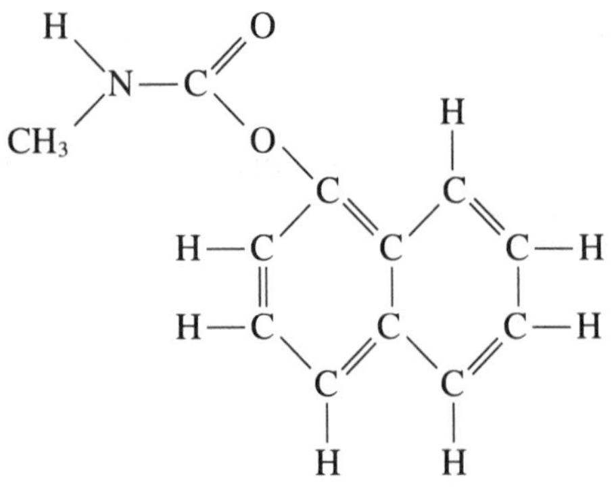

The mode of action of carbamate pesticides, like *Carbyl,* on insects and vertebrates is to inhibit the enzyme cholinesterase. Some carbamate pesticides also function as fungicides and herbicides.

Urea Pesticides

The molecular structure of urea is denoted as follows:

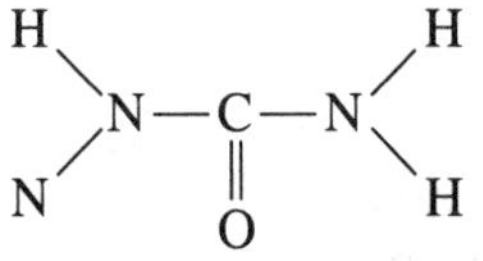

The urea pesticides are derivatives of urea, that is, substances in which at least one of the hydrogen atoms in urea has been chemically replaced by another atom or combination of atoms.

Most urea pesticides are used as herbicides. An example of such a herbicide is *Linuron,* whose molecular structure is

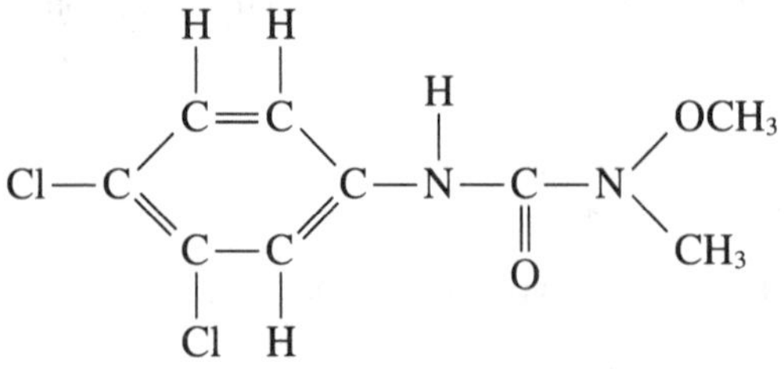

The urea pesticides act as herbicides by inhibiting photosynthesis in plants.

9-15 RESPIRATORY PROTECTION FROM TOXIC SUBSTANCES

Respiratory protection is often necessary when fighting fires or responding to incidents involving toxic substances. Respiratory protection may be provided through the use of devices called *respirators*. There are respirators that filter gases, vapors, and particulates from the ambient atmosphere, and other respirators that supply clean air to the user that is independent of the ambient atmosphere. The use of respirators in the workplace is regulated by laws enforced by the Occupational Safety and Health Administration.

Respiratory protective devices are commercially available in two general categories: *air purifying* and *atmosphere supplying*. An air-purifying respirator is either a quarter-, half-, or full facepiece, to which twin cartridges are attached. Cartridges are available with different filters designed to remove specified toxins from the ambient atmosphere. When the user of this respirator inhales, as illustrated in Fig. 9-21, negative pressure with respect to the surrounding atmospheric pressure is created in the facepiece. This forces air to enter through the cartridges, where it is

Figure 9-21   A negative-pressure, air-purifying, twin-cartridge respirator. The cartridges contain both a particulate filter and vapor- or gas-removing sorbents that remove specific substances only from the ambient atmosphere. Such respirators must only be used in atmospheres where there is at least 19% oxygen by volume and where contaminants and their concentrations are known. (Courtesy of Mine Safety Appliances Co., Pittsburgh, Pennsylvania)

purified. Air-purifying respirators should only be used when entering atmospheres where the concentration of a contaminant is below its IDLH level.

An atmosphere-supplying respirator is designed to provide breathable air to the user for a period ranging from 5 minutes to several hours. One type recirculates self-generating oxygen. The moisture and carbon dioxide in the user's breath are absorbed by potassium superoxide contained in a canister. Chemical action of the moisture and the superoxide causes oxygen to be generated. The chemical phenomenon is essentially the decomposition of potassium superoxide. The oxygen enters a breathing bag for inhalation and the exhaled breath repeats the cycle. The chemical reactions involved in this process are represented by the following equations:

$$4KO_2(s) \longrightarrow 2K_2O(s) + 3O_2(g)$$

$$K_2O(s) + CO_2(g) \longrightarrow K_2CO_3(s)$$

This type of atmosphere-supply respirator was formerly used by the military and is still employed in mines to escape from a noxious or poisonous atmosphere.

Another type of atmosphere-supplying respirator uses a hose mask. Clean air is drawn, generally from compressed air cylinders, through a hose into a facepiece by means of a blower. Air is forcibly delivered to the user. The air pressure inside the

Figure 9-22 Pressure-demand, self-contained breathing apparatus provides the best degree of protection when entering an atmosphere contaminated by toxic substances. It is used where oxygen deficiency is an issue of concern or when hazardous contaminants are present. (Courtesy of Mine Safety Appliances Co., Pittsburgh, Pennsylvania)

facepiece is thus greater than the pressure of the ambient air outside the mask. This type of respirator provides an extra dimension of protection over other respirators. Even if leaks exist between the facepiece and the user's face, the greater pressure inside the mask constantly forces air outward. This prevents the inward movement of contaminated air.

A special type of hose mask is the self-contained breathing apparatus. This is designed so that the user has the freedom to move from place to place without the normal confinement provided by a long hose attached some distance away to an air supply. Of the various types commercially available, the best protection to the user is afforded by a type that operates on the pressure-demand mode; this type is shown in Fig. 9-22. During routine firefighting and rescue work, the only device that provides adequate protection to the user is the positive-demand, self-contained breathing apparatus.

In this type of respiratory device, positive air pressure is maintained within the facepiece at all times during use with respect to the ambient atmospheric pressure. The pressure developed inside the facepiece operates two valves automatically, the admission valve and the exhalation valve. When the admission valve opens, clean air

Figure 9-23 Two individuals from an emergency-response team investigating an abandoned hazardous waste disposal site. Each individual is wearing a commercially available total-encapsulating suit. Under the suits, each individual is wearing self-contained breathing apparatus. The individual in the foreground is carrying a combustible gas and oxygen alarm to check for the presence of flammable and combustible gases in the nearby atmosphere. (Courtesy of Mine Safety Appliances Co., Pittsburgh, Pennsylvania)

enters the facepiece from the cylinder. When the admission valve closes, the exhalation valve opens, which allows exhaled air to be released. During operation, these two valves open and close, one after the other, allowing the user to concentrate on emergency-response duties. The air supply is limited from 5 to 30 minutes. When it becomes nearly exhausted, an alarm sounds, which informs the user that the air supply will be totally consumed in a preestablished time.

Finally, on occasion, it is sometimes essential for emergency-response personnel to wear total-encapsulating suits, like either of those illustrated in Fig. 9-23. Total-encapsulating suits are not in themselves respiratory protection devices; thus, they must be worn with a self-contained breathing apparatus. The advantage of a total-encapsulating suit is that it helps protect the wearer from exposure to hazardous materials by skin contact. Such suits are recommended to be worn when responding to hazardous material situations in which the materials at issue are unknown; this is typically the case when individuals are called upon to investigate sites where chemical wastes have been illegally disposed or abandoned. Such suits are also recommended when responding to situations involving highly toxic materials, particularly those which are known to cause adverse health effects by absorption through the skin.

REVIEW EXERCISES

Basic Elements of Toxicology

9.1. To determine the cumulative exposure to an air contaminant for an 8-hour workshift, OSHA requires use of the following formula:

$$E = \frac{(C_a T_a + C_b T_b + \cdots + C_n T_n)}{8}$$

E is the equivalent exposure for the workshift, C is the concentration during any period of time T where the concentration remains constant, and T is the duration of time in hours of the exposure at the concentration C. OSHA further stipulates that the value of E shall not exceed the 8-hour time-weighted average limit set by the agency for the substance at issue. If a worker in the chemical industry is exposed to ammonia in the following concentrations for the indicated time periods during a single workday, are the combined exposures acceptable to OSHA: (1) 4 hours exposure at 125 ppm; (2) 2 hours exposure at 75 ppm; and (3) 2 hours exposure at 25 ppm? [OSHA stipulates the 8-hour time-weighted average limit of exposure to ammonia as 50 ppm (Title 29 *Code of Federal Regulations*, §1910.1000).]

9.2. Distinguish between the local and systemic effects most likely to be experienced from inhalation of hydrogen chloride vapor.

9.3. An individual is often afflicted with pneumonia within 1 to 2 weeks following inhalation of beryllium compounds. However, at later periods, this same exposure may cause excess connective tissue to develop in the lungs; this latter condition is called *fibrosis*.

Classify pneumonia and fibrosis as either an acute or chronic effect resulting from exposure to beryllium.

9.4. Which factor affecting toxicity is illustrated by each of the following observations?

   (a) In behavioral studies, rats exposed to 4000 ppm by volume of acetone for 4 hours/day, 5 days/week for two weeks showed modified avoidance and escape behavior after one exposure, but no change after subsequent exposures.

   (b) Barium sulfate and barium chloride are soluble to the extent of 0.00028 g and 33 g per 100 g of water, respectively, at 25°C. Barium sulfate is often administered orally to increase contrast in x-ray photographs of the gastrointestinal tract with no ill side effects, whereas ingestion of barium chloride causes a strong stimulant action on smooth, cardiac, and skeletal muscle.

   (c) The probability of acquiring cancer from exposure to chloroform vapor increases as a function of the concentration to which a population has been exposed.

   (d) Individuals with heart ailments are more prone to experience discomfort from inhaling nitric oxide than is the population as a whole.

   (e) Zinc is an essential nutrient in our diets, but ingestion of high concentrations of zinc can cause fever, vomiting, stomach cramps, and diarrhea.

   (f) Arsenic trioxide and arsenic pentoxide pose different adverse health effects on humans.

9.5. The International Agency for Research on Cancer classifies acrylonitrile and carbon tetrachloride as a "suspect human carcinogen" and "human carcinogen," respectively. What do these terms imply insofar as acquiring cancer is concerned?

Measurement of Toxicity

9.6. The average weight of a 5-year-old boy is 45 lb. Estimate the lethal dose of aspirin in ounces for the boy. (The LD_{50} value for aspirin is 1750 mg/kg.)

9.7. John Doe weighs 190 lb. He mistakenly swallows a water solution containing 5 oz of phenol. Does this amount exceed the lethal dose? (The LD_{50} for phenol is 530 mg/kg.)

Smoke and the Fire Gases

9.8. Why is inhalation of an atmosphere containing a relatively low concentration of carbon monoxide and smoke likely to be more dangerous to one's health compared to inhalation of an atmosphere containing the same carbon monoxide concentration but in the absence of smoke?

9.9. Which of the following situations is more likely to produce the larger amount of carbon monoxide: (1) a burning pile of cardboard, 13 ft × 13 ft × 13 ft in size: or (2) the combustion of 1 gallon of gasoline uniformly spilled on a surface 13 ft × 13 ft in area?

9.10. The Center for Disease Control in Atlanta, Georgia, recommends regular cleaning and adjustment of air inlets for all fuel-burning devices used indoors. What is the most likely reason for this recommendation?

9.11. By what physical means may firefighters and others be warned of their overexposure to carbon monoxide?

9.12. Why does overexposure to carbon monoxide affect the cardiovascular system, while overexposure to carbon dioxide does not?

9.13. According to the National Oceanic and Atmospheric Administration's Air Resources Laboratory, the atmospheric level of carbon dioxide increased to 345 ppm by volume in the second half of the 1980s compared to 290 ppm a century ago. What single factor is most likely the cause of this atmospheric increase in the concentration of carbon dioxide?

9.14. When Dry Ice is transported on ocean vessles, USDOT requires its packaging to be marked as follows: "Stow clear of living quarters." What is the most likely reason that USDOT regulates the stowage of Dry Ice in this manner?

9.15. For each of the following pairs of gases individually contained in steel cylinders, identify the one from which all sources of ignition should be absolutely eliminated: (1) carbon monoxide or carbon dioxide, (2) hydrogen sulfide or sulfur dioxide, or (3) nitric oxide or ammonia?

9.16. Other than carbon monoxide and carbon dioxide, which fire gas is most likely to be encountered in each of the following fire-response situations?
  (a) A burning warehouse that stores clothing made from polyacrylonitrile.
  (b) An aircraft interior whose seats are cushioned with polyurethane.
  (c) A smoldering pile of tires made from vulcanized rubber.

9.17. Although hydrogen sulfide possesses a characteristic foul-smelling odor, why is reliance on the sense of smell not recommended for detecting the presence of this gas?

9.18. Why are low-sulfur coal and oil generally more expensive than high-sulfur coal and oil?

9.19. Why does overexposure to nitric oxide affect the cardiovascular system, while overexposure to nitrous oxide does not?

9.20. Why is it impossible to use water fog or water spray to disperse ammonia vapors when directed from an upwind direction?

Toxic Heavy Metals

9.21. To protect the environment, what action should be taken when using water to fight a fire in a transportation mishap incident involving a consignment of lead arsenate?

9.22. Why is the concentration of lead higher in the top 6 inches of undisturbed soil along the shoulders of a highway than in soil that is not near a roadway?

9.23. Faced with the fact that her two children were sick, a woman prepared and stored fresh orange juice in a ceramic pitcher originally purchased in Mexico. With therapeutic intent, she then fed more and more orange juice to her sick children. Based only on this limited information, what was the most likely reason the children subsequently suffered from permanent brain damage?

9.24. Cinnabar is a mineral containing mercuric sulfide, a compound soluble in water only to a very small extent. Why is the inadvertent consumption of ground cinnabar not considered a health hazard?

Asbestos

9.25. For the past several decades, Lake Superior has been plagued by asbestoslike fibers that originated in local taconite mining tailings. Communities once drew drinking water from the lake with virtually no treatment. Now, however, the state of Minnesota recommends that they turn to using bottled water and to install filtration plants. What is the most likely reason the state recommends these practices?

9.26. Assign appropriate numbers to each quadrant of a NFPA 704 symbol representing asbestos.

9.27. Why is it that attorneys defending clients in asbestos-related damage suits often desire to establish that their clients have acquired mesothelioma?

Organic Pesticides

9.28. Approximately half of the pesticides listed in Table 9-11 are organochlorine pesticides. What is the most likely overall reason that USEPA canceled their registration?

9.29. Estimate the lethal dose of parathion in grams for a person who weighs 175 lb. (The LD_{50} for parathion is 8 mg/kg.)

9.30. When responding to a fire involving pesticides, why is it generally a wise practice to first contact the manufacturer of the pesticide *before* extinguishing the fire?

Respiratory Protection

9.31. What do the terms negative pressure and positive pressure mean in connection with respiratory protection?

9.32. What type of respiratory protection, if any, is normally recommended for use by the following workers during the indicated work-related activity?

    (a) A building custodian who occasionally encounters crumbling asbestos-covered pipes during the workday.

    (b) Medical personnel responding to a transportation mishap involving leaking hydrogen sulfide.

    (c) Firefighters responding to a pesticide warehouse fire.

    (d) An environmental engineer who must turn off the valve of a tank from which lead-contaminated wastewater is leaking.

10

Oxidation–Reduction Phenomena

In Sec. 4-4 we first observed that simple chemical reactions may be classified by dividing them into two general groups: (1) those in which the oxidation numbers of all component elements remain the same, and (2) those in which the oxidation numbers of some component elements change during the course of the reaction. The reactions in this latter group constitute the subject matter of this chapter; they are called *oxidation–reduction reactions,* or *redox* reactions. The chemical participants in a redox reaction are called *oxidizing agents* and *reducing agents.*

Many controlled redox reactions provide beneficial uses to society in such ways as the following: the combustion of fuels, the chlorination of water, the explosion of dynamite, the bleaching of fabrics, and the flare of fireworks. Furthermore, the energy of redox reactions may often be harnessed to provide still further benefits. For example, batteries, dry cells, and fuel cells are examples of simple devices in which specific redox reactions provide a portable supply of electrical energy. But oxidizing and reducing agents are chemically incompatible substances. When reactions occur between them in an uncontrolled fashion, relatively large amounts of energy may be generated, which could subsequently cause fire, explosion, and the loss of life. This potential source of destructiveness emphasizes the need to examine the oxidation–reduction phenomenon as an element of the study of hazardous materials.

USDOT directly regulates the transportation of a subclass of oxidizing agents called oxidizers (Sec. 5-7). USDOT defines an oxidizer as a substance that yields oxygen to stimulate the combustion of organic matter.

Finally, we observed earlier (Sec. 2-11) ways by which the property of *ignitability* may be used to characterize RCRA hazardous wastes (Sec. 1-5). Another way is when the waste is an oxidizer; that is, a chemical waste exhibits the characteristic of ignitability when the waste conforms to the USDOT definition of an oxidizer. USEPA regulates the treatment, storage, and disposal of such chemical wastes.

10-1 OXIDATION NUMBER

The *oxidation number* is a term used to reflect the combining capability of one ion for another ion or of one atom for another atom in a given substance. It is usually a positive or negative whole number, or zero. For instance, in sodium chloride (NaCl), the oxidation numbers of the sodium and chloride ions are $+1$ and -1, respectively. Since all chemical substances are electrically neutral, the algebraic sum of the oxidation numbers in a substance must numerically equal zero.

In practice, an oxidation number is assigned to the constituents of substances by applying certain rules. For our purposes, the following six rules are appropriate:

1. The oxidation number of an element in an uncombined state is zero. This means that the atoms in substances represented as H_2, O_2, Na, Mg, and any other element are defined to have an oxidation number of zero.
2. The algebraic sum of the oxidation numbers in any substance is zero.
3. The hydrogen in most hydrogen-containing compounds possesses an oxidation number of $+1$. [An exception is the hydrogen in a metallic hydride (Sec. 8-4), which possesses an oxidation number of -1.]
4. The oxygen in most oxygen-containing compounds possesses an oxidation number of -2. [An exception is the oxygen in a metallic peroxide (Sec. 8-6), each atom of which possesses an oxidation number of -1.]
5. The oxidation number of a monatomic ion (having one atom) is the same as its net ionic charge. The charges of some common ions were previously listed in Table 3-7. Thus, the oxidation number of sodium is $+1$; of magnesium, $+2$; of chlorine in the chloride ion, -1; and of sulfur in the sulfide ion, -2.
6. The algebraic sum of the oxidation numbers in any polyatomic ion is equal to its ionic charge.

Let's see how these rules may be used to determine the oxidation number of each element in sodium chlorate ($NaClO_3$), an oxidizing agent used in some fireworks. First, the substance is composed of two ions, the sodium ion (Na^+) and the chlorate ion (ClO_3^-). Rule 5 informs us that the oxidation number of sodium is $+1$, and rule 4 indicates that the oxidation number of oxygen is -2. But the oxidation number of chlorine must be determined through the use of simple arithmetic.

We may determine the oxidation number of chlorine by using either rule 2 or 6.

1. Using rule 2,

$$0 = +1 + x + [3 \times (-2)]$$

$$x = +5$$

2. Using rule 6,

$$-1 = x + [3 \times (-2)]$$

$$x = +5$$

Table 10-1 provides additional examples of oxidation numbers that have been determined through the use of these rules.

TABLE 10-1   EXAMPLES OF OXIDATION NUMBERS

Element	Oxidation number	Chemical formula of representative compound
F	-1	HF
O	-2	SO_2
	-1	H_2O_2
N	-3	NH_3
	-2	N_2H_4
	-1	NH_2OH
	$+1$	N_2O
	$+2$	NO
	$+3$	HNO_2
	$+4$	NO_2
	$+5$	HNO_3
Cl	-1	HCl
	$+1$	HClO
	$+3$	$HClO_2$
	$+5$	$HClO_3$
	$+7$	$HClO_4$
S	-2	H_2S
	$+4$	H_2SO_3
	$+6$	H_2SO_4
P	-3	AlP
	$+3$	H_3PO_3
	$+5$	H_3PO_4

10-2 OXIDATION–REDUCTION REACTION

A redox reaction occurs between an oxidizing agent and a reducing agent. The equation illustrating a redox reaction is thus written in generalized form as

$$\text{Oxidizing agent + reducing agent} \longrightarrow \text{products}$$

Let's consider a specific example of a redox reaction that occurs as a consequence of an electron-transfer mechanism. Iron(III) chloride reacts with tin(II) chloride, forming iron(II) chloride and tin(IV) chloride. The equation illustrating this reaction is written as follows:

$$2FeCl_3(aq) + SnCl_2(aq) \longrightarrow 2FeCl_2(aq) + SnCl_4(aq)$$

To appreciate the mechanism of a redox reaction, let's first determine the oxidation number of the elements in the participating compounds. As monatomic ions, the oxidation number of each chloride, iron(II), and iron(III) ion is -1, $+2$, and $+3$, respectively. Similarly, tin(II) and tin(IV) possess oxidation numbers of $+2$ and $+4$, respectively. Clearly, in this redox reaction the iron ions and tin ions simply transfer electrons and thus change their oxidation numbers. Such a simultaneous increase and decrease in oxidation number always accompanies oxidation and reduction, respectively.

In terms of the net process, the chloride ions do not participate in this redox reaction; we say they are *spectator ions*. Eliminating them, the following ionic equation may be written:

$$2Fe^{3+}(aq) + Sn^{2+}(aq) \longrightarrow 2Fe^{2+}(aq) + Sn^{4+}(aq)$$

The ionic equation allows us to note more clearly the nature of the overall phenomenon, which may be summarized as follows:

1. Tin(II) ions become tin(IV) ions. The oxidation number *increases from* $+2$ to $+4$; the tin(II) ions are *oxidized*. Tin(II) chloride is the *reducing agent*.
2. Iron(III) ions become iron(II) ions. The oxidation number *decreases* from $+3$ to $+2$; the iron(III) ions are *reduced*. Iron(III) chloride is the *oxidizing agent*.

Sometimes we illustrate oxidation and reduction by individual equations like the following:

$$Fe^{3+}(aq) + e^- \longrightarrow Fe^{2+}(aq)$$
$$Sn^{2+}(aq) \longrightarrow Sn^{4+}(aq) + 2e^-$$

Each of these equations illustrates the independent steps of a redox reaction; they are called *half-reactions*. One half-reaction represents oxidation, while the other represents reduction. Processes represented by half-reactions always occur simultaneously.

TABLE 10-2 SOME COMMON OXIDIZING AGENTS

Oxidizing agent	Element that changes oxidation number	Oxidation number		Equation illustrating half-reaction
		in reactant	in product	
Sodium peroxide	Oxygen	−1	−2	$Na_2O_2(aq) + 2H_2O(l) + 2e^-$ $\longrightarrow 2Na^+(aq) + 4OH^-(aq)$
Metallic hypochlorites	Chlorine	+1	−1	$ClO^-(aq) + 2H^+(aq) + 2e^-$ $\longrightarrow Cl^-(aq) + H_2O(l)$
Metallic chlorates	Chlorine	+5	−1	$ClO_3^-(aq) + 6H^+(aq) + 6e^-$ $\longrightarrow Cl^-(aq) + 3H_2O(l)$
Nitric acid (concentrated)	Nitrogen	+5	+4	$NO_3^-(aq) + 2H^+(aq) + e^-$ $\longrightarrow NO_2(g) + H_2O(l)$
Nitric acid (dilute)	Nitrogen	+5	+2	$NO_3^-(aq) + 4H^+(aq) + 3e^-$ $\longrightarrow NO(g) + 2H_2O(l)$
Metallic peroxydisulfates	Sulfur	+7	+6	$S_2O_8^{2-}(aq) + 2e^-$ $\longrightarrow 2SO_4^{2-}(aq)$

Several other examples of oxidation–reduction phenomena are provided in Table 10-2.

10-3 COMMON HAZARDOUS FEATURES OF OXIDIZERS

Oxidizing agents are relatively powerful chemical substances. Hence, they readily react with a large group of other substances, many of which are common even to the normal household. For instance, oxidizing agents react with certain cleaning agents, fuels, lubricants, greases, and oils.

The relative strength of a given oxidizing agent varies under different reaction conditions. Nevertheless, the relative strength of oxidizing agents may be roughly established by means of the listing in Table 10-3. In this series, the oxidizing agents are arranged according to their decreasing oxidizing power. This means that any substance on this list is a stronger oxidizing agent than the substances whose names are listed below it. While Table 10-3 is not meant to be all-inclusive, it is instructive to note that many substances are more reactive as oxidizing agents than oxygen itself. In turn, these substances possess a greater degree of hazard than an equal quantity of oxygen.

The degree of hazard of oxidizers may also be evaluated by consideration of the NFPA classification for oxidizers.* NFPA identifies the following four classes of oxidizers:

Class 1 oxidizer: An oxidizing material whose primary hazard is that it may increase the burning rate of combustible material with which it comes in contact.

** NFPA 43A-1980, Code for the Storage of Liquid and Solid Oxidizing Materials.

TABLE 10-3   RELATIVE STRENGTH OF OXIDIZING AGENTS[a]

Fluorine
Ozone
Hydrogen peroxide
Hypochlorous acid
Metallic chlorates[b]
Lead(II) dioxide
Metallic permanganates[b]
Metallic dichromates[b]
Nitric acid (concentrated)
Chlorine
Sulfuric acid (concentrated)
Oxygen
Metallic iodates
Bromine
Iron(III) (Fe^{3+}) compounds
Iodine
Sulfur
Tin(IV) (Sn^{4+}) compounds

[a] Listed in descending order of oxidizing power.
[b] In an acidic environment.

Class 2 oxidizer: An oxidizing material that will moderately increase the burning rate or that may cause spontaneous ignition of combustible material with which it comes in contact.

Class 3 oxidizer: An oxidizing material that will cause a severe increase in the burning rate of combustible material with which it comes in contact or that will undergo vigorous self-sustained decomposition when catalyzed or exposed to heat.

Class 4 oxidizer: An oxidizing material that can undergo an explosive reaction when catalyzed or exposed to heat, shock, or friction.

Thus, the relative degree of hazard increases in the following order for oxidizers: $4 > 3 > 2 > 1$. Table 10-4 illustrates examples of oxidizers in each NFPA class.

Relatively large amounts of energy are typically evolved when redox reactions occur. Consider the combustion reaction alone. Sufficient energy is liberated when natural gas burns to use beneficially, for example, for heating and cooking purposes. But the heat evolved from a redox reaction could be absorbed by nearby combustible materials; then the ignition of these materials may occur. Sometimes even a small trace of an oxidizing agent is enough to cause the spontaneous ignition of materials like sulfur, charcoal, and turpentine. Hence, as a general practice it is a sound policy to store oxidizing agents in areas removed from ignitable and combustible matter.

The oxidizing agents containing oxygen in their composition are inherently unstable when heated, thereby decomposing by evolving oxygen. This is illustrated

TABLE 10-4   SOME TYPICAL OXIDIZERS BY NFPA CLASSIFICATION

Class 1

Aluminum nitrate
Ammonium persulfate
Barium chlorate
Barium nitrate
Calcium chlorate
Calcium nitrate
Calcium peroxide
Cupric nitrate
Hydrogen peroxide solutions, over 8% but not
   exceeding 27.5% concentration by mass
Lead nitrate
Lithium hypochlorite
Lithium peroxide
Magnesium nitrate
Magnesium perchlorate
Magnesium peroxide
Nickel nitrate
Nitric acid, 70% concentration or less
Perchloric acid, less than 60% by mass

Potassium dichromate
Potassium nitrate
Potassium persulfate
Silver nitrate
Sodium carbonate peroxide
Sodium dichloro-*s*-triazinetrione dihydrate
Sodium dichromate
Sodium nitrate
Sodium nitrite
Sodium perborate
Sodium perborate tetrahydrate
Sodium perchlorate monohydrate
Sodium persulfate
Strontium chlorate
Strontium nitrate
Strontium peroxide
Thorium nitrate
Uranium nitrate
Zinc chlorate
Zinc peroxide

Class 2

Calcium hypochlorite, 50% or less by mass
Chromium trioxide (chromic acid)
Halane (1,3-dichloro-5,5-dimethyl hydantoin)
Hydrogen peroxide, 27.5% to 52%
   concentration by mass
Nitric acid, more than 70% concentration

Potassium permanganate
Sodium chlorite, 40% or less
Sodium peroxide
Sodium permanganate
Trichloro-*s*-triazinetrione (trichloroisocyanuric
   acid)

Class 3

Ammonium dichromate
Calcium hypochlorite, over 50% by mass
Hydrogen peroxide, 52% to not more than
   91% concentration by mass
Mono-(trichloro)tetra-(monopotassium
   dichloro)-penta-*s*-triazinetrione
Perchloric acid solutions, 60% to 72.5% by
   mass

Potassium bromate
Potassium chlorate
Potassium dichloro-*s*-triazinetrione (potassium
   dichloroisocyanurate)
Sodium chlorate
Sodium chlorite, over 40% by mass
Sodium dichloro-*s*-triazinetrione (sodium
   dichloroisocyanurate)

Class 4

Ammonium perchlorate
Ammonium permanganate
Guanidine nitrate
Hydrogen peroxide solutions, more than 91%
   by mass

Perchloric acid solutions, more than 72.5% by
   mass
Potassium superoxide

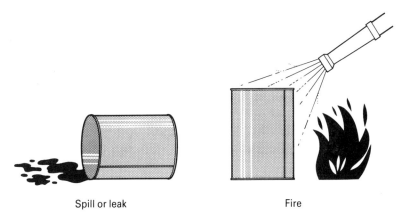

Spill or leak   Fire

Figure 10-1   Some general emergency procedures for mishaps involving oxidizers.

by the thermal decomposition of potassium chlorate, which is noted by the following equation:

$$2KClO_3(s) \longrightarrow 2KCl(s) + 3O_2(g)$$

Hence, the oxidizers listed in Table 10-3 do not burn in the traditional sense; but they are capable of supporting normal combustion by providing a supply of oxygen. During fires, the presence of oxidizers may be particularly hazardous. Oxidizers may serve as potential sources of oxygen for sustaining combustion, even when the atmospheric oxygen supply has been totally depleted. This means that an oxidizing agent may replace atmospheric oxygen as an essential component of the fire tetrahedron (Fig. 5-6).

As Fig. 10-1 illustrates, most fires supported by liquid or solid oxidizers may be easily extinguished by applying flooding amounts of water to them. The manner by which water effectively extinguishes such fires varies with the specific nature of the oxidizer. But, in general, these oxidizers dissolve in the water, thereby reducing their concentration; this usually eliminates their chemical reactivity.

Oxidizers supply oxygen when they thermally decompose. However, before many oxidizers decompose, they first melt. This hot molten material may spread to adjoining areas, where it may cause any combustible material to burn that it contacts. While water is generally used on fires involving oxidizers, water should not be applied to bulk quantities of molten oxidizers. This causes steam to be rapidly generated, which may splatter the molten material. Whenever practicable, fires involving bulk quantities of molten oxidizers should be smothered with sand.

10-4 HYDROGEN PEROXIDE

35% to 52%

Hydrogen peroxide is the most common peroxide in commerce; its chemical formula is H_2O_2. Pure hydrogen peroxide and its aqueous solutions resemble water in physical appearance, but possess a slightly sharp odor. To the lay-person the most

52% or
greater

common form of hydrogen peroxide is an aqueous solution containing approximately 1% to 3% hydrogen peroxide by volume in water. It is used as a mild antiseptic. Industrially, more concentrated solutions are used to bleach cotton, wool, straw, leather, gelatin, and various waxes and oils. In the chemical industry, hydrogen peroxide is often the raw material used for synthesizing other metallic and organic peroxides.

Most hydrogen peroxide in use today is an aqueous solution that was produced by the electrolytic oxidation of a sulfuric acid solution of peroxydisulfuric acid, $H_2S_2O_8$, followed by hydrolysis. The equations illustrating this production process are

$$2H_2SO_4(aq) \longrightarrow H_2S_2O_8(aq) + H_2(g)$$

$$H_2S_2O_8(aq) + 2H_2O(l) \longrightarrow 2H_2SO_4(aq) + H_2O_2(aq)$$

The resulting mixture of hydrogen peroxide and sulfuric acid is distilled, typically generating an aqueous solution of hydrogen peroxide whose strength is approximately 30% by volume. Using special techniques, additional water may be removed from this distillate, resulting in strengths of hydrogen peroxide up to 99% by volume. The greater the concentration of hydrogen peroxide, the greater is the relative hazard of the solution.

Hydrogen peroxide is an inherently unstable substance: it slowly decomposes spontaneously as noted by the following equation:

$$2H_2O_2(aq) \longrightarrow 2H_2O(l) + O_2(g)$$

This decomposition is catalyzed by sunlight. An aqueous solution of hydrogen peroxide containing 8% by mass H_2O_2 is completely decomposed after ten-month exposure to light, whereas a similar solution kept in darkness for the same length of time is essentially unaltered. Usually this decomposition occurs so slowly that it is virtually imperceptible, especially in solutions whose strength is less than 30% by volume. However, concentrated hydrogen peroxide solutions decompose rapidly. Concentrated solutions may become so strongly heated during decomposition that they completely vaporize. To prevent the decomposition from posing an explosive hazard, most commercial forms of hydrogen peroxide, regardless of their concentration, are stabilized with a small amount of sodium pyrophosphate ($Na_4P_2O_7$). The stabilizer acts as a catalyst, retarding the decomposition, but not totally preventing its occurrence.

As expected, the rate of decomposition of hydrogen peroxide increases with a rise in temperature. When heated to 291°F (144°C), hydrogen peroxide decomposes violently. Certain metals also catalyze the decomposition, especially iron, steel, lead, copper, brass, bronze, chromium zinc, manganese, and silver.

The fact that hydrogen peroxide is a potential source of oxygen should not be regarded lightly. Solutions exceeding 50% in hydrogen peroxide by volume may cause the spontaneous ignition of combustible material on contact. Reactions occur rapidly in which hydrogen peroxide participates, and they evolve considerable heat.

Figure 10-2 Bulk quantities of hydrogen peroxide are often transported as 70% solutions by mass for dilution to lower concentrations on delivery. The transportation of bulk quantities is generally accomplished in either tank trucks or rail tank cars, as shown here. Thereafter, the substance must be stored in tanks built from compatible materials that have been properly designed and thoroughly passivated, like high purity aluminum. (Courtesy of FMC Corporation, Philadelphia, Pennsylvania)

Not only may hydrogen peroxide act as a strong oxidizing agent, but it may also act as a weak reducing agent. As an oxidizing agent, the oxidation number of oxygen changes from -1 to -2; but as a reducing agent, the oxidation number changes from -1 to 0. For instance, hydrogen peroxide is sometimes used to counteract excess chlorine in drinking water supplies, in which case it acts as a reducing agent:

$$H_2O_2(aq) + Cl_2(aq) \longrightarrow 2HCl(aq) + O_2(g)$$

When hydrogen peroxide is employed as a reducing agent, oxygen is always liberated. Since hydrogen peroxide may act as a either an oxidizing or reducing agent, its self-decomposition occurs when one H_2O_2 molecule oxidizes another one, which is simultaneously reduced.

Exposure to hydrogen peroxide may constitute a serious health hazard. Hydrogen peroxide in a solution whose strength exceeds 30% by volume is especially corrosive to skin. In the workplace, OSHA recommends a permissible exposure limit to hydrogen peroxide of only 1 ppm.

USDOT regulates the transportation of four hydrogen peroxide solutions as oxidizers that exist in the following concentration ranges: 8% to 20%, 20% to 40%, 40% to 60%, and over 60%. Containers of hydrogen peroxide having a concentration from 8% to 20% are labeled OXIDIZER. Containers of other concentrations are labeled OXIDIZER and CORROSIVE. When required, the transport vehicles are placarded OXIDIZER, as illustrated in Fig. 10-2.

10-5 METALLIC HYPOCHLORITES, CHLORITES, CHLORATES, AND PERCHLORATES

There are four oxychlorinated acids that are strong oxidizing agents. Their names and formulas, as well as the names and formulas of their corresponding sodium salts, are as follows:

Hypochlorous acid, HClO	Sodium hypochlorite, NaClO
Chlorous acid, $HClO_2$	Sodium chlorite, $NaClO_2$
Chloric acid, $HClO_3$	Sodium chlorate, $NaClO_3$
Perchloric acid, $HClO_4$	Sodium perchlorate, $NaClO_4$

The chlorine atom assumes an oxidation number of $+1$, $+3$, $+5$, and $+7$ in each compound and column by descending order.

Hypochlorous acid, chlorous acid, and chloric acid are known only in aqueous solution. In terms of their ability to ionize, hypochlorous acid and chlorous acid are weak acids, but chloric acid and perchloric acid are strong acids. However, in terms of their oxidizing potential, each is regarded as a strong oxidizing acid.

Metallic Hypochlorites

Either sodium or calcium hypochlorite is the active component of many commercial bleaching and sanitation agents. For instance, common laundry bleach is a 3% to 5% percent aqueous solution of sodium hypochlorite. These metallic hypochlorites effectively bleach cloth because of their ability to react with atmospheric carbon dioxide, producing hypochlorous acid, which photochemically decomposes to oxygen. The oxygen generated by the photodecomposition is believed to bleach fabric; chlorine only plays an indirect role. These reactions are illustrated by the following equations:

$$2NaClO(aq) + H_2O(l) + CO_2(g) \longrightarrow Na_2CO_3(aq) + 2HClO(aq)$$

$$2HClO(aq) \longrightarrow 2HCl(aq) + O_2(g)$$

Sodium hypochlorite is also employed as a disinfectant and deodorant in treating water supplies, sewage effluent, swimming pools, and dairies.

When involved in fires, metallic hypochlorites support combustion by generating oxygen. When heated, calcium hypochlorite decomposes to calcium chloride and oxygen, as the following equation illustrates:

$$Ca(ClO)_2(s) \longrightarrow CaCl_2(s) + O_2(g)$$

There are three different chemical types of bleaching powder containing calcium hypochlorite, which are discussed independently next.

Mixed lime and calcium hypochlorite.   This bleach is a solid obtained by heating calcium hydroxide (lime) with chlorine, as the following equation notes:

$$3Ca(OH)_2(s) + 2Cl_2(g) \longrightarrow Ca(ClO)_2 \cdot H_2O(s) + CaCl_2 \cdot Ca(OH)_2 \cdot H_2O(s)$$

This substance tends to decompose when only slightly wet and does not store well, which can be a hazard during transportation. Hence, it is used only infrequently.

Chloride of lime.   This bleach is also a solid, produced by controlling the reaction previously noted so that 1 mol of calcium hydroxide reacts with 1 mol of chlorine, as the following equation notes:

$$Ca(OH)_2(s) + Cl_2(g) \longrightarrow CaCl(ClO)(s) + H_2O(l)$$

The product of this reaction is not the true calcium hypochlorite; it is called *chloride of lime* or *calcium chlorohypochlorite*. However, when dissolved in water, the true calcium hypochlorite forms as follows:

$$2CaOCl_2(aq) \longrightarrow CaCl_2(aq) + Ca(ClO)_2(aq)$$

Calcium hypochlorite.   The third bleaching powder is the actual calcium hypochlorite. Other than its use for laundering, this form is also commonly employed by municipalities for sanitizing drinking water and disinfecting domestic and

Nonfire

municipal swimming pools. This form of calcium hypochlorite is sometimes known commercially by the tradename *high-test hypochlorite* (HTH), illustrated in Fig. 10-3.

Bleaching agents are chemically incompatible with mineral acids, generating chlorine. For instance, calcium hypochlorite reacts with hydrochloric acid, as the following equation illustrates:

$$Ca(ClO)_2(s) + 4HCl(aq) \longrightarrow CaCl_2(aq) + 2H_2O(l) + Cl_2(g)$$

This particular reaction is used as the basis of establishing the amount of *available chlorine* in a chlorine-containing oxidizer. The latter term refers to the amount of

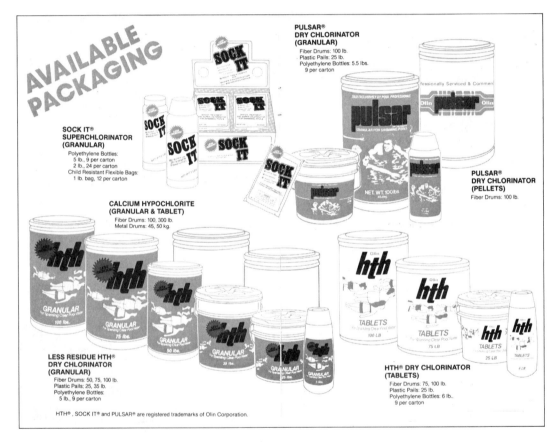

Figure 10-3 Commercially available packaging for calcium hypochlorite, an oxidizing agent. Calcium hypochlorite is commonly used for disinfecting and sanitizing swimming pool water and potable water, as well as for sewage treatment and other industrial sanitation applications. When properly stored and handled, calcium hypochlorite is a stable chemical substance. However, it is chemically incompatible with numerous other substances. Such chemical reactions are exothermic; they often liberate chlorine, and they may cause a very intense fire. (Courtesy of Olin Corporation, Stamford, Connecticut; HTH^R is the registered trademark of Olin Corporation)

chlorine that is evolved when acid is mixed with a chlorine-containing oxidizer; thus, "99.2% available chlorine" means that the the material at issue has the effectiveness of 99.2% by mass of chlorine. Substances having 99.2% available chlorine are useful for sanitizing water; commercial laundry bleaches typically have approximately 35% available chlorine.

USDOT regulates the transportation of three forms of calcium hypochlorite as oxidizers: a hydrated form, containing from 5.5% but not more than 10% water by mass; a dry form with more than 10% but not more than 39% available chlorine; and a dry form with more than 39% available chlorine. Containers of these forms of calcium hypochlorite are labeled OXIDIZER. When required, their transport vehicles are placarded OXIDIZER.

USDOT also regulates the transportation of two forms of metallic hypochlorite solutions. Hypochlorite solutions with more than 5% but less than 16% available chlorine are regulated as corrosive materials; so also are hypochlorite solutions having not less than 16% available chlorine. Containers of these hazardous materials are labeled CORROSIVE; when required, their transport vehicles are placarded CORROSIVE.

Metallic Chlorites

These substances are potentially important bleaching agents. In practice, however, sodium chlorite is the only metallic chlorite employed for bleaching purposes. Sodium chlorite is commercially available in an 80% aqueous solution by mass, having about 125% available chlorine. It is used to bleach paper pulp and textiles.

Solid sodium chlorite is chemically incompatible with finely divided magnesium and aluminum. When these combinations of substances are ignited, the redox reaction occurs with explosive violence. The following equations illustrate the respective reactions:

$$2Mg(s) + NaClO_2(s) \longrightarrow 2MgO(s) + NaCl(s)$$

$$4Al(s) + 3NaClO_2(s) \longrightarrow 2Al_2O_3(s) + 3NaCl(s)$$

Such chemical mixtures are potentially useful in fireworks. When exposed to intense sources of heat, sodium chlorite decomposes to form sodium chloride and oxygen, as follows:

$$NaClO_2(s) \longrightarrow NaCl(s) + O_2(g)$$

USDOT regulates the transportation of sodium chlorite and its aqueous solution having more than 5% available chlorine as an oxidizer and corrosive material, respectively. Containers of sodium chlorite are labeled OXIDIZER; when required, their transport vehicles are placarded OXIDIZER. When regulated, containers of sodium chlorite solutions are labeled CORROSIVE; when required, their transport vehicles are placarded CORROSIVE.

Metallic Chlorates

These compounds are also commercially important, especially as components of certain fireworks, gunpowder, flares, herbicides, and plant defoliants.

Sodium chlorate is very sensitive to impact and friction and thus initiates fire quite easily. It is a very powerful oxidizing agent; hence, sodium chlorate is incompatible with many materials. The redox reactions between sodium chlorate and either hot charcoal or sulfur occur spontaneously as the following equations note:

$$3C(s) + 2KClO_3(s) \longrightarrow 2KCl(s) + 3CO_2(g)$$

$$3S_8(s) + 16KClO_3(s) \longrightarrow 16KCl(s) + 24SO_2(g)$$

Furthermore, sodium chlorate reacts with finely divided metals with explosive violence. For instance, the following equation represents the reaction between sodium chlorate and finely divided aluminum:

$$2Al(s) + KClO_3(s) \longrightarrow Al_2O_3(s) + KCl(s)$$

In addition, sodium chlorate causes certain organic compounds to burst into flame spontaneously.

Sodium chlorate undergoes thermal decomposition quite readily. When heated to just its melting point, it forms sodium perchlorate and sodium chloride.

$$4NaClO_3(s) \longrightarrow 3NaClO_4(s) + NaCl(s)$$

When the temperature is further raised, the perchlorate decomposes into sodium chloride and oxygen.

$$NaClO_4(s) \longrightarrow NaCl(s) + 2O_2(g)$$

The thermal decomposition of sodium chlorate is catalyzed by minute quantities of iron(III) oxide or manganese(IV) oxide.

USDOT regulates the transportation of metallic chlorates as oxidizers. Containers of metallic chlorates are labeled OXIDIZER; when required, their transport vehicles are placarded OXIDIZER.

Metallic Perchlorates

These compounds are commercially useful in much the same manner as metallic chlorates are. However, when used in fireworks, they are considered more reliable, since they are not as frequently associated with premature reactions. This is due to the fact that metallic perchlorates are relatively stable substances when compared to the metallic hypochlorites, chlorites, or chlorates. Hence, they are safer to use than the other three and relatively less hazardous to store and transport; but metallic perchlorates still must be handled very carefully.

Potassium perchlorate is a strong oxidizing agent. When mixed with either charcoal or sulfur, for example, spontaneous ignition occurs, as the following equations represent:

$$2C(s) + KClO_4(s) \longrightarrow KCl(s) + 2CO_2(g)$$

$$S_8(s) + 4KClO_4(s) \longrightarrow 4KCl(s) + 8SO_2(g)$$

Potassium perchlorate is often used along with either magnesium or aluminum in fireworks. Magnesium powder enhances the brilliance of fireworks, and the addition of coarse aluminum flakes produces luminous tails; these pyrotechnic effects are illustrated in Fig. 10-4. The corresponding redox reactions are illustrated by the following equations:

$$4Mg(s) + KClO_4(s) \longrightarrow 4MgO(s) + KCl(s)$$

$$8Al(s) + 3KClO_4(s) \longrightarrow 4Al_2O_3(s) + 3KCl(s)$$

Figure 10-4 Fireworks are beautiful pyrotechnic displays resulting from the occurrence of certain oxidation–reduction reactions. These fireworks were exhibited in Chicago, Illinois. But the substances that produce fireworks are hazardous materials and must always be stored, transported, and handled as oxidizers and displayed only by experienced, knowledgeable professionals. (Copyrighted 1987, Chicago Tribune Company, all rights reserved, used with permission)

USDOT regulates the transportation of metallic perchlorates as oxidizers. Containers of metallic perchlorates are labeled OXIDIZER; when required, their transport vehicles are placarded OXIDIZER.

10-6 AMMONIUM COMPOUNDS

All compounds containing the ammonium ion (NH_4^+) are thermally unstable. When heated, they decompose in either of the following two ways:

1. Ammonium compounds that are not oxidizing agents decompose to form ammonia. For instance, ammonium chloride thermally decomposes at temperatures less than 350°F (167°C) to form ammonia and hydrogen chloride, as the following equation notes:

$$NH_4Cl(s) \longrightarrow NH_3(g) + HCl(g)$$

2. Ammonium compounds that are oxidizing agents may also decompose to form ammonia; but, more generally, they form nitrogen, oxygen, and either metallic or nonmetallic oxides. Examples of such compounds are ammonium nitrate and ammonium perchlorate. The chemical equations in Table 10-5 illustrate the thermal decomposition of some ammonium compounds that are oxidizing agents. These decompositions pose a unique concern, since they may occur explosively.

In the modern world, ammonium compounds are likely to be frequently encountered. Ammonium sulfate is used as a fertilizer in larger amounts than any other ammonium compound. But, more generally, ammonium nitrate is usually regarded to be the most commercially important ammonium compound. This substance is used as a fertilizer and as a component of certain explosives.

Ammonium nitrate is most frequently encountered as a fertilizer-grade material; it is illustrated in Fig. 10-5. Ordinarily, ammonium nitrate may be safely transported, stored, and used as long as certain precautionary measures are taken. Nonetheless, incidents are known in which ammonium nitrate has exploded violently. Consequently, precaution should always be taken when handling this hazardous material.

Why is it that ammonium nitrate is a potential explosion risk? To answer this question, we must examine the chemistry of ammonium nitrate at the temperatures likely to be achieved in a major fire. Figure 10-6 illustrates that from 175° to 200°F (80° to 93°C), ammonium nitrate decomposes to ammonia and nitric acid. This reaction occurs endothermically, as the following equation represents:

$$NH_4NO_3(s) \longrightarrow NH_3(g) + HNO_3(g)$$

At approximately 330°F (166°C), ammonium nitrate melts. Thus, molten pools of ammonium nitrate are usually encountered when ammonium nitrate is involved in a

TABLE 10-5 SOME POTENTIALLY HAZARDOUS AMMONIUM COMPOUNDS

Oxidizer		Equation illustrating thermal decomposition
Ammonium bromate	$2NH_4BrO_3(s)$ $\longrightarrow$	$2NH_4Br(s) + 3O_2(g)$
Ammonium chlorate	$2NH_4ClO_3(s)$ $\longrightarrow$	$2NH_4Cl(s) + 3O_2(g)$
Ammonium dichromate	$(NH_4)_2Cr_2O_7(s)$ $\longrightarrow$	$Cr_2O_3(s) + 4H_2O(g) + N_2(g)$
Ammonium nitrate[a]	$NH_4NO_3(s)$ $\longrightarrow$	$N_2O(g) + 2H_2O(g)$
Ammonium nitrite	$NH_4NO_2(s)$ $\longrightarrow$	$N_2(g) + 2H_2O(g)$
Ammonium perchlorate	$2NH_4ClO_4(s)$ $\longrightarrow$	$N_2(g) + Cl_2(g) + 4H_2O(g) + 2O_2(g)$
Ammonium permanganate	$2NH_4MnO_4(s)$ $\longrightarrow$	$2MnO(s) + N_2(g) + 4H_2O(g) + O_2(g)$
Ammonium peroxydisulfate	$3(NH_4)_2S_2O_8(s)$ $\longrightarrow$	$4NH_3(g) + N_2(g) + 6SO_2(g) + 6H_2O(g) + 3O_2(g)$

[a] See text also.

Figure 10-5   A farm distribution center where ammonium nitrate can be stored in
bulk as well as in bags. Ammonium nitrate is an important source of nitrogen for
crops and thus serves as an important fertilizer. (Courtesy of the Fertilizer Institute,
Washington, D.C.)

fire. The explosion risk becomes prevalent when molten ammonium nitrate is heated
to still higher temperatures. At temperatures exceeding 410°F (212°C), ammonium
nitrate undergoes decomposition into nitrous oxide and water vapor. This decomposition occurs exothermically as Fig. 10-7 shows; the decomposition is represented
by the following equation:

$$NH_4NO_3(s) \longrightarrow N_2O(g) + 2H_2O(g)$$

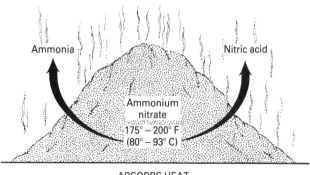

ABSORBS HEAT
NO RUNAWAY REACTION

Figure 10-6   The chemical phenomenon
that occurs when ammonium nitrate is
heated in the range of temperatures from
175° to 200°F (80° to 93°C). (Courtesy
of the Fertilizer Institute, Washington, D.C.)

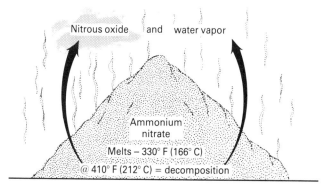

SPEED INCREASES AS
TEMPERATURE INCREASES

Figure 10-7   The chemical phenomenon that occurs when ammonium nitrate is heated to its decomposition temperature, 410°F (212°C). (Courtesy of the Fertilizer Institute, Washington, D.C.)

In conventional storage facilities and transportation vehicles, these vapors produced by decomposing ammonium nitrate freely escape into the atmosphere. For this reason, pressure does not buildup and an explosion cannot occur. However, when the decomposition products are tightly confined in some way, explosion is imminent. Such is the situation in the tightly constructed hold of a ship.

It was in this latter fashion that ammonium nitrate caused one of the worst maritime incidents associated with a chemical substance. In 1947, the *S.S. Grandcamp,* a French cargo ship docked in Texas City, Texas, caught fire with 2200 tons of fertilizer-grade ammonium nitrate on board, in addition to 1500 tons of fuel oil. The heat caused the ammonium nitrate to decompose. In an attempt to smother the fire, the hatches were closed. In addition, steam was applied throughout the hold. These features of this infamous chemical tragedy are illustrated in Fig. 10-8.

Under such confinement, neither heat nor pressure can dissipate from a ship's hold. Furthermore, the fire does not smother; the combustion continues, supported by nitrous oxide instead of atmospheric oxygen. The rate of the decomposition reactions escalates out of control at these elevated temperatures, causing the internal pressure to reach an intolerable level. The application of steam increases the heat and pressure within the hold. When the internal pressure cannot dissipate, which occurred on board the *Grandcamp,* the remaining material explodes. The self-decomposition reaction is represented by the following equation:

$$2NH_4NO_3(s) \longrightarrow 2N_2(g) + 4H_2O(g) + O_2(g)$$

Certain substances are known to increase the rate at which ammonium nitrate decomposes. Copper-containing compounds, for instance, catalyze the self-decomposition of ammonium nitrate. Furthermore, mixtures of ammonium nitrate with either sulfur, charcoal, or powdered metals are chemically incompatible. For this rea-

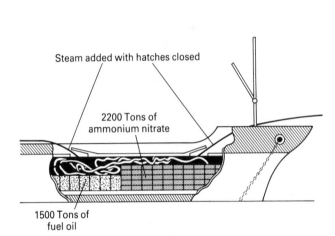

Steam added with hatches closed

2200 Tons of
ammonium nitrate

1500 Tons of
fuel oil

Figure 10-8 On April 16, 1947, 2200 tons of ammonium nitrate was stored on board the *S.S. Grandcamp* along with 1500 tons of fuel oil. While docked at Texas City, Texas, a fire was detected in the hold. An order was given to close the hatches and to apply steam throughout the hold. Under these conditions, the fire could not extinguish. Instead, the internal pressure built up at an uncontrollable rate, causing the cargo vessel to explode. Six hundred people were killed and another 3500 were injured. The property damage was comparable to that experienced during a major wartime bombing incident. The total property loss was estimated as $33 million based on 1947 costs.

son, ammonium nitrate should be stored in a manner whereby it is segregated from other chemical substances.

There are several lessons to be learned from the *Grandcamp* disaster. First, ammonium nitrate should always be stored in well-ventilated structures like that shown in Fig. 10-9 and segregated from combustible materials and sources of heat. This prevents pressure buildup, if the material becomes involved in a fire. When fighting fires involving ammonium nitrate, firefighters should always first ventilate these structures. One manner by which ventilation may be accomplished is to puncture multiple holes in the roofs.

While water generally extinguishes fires involving oxidizing agents, water was never applied to the fire on board the *Grandcamp*. When ammonium nitrate becomes involved in fire, it should be cooled as soon as possible by penetrating the material with large volumes of water.

Finally, although nitrous oxide is nontoxic, it may readily oxidize to other nitrogenous gases that are toxic, like nitric oxide and nitrogen dioxide (Sec. 9-9). Special precaution needs to be taken to prevent inhaling these gases. Hence, when it is necessary during firefighting to enter an area where ammonium nitrate is stored, firefighters should use positive-demand, self-contained breathing apparatus.

Today, ammonium nitrate is no longer used in its pure form as a fertilizer. Instead, it is mixed with ammonium sulfate or calcium carbonate. This tends to dilute the ammonium nitrate, which also reduces its risk of explosion. However, pure ammonium nitrate is still used by the explosives industry along with the *ammonia dynamites* (Sec. 13-6); these are explosives composed of approximately 32% by weight ammonium nitrate, nitroglycerin, and other substances.

USDOT regulates the transportation of various formulations of ammonium nitrate, generally as oxidizers; ammonium dichromate, ammonium perchlorate, and ammonium permanganate are also regulated as oxidizers. When their transportation is approved by USDOT, containers of these ammonium compounds are labelled OX-

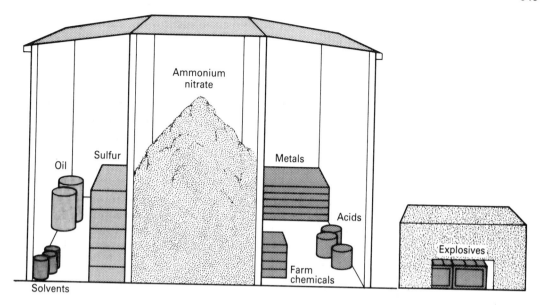

Figure 10-9 Recommended storage for bulk quantities of ammonium nitrate. When ammonium nitrate is stored in bulk, the storage bin should be clean and free of contaminants. The pile within the bin should be sized so that all material in the pile is moved periodically. The bin should be separated by approved fire walls from flammable and combustible materials, corrosive materials, other oxidizers, and substances with which the ammonium nitrate may chemically react. Explosives or blasting agents should never be stored in the same building as ammonium nitrate. (Courtesy of the Fertilizer Institute, Washington, D.C.)

IDIZER; when required, their transport vehicles are placarded OXIDIZER. It is forbidden to transport ammonium bromate, ammonium chlorate, ammonium fulminate, and ammonium nitrite.

10-7 OXIDIZING AGENTS CONTAINING CHROMIUM

Chromium assumes the oxidation state of +6 in four compounds: metallic chromates, metallic dichromates, chromium trioxide, and chromyl chloride. When the metallic ion is colorless, metallic chromates and dichromates are yellow and orange compounds, respectively. Potassium chromate may be produced from potassium dichromate, and vice versa, as the following equations note:

$$K_2Cr_2O_7(aq) + 2KOH(aq) \longrightarrow 2K_2CrO_4(aq) + H_2O(l)$$

$$2K_2CrO_4(aq) + 2HCl(aq) \longrightarrow K_2Cr_2O_7(aq) + H_2O(l)$$

Metallic dichromates are considered *strong* oxidizing agents, particularly when dissolved in diluted acid. Potassium dichromate is often industrially encountered. It is used to electroplate metallic chromium, in dyeing and printing, as a component of explosives, and for other uses. But it is also incompatible with a great many substances. For example, the following equation illustrates the incompatibility of potassium dichromate and hydrochloric acid:

$$K_2Cr_2O_7(aq) + 14HCl(aq) \longrightarrow 2CrCl_3(aq) + 2KCl(aq) + 2Cl_2(g) + 7H_2O(l)$$

When heated, metallic dichromates release some of their oxygen, but only at very elevated temperatures. For instance, potassium dichromate is thermally decomposed at approximately 900°F (483°C), as the following equation notes:

$$4K_2Cr_2O_7(s) \longrightarrow 2Cr_2O_3(s) + 4K_2CrO_4(s) + 3O_2(g)$$

The acids associated with chromates and dichromates exist only in aqueous solution. When the water is evaporated from them, only *chromium(VI) oxide* remains, which is also known as either *chromium trioxide, chromium anhydride,* or, most commonly, *chromic acid*. Its chemical formula is CrO_3. Chromic acid is a red solid prepared industrially by adding concentrated sulfuric acid to a saturated solution of potassium dichromate, as the following equation notes:

$$K_2Cr_2O_7(aq) + 2H_2SO_4(conc) \longrightarrow 2KHSO_4(aq) + H_2O(l) + 2CrO_3(s)$$

It is widely used industrially for many purposes, such as cleaning metal and glass surfaces.

Like potassium dichromate itself, chromic acid is incompatible with many substances. When ethyl alcohol is dropped on chromic acid, it readily catches fire. The alkali metals, sulfur, phosphorus, lubricating oils, and greases are readily oxidized.

Another similar compound is *chromyl chloride,* also called *chromium oxychloride*. This is a red liquid that forms when a mixture of concentrated hydrochloric and sulfuric acids is added to a saturated solution of potassium dichromate. The chemical formula of chromyl chloride is CrO_2Cl_2. It is used industrially in much the same manner as chromic acid, although not as commonly.

USEPA regards all compounds of chromium as toxic, although the compounds of major toxicological importance in higher organisms are only those in which the oxidation state of chromium is +6. Such compounds are carcinogenic and produce kidney damage in animals and humans. USDOT regulates the transportation of chromic acid as either an oxidizer or corrosive material, depending on its state of matter.

10-8 METALLIC PERMANGANATES

Metallic permanganates are compounds containing manganese in the +7 oxidation state. When the metallic ion is colorless, these permanganates are intensely purple. The most industrially popular metallic permanganates are sodium and potassium permanganate.

Dilute solutions of potassium permanganate are used to cure dermatitis having a bacterial or fungal origin. Concentrated solutions are sometimes used industrially to remove objectionable matter from process wastes by oxidation. The effluent gases from many industrial sources are also often eliminated or improved through the use of metallic permanganates.

When exposed to a source of heat, potassium permanganate evolves some of its oxygen. The associated chemical reactions are a function of temperature and are illustrated by the following equations:

$$2KMnO_4(s) \longrightarrow K_2MnO_4(s) + MnO_2(s) + O_2(g) \quad (T < 200°C)$$

$$4KMnO_4(s) \longrightarrow 2K_2MnO_3(s) + 2MnO_2(s) + 3O_2(g) \quad (T > 200°C)$$

USDOT regulates the transportation of metallic permanganates as oxidizers. Containers of metallic permanganates are labeled OXIDIZER; when required, their transport vehicles are placarded OXIDIZER.

10-9 METALLIC NITRITES AND NITRATES

The metallic nitrites and nitrates are important oxidizing agents containing nitrogen in the +3 and +5 oxidation states, respectively. They are salts of nitrous and nitric acid, respectively. The commercial uses of these compounds usually vary with the identity of the metallic ion. Until very recently, sodium nitrite and sodium nitrate were added to raw meat, like bacon, in order to preserve its color.

Metallic nitrites may act as either oxidizing or reducing agents. Metallic nitrites are oxidized to metallic nitrates and reduced to nitric oxide. The following equations are illustrative of this chemical behavior:

$$NaNO_2(aq) + Na_2O_2(aq) + H_2O(l) \longrightarrow NaNO_3(aq) + 2NaOH(aq)$$

$$2NaNO_2(aq) + Na_2SO_3(aq) + 2HCl(aq) \longrightarrow$$
$$Na_2SO_4(aq) + 2NaCl(aq) + 2NO(g) + H_2O(l)$$

Metallic nitrites are generally stable to heat. However, at relatively high temperatures, sodium nitrite decomposes, as the following equation illustrates:

$$2NaNO_2(s) \longrightarrow NO_2(g) + NO(g) + Na_2O(s)$$

Nonfire

Metallic nitrates react only as oxidizing agents. In the presence of acids, they may be converted to either nitric oxide, nitrogen dioxide, or ammonia. Typical of such chemical behavior are the chemical reactions of zinc with nitric acid (Sec. 7-6).

Fire

When heated, metallic nitrates decompose in various ways. The alkali metals decompose to form nitrites and oxygen; but the heavy metal nitrates decompose to form the respective metallic oxide, nitrogen dioxide, and oxygen. The following equations illustrate the thermal decomposition of sodium nitrate, silver nitrate, and lead(II) nitrate:

$$2NaNO_3(s) \longrightarrow 2NaNO_2(s) + O_2(g)$$

$$2AgNO_3(s) \longrightarrow 2Ag(s) + 2NO_2(g) + O_2(g)$$

$$2Pb(NO_3)_2(s) \longrightarrow 2PbO(s) + 4NO_2(g) + O_2(g)$$

As noted earlier, sodium nitrate and sodium nitrite have been used as additives in meat products to retain the pink color associated with raw meat. Sodium nitrite may be converted under proper enzymatic action to substances called *nitrosamines,* compounds with the $-N-N=O$ grouping of atoms. Nitrosamines are potent carcinogenic substances. The use of sodium nitrate in meats is also a matter of concern, since stomach bacteria are capable of converting metallic nitrates to nitrites.

The presence of nitrates in public water supplies is also a matter of some concern. Repeated consumption of nitrates in drinking water may cause methemoglobinemia (Sec. 9-9), particularly in infants.

USDOT regulates the transportation of metallic nitrites and metallic nitrates as oxidizers. Containers of metallic perchlorates are labeled OXIDIZER; when required, their transport vehicles are placarded OXIDIZER.

10-10 CHEMISTRY OF MATCHES

Matches served as one of the earliest commercial applications that invoked the use of redox reactions as an instant source of fire. Even today, the average American lights about nine matches per day. It is appropriate to review the chemistry associated with the burning of matches, as it involves the chemical principles that cause oxidizers to be regarded as hazardous materials.

Two kinds of matches, illustrated in Fig. 10-10, are available in the United States: *strike-anywhere* matches and *safety* matches. The head of a strike-anywhere match consists of a mixture of tetraphosphorus trisulfide, sulfur, lead(IV) oxide, powdered glass, and glue. This mixture is covered with paraffin wax and then mounted on a small stick of wood. Tetraphosphorus trisulfide is a substance having a relatively low ignition temperature. As the match is struck against a hard surface, friction causes ignition of the trisulfide; the evolved heat then causes the entire match head to burn. This initial ignition is illustrated by the following equation:

$$P_4S_3(s) + 8O_2(g) \longrightarrow P_4O_{10}(s) + 3SO_2(g)$$

Thereafter, the sulfur ignites. As Fig. 10-10 notes, the lead(IV) oxide acts as the oxidizer and replaces atmospheric oxygen as the third leg of the fire triangle.

Safety matches contain different components than strike-anywhere matches. They consist of a mixture of antimony sulfide, sulfur, and potassium chlorate held to a piece of cardboard by means of glue. The surface on which this mixture is struck consists of red phosphorus and powdered glass. This latter mixture is particularly sensitive. When the match head is rubbed on this surface, a redox reaction occurs, as the following equation illustrates:

$$3S_8(s) + 16KClO_3(s) \longrightarrow 16KCl(s) + 24SO_2(g)$$

The energy evolved from this reaction causes the antimony trisulfide to ignite.

$$2Sb_2S_3(s) + 9O_2(g) \longrightarrow 2Sb_2O_3(g) + 6SO_2(g)$$

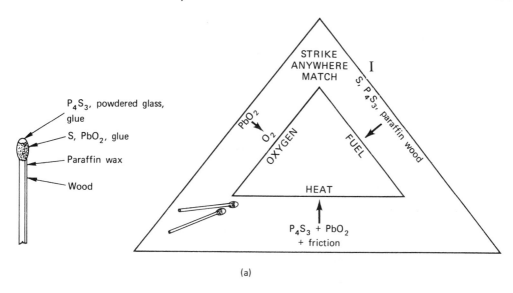

(a)

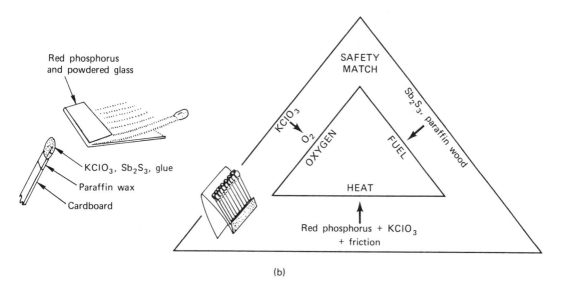

(b)

Figure 10-10   (a) A strike-anywhere match ignites when the surface of the match head is rubbed against a hard surface. (b) A safety match ignites when the surface of a match head is struck against a surface containing red phosphorus and powdered glass. In each instance a unique oxidation–reduction reaction occurs, which produces a flame. Fire triangles are also noted that identify the fuel, the source of oxygen, and the source of heat.

USDOT regulates the transportation of strike-anywhere and safety matches as flammable solids. Containers of these matches are labeled FLAMMABLE SOLID, as illustrated in Fig. 10-8; when required, their transport vehicles are placarded FLAMMABLE SOLID.

10-11 HYDRAZINE

Hydrazine is a colorless liquid resembling water in appearance and possessing a weak, ammonialike odor. Its chemical formula is N_2H_4. It is industrially prepared by reacting liquid ammonia and sodium hypochlorite:

$$2NH_3(l) + NaClO(s) \longrightarrow N_2H_4(l) + NaCl(s) + H_2O(g)$$

Commercially, hydrazine is marketed as the anhydrous liquid and as aqueous solutions.

Anhydrous hydrazine is a very powerful reducing agent. It is best known in connection with its use as a high-energy rocket fuel, but it is also used to produce certain agricultural chemicals, plastics, and drugs. Hydrazine is chemically incompatible with all oxidizing agents, many of which react explosively on contact with hydrazine. Such mixtures of oxidizing and reducing agents that react explosively on contact are said to represent a *hypergolic mixture*. For instance, hydrazine and hydrogen peroxide are a hypergolic mixture; it explodes, producing nitrogen and water vapor.

Figure 10-11 The use of protective clothing is highly recommended when handling containers of anhydrous hydrazine or when removing spills of this substance. Anhydrous hydrazine is very toxic by skin absorption; hence, the protective clothing must be of a material that is resistant to penetration by hydrazine. To avoid inhalation of hydrazine vapors, the use of a positive-pressure respirator is also recommended. Emergency-response personnel should be wary of the potential flammability of anhydrous hydrazine, since it forms explosive mixtures with air over a relatively wide range, 4% to 100% by volume. (Courtesy of Mine Safety Appliances Co., Pittsburgh, Pennsylvania)

$$N_2H_4(l) + 2H_2O_2(l) \longrightarrow N_2(g) + 4H_2O(g)$$

Anhydrous hydrazine is also a flammable liquid. Its flash point is 126°F (52°C). Mixed with air, the vapors of hydrazine explosively burst into flame. The oxidation is represented by the following equation:

$$N_2H_4(g) + O_2(g) \longrightarrow N_2(g) + 2H_2O(g)$$

This combustion is highly exothermic, yielding 148.6 kcal/mol (590 Btu/mol; 622 kJ/mol). This relatively large amount of heat makes hydrazine useful as a rocket fuel. Uncontrolled fires involving anhydrous hydrazine are readily extinguished with flooding amounts of water.

Hydrazine is also a particularly toxic substance, whether by inhalation, ingestion, or skin absorption. Its TLV is only 1 ppm by absorption through the skin. Protective clothing, like that illustrated in Fig. 10-11, should be employed. OSHA reg-

Figure 10-12 The launch of the Pegasus III satellite from the space complex at Cape Kennedy, Florida. A mixture of anhydrous hydrazine and liquid oxygen was used as the propellant. (Courtesy of the National Aeronautics & Space Administration, Washington, D.C.)

ulates the exposure of workers to hydrazine; the permissible exposure limit is only 1 ppm.

USDOT regulates the transportation of aqueous solutions of hydrazine containing more than 64% hydrazine as flammable liquids. Containers of these solutions are labeled FLAMMABLE LIQUID, POISON, and CORROSIVE; when required, their transport vehicles are placarded FLAMMABLE LIQUID. USDOT regulates the transportation of aqueous solutions of hydrazine containing less than 64% hydrazine as corrosive materials. Containers of these solutions are labeled CORROSIVE and POISON; when required, their transport vehicles are placarded CORROSIVE.

Organic derivatives of hydrazine are also used industrially. These are compounds in which one or more of the hydrogen atoms have been replaced with organic groups. *Unsymmetrical dimethyl hydrazine,* for example, is a compound in which two hydrogen atoms bonded to the same nitrogen atom have been replaced by methyl ($-CH_3$) groups. Hence, its chemical formula is $(CH_3)_2-N-NH_2$. This hazardous material has been used with liquid oxygen as the propellant of the Gemini lunar flight shown in Fig. 10-12.

USDOT regulates the transportation of *unsymmetrical* dimethyl hydrazine as a flammable liquid. Containers of this substance are labeled FLAMMABLE LIQUID, POISON, and CORROSIVE; when required, their transport vehicles are placarded FLAMMABLE LIQUID.

REVIEW EXERCISES

Basic Concepts Regarding Oxidation–Reduction

10.1. What is the oxidation number of each underlined element in the following chemical formulas?

(a) $\underline{Na}_2O$	(b) $Na_2\underline{O}_2$	(c) $N\underline{O}$
(d) $H_3\underline{P}O_4$	(e) $Mn\underline{O}_2$	(f) $KCl\underline{O}_4$
(g) $Ca\underline{H}_2$	(h) $\underline{Hg}_2Cl_2$	(i) $\underline{O}_2$
(j) $H_2\underline{S}O_4$	(k) $\underline{Hg}S$	(l) $\underline{Na}BrO_3$
(m) $\underline{Cr}Cl_3$	(n) $\underline{Pb}O_2$	(o) $H_2\underline{S}O_3$

10.2. The following equation notes the reaction between potassium dichromate, perchloric acid, and hydroiodic acid:

$$K_2Cr_2O_7(aq) + 8HClO_4(aq) + 6HI(aq) \longrightarrow$$
$$2Cr(ClO_4)_3(aq) + 2KClO_4(aq) + 3I_2(aq) + 7H_2O(l)$$

In terms of this equation, identify the following:
(a) The oxidation numbers of all elements
(b) The element oxidized
(c) The element reduced
(d) The oxidizing agent
(e) The reducing agent

10.3. Using Table 10-3, identify the substance in the following pairs that possesses the greater oxidizing potential:

    (a) Potassium chlorate or concentrated sulfuric acid

    (b) Concentrated sulfuric acid or concentrated nitric acid

    (c) Oxygen or chlorine

    (d) Chlorine or bromine

    (e) Hydrogen peroxide or sodium dichromate

10.4. Which of the following mineral acids are oxidizing acids: sulfuric acid; hydrochloric acid; nitric acid; phosphoric acid; perchloric acid; and hydrofluoric acid?

10.5. When an industry uses oxidizers, safety officials often recommend their storage in a cooled, ventilated building separated from the industry's major structures. What is the technical basis for this recommendation?

10.6. Why are smothering-type fire extinguishers, like carbon dioxide, generally ineffective when used on fires supported by oxidizers other than atmospheric oxygen?

10.7. Stipulate whether the following incidents constitute a potentially hazardous situation:

    (a) An organochlorine insecticide is inadvertently mixed with calcium hypochlorite bleach.

    (b) Waste barium peroxide is disposed of in a municipal sewer.

    (c) A spill of solid sodium chlorate is mixed with sawdust and the resulting residue disposed of in an open container with ordinary office wastes.

    (d) A spill of solid sodium chlorate is mixed with sawdust and the resulting residue is disposed of in a closed container with ordinary office wastes.

    (e) Five-gallon containers of aqueous sodium hypochlorite, en route to a dealer from its manufacturer, are exposed to strong sunlight.

    (f) Steel drums containing a residue of an oxidizer are used to collect waste paper.

    (g) Cardboard fiber drums containing potassium chlorate are exposed to the heat from a fire.

    (h) Sulfur is shoveled into a pile near sacks containing ammonium nitrate.

    (i) Charcoal is mixed with the swimming pool disinfectant, sodium dichloro-*s*-triazinetrione.

    (j) Chemicals to be used for chlorinating a domestic swimming pool, sodium dichloro-*s*-triazinetrione and muriatic acid, are stored in a common cupboard.

Hydrogen Peroxide

10.8. Why is hydrogen peroxide usually stored in dark brown glass bottles?

10.9. When hydrogen peroxide containers are shipped on board watercraft, USDOT requires that the containers be shaded from radiant heat, stowed separately from metallic permanganates, and stowed away from powdered metals. What is the most likely reason USDOT stipulates these special requirements?

10.10. When bulk quantities of hydrogen peroxide spill during a transportation mishap, why should the runoff be prevented from draining into sewers?

10.11. Water is employed to extinguish a fire in a warehouse that is used to store a variety of chemical substances, including sodium peroxide. When moving the drums of sodium peroxide, several drums are punctured, causing their contents to spill on the underlying concrete floor. Some water contacts the sodium peroxide, but not in flooding amounts. Why is this likely to intensify the ongoing fire?

Metallic Hypochlorites, Chlorites, Chlorates, and Perchlorates

10.12. Why do we say that chlorine bleaches cotton *indirectly?*

10.13. Hypochlorous acid has a distinctive sanitizing odor. Why is this odor often prevalent on freshly laundered clothing?

10.14. When the bleaching powder called chloride of lime is mixed with sulfuric acid, chlorine is produced, as the following equation illustrates:

$$CaCl(ClO)(s) + H_2SO_4(aq) \longrightarrow CaSO_4(s) + H_2O(l) + Cl_2(g)$$

Show that chlorine is both oxidized and reduced in this reaction.

10.15. Why is each of the following often a component of incendiary bombs?
  (a) An oxidizer
  (b) Finely divided magnesium or aluminum

10.16. A firecracker usually consists of a delicately balanced mixture of sulfur, potassium perchlorate, and flaked aluminum, loosely packed in cardboard tubes. What specific role is played by each component when the firecracker is ignited?

Ammonium Compounds

10.17. When used in fireworks as an oxidizer, ammonium perchlorate has the advantage of only producing colorless gaseous products. Write the equation that illustrates the thermal decomposition of ammonium perchlorate.

10.18. Why is it generally recommended to segregate ammonium nitrate from other oxidizers during storage and transportation?

Oxidizing Agents Containing Chromium

10.19. To protect the environment, what action should be taken when using water to fight a fire involving a consignment of combustible material and containers of sodium chromate?

10.20. USDOT regulates the transportation of potassium chromate and potassium dichromate as ORM-E materials. What does this hazard class assignment tell us about the thermal stability of these two substances?

Metallic Permanganates

10.21. Why do safety officials often recommend the storage of metallic permanganates in areas not used for storage of concentrated sulfuric acid?

Metallic Nitrites and Nitrates

10.22. Suggest a simple chemical test that serves to differentiate between samples of sodium nitrite and sodium nitrate.

10.23. Metallic nitrites are occasionally identified in polluted drinking water supplies. What specific adverse health effect is likely to result from the consumption of such water?

Matches

10.24. Identify the specific role served by each of the following components of the strike-anywhere match: (a) tetraphosphorus trisulfide: (b) sulfur; (c) lead(IV) oxide; (d) glass.

10.25. Identify the specific role served by each of the following components of the safety match: (a) potassium chlorate; (b) antimony trisulfide; (c) sulfur.

Hydrazine

10.26. Two groups of emergency-response personnel disagree over the type of respiratory protection to be worn when responding to an incident involving anhydrous hydrazine. The respective groups recommend use of respirators based on the principles of negative pressure and positive pressure, respectively. Which group is correct?

10.27. Write the Lewis structures for *symmetrical* and *unsymmetrical* dimethyl hydrazine.

11

Chemistry of Some Hazardous Organic Compounds

Organic compounds are components of all the familiar commodities that our technological world requires. Examples of such commodities are heating and motor fuels, cleaning solvents, adhesives, plastics, resins, fibers, paints, varnishes, refrigerants, propellants, aerosols, textiles, and explosives, among many others. The relative commonplace of these materials underscores the need to study organic chemistry in some detail.

From a safety perspective, the principal concern about organic compounds is that they are usually flammable or combustible substances, with few exceptions. When ignited, the flammable gaseous organic compounds burn explosively in air, and most possess relatively wide flammable ranges. Furthermore, many flammable liquid organic compounds evaporate easily at room conditions; their vapors also ignite readily.

Flammability is generally the principal hazard of organic compounds. The uncontrolled burning of even small quantities (one gallon or less) of liquid organic compounds always comprises a dangerous situation. This is due to the heat evolved from their fires, which when radiated to other nearby flammable and combustible materials is usually sufficient to ignite them as well.

A secondary concern about organic compounds is their ability to cause a range of detrimental health effects. While many organic compounds are nontoxic (in fact, many are essential for life itself), the common industrial organic compounds are often associated with varying degrees of toxicity. In humans, some of these compounds damage the liver, kidneys, and heart, other depress the central nervous system, and several even cause cancer.

11-1 WHAT ARE ORGANIC COMPOUNDS?

The molecules of all organic compounds have one common feature: one or more carbon atoms that invariably covalently bond to other atoms; that is, pairs of electrons are shared between atoms (Sec. 3-10). On the one hand, a carbon atom may share electrons with other nonmetallic atoms. As Fig. 11-1 notes, methane, carbon tetrachloride, carbon monoxide, and carbon disulfide are compounds having molecules in which the carbon atom is bonded to other nonmetallic atoms.

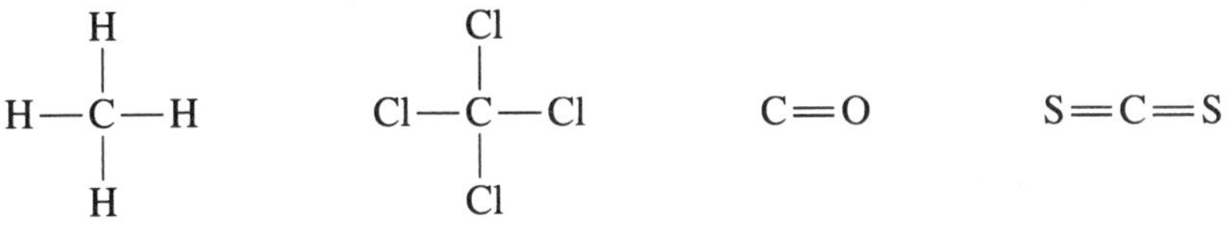

Methane   Carbon tetrachloride   Carbon monoxide   Carbon disulfide

Figure 11-1 Carbon atoms may share their electrons with the electrons of other nonmetallic atoms, like hydrogen, chlorine, oxygen, and sulfur. The compounds that result from such electron sharing are methane, carbon tetrachloride, carbon monoxide, and carbon disulfide, respectively.

On the other hand, carbon atoms may also share electrons with other carbon atoms. When the molecular structures of such compounds are examined, we find that two carbon atoms may share electrons in such a manner that they form either of the following: carbon–carbon single bonds (C—C); carbon–carbon double bonds (C=C); and carbon–carbon triple bonds (C≡C). Each bond written here as a dash (−) is a shared pair of electrons. Figure 11-2 illustrates the bonding in molecules of ethane, ethylene, and acetylene, compounds having molecules with only two carbon atoms. Molecules of ethane possess carbon–carbon single bonds; molecules of ethylene possess carbon–carbon double bonds; and molecules of acetylene possess carbon–carbon triple bonds.

Covalent bonds between carbon atoms in molecules of more complex organic compounds may be linked into chains, including branched chains, or into rings.

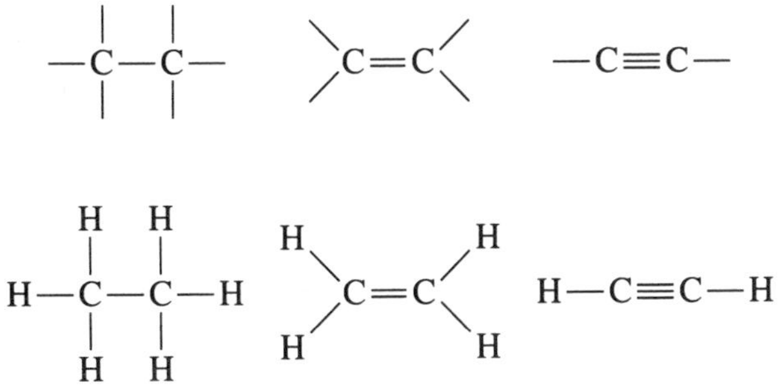

Figure 11-2 Two carbon atoms may share their own electrons in either of the three ways noted. This results in the formation of carbon–carbon single bonds, carbon–carbon double bonds, and carbon–carbon triple bonds. When these carbon atoms further bond to hydrogen atoms, the resulting compounds are ethane, ethene, and acetylene, respectively.

Consider just the carbon skeleton of the two molecules noted as (a) and (b) below:

$$-C-C-C-C-C-C- \qquad -C-C-C-C-C-C-$$

(a)   (b)

In (a), the carbon atoms are bonded to one another in a continuous chain, whereas in (b) the carbon atoms are bonded to one another in a branched pattern. This means that carbon atoms may bond to other carbon atoms in long chains, but may also bond to groups of other carbon atoms as side chains attached to the main chain.

Aside from being bonded to other carbon atoms, each carbon atom in either a straight chain or branched chain arrangement is always bonded to one or more atoms of hydrogen, oxygen, nitrogen, sulfur, or the halogens. For instance, when the carbon atoms in the compound having the carbon skeleton (a) are bonded to hydrogen atoms, the molecular structure [that is, the Lewis structure (Sec. 3-10)] of the compound may be written as follows:

$$
\begin{array}{c}
\ \ \ H \ \ \ H \ \ \ H \ \ \ H \ \ \ H \ \ \ H \\
\ \ \ | \ \ \ | \ \ \ | \ \ \ | \ \ \ | \ \ \ | \\
H-C-C-C-C-C-C-H \\
\ \ \ | \ \ \ | \ \ \ | \ \ \ | \ \ \ | \ \ \ | \\
\ \ \ H \ \ \ H \ \ \ H \ \ \ H \ \ \ H \ \ \ H
\end{array}
$$

When the carbon atoms in (b) are bonded to hydrogen atoms, the molecular structure of the compound may be written as follows:

$$
\begin{array}{c}
\ \ \ H \ \ \ H \ \ \ H \ \ \ H \ \ \ H \ \ \ H \\
\ \ \ | \ \ \ | \ \ \ | \ \ \ | \ \ \ | \ \ \ | \\
H-C-C-C-C-C-C-H \\
\ \ \ | \ \ \ | \ \ \ | \ \ \ | \\
\ \ \ H \ \ \ H \ \ H \ \ \ H \\
\ \ \ \ \ \ \ \ | \ \ \ \ \ \ \ \ \ | \\
\ \ \ \ \ H-C-H \ \ H-C-H \\
\ \ \ \ \ \ \ \ | \ \ \ \ \ \ \ \ \ | \\
\ \ \ \ \ \ H \ \ \ \ \ \ \ \ H
\end{array}
$$

This illustrates that many organic compounds consist of continuous or branched chains of carbon atoms. The actual number of such compounds appears to be almost unlimited. Over 1 million organic compounds are already known. This abundance may be cause for some initial concern among students. How can anyone know the properties of so many substances? Fortunately, it is possible to systematically organize organic compounds into a number of groups. The properties of the members of each group are largely dictated by certain atoms or groups of atoms. These are called *functional groups* and are identified later in Sec. 11-7. The study of organic chemistry then becomes the study of compounds possessing these functional groups.

We shall begin the study of organic chemistry by examining the simplest organic compounds, the *hydrocarbons;* these are compounds whose molecules are composed only of carbon and hydrogen atoms. Such compounds are of particular significance to firefighters, as all of them are flammable or combustible. They may be either gaseous, liquid, or solid substances at room conditions. On complete combustion, they form carbon dioxide and water vapor.

All hydrocarbons are broadly divided into two groups: aliphatic and aromatic hydrocarbons. We shall note some of their features in the following two sections of this chapter.

11-2 ALIPHATIC HYDROCARBONS

Aliphatic hydrocarbons are those that can be characterized by the chain arrangements of their constituent carbon atoms. They are divided into the following series, which we shall examine independently.

Alkanes

The *alkanes* are hydrocarbons whose molecules have only carbon–carbon single bonds. The general chemical formula of an alkane is C_nH_{2n+2}, where n is a nonzero integer. In other words, when the number of carbon atoms is only one, the number of hydrogen atoms must be four; the corresponding compound is methane (CH_4). When the number of carbon atoms is two, the number of hydrogen atoms is six; the corresponding compound is ethane (C_2H_6). The alkanes are called *saturated hydrocarbons* since each bonding electron from the carbon atoms is shared with a valence electron from an atom other than carbon. Several examples of alkanes having carbon atoms from one to eight are noted in Table 11-1. The names of these simple alkanes and their formulas should be memorized.

Table 11-1 also provides the Lewis structures of these simple alkanes; for simplicity, only the straight chain arrangements have been noted. But beginning with C_4H_{10}, the molecular structure may be correctly written in a variety of ways. There are two ways to write the Lewis structure for the formula C_4H_{10}, as follows:

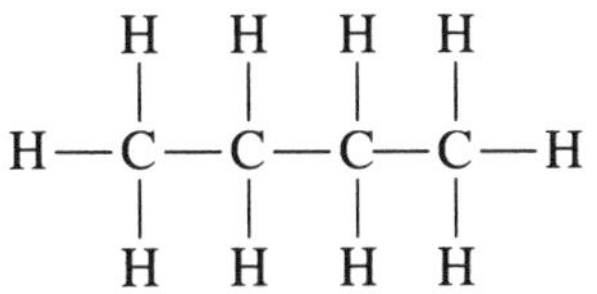

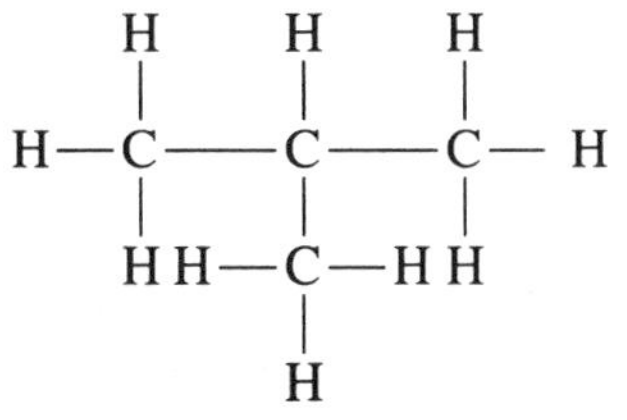

In the first structure, the carbon atoms are bonded to one another in a continuous chain. In the second structure, only three of them are so bonded; the fourth carbon atom is bonded to the carbon atom in the middle of the chain. This means that there are two distinctly different compounds having the formula C_4H_{10}. To distinguish them by name, the compound having the carbon atoms bonded in a continuous chain

TABLE 11-1 SIMPLE ALKANES HAVING A CONTINUOUS CHAIN OF CARBON ATOMS

Name	Molecular formula	Lewis structure	Condensed formula
Methane	CH_4	$\begin{array}{c} H \\ | \\ H-C-H \\ | \\ H \end{array}$	CH_4
Ethane	C_2H_6	$\begin{array}{c} H\ \ H \\ |\ \ \ | \\ H-C-C-H \\ |\ \ \ | \\ H\ \ H \end{array}$	CH_3CH_3
Propane	C_3H_8	$\begin{array}{c} H\ \ H\ \ H \\ |\ \ \ |\ \ \ | \\ H-C-C-C-H \\ |\ \ \ |\ \ \ | \\ H\ \ H\ \ H \end{array}$	$CH_3CH_2CH_3$
Butane	C_4H_{10}	$\begin{array}{c} H\ \ H\ \ H\ \ H \\ H-C-C-C-C-H \\ H\ \ H\ \ H\ \ H \end{array}$	$CH_3CH_2CH_2CH_3$
Pentane	C_5H_{12}	$\begin{array}{c} H\ \ H\ \ H\ \ H\ \ H \\ H-C-C-C-C-C-H \\ H\ \ H\ \ H\ \ H\ \ H \end{array}$	$CH_3CH_2CH_2CH_2CH_3$
Hexane	C_6H_{14}	$\begin{array}{c} H\ \ H\ \ H\ \ H\ \ H\ \ H \\ H-C-C-C-C-C-C-H \\ H\ \ H\ \ H\ \ H\ \ H\ \ H \end{array}$	$CH_3CH_2CH_2CH_2CH_2CH_3$
Heptane	C_7H_{16}	$\begin{array}{c} H\ \ H\ \ H\ \ H\ \ H\ \ H\ \ H \\ H-C-C-C-C-C-C-C-H \\ H\ \ H\ \ H\ \ H\ \ H\ \ H\ \ H \end{array}$	$CH_3CH_2CH_2CH_2CH_2CH_2CH_3$
Octane	C_8H_{18}	$\begin{array}{c} H\ \ H\ \ H\ \ H\ \ H\ \ H\ \ H\ \ H \\ H-C-C-C-C-C-C-C-C-H \\ H\ \ H\ \ H\ \ H\ \ H\ \ H\ \ H\ \ H \end{array}$	$CH_3CH_2CH_2CH_2CH_2CH_2CH_2CH_3$

is called *n*-butane, while the compound having the branched structure is named *iso*butane.

Two or more compounds that have the same molecular formula, but different structural arrangements of their atoms, are called *structural isomers*. One way by which isomers are named is to use an *n*- (for *normal*) in front of the name of the compound that contains a continuous chain of carbon atoms, and *iso*- for the compound that has a —CH_3 group bonded to a carbon atom next to the terminal (end) carbon atom. This system of nomenclature is generally referred to as the *common system*. Thus, as noted earlier, the two isomers of butane are named *n*-butane and *iso*butane, respectively.

Although we may write a Lewis formula for any alkane, it is generally convenient to condense the structure by simply writing the symbols of the atoms next to the symbol of the carbon atom to which they are bonded. For an alkane, we write the symbols of the hydrogen atoms next to the symbols of the carbon atoms to which they are bonded. Such formulas are called *condensed formulas*. They convey the same information as the more complete Lewis structure, but the dashes are either entirely omitted, or used only in a limited fashion. Condensed formulas for *n*-butane and *iso*butane are noted below:

$$CH_3CH_2CH_2CH_3 \qquad \begin{array}{c} CH_3—CH—CH_3 \\ | \\ CH_3 \end{array}$$

When using the common system of nomenclature, not all isomers are named as easily as those of butane. Thus, a second system of nomenclature was devised that is both simple and systematic. It was adopted by the International Union of Pure and Applied Chemistry (IUPAC) and is called the *IUPAC system of nomenclature*. This system applies the following rules to the naming of alkanes from known molecular structures:

1. Find the longest continuous chain of carbon atoms in the structure. This is called the *main chain*. It is not necessary that the main chain be written horizontally in order to be continuous.

2. Assign numbers to each carbon atom in the main chain starting from the end that will give the groups attached to the chain the smaller numbers. In alkanes, the groups are called *alkyl substituents;* their general chemical formula is C_nH_{2n+1}. The names of some common alkyl substituents are provided in Table 11-2; they should be memorized.

3. Designate the position of each substituent by the number of the carbon atom along the main chain to which it is attached.

4. Name the substituents alphabetically (ethyl before methyl, and so on) and place the names of the substituents as prefixes on the name of the main chain.

5. If several substituents occur in the same compound, indicate the number of identical groups by the use of the following prefixes before the name of the

TABLE 11-2   SOME COMMON ALKYL SUBSTITUENTS

Name[a]	Chemical formula
Methyl	CH_3-
Ethyl	CH_3CH_2- or C_2H_5-
n-Propyl	$CH_3CH_2CH_2-$ or C_3H_7-
Isopropyl	$\begin{array}{c} H \\ \mid \\ CH_3-C- \\ \mid \\ CH_3 \end{array}$ or $(CH_3)_2CH-$
n-Butyl	$CH_3CH_2CH_2CH_2-$ or C_4H_9-

[a] Alkyl groups are named by replacing the -ane suffix from the name of the corresponding hydrocarbon with -yl.

substituent: di- for two identical groups, tri- for three identical groups, tetra- for four identical groups, and so on.

Table 11-3 illustrates the use of these rules in naming several alkanes by the IUPAC system of nomenclature.

Alkanes are also known in which the first and last carbon atoms in a continuous chain are joined to each other so that a cyclic arrangement of carbon atoms exists. These compounds are called *cycloalkanes*. Their general chemical formula is C_nH_{2n}. They are named by placing the prefix *cyclo-* in front of the name of the parent hydrocarbon: cyclopropane, cyclobutane, cyclopentane, and so on. Sometimes the structural formula is abbreviated by writing a triangle, square and pentagon for cyclopropane, cyclobutane, and cyclopentane, respectively, as noted in Table 11-4.

Alkenes

Hydrocarbons whose molecules contain one or more carbon–carbon double bonds are called *alkenes,* or olefins. Since each alkene is deficient in hydrogen relative to its corresponding alkane, they are said to be *unsaturated hydrocarbons*. The general chemical formula of this group of organic compounds is C_nH_{2n}. The simplest alkene is named ethene, or ethylene. Its molecular formula is C_2H_4, and its Lewis structure is the following:

$$\begin{array}{ccc} H & & H \\ \diagdown & & \diagup \\ & C=C & \\ \diagup & & \diagdown \\ H & & H \end{array}$$

The presence of the carbon–carbon double bond restricts the movement of the atoms

TABLE 11-3 EXAMPLES OF NAMING ALKANES

Chemical formula	Name
$CH_3CH_2CH_2CH_2CH_3$	*n*-Pentane
$CH_3CH_2CHCH_3$	3-Methylheptane

$$
\begin{array}{l}
CH_3CH_2\overset{\displaystyle |}{C}HCH_3 \\
\quad\ \ |\\
\quad\ \ CH_2 \\
\quad\ \ |\\
\quad\ \ CH_2 \\
\quad\ \ |\\
\quad\ \ CH_2 \\
\quad\ \ |\\
\quad\ \ CH_3
\end{array}
$$

$CH_3\!-\!\overset{\displaystyle CH_3}{\underset{\displaystyle CH_3}{C}}\!-\!CH_2CH_3$	2, 2-Dimethylbutane
$CH_3\underset{\displaystyle CH_3}{C}HCH_2\underset{\displaystyle \underset{\displaystyle CH_3}{CH_2}}{C}HCH_3$	2, 4-Dimethylhexane
$CH_3\underset{\displaystyle CH_3CH_2CHCH_2CH_3}{C}H\overset{\displaystyle CH_3}{C}HCH_2CH_2CH_3$	3-Ethyl-4-isopropylheptane

TABLE 11-4 SOME SIMPLE CYCLOALKANES

Chemical formula	Notation	Name
CH_2 $CH_2\!-\!CH_2$	△	Cyclopropane
$CH_2\!-\!CH_2$ $CH_2\!-\!CH_2$	□	Cyclobutane
CH_2 $CH_2\quad CH_2$ $CH_2\!-\!CH_2$	⬠	Cyclopentane

in this molecule. This means that the six atoms in ethylene spend their average time in a plane, such as in the plane of this page.

Ethene and propene do not possess structural isomers; that is, the following ways of representing propene are identical:

$$CH_2\!\!=\!\!CHCH_3 \qquad CH_3CH\!\!=\!\!CH_2$$

Either formula is the other one turned around end for end. However, in C_4H_8, the double bond may be positioned differently, as follows:

$$CH_2\!\!=\!\!CHCH_2CH_3 \qquad CH_3CH\!\!=\!\!CHCH_3$$

These formulas represent the two structural isomers of butene.

Alkenes are named much like alkanes, except that we use the *-ene* suffix to identify them. Hence, the first several members of the alkene series are named ethene, propene, butene, and pentene. In the IUPAC system, we also indicate the position of the double bond in the main chain. This is accomplished by using the following simple rules:

1. Number the main chain of continuous carbon atoms consecutively by beginning at the end that is nearer to the double bond.
2. Indicate the position of the double bond by the appropriate numerical prefix.

For example, using these rules, $CH_2\!\!=\!\!CHCH_2CH_3$ is named 1-butene; $CH_3CH\!\!=\!\!CHCH_3$ is named 2-butene.

The relative rigidity of the carbon–carbon double bond leads to a new kind of isomerism, called *geometrical isomerism*. This phenomenon arises when the two substituents are different on each carbon atom that comprises the carbon–carbon double bond. For example, we may write two Lewis structures for 2-butene, as follows:

trans-2-butene   *cis*-2-butene

(Note that the substituents on each carbon atom are different: hydrogen and the methyl group.) These two compounds have exactly the same atoms bonded to the same atoms, so they are not structural isomers. Yet the arrangement of these atoms differs in space. When two compounds have the same structural formula, but differ by the spatial arrangement of their atoms, they are called *geometrical isomers*.

Geometrical isomers are named by using either of the prefixes *trans-* (meaning on the opposite side of the double bond) or *cis-* (meaning on the same side of the bond). Thus, the two geometrical isomers of 2-butene are named *trans*-2-butene and *cis*-2-butene, respectively.

Alkynes

Hydrocarbons possessing one or more carbon–carbon triple bonds are called *alkynes*. Like alkenes, they are unsaturated hydrocarbons. The general chemical formula of an alkyne is C_nH_{2n-2}. The simplest member of this series has the formula C_2H_2, which is named ethyne or acetylene. Its Lewis structure is:

$$H-C\equiv C-H$$

In the IUPAC system, alkynes are named by replacing the *-ane* or *-ene* suffix on the associated alkane or alkene, respectively, with the suffix *-yne*.

Like alkenes, alkynes may have structural isomers, like 1-butyne and 2-butyne:

$$H-C\equiv C-CH_2CH_3 \qquad CH_3-C\equiv C-CH_3$$
$$\text{1-butyne} \qquad\qquad\qquad \text{2-butyne}$$

When naming alkynes by the IUPAC system, a numerical prefix is used to indicate the position of the triple bond in the main chain of continuous carbon atoms. Thus, 1-butyne is the alkyne having the molecular formula C_4H_6 and having the carbon–carbon triple bond between a terminal carbon atom and the one immediately adjacent to it; on the other hand, 2-butyne is the C_4H_{10} isomer that has the triple bond between two nonterminal carbon atoms.

11-3 AROMATIC HYDROCARBONS

Aromatic hydrocarbons are those containing one or more rings of carbon atoms. These compounds are typified by benzene, the simplest aromatic hydrocarbon, whose molecular formula is C_6H_6. The molecular structure of benzene is shown by either formula drawn below to the left, but it is more commonly represented by a hexagon with a circle inside as shown below on the right:

Aromatic and aliphatic hydrocarbons are thus differentiated by the fact that their molecular structures either resemble benzene or do not resemble benzene, respectively. Hereafter we shall represent benzene by the hexagonal structural formula with the inscribed circle. The symbols of the carbon atoms are not written as part of the hexagon, but it is understood that a carbon atom occupies each corner of the hexagon, where it is bonded to a hydrogen atom.

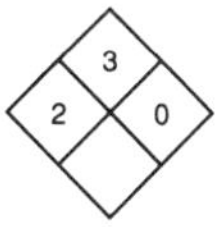

 Benzene itself is a clear, colorless, water-insoluble liquid that vaporizes readily at room conditions. When exposed to an ignition source, a mixture of benzene vapor and air burns readily with a very sooty flame. USDOT regulates its transportation as a flammable liquid. Containers of benzene are labeled FLAMMABLE LIQUID; when required, their transport vehicle is placarded FLAMMABLE.

Benzene was formerly a very popular industrial solvent, but no longer. Benzene is now recognized as a human carcinogen, causing leukemia in exposed individuals. Worker exposure to benzene is regulated by OSHA at 10 ppm as an 8-hour, time-weighted average.

 When any of the six hydrogen atoms in benzene is substituted with a methyl group (—CH₃), the resulting compound is commonly called *toluene*. Its molecular structure is represented as follows:

 Two hydrogen atoms in benzene may also be substituted with methyl groups. It is possible to position them in three different ways; that is, structural isomerism exists. The resulting compounds are said to be the isomers of *xylene*. Their molecular structures are

ortho-xylene   *meta*-xylene   *para*-xylene

As noted, the prefixes *ortho-, meta-* and *para-* are employed to distinguish these structural isomers from each other; this is the "common" system of nomenclature. *Ortho-,* means "straight ahead," *meta-* means "beyond," and *para-* means "opposite." Figure 11-3 illustrates two arbitrary substituents, A and B, on the ben-

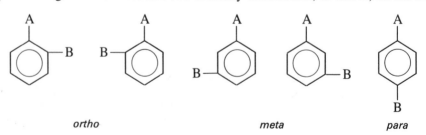

ortho   meta   para

Figure 11-3   The *ortho, meta,* and *para* positions of B, an arbitrary substituent, relative to a fixed position on the benzene ring of another arbitrary substituent, A. (A and B may be identical or different substituents.) The two configurations noted for the ortho and meta positions are equivalent ways of denoting the same compound.

zene ring; B is in either the ortho, meta- or para- position relative to the position of A. When naming such derivatives of benzene, these prefixes are generally shortened to the italicized letters *o-*, *m-* and *p-*, respectively.

In the IUPAC system, when two or more alkyl groups are bonded to the benzene ring, the resulting hydrocarbon may be named by listing the alkyl groups alphabetically and numbering the first one with a 1. Each of the six carbon atoms in the benzene ring is then numbered consecutively from 1 to 6. In this manner, ortho-xylene is named 1,2-dimethylbenzene; meta-xylene is named 1,3-dimethylbenzene; and para-xylene is named 1,4-dimethylbenzene.

At room conditions, toluene and the isomers of xylene are clear, colorless, water-insoluble liquids that vaporize readily. Mixtures of the vapors of these substances with air burn readily when exposed to a source of ignition. USDOT regulates their transportation as flammable liquids. They are stored in metal or glass containers labeled FLAMMABLE LIQUID, as shown in Fig. 11-4. When required, their transport vehicles are placarded FLAMMABLE.

Toluene and the xylene isomers are toxic. In humans, acute exposure to these substances depresses the central nervous system and causes narcosis. Unlike benzene, however, these substances are not known to be carcinogenic. Worker exposure to toluene and the isomers of xylene is regulated by OSHA at 200 ppm (8-hour, time-weighted average) and 100 ppm, respectively.

Figure 11-4 Toluene is one of the most commonly encountered industrial organic solvents. Chemical laboratories often dispense it from a 5-gallon metal container, like that illustrated here. When offered for transportation, the FLAMMABLE LIQUID label is affixed to the container. (Courtesy of J. T. Baker, Inc., Phillipsburg, New Jersey)

Sometimes, individual aromatic hydrocarbons may be named by identifying the benzene ring itself. In such instances, C_6H_5— is regarded as a substituent of a parent molecule and is named the *phenyl* group. For instance, the compound whose molecular structure follows is named phenylethylene:

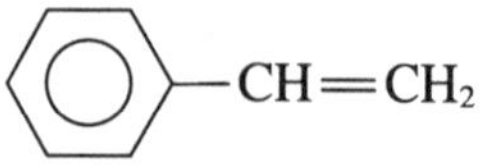

It is also called styrene (Sec. 12-1). The phenyl group and other similar aromatic hydrocarbon groups of atoms are called *aryl groups*. Aside from the phenyl group, the only other aryl group that we shall encounter is the benzyl group, $C_6H_5CH_2$—. Table 11-5 illustrates additional examples of naming aromatic hydrocarbons.

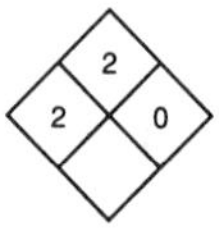

Another group of aromatic hydrocarbons exists in which two or more benzene rings are mutually bonded. They are called *polynuclear aromatic hydrocarbons* (PAHs). Three important PAHs are naphthalene, anthracene, and phenanthrene, whose molecular structures are shown below:

naphthalene   anthracene   phenanthrene

Each substance is a white to colorless solid, but only naphthalene has commercial importance, primarily as a moth repellant and fungicide and as an intermediate in the chemical industry. USDOT regulates the transportation of molten naphthalene as a flammable solid.

PAHs are likely to be encountered inadvertently, since they are frequently produced during fires. For instance, PAHs have been detected in forest and prairie fires. This can be a point of concern, as some PAHs are carcinogenic, causing tumors at the site of application as well as systemically.

11-4 FUNCTIONAL GROUPS

Many organic compounds are known in which one or more hydrogen atoms in a hydrocarbon have been substituted with another atom or group of atoms. The atom or group of atoms that substitutes for the hydrogen atom is called the *functional group*.

Table 11-6 lists the important functional groups of organic molecules. The symbols R and R′ represent any arbitrary alkyl or aryl group.

TABLE 11-5 NAMING AROMATIC HYDROCARBONS

Chemical formula	Name
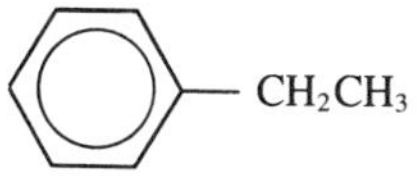 ⬡—CH_2CH_3	Ethylbenzene
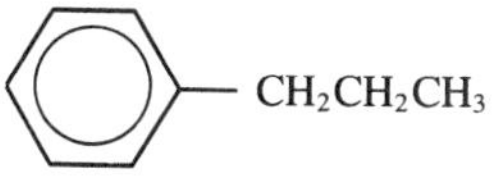 ⬡—$CH_2CH_2CH_3$	n-Propylbenzene
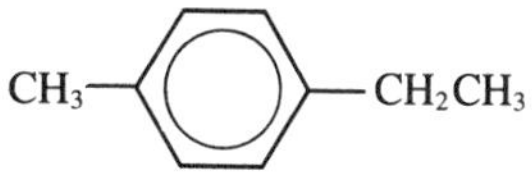 CH_3—⬡—CH_2CH_3	p-Ethyltoluene
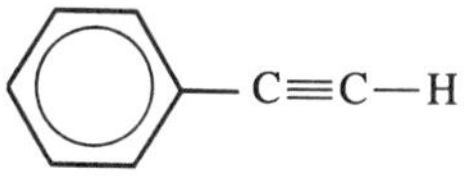 ⬡—$C{\equiv}C{-}H$	Phenylacetylene
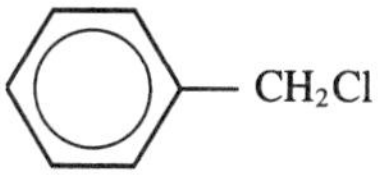 ⬡—CH_2Cl	Benzyl chloride
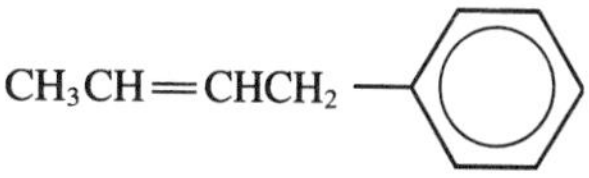 $CH_3CH{=}CHCH_2$—⬡	1-Phenyl-2-butene

TABLE 11-6 SOME IMPORTANT FUNCTIONAL GROUPS IN ORGANIC
MOLECULES

Name	General formula	Functional group
Haloalkane or alkyl halide or aryl halide	R—X	—X (X = halogen)
Alcohol	R—OH	—OH, hydroxyl
Ether	R—O—R′	—O—
Aldehyde	R—C(=O)H	—C(=O)H
Ketone	R—C(=O)—R′	—C(=O)—, carbonyl
Acid	R—C(=O)OH	—C(=O)OH, carboxyl
Ester	R—C(=O)OR′	—C(=O)O—
Hydroperoxides	H—O—O—R	—O—O—, peroxo
Peroxides	R—C(=O)—O—O—C(=O)—R′	—O—O—

11-5 PETROLEUM AND ITS PRODUCTS

Petroleum is a highly complex liquid mixture, mainly consisting of a variety of hydrocarbons, each of whose molecules has from 3 to 60 carbon atoms. Deposits of petroleum occur naturally in certain areas around the world and are located in oil-sand under folds in overlying rock. To obtain the petroleum, wells are drilled through the rock to the oil-bearing stratum, through which the petroleum is then pumped to the earth's surface. In this form, it is usually called *crude petroleum*.

Scientists believe that petroleum originated from the partial decomposition of animals and plants that lived in the sea millennia ago. Changes in the position of Earth's crust buried these materials underground, sometimes at great depths. The accompanying pressure then caused them to partially decompose, resulting in the production of petroleum.

The various components of petroleum are individual compounds possessing different boiling points. Thus, when crude petroleum is heated at specified temperatures, mixtures of its volatile components may be removed one by one. This process is referred to as *fractional distillation*. Petroleum fractions are obtained in certain temperature ranges and are thereby collected in separate receivers. The components of these fractions are not pure compounds, but mixtures of individual hydrocarbons. One of the principal products resulting from the fractional distillation of petroleum, whether directly or indirectly, is gasoline, followed by other fuel oils, jet oils, lubricants, asphalt, and various other products. Some physical properties of several of these fuels are noted in Table 11-7. The production of petroleum products comprises the major business of the petroleum industry, the largest chemical industry in the world. Figure 11-5 illustrates just one facility in the United States where petroleum products are stored prior to distribution.

TABLE 11-7 PHYSICAL PROPERTIES OF SOME COMMON FUELS

	Gasoline	Kerosene	Fuel oil #4	Jet fuel (JP-6)
Boiling point	100°–400°F (38°–204°C)	338°–572°F (170°–300°C)		250°F (121°C)
Specific gravity	0.8	0.81	<1	0.8
Vapor density	3.0–4.0	4.5		1
Flash point	−45°F (−43°C)	100°–150°F (38°–66°C)	130°F (54°C)	100°F (38°C)
Autoignition temperature	536°–853°F (280°–456°C)	444°F (229°C)	505°F (263°C)	435°F (224°C)
Lower explosive limit	1.4%	0.7%		0.6%
Upper explosive limit	7.6%	5%		3.7%

A number of commercially important products are derived from the fractional distillation of petroleum, either directly or by blending certain fractions together. Three important petroleum products are the following:

1. *Petroleum ether** or *petroleum naphtha* is a mixture of hydrocarbons whose molecules mainly have five, six, or seven carbon atoms. It primarily distills between 95° and 194°F (35° and 90°C).

2. *Gasoline* is a mixture of hydrocarbons whose molecules primarily have five to nine carbon atoms. Its approximate boiling point range is from 100° to 400°F (38° − 204°C).

3. *Kerosene* is a mixture of heavier hydrocarbons than those found in gasoline.

---

*Petroleum ether is a generic term and is not associated with the class of organic compounds called ethers.

Figure 11-5 The Los Angeles Texaco refinery at Wilmington, California, where numerous floating-roof tanks are used to store petroleum products prior to their commercial distribution. Over 1 billion gallons of petroleum products may be stored at such locations at any time. (Courtesy of Texaco, Inc., Bellaire, Texas; photographer Thomas Carroll)

Decane ($C_{10}H_{22}$) and dodecane ($C_{12}H_{26}$) are commonly found in kerosene. Its approximate boiling point range is $338° - 572°F$ ($170° - 300°C$).

Other products may be separated from the very highest boiling petroleum fractions, like fuel oil (including diesel oil), lubricating oil, greases, petroleum jelly, paraffin wax, asphalt, and petroleum coke. These petroleum products typically contain organosulfur compounds in addition to high-molecular-weight hydrocarbons.

Aside from these products that result directly from crude petroleum, still others may be obtained by subjecting petroleum fractions to high temperatures under pressure, generally in the presence of a catalyst. This process, called *cracking*, breaks up complex molecules into simpler ones. In the petroleum industry, the components of higher boiling fractions are often decomposed, or "cracked," into other substances having lower boiling points, which may be blended to make gasoline.

Certain petroleum fractions may be further subjected to distillation, resulting in the production of a host of individual substances. These substances that are derivable from petroleum but not ordinarily used as fuels are collectively called *petrochemicals*. They serve as indispensable raw materials for the plastics, rubber, and synthetic fiber industries, among others. Roughly 175 petrochemicals are derived from petroleum. A typical one is ethylene, which is converted to products like ethylene glycol, a common antifreeze.

The individual components of certain petroleum fractions may also be chemically united. One such process, called *alkylation,* is a method by which an alkane and an alkene are selectively combined in the gaseous phase. When the alkane is isobutane and the alkene is isobutene, the product is isooctane as the following equation illustrates:

$$CH_3-\underset{\underset{CH_3}{|}}{CH}-CH_3(g) \; + \; \underset{\underset{CH_3}{\diagup}}{\overset{CH_3\diagdown}{}}C{=}CH_2(g) \; \longrightarrow \; CH_3-\underset{\underset{CH_3}{|}}{\overset{\overset{CH_3}{|}}{C}}-CH_2-\underset{\underset{CH_3}{|}}{\overset{\overset{H}{|}}{C}}-CH_3(g)$$

The name isooctane is actually a misnomer; the correct name is 2,2,4-trimethylpentane. Hydrocarbons having molecules with highly branched structures, like isooctane, are highly desirable as components of gasoline fuels, since they reduce knocking, the pinging noise that results when the mixture of gasoline vapor and air burns within the cylinders of a motor vehicle.

Alkylation processes using other alkanes and alkenes are employed on a grand scale to produce a variety of useful chemical products. For this reason, refineries often store and transport large volumes of such feedstocks. Figure 11-6 illustrates refrigerated spherical storage tanks that are sometimes used specifically for storing alkylation feedstock chemicals.

11-6 SIMPLE HYDROCARBONS

Certain simple hydrocarbons are occasionally encountered as components of smokestack emissions from many industrial and commercial facilities. Sometimes we are unaware of their presence in the atmosphere, since they are colorless, odorless compounds and are typically encountered in relatively small concentrations. No scientific evidence suggests that inhalation of these pollutants directly harms us, but some hydrocarbon pollutants adversely affects the quality of air in another way. By chemically interacting with sunlight, they contribute to the formation of atmospheric ozone (Sec. 6-1).

Under the Clean Air Act, USEPA regulates nonmethane hydrocarbon emissions from factories and other plants. The relevant standards are considered supplementary to the standards for ozone; thus, the standards for nonmethane hydrocarbans and ozone are enforced together. The primary and secondary standards for nonmethane hydrocarbons are identical: 160 μg/m^3 (0.24 ppm) measured for the 3-

Figure 11-6   These refrigerated spheroidal storage tanks resemble giant silver basketballs. Each tank holds the equivalent of 25,000 barrels of alkylation feedstock chemicals. A thick blanket of glass wool covered by aluminum sheathing insulates the shell of each tank. An adjacent refrigeration plant keeps the alkylate feedstock in the liquid state. (Courtesy of Marathon Oil Co., Robinson, Illinois)

hour period between 6 and 9 A.M. The primary and secondary standards for ozone are also identical: 235 μg/^{3} (0.12 ppm) for 1 hour.

Some simple hydrocarbons are commonly used domestically and industrially, primarily as fuels. It is appropriate to review them independently.

Methane

Methane is the simplest hydrocarbon; its chemical formula is CH_4. It is a component of *natural gas,* the common heating fuel, in which its concentration may equal 97% by volume. Thus, the chemistry of natural gas is principally the chemistry of methane. This gas also occurs naturally in the atmosphere of coal mines, where it is called *firedamp*. It is also encountered when the mud at the bottom of stagnant pools is disturbed; here it is called *marsh gas*. In each of these instances, methane results naturally from the decay and alteration of animal and plant remains. But methane may be produced artificially as well. It is produced by cracking petroleum and by subjecting coal to high temperatures in the absence of air.

Methane is an odorless, colorless, and tasteless gas and is only slightly soluble in water. Commercial natural gas, however, possesses a slight odor of hydrogen sulfide or an organosulfur compound, which is intentionally added to assist in the detection of gas leaks. It burns with a slightly luminous flame and possesses a relatively high heat of combustion, 213 kcal/mol (845 Btu/mol; 891 kJ/mol). Methane is nontoxic, but when inhaled, it acts as an asphyxiant. Thus, an atmosphere of methane is a direct threat to life only by suffocation.

Like petroleum, natural gas occurs in underground, porous reservoirs. When encounterd naturally, it may or may not be accompanied by crude petroleum. Natural gas is very desirable as a fuel for several reasons: It is easy to transport, cost effective, yields a relatively high heat value, and burns without adversely affecting our environment. The use of natural gas is now so popular that the fuel is supplied by pipeline to most cities of the nation directly from well fields; the terminal of such a pipeline is shown in Fig. 11-7.

Natural gas may also be liquefied; in this form it is called *liquefied natural gas* (LNG). It is transported as either a compressed gas or cryogenic liquid. USDOT regulates the transportation of methane as a flammable gas. Methane containers are labeled FLAMMABLE GAS; when required, their transport vehicles are placarded FLAMMABLE GAS.

Liquefied Petroleum Gas

Liquefied propane, butane, or a mixture of these compounds is often called *liquefied petroleum gas* (LPG); it is also commonly known as *bottled gas*. It is produced as a by-product from rectifiers treating natural gas. While propane or butane is the main constituent, lesser amounts of ethane, ethene, propene, butene, isobutane, isobutene, and isopentene may also be present.

At ambient conditions, propane and butane are colorless, odorless, and tasteless gases, but under moderate pressure, they are readily liquefied. When ordinarily encountered, however, a small amount of a mercaptan has been added to bottled gas to warn users of leaks; a *mercaptan* is a detestable-smelling organosulfur compound (Sec. 6-7). Both propane and butane possess heating values similar to that of methane and thus are highly desirable as domestic and industrial fuels. Some important physical properties of these two hydrocarbons are noted in Table 11-8.

Propane and butane are often transferred by pipeline or shipped under pressure in rail tank cars from petroleum gas wells or refineries where they were produced. These liquids are then typically transferred to storage tanks like that shown in Fig. 11-8; from here, they are often transferred to smaller storage vessels (that is, they are bottled). The bottled gas is then delivered to locations where natural gas cannot

Figure 11-7   At this point near Houston, Texas, products from Phillips Petroleum Company's Sweeny refinery are transported by pipeline to distant locations, like New York. In 1964, a 1531-mile pipeline was completed that linked the Sweeny refinery with the New York City harbor area. (Courtesy of Phillips Petroleum Company, Borger, Texas)

be directly supplied economically, such as to rural areas. The cylinders are always stored outside buildings and connected by means of copper tubing to heaters and stoves inside. While stored as the liquid, the fuel is generally gaseous in the tubing and is delivered to its outlet as the gas.

USDOT regulates the transportation of propane, butane, and liquefied

TABLE 11-8 PHYSICAL PROPERTIES OF PROPANE AND BUTANE

	Propane	Butane
Boiling point	−49°F (−45°C)	31°F (−0.5°C)
Melting point	−305°F (−187°C)	−216°F (−138°C)
Specific gravity	0.58	0.60
Vapor density (air = 1)	1.56	2.04
Flash point	−156°F (−104°C)	(−76°F) (−60°C)
Autoignition temperature	874°F (468°C)	761°F (405°C)
Lower explosive limit	2.2%	1.90%
Upper explosive limit	9.5%	8.5%

Figure 11-8   At this terminal in Morris, Illinois, butane is stored in a spheroidal tank prior to distribution to customers. (Courtesy of Chicago Bridge and Iron Company, Oakbrook, Illinois)

petroleum gas as flammable gases. Containers of these substances are labeled FLAMMABLE GAS; when required, their transport vehicles are placarded FLAMMABLE GAS.

Acetylene

From an industrial viewpoint, acetylene is our most important alkyne. Its chemical formula is C_2H_2. At one time, acetylene was manufactured by the hydrolysis of calcium carbide (Sec. 8-7), but today it is mainly produced by cracking methane, as the following equation illustrates:

$$2CH_4(g) \longrightarrow C_2H_2(g) + 3H_2(g)$$

Dissolved in acetone in closed cylinder

Acetylene is a colorless gas, possessing an ethereal odor when pure. However, industrial-grade acetylene that has been prepared by hydrolyzing calcium carbide often possesses a garliclike odor. This odor is caused by the presence of phosphine as an impurity in industrial grade acetylene.

Acetylene is commonly used as the fuel in the oxyacetylene torch illustrated in Fig. 11-9. Mixed with oxygen, it burns to achieve temperatures as high as 5400°F (3300°C). The combustion yields 312.4 kcal/mol (1240 Btu/mol; 1307 kJ/mol), which is more than adequate heat to weld and cut steel and clad metals.

Heat sufficient to weld steel is also adequate to trigger the ignition of many other commonly encountered flammable and combustible materials. Consequently, when acetylene is being used for welding and other similar purposes, the high heat of

Figure 11-9 Upon combustion, a mixture of acetylene with air or oxygen produces very high temperatures. This feature accounts for the popular use of acetylene for cutting and welding metals. Caution should always be exercised, however, when using acetylene for such purposes. The relatively large amount of heat produced by burning acetylene may be conducted or radiated to nearby flammable and combustible materials, thereby causing their unintended ignition.

combustion should always be a matter of concern. Transmission of this heat from the combustion area to nearby flammable and combustible materials may result in a fire of the greatest magnitude.

Acetylene possesses an additional hazard: It is an innately unstable substance. Thus, if compressed in steel cylinders under great pressure, acetylene could explode. To counteract this likelihood, special methods are employed to safely contain acetylene in steel cylinders. Some structural features of a cylinder used for the containment of acetylene are illustrated in Fig. 11-10. A solvent, generally acetone, is first absorbed into a porous filler inside the cylinder, like asbestos. Then acetylene is charged into the cylinder, where it dissolves in the acetone. Dissolving acetylene into acetone increases the amount that may be safely compressed into a cylinder. One volume of acetone dissolves 25 volumes of acetylene at 1 atm and 300 volumes at 12 atm.

USDOT regulates the transportation of acetylene as a flammable gas. When cylinders containing acetylene are to be transported, USDOT regulates the nature of its cylinders, the porous filling, and solvent and further requires that the cylinder pressure not exceed 250 psig at 70°F. Acetylene containers are labeled FLAMMABLE GAS; when required, their transport vehicles are placarded FLAMMABLE GAS.

Pure acetylene is not toxic, but in low concentrations it acts as an asphyxiant.

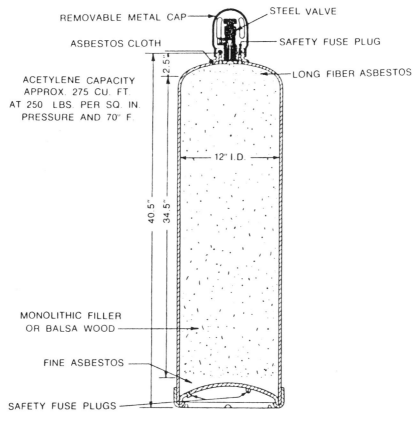

REMOVABLE METAL CAP
STEEL VALVE
ASBESTOS CLOTH
SAFETY FUSE PLUG
LONG FIBER ASBESTOS
2.5"
ACETYLENE CAPACITY
APPROX. 275 CU. FT.
AT 250 LBS. PER SQ. IN.
PRESSURE AND 70° F.
12" I.D.
40.5"
34.5"
MONOLITHIC FILLER
OR BALSA WOOD
FINE ASBESTOS
SAFETY FUSE PLUGS

Figure 11-10   The cutaway of a typical acetylene cylinder, illustrating its components. The monolithic filler or balsa wood is a porous material that is charged with a suitable solvent, like acetone.

Proportions by volume of 40% or more with oxygen were once used in anesthesia, although this practice is now obsolete. On the other hand, it is important to recall that industrial-grade acetylene may pose a health hazard due to the presence of phosphine.

11-7 HALOGENATED HYDROCARBONS

A group of organic compounds may be derived by substituting at least one hydrogen atom in a hydrocarbon with a halogen atom; such compounds are called *halogenated hydrocarbons*. If the halogen is chlorine, the compounds are called *chlorinated hydrocarbons*. To illustrate, let's consider the chlorinated derivatives of methane with the following Lewis structures:

$$
\begin{array}{cccc}
\text{H} & \text{H} & \text{H} & \text{Cl} \\
| & | & | & | \\
\text{H}-\text{C}-\text{Cl} \quad & \text{Cl}-\text{C}-\text{Cl} \quad & \text{Cl}-\text{C}-\text{Cl} \quad & \text{Cl}-\text{C}-\text{Cl} \\
| & | & | & | \\
\text{H} & \text{H} & \text{Cl} & \text{Cl}
\end{array}
$$

Notice as we proceed across the page from left to right that each structure identifies a substance in which one, two, three, and four hydrogen atoms have been respectively substituted with chlorine atoms. These compounds are named methyl chloride, methylene chloride, chloroform, and carbon tetrachloride, respectively.

In the IUPAC system, the halogen atoms are named as substituents of the compound having the longest continuous chain of carbon atoms. The halogen atoms are named as substituents by replacing the *-ine* suffix on the name of the halogen with *-o*. Thus the chlorinated derivatives of methane are named chloromethane, dichloromethane, trichloromethane, and tetrachloromethane, respectively. Other examples of halogenated hydrocarbons are Halon agents, fire extinguishers (Sec. 4-9), and organochlorine pesticides (Sec. 9-14).

When a halogen atom is substituted for a hydrogen atom in a hydrocarbon, the resulting substance attains partial or complete nonflammability. This is illustrated by comparing the flash points of the chlorinated derivatives of methane. The flash point of methane itself is −306°F (−188°C); the flash point of chloromethane is below 32°F (0°C), and its flammable range is only from 10.7% to 11.4% by volume; however, chloroform and carbon tetrachloride are nonflammable substances. When all the hydrogen atoms have been substituted with halogen atoms in any arbitrary hydrocarbon, the resulting compound is nonflammable.

This should not be taken to mean that such substances are entirely safe under fire conditions. While they do not readily burn as individual substances, the vapor of most halogenated hydrocarbons either thermally decomposes or burns at elevated temperatures. With an auxiliary fuel, chlorinated hydrocarbons burn in air, forming phosgene and hydrogen chloride.

Many of the simpler halogenated hydrocarbons are clear, colorless liquids that readily vaporize at room conditions. They have properties that make them highly desirable as cleaning solvents, degreasers, aerosol propellants, blowing agents, refrigerants, and fire extinguishing agents. Hence, such substances are commonly encountered in manufacturing and process industries. Several of these compounds are noted in Table 11-9, which lists the halogenated hydrocarbons having only two carbon atoms per molecule.

While halogenated hydrocarbons are often associated with desirable industrial properties, repeated exposure to them may cause adverse health problems. For example, the toxic effects of chloroform include depression of the central nervous system; skin, eye, and gastrointestinal irritation; and damage to the liver, kidneys, and heart. Carbon tetrachloride causes liver and kidney damage in animals and humans and is a suspect human carcinogen. Many other halogenated hydrocarbons pose health risks, as Table 11-9 illustrates.

TABLE 11-9 SOME SIMPLE CHLORINATED HYDROCARBONS

	Compound	Condensed formula	Adverse health effects
	1, 1-Dichloroethane	CH_3CHCl_2	Inhalation exposure to high concentrations causes central nervous system depression in humans; may cause liver and kidney damage and retard fetal development.
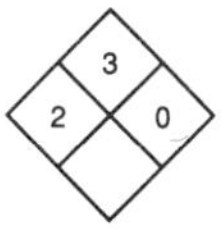	1, 2-Dichloroethane	CH_2ClCH_2Cl	Carcinogenic in animals; mutagenic in bacterial test systems; suspect human carcinogen.
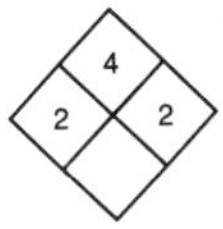	1, 1-Dichloroethene	$CH_2{=}CCl_2$	Exposure to high concentrations causes liver and kidney damage.
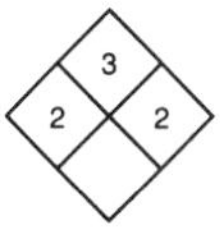	*cis/trans*-1, 2-Dichloroethene	$CHCl{=}CHCl$	Chronic inhalation of the *trans* isomer causes liver degeneration; acute exposure to high concentrations adversely affects the central nervous system.
	1, 1, 1-Trichloroethane	CH_3CCl_3	Inhalation exposure to high concentrations depresses the central nervous system, affects the cardiovascular system, and damages the lungs, liver, and kidneys in animals and humans.
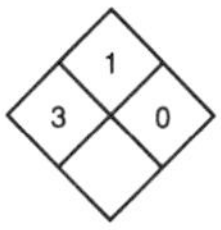	1, 1, 2-Trichloroethane	$CH_2ClCHCl_2$	Induces liver tumors in mice; causes liver and kidney damage in dogs.
	Trichloroethylene	$CHCl{=}CCl_2$	Induces liver carcinomas in mice; chronic inhalation of high concentrations causes liver, kidney, and neural damage in animals.
	1, 1, 2, 2-Tetrachloroethane	$CHCl_2CHCl_2$	Oral administration induces liver tumors in mice; acute and chronic exposure damages the liver, kidneys, and central nervous system; acute exposure may be fatal.
	Tetrachloroethylene	$CCl_2{=}CCl_2$	Induces liver tumors in mice; also causes liver, kidney, and central nervous system damage.

USDOT regulates the transportation of halogenated hydrocarbons, generally as poisonous materials. Containers of substances like 1,1,1-trichloroethane, trichloroethylene and tetrachloroethylene are labeled KEEP AWAY FROM FOOD. However, their transport vehicles are not placarded.

Fluorocarbons

A special class of halogenated hydrocarbons are the *fluorocarbons* (also called *chlorofluorocarbons* or *chlorofluoromethanes*). The general chemical formula of members of this class of compounds is CF_nCl_{n-x}, *where n and x are whole numbers* less than 4. Many of such compounds are known by the trademark Freon. Several industrially important fluorocarbons are identified in Table 11-10. They are most commonly encountered as refrigerants and, until relatively recently, as aerosols.

As a class of organic compounds, the fluorocarbons are relatively inert. For instance, they do not readily burn; nor do they react with acids or other common laboratory reagents. But when released to the environment, fluorocarbons disperse and migrate upward to the ozone layer (Sec. 6-1), where scientists fear they are causing damage to the environment.

A fluorocarbon molecule is capable of absorbing ultraviolet radiation from the sun. Upon doing so, a carbon-to-chlorine bond is ruptured in the molecule, resulting in the production of chlorine atoms ($Cl\cdot$), as the following equation notes:

$$CF_nCl_{n-x}(g) \longrightarrow CF_nCl_{3-x}\cdot(g) + Cl\cdot(g)$$

In the ozone layer, these chlorine atoms are exposed to ozone molecules, with which they may unite by a stepwise process that forms molecular oxygen. This process is illustrated by the following equations:

$$O_3(g) + Cl\cdot(g) \longrightarrow ClO\cdot(g) + O_2(g)$$

$$ClO\cdot(g) + O\cdot(g) \longrightarrow Cl\cdot(g) + O_2(g)$$

Notice that the overall process results in the production of oxygen.

$$O\cdot(g) + O_3(g) \longrightarrow 2O_2(g)$$

By this mechanism, ozone is consumed, permitting ultraviolet radiation to penetrate to the troposphere, the part of the atmosphere in which we live.

As noted in Sec. 6-1, scientists have speculated that depletion of the ozone layer could cause serious environmental problems. This speculation was further enhanced when American scientists identified a depletion hole in the ozone layer at an altitude of 12 to 20 km over Antarctica. Concern over the possible depletion of the ozone layer led the U.S. Environmental Protection Agency and the U.S. Food and Drug Administration to ban the use of fluorocarbons as aerosol propellants in the United States. In the interest of worldwide environmental protection, 24 nations

TABLE 11-10 SOME INDUSTRIALLY IMPORTANT FLUOROCARBONS

Trademark	Chemical name	Chemical formula
Freon-21	Dichloromonofluoromethane	$\begin{array}{c} \text{H} \\ \mid \\ \text{Cl}-\text{C}-\text{Cl} \\ \mid \\ \text{F} \end{array}$
Freon-12	Dichlorodifluoromethane	$\begin{array}{c} \text{F} \\ \mid \\ \text{Cl}-\text{C}-\text{Cl} \\ \mid \\ \text{F} \end{array}$
Freon-22	Monochlorodifluoromethane	$\begin{array}{c} \text{H} \\ \mid \\ \text{F}-\text{C}-\text{Cl} \\ \mid \\ \text{F} \end{array}$
Freon-13	Monochlorotrifluoromethane	$\begin{array}{c} \text{F} \\ \mid \\ \text{Cl}-\text{C}-\text{F} \\ \mid \\ \text{F} \end{array}$
Freon-11; Freon MF	Trichloromonofluoromethane	$\begin{array}{c} \text{Cl} \\ \mid \\ \text{Cl}-\text{C}-\text{Cl} \\ \mid \\ \text{F} \end{array}$
Freon-113; Freon TF	1, 1, 2-Trichloro-1, 2, 2-trifluoroethane	$\begin{array}{c} \text{F}\quad\text{F} \\ \mid\quad\mid \\ \text{Cl}-\text{C}-\text{C}-\text{F} \\ \mid\quad\mid \\ \text{Cl}\;\;\text{Cl} \end{array}$
Freon-114	1, 2-Dichloro-1, 1, 2, 2-tetrafluoroethane	$\begin{array}{c} \text{F}\quad\text{F} \\ \mid\quad\mid \\ \text{Cl}-\text{C}-\text{C}-\text{Cl} \\ \mid\quad\mid \\ \text{F}\quad\text{F} \end{array}$

signed an agreement in 1988 to halve production and the use of fluorocarbons in their countries by 1999.

USDOT regulates the transportation of most fluorocarbons as nonflammable gases. Containers are labeled NONFLAMMABLE GAS; when required, their transport vehicles are placarded NONFLAMMABLE GAS.

Polychlorinated Biphenyls

Another important group of halogenated hydrocarbons are the *polychlorinated biphenyls,* commonly denoted as *PCBs.* The Lewis structure of the biphenyl molecule is

PCBs are chlorinated derivatives of this hydrocarbon, that is, substances in which one or more hydrogen atoms have been replaced by a chlorine atom. There are several structural PCB isomers, but their properties are relatively similar. PCBs are generally encountered industrially as a mixture of some of these isomers.

Most PCBs are mobile liquids at room conditions, although some are viscous and sticky. Like the fluorocarbons, these compounds are resistant to chemical attack by nearly all common substances. But, unlike the simpler halogenated hydrocarbons, they are stable when exposed to elevated temperatures. Some PCBs can be heated to their minimum boiling point of 513°F (267°C) without undergoing decomposition. This combination of properties is ideal for certain industrial purposes requiring high temperatures. In particular, PCBs have been selected as the insulating fluids for many electrical transformers. For years, they were commonly used in conjunction with the operation of electrical equipment (transformers, circuit breakers, capacitors, and the like), heat transfer systems, and hydraulic systems. Today, PCBs still are used in certain pieces of electrical equipment, especially transformers. A typical transformer contains 200 gal of fluid with a PCB concentration of approximately 50% to 60% by volume, but some transformers hold as much as 1000 gal of liquid PCBs.

In the 1960s, it was discovered that PCBs and by-products resulting from their manufacture may cause a number of adverse health problems: cancer, birth defects, liver damage, acne, impotence, and death. An incident that dramatically illustrates the toxicological problems associated with PCBs occurred in 1968 in Japan. Over 1000 people consumed rice bean oil that had been contaminated with PCBs leaking from heat transfer pipes used in processing the oil. These individuals developed acne, brown pigmentation of the skin and nails, distinctive hair follicles, increased eye discharge, eyelid swelling, and gastrointestinal disorders.

Once ingested, PCBs distribute themselves into various receptor tissues of the thymus, lungs, spleen, kidneys, liver, brain, muscle, and testes. They remain stored in these body organs, where they disrupt normal biological functions and cause pathological changes. PCBs are the cause of a variety of neurobehavioral disorders and birth abnormalities. They are animal carcinogens and, in addition, sometimes enhance the cancer-causing capability of certain other carcinogens.

When improperly disposed of or discharged to the environment, PCBs can be particularly hazardous substances. Since they are so stable, PCBs persist for long periods of time in the environment, passing from animal to animal by means of the food chain; thus, they may adversely affect many varieties of animal life. In birds alone, PCBs have caused eggshell thinning, lower egg production, decreased ability of eggs to hatch, deformities in birds that did hatch, and reduction in growth and survival.

In the United States, USEPA banned the production and manufacture of PCBs in 1979. Systems employing the use of PCBs are permitted to continue operating, but only under specific regulatory conditions. Generally, PCBs may be used only in totally enclosed systems, which must be marked as shown in Fig. 11-11 to identify that PCBs are present. Use of PCB-containing waste oil as a sealant, coating, or dust control agent is totally prohibited. Furthermore, USEPA highly regulates the storage and disposal of PCBs. In fact, no other single substance is as highly regulated by environmental laws.

Figure 11-11   Examples of warnings required by USEPA on equipment containing polychlorinated biphenyls (PCBs).

USEPA has also put certain controls on facilities that use PCB-containing transformers; these controls directly affect the firefighting profession. When these electrical transformers catch fire, the resulting soot and smoke may be inhaled by firefighters or other innocent individuals. Realizing that such fires expose these people to a substantial health risk, USEPA now requires the registration of PCB-containing transformers with the appropriate local fire department. This registration gives firefighters the knowledge as to where such transformers are located in their areas of jurisdiction. USEPA also requires the isolation of these transformers from flammable and combustible materials, as well as from locations near ventilation equipment and ductwork inside buildings.

Aside from PCBs themselves, fires involving these substances pose yet another health risk. The incomplete combustion of PCBs may form other toxic compounds, like dioxin (Sec. 11-9). Hence, a PCB transformer fire may easily represent one of the most dangerous fires to which firefighters are exposed. Fortunately, these fires are not common occurrences. Nevertheless, firefighters should be prepared to respond to fires involving PCBs by having available proper respiratory devices and protective clothing.

When transported, USDOT regulates PCBs as a miscellaneous hazardous material. Furthermore, exposure to PCBs is regulated in the workplace. NIOSH recommends 1.0 $\mu g/m^3$ as the maximum permissible exposure concentration. The TLV is 0.5 mg/m^3 as a weighted average.

11-8 ALCOHOLS

The *alcohols* are organic compounds derived from the hydrocarbons in which at least one hydrogen atom has been substituted with the hydroxy (—OH) group. Two alcohols that are frequently encountered commercially have the common names methyl alcohol and ethyl alcohol. In the IUPAC system, simple alcohols are named by replacing the *-e* in the name of the corresponding alkaline with *-ol*; hence, methyl alcohol and ethyl alcohol are named methanol and ethanol, respectively. Another important group of alcohols are the *phenols,* the parent compound of which is itself named *phenol.* Some of the common properties of methanol, ethanol, and phenol are noted in Table 11-11; their chemical formulas are

$$CH_3OH \qquad CH_3CH_2OH$$
methanol   ethanol   phenol

The simple alcohols are flammable or combustible substances, forming carbon dioxide and water on complete combustion. Methanol and ethanol burn with a pale blue flame without the accompaniment of soot formation. These substances are also infinitely soluble in water; thus water is an effective extinguisher of alcohol fires.

TABLE 11-11 PHYSICAL PROPERTIES OF SOME SIMPLE ALCOHOLS

	Methanol	Ethanol	Phenol
Boiling point	149°F (65°C)	174°F (79°C)	358°F (181°C)
Melting point	−144°F (−98°C)	−173°F (−114°C)	104°F (40°C)
Specific gravity	0.79	0.79	1.07
Vapor density (air = 1)	1.11	1.59	3.24
Flash point	54°F (12°C)	54°F (12°C)	175°F (79°C)
Autoignition temperature	867°F (464°C)	793°F (423°C)	1319°F (715°C)
Lower explosive limit	6%	3.3%	1.5%
Upper explosive limit	36.5%	19%	—
TLV	200 ppm	1000 ppm	5 ppm (skin)

Methanol

Methanol was once called *wood alcohol,* since it can be produced along with other substances by heating wood in the absence of air. But today large quantities of methanol are produced by the high-temperature, high-pressure hydrogenation of carbon monoxide in the presence of an appropriate catalyst. This chemical reaction is represented by the following equation:

$$CO(g) + 2H_2(g) \longrightarrow CH_3OH(g)$$

Methanol is a clear, colorless liquid that is totally miscible with water. It is used industrially as a solvent for shellac, gums, and other substances; for the preparation of formaldehyde; and for the preparation of certain dyes. When ingested, as little as one-half ounce (30 mL) can cause death, and lesser amounts have been known to cause irreversible blindness.

Ethanol

Ethanol is the alcohol commonly encountered in wine, beer, whiskey, brandy, and similar beverages. It has also been known as *grain alcohol,* since it may be prepared by the yeast fermentation of carbohydrates, such as grains. Ethanol is a clear, colorless liquid that is totally miscible with water. When pure ethanol is mixed with water, it forms a mixture with 5% water that boils at 79°C (174°F), which is the ethyl alcohol that is ordinarily encountered. Commercially, pure ethanol, without water, is called *absolute alcohol.*

The concentration of ethanol in alcoholic solutions is frequently expressed in *proof.** The percentage by volume is about half the proof of an alcoholic solution at room temperature. Thus, absolute alcohol is 200 proof; 95% alcohol by volume is approximately 190 proof.

Almost 90% of the alcohol produced throughout the world is made by the vapor catalyzed hydrolysis of ethylene, as the following equation notes:

$$CH_2{=}CH_2(g) + H_2O(g) \longrightarrow CH_3CH_2OH(g)$$

Its manufacture is strictly controlled, owing to the high tax placed on its sale.

Ethanol is frequently encountered industrially as *denatured alcohol.* This is ethanol to which a denaturant has been added, like methanol, pyridine, or aviation gasoline, which causes the ethanol to become unpalatable or even poisonous. While

---

*Long, long ago, people believed spirits caused the physiological effects observed from consuming alcohol. The procedure that was devised to test for the presence of spirits in whiskey and other alcoholic beverages consisted of pouring a sample over gunpowder and igniting the resulting mixture. If an explosion or fire resulted, either event was taken as "proof" that spirits were actually present in the sample, whereas the lack of either incident indicated that it contained too much water. This qualitative "proof" was later quantified as noted herein.

the sale of absolute alcohol is closely regulated, denatured alcohol may be sold without tax.

Ethanol is also the substance used in *gasohol,* the alcohol-augmented motor fuel. The idea of using ethanol in motor fuels was first implemented during the oil crisis of the mid-1970s when gasohol was introduced in the United States as a fuel extender. However, when used in motor vehicles, gasohol clogged fuel injectors, corroded fuel lines, and ignited poorly during cold weather; hence, its early use was poorly received. These problems were resolved, however, by mixing gasohol with detergents and other fuel additives. The use of gasohol is now almost obsolete, but since gasohol provides more octane power and generates less carbon monoxide than standard gasoline, it holds some promise for future use.

Ethanol acts as a depressant of the central nervous system. Consumers become "intoxicated" on consuming ethanol; that is, they first feel relaxed, then lightheaded, and ultimately unable to control their faculties. Ethanol is not ordinarily considered a toxic substance, except when it is excessively consumed, and especially when excessive consumption occurs during a relatively short time period.

The adverse health effects caused by the consumption of ethanol depend directly on the amount that has been absorbed into the bloodstream, as Fig. 11-12 illustrates; a concentration in excess of 5.0% by volume may cause death. This concentration may be achieved quite readily by overconsumption of alcohol, since alcohol passes directly through the stomach wall into the bloodstream, unlike most other substances. Individuals on medications and pregnant women must be especially wary of drinking alcoholic beverages. Ethanol in combination with barbiturates, tranquilizers, or narcotics can produce synergistic effects, including coma and death. Women who drink excessively during pregnancy are the third most common cause of mentally handicapped children in the Western world. Moderate or social drinking may also be harmful to the developing fetus. The U.S. Surgeon General and Britain's Royal College of Psychiatrists uncompromisingly advise total abstinence from alcohol during pregnancy.

USDOT regulates the transportation of ethanol and alcoholic beverages as flammable liquids. Containers are labeled FLAMMABLE LIQUID; when required, their transport vehicle are placarded FLAMMABLE.

Phenols

The *phenols* are a group of aromatic alcohols; they may be regarded as hydroxy derivatives of benzene. The parent compound of this class of alcohols is also named *phenol* or *hydroxybenzene.* It is also sometimes called *carbolic acid,* a name not to be confused with carbonic acid, the nonhazardous acid in carbonated beverages. Phenol is a colorless to white crystalline solid that often darkens to red if exposed to light, but phenol readily absorbs atmospheric moisture and thus is often encountered as a liquid. Phenol possesses a sharp, distinctive odor.

Phenol is an important industrial chemical, since a number of pharmaceutical products and phenolic resins are manufactured from it. The most common example

Figure 11-12   The effect on an average adult of the ethyl alcohol concentration in the blood. (Adapted from an illustration by Joe Orlando from *Chemistry*, by Harvey A. Yablonsky, copyright 1975 by Thomas Y. Crowell; used by permission of the publisher)

of the latter is the phenol-formaldehyde resin (see Table 12-2). Phenol and phenol derivatives are common ingredients of mouthwashes, gargles, and sprays; a solution of 1 part of phenol in 850 parts of water by weight prevents multiplication of certain bacteria.

While phenols burn, their flammability is not generally a significant hazard, since all phenols possess flash points greater than 172°F (78°C). The primary hazard associated with phenols is their poisonous nature. Phenol is highly toxic in humans by inhalation and ingestion. It is an eye, nose, and throat irritant; following dermal, oral, or inhalation exposure, phenol may cause systemic damage to the nervous system in humans.

Most likely for these reasons, USDOT regulates the transportation of phenol as a poisonous material. Containers are labeled POISON; when required, transport vehicles are placarded POISON. Worker exposure to phenol is regulated by OSHA at 5 ppm for contact with skin tissue. The TLV for phenol is 19 mg/m^3.

A group of phenols, called *cresols,* may be regarded as hydroxy derivatives of xylene in which a methyl group has been substituted with a hydroxy group. Thus, there are three cresols, whose formulas and names follow:

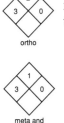

ortho

meta and
para

CH$_3$ — OH

ortho-cresol

CH$_3$ — OH

meta-cresol

CH$_3$

OH

para-cresol

A mixture of the cresol isomers is called *cresylic acid.* Like phenol itself, the cresols are used as disinfectants and for the production of phenolic resins.

As a group of substances, the cresols are highly irritating to the skin, mucous membranes, and eyes, may impair liver and kidney function, and may disturb the central nervous system. When applied to the skin of mice, the cresol isomers promote the formation of skin tumors. USDOT regulates the transportation of cresols as poisonous materials. Containers of cresol are labeled POISON; when required, transport vehicles are placarded POISON. Worker exposure to the cresol isomers is also regulated by OSHA at 5 ppm for contact with the skin. The TLV is 22 mg/m^3.

11-9 ETHERS

An *ether* is an organic compound whose molecules possess an oxygen atom bridged between two alkyl or aryl groups. Thus, the ethers have the general chemical formula R—O—R'. The common names of the ethers are determined by noting the names of the groups bonded to the oxygen atom, followed by the word ether. Hence, for example, the substance having the chemical formula CH_3—O—C_2H_5 is named methyl ethyl ether. In the IUPAC system, ethers are named by replacing the -*yl* suffix of the alkyl group with -*oxy*, and naming the ether as a substituted alkane. Thus, methyl ethyl ether is also named methoxyethane.

The simple ethers are highly volatile and constitute dangerous fire and explosion hazards. This is illustrated by noting in Table 11-12 the general properties of the most commonly used ether, diethyl ether. This is a clear, colorless liquid that many individuals have encountered in medical clinics or hospitals where it is frequently used as an anesthetic; in such environments it is commonly known as *ether*. Diethyl ether vaporizes readily at room conditions and is highly flammable. Note that its flash point is −49°F (−45°C) and its flammable range extends from 1.85% to 48% by volume. Note, too, that the vapor of diethyl ether is two and one-half times heavier than air. Such vapors can flow toward low areas until they reach an ignition source and ignite, flashing back to their initial source. Figure 11-13 illustrates this dangerous feature of diethyl ether.

TABLE 11-12 PHYSICAL PROPERTIES OF DIETHYL ETHER

Boiling point	94°F (34°C)
Melting point	−189°F (−123°C)
Specific gravity	0.71
Vapor density (air = 1)	2.55
Flash point	−49°F (−45°C)
Autoignition temperature	356°F (180°C)
Lower explosive limit	1.85%
Upper explosive limit	48%
TLV	400 ppm

The flammability of diethyl ether is also a property of other simple ethers. But, as ethers become more complex in molecular structure, their flash points correspondingly rise; these liquid ethers are combustible liquids. For instance, certain mono- and dialkyl ethers of ethylene glycol are combustible liquids, but not flammable liquids. These ethers are commonly employed as solvents; they are often referred to by the common name of Cellosolve, the trademark for these substances.

Figure 11-13   Vapors of diethyl ether are heavier than air (vapor density = 2.55), as is illustrated by this experiment. A cloth saturated with the ether is placed at the top of a trough that has been arranged at a 45° angle. A lighted candle is positioned near its lower end, as shown at the left. Ether vapors slowly move down the trough. The candle flame serves as an ignition source. Fire rushes upward to the source of the vapors, as noted at the right.

Some examples are noted in Table 11-13 along with their important physical properties.

Aside from their flammability, liquid ethers are typically hazardous in another way: They often contain dissolved organic peroxides (Sec. 11-14), which are potentially explosive substances. Organic peroxides form when ethers react with dissolved atmospheric oxygen, a reaction that is catalyzed by light. For this reason, liquid ethers are often stored in metal cans rather than glass bottles. The reaction with atmospheric oxygen occurs over time; hence, ethers purchased a year ago are likely to be more hazardous than those recently acquired. Since ethers typically evaporate with ease, organic peroxides become concentrated in the final volume remaining in its container. This means that the likelihood of an explosive reaction is greater in containers having only a small liquid residue.

The following ethers are particularly susceptible to organic peroxide formation:

· Diethylene ether
· Diethylene glycol dimethyl ether
· Diethyl ether
· Ethylene glycol dimethyl ether
· Isopropyl ether
· Tetrahydrofuran

Manufacturers of these ethers typically include warning statements on the label of

TABLE 11-13 SOME COMMON ETHERS OF ETHYLENE GLYCOL

Trademark	Name	Chemical formula
Butyl Cellosolve	Ethylene glycol monobutyl ether (2-butoxyethanol)	$C_4H_9-O-CH_2CH_2OH$
Butyl Cellosolve acetate	Ethylene glycol monobutyl ether acetate	$CH_3COO-CH_2CH_2-O-C_4H_9$
Cellosolve acetate	Ethylene glycol monoethyl ether acetate	$CH_3COO-CH_2CH_2-O-C_2H_5$
Cellosolve solvent	Ethylene glycol monoethyl ether (2-ethoxyethanol)	$C_5H_5-O-CH_2CH_2OH$
Dibutyl Cellosolve	Ethylene glycol dibutyl ether	$C_4H_9-O-CH_2CH_2-O-C_4H_9$
Methyl Cellosolve	Ethylene glycol monomethyl ether (2-methoxyethanol)	$CH_3-O-CH_2CH_2OH$
Methyl Cellosolve acetate	Ethylene glycol monomethyl ether acetate	$CH_3COO-CH_2CH_2-O-CH_3$
Phenyl Cellosolve	Ethylene glycol monophenyl ether (2-phenoxyethanol)	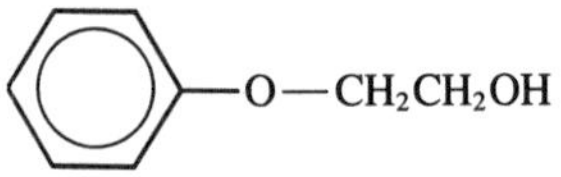 ⬡$-O-CH_2CH_2OH$

their products like the following: "Discard thirty days after opening or after one year if unopened."

Exposure to the common ethers does not generally lead to adverse health effects in humans, although inhalation of ether vapors may cause asphyxiation.

USDOT regulates the transportation of most ethers as flammable liquids. Containers of diethyl ether, for instance, are labeled FLAMMABLE LIQUID; when required, their transport vehicle are placarded FLAMMABLE.

Dioxin

A compound often regarded today as one of the most toxic of all substances is a chlorinated cyclic ether named 2,3,7,8-tetrachlorodibenzo-p-dioxin. It is often simply called dioxin or TCDD and has the following molecular structure:

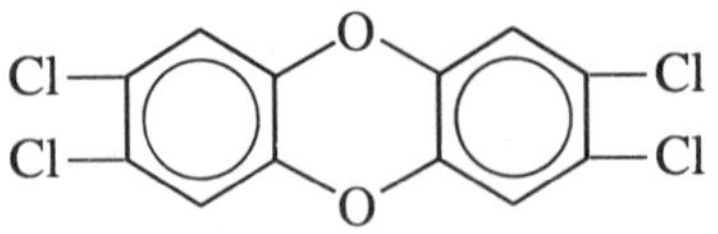

The toxicity of this substance is demonstrated in Table 11-14 by comparing the LD_{50} value for dioxin with the LD_{50} values of some other common poisons.

Dioxin has been a contaminant in a number of phenolic compounds formerly manufactured as broad spectrum herbicides, like 2,4-dichlorophenoxyacetic acid and

TABLE 11-14 RELATIONSHIP OF LETHALITY OF
DIOXIN AND OTHER POISONS

Substance	Animal	LD_{50}
Botulinum toxin A	Mouse	3.3×10^{-17}
Tetanus toxin	Mouse	1.0×10^{-15}
Diphtheria toxin	Mouse	4.2×10^{-12}
Dioxin	Guinea pig	3.1×10^{-9}
Bufotenine[a]	Cat	5.2×10^{-7}
Curare[b]	Mouse	7.2×10^{-7}
Strychnine	Mouse	1.5×10^{-6}
Sodium cyanide	Mouse	2.0×10^{-4}

[a] Bufotenine is a toxin derived from the skin glands of certain toads.

[b] Curare is a mixture of several toxins derived from certain species of South American trees.

2,4,5-trichlorophenol. The toxicological effects caused by exposure to this substance is an active area of modern research. Yet a variety of adverse health effects has been already attributed to exposure to extremely low concentrations of dioxin, concentrations as low as the range of nanograms per liter. These effects include neurobehavioral disorders, reproductive disorders, chloracne, and cancer. Dioxin also is suspected of affecting the immune system, thereby weakening the body's defenses toward contracting diseases.

Dioxin is an extremely stable substance; it begins to thermally decompose only when heated to its boiling point, 930°F (500°C). Hence, when dioxin is present as a contaminant on wild vegetation, it constitutes a component that is passed from one animal to another in the food chain. Dioxin readily dissolves in the fatty tissues of such animals and may thereafter be bioaccumulated.

In the United States, dioxin inadvertently entered our natural environment through the use of related herbicides and by the improper disposal of chemical wastes associated with phenolic herbicide production. Over 30 such sites have already been identified in Missouri alone where dioxin has been detected in the soil. Left unattended, this substance could remain for decades to pose a threat to public health and the environment. Fortunately, however, USEPA is actively pursuing the cleanup of sites contaminated by dioxin. Nonetheless, environmental problems associated with this pollutant are certain to remain for decades.

The best known dioxin-contaminated herbicide is *Agent Orange,* which the U.S. military used extensively during the Vietnam encounter as a defoliant. On occasion, individuals were either inadvertently sprayed or otherwise came in contact with this herbicide; later they suffered health effects thought to have been caused by it, such as mental disorders. Today, considerable controversy still exists over the role that Agent Orange may have contributed to the adverse health problems now exhibited in some Vietnam veterans.

11-10 ALDEHYDES AND KETONES

The *aldehydes* and *ketones* are organic compounds whose molecules have a carbonyl group ($-C=O$). In aldehydes, the carbonyl group is located at the end of a chain of carbon atoms; in ketones, it is located at a nonterminal position. Thus, aldehydes have the general chemical formula $R-CHO$, whereas ketones have the formula $R-CO-R'$.

Gas

The simplest aldehyde is *formaldehyde* or *methanal*. Its chemical formula is HCHO. Formaldehyde is a colorless gas at room conditions, possessing a pungent characteristic odor. Large quantities are used for preparation of certain plastics and resins, like formaldehyde-phenol resins (Sec. 12-5). Formaldehyde is also commonly encountered as an approximately 40% aqueous solution containing 5% to 12% methanol called *formalin*. This solution is widely used as a disinfecting, sterilizing, and embalming agent. Its flash point depends on the amount of water and methanol present. Formalin containing 37% water and 15% methanol flashes at 122°F (50°C); methanol-free formalin flashes at 185°F (85°C). Formalin is also used by the apparel and furniture industries to make cloth wrinkle-resistant.

Water solution

Formaldehyde is moderately toxic by inhalation and skin contact. Inhalation causes respiratory problems ranging from mild irritation of the throat and bronchial passageways to bronchopneumonia; it may also produce localized sores in the nose, throat, and lungs. It also irritates skin tissue and causes allergic dermatitis in susceptible individuals. A concentration in air of only 4 to 5 ppm causes the eyes to tear. In rats, formaldehyde produces malignant nasal tumors, making it a suspect carcinogen.

USDOT regulates the transportation of formaldehyde solutions possessing a flash point less than 141°F (60.5°C) as flammable liquids. Containers are labeled FLAMMABLE LIQUID; when required, their transport vehicles are placarded FLAMMABLE.

The simplest ketone is *acetone* or *dimethyl ketone*. Its chemical formula is $CH_3-CO-CH_3$. Acetone is a clear, colorless, water-soluble liquid with a sweetish odor. It is also a highly volatile liquid at room conditions. Since its flash point is 15°F (−9°C) and since acetone possesses a flammable range from 2.6% to 12.8% by volume in air, acetone is considered a dangerous fire and explosion risk. USDOT regulates its transportation as a flammable liquid.

Commercially, acetone is largely employed as a solvent; for instance, it is commonly encountered in varnishes, lacquers, and fingernail polish.

Low acute toxicity is associated with acetone, and no chronic health hazards have been associated with it. Worker exposure to acetone is limited by OSHA to 1000 ppm.

MEK

Other simple ketones with properties similar to acetone are methyl ethyl ketone (MEK; butanone) and methyl isobutyl ketone (MIBK; 4-methyl-2-pentanone). Their chemical formulas are $CH_3-CO-C_2H_5$ and $(CH_3)_2CHCH_2-CO-CH_3$,

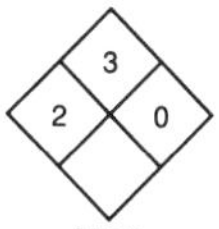

MIBK

respectively. These ketones are liquids at room conditions and are also commonly encountered as solvents. They are considered dangerous fire risks.

USDOT regulates the transportation of the simple ketones as flammable liquids. Containers are labeled FLAMMABLE LIQUID; when required, their transport vehicles are placarded FLAMMABLE.

At low doses, methyl ethyl ketone and methyl isobutyl ketone are associated with relatively low toxicities. But at higher doses, methyl ethyl ketone adversely affects the nervous system and causes irritation of the eyes, nose, and skin in humans; it also retards fetal development in rats. Methyl isobutyl ketone causes nausea, headaches, vomiting, and eye irritation in humans and damages the kidneys in rats. Worker exposure to methyl ethyl ketone and methyl isobutyl ketone is limited under OSHA to 200 ppm and 100 ppm, respectively.

11-11 ORGANIC ACIDS

Organic acids are organic compounds containing the carboxyl group, —COOH; hence, they are sometimes called *carboxylic acids*. The general chemical formula of organic acids is R—COOH. The simplest organic acid is formic acid (methanoic acid), whose chemical formula is HCOOH. In Sec. 7-12, we noted the properties of acetic acid (ethanoic acid), the most commonly encountered organic acid. The common organic acids and their chemical formulas are the following:

Acetic acid (ethanoic acid), CH_3COOH
Propionic acid (propanoic acid), CH_3CH_2COOH
Butyric acid (butanoic acid), $CH_3CH_2CH_2COOH$
Valeric acid (pentanoic acid), $CH_3CH_2CH_2CH_2COOH$

Like acetic acid, these simple organic acids are clear, colorless, water-soluble liquids with characteristic odors. Formic acid, acetic acid, and propionic acid possess pungent but not disagreeable odors. However, the odors of butyric acid and valeric acids are highly disagreeable.

Organic acids are also ignitable or combustible liquids. The flash points of acids more complex than acetic acid increase as the number of carbon atoms increases per molecule. Beginning with formic acid and proceeding to valeric acid, the flash points are respectively the following: 156°F (69°C), 110°F (43°C), 130°F (56°C), 170°F (77°C), and 205°F (96°C).

USDOT regulates the transportation of these simple organic acids as corrosive materials. Containers are labeled CORROSIVE; when required, their transport vehicles are placarded CORROSIVE.

The organic acids are generally associated with a low to moderate toxicity. Thus, they are not regarded as posing health hazards.

11-12 ESTERS

Organic compounds having the general chemical formula R—CO—OR′ are called *esters*. They are produced when organic acids react with alcohols. For example, the ester ethyl acetate results when ethanol reacts with acetic acid, as noted by the following equation:

$$C_2H_5OH(aq) + CH_3COOH(aq) \longrightarrow CH_3-CO-OC_2H_5(aq) + H_2O(l)$$
    ethanol   acetic acid   ethyl acetate   water

Esters are named by changing the −*ic* suffix of the organic acid to −*ate*, preceded by the name of the alkyl or aryl group of the alcohol.

The simple esters are clear, colorless liquids, many of which possess fruity odors. In fact, many esters occur naturally in apples, pineapples, bananas, and oranges and are responsible for the odor of these fruits. Synthetic esters are used commonly as food additives.

A commonly encountered industrial ester is ethyl acetate. It is frequently used as a solvent. It is a clear, colorless liquid with a fragrant odor. Its flash point is only 24°F (−5°C); consequently, it is considered a dangerous fire and explosion risk. USDOT regulates the transportation of ethyl acetate as a flammable liquid. Containers are labeled FLAMMABLE LIQUID; when required, their transport vehicles are placarded FLAMMABLE.

In humans, ethyl acetate vapor causes only mild irritation of the nose, eyes, and throat. But animals exposed by inhalation to relatively high concentrations (equal to or greater than 6000 mg/m³) of ethyl acetate vapor exhibit pulmonary edema, hemorrhaging of the respiratory tract, and degeneration of vital organs. Worker exposure to ethyl acetate is limited to 400 ppm by OSHA.

11-13 PEROXO-ORGANIC COMPOUNDS

Two classes of organic compounds are characterized by the presence of the peroxy group (—O—O—): *organic hydroperoxides* and *organic peroxides*. These compounds may be considered as derivatives of hydrogen peroxide (Sec. 10-4), in which either one or both of the hydrogen atoms have been substituted with one or more alkyl or aryl groups. Hence, the general chemical formulas of these peroxo-organic compounds are

$$H-O-O-R \quad \text{and} \quad R-\underset{\underset{O}{\|}}{C}-O-O-\underset{\underset{O}{\|}}{C}-R'$$

respectively.

The peroxo-organic compounds are commonly used industrially to induce free-radical polymerization (Sec. 12-1), a chemical reaction essential to the production of plastics. They are either liquids or solids, although these substances are often encountered dissolved in water or an appropriate organic solvent.

Like inorganic peroxides, the peroxo-organic compounds are oxidizing agents containing active oxygen within their molecular structures; thus, they are capable of supporting combustion. These compounds are also flammable substances. The combination of flammability and reactive oxygen in peroxo-organic compounds results in fires that burn more furiously and intensely than other flammable substances. But their main hazard is a potential to undergo rapid, auto-accelerated decomposition. Almost all peroxo-organic compounds are intrinsically unstable. Some are extremely sensitive to friction, heat, or shock and cannot be safely handled unless diluted. Notwithstanding their instability, the decomposition of commercially available peroxo-organic compounds is avoided by storing them at proper temperatures. Thus, they are safe to handle in relatively small quantities.

Peroxo-organic compounds are chemically incompatible with a number of other chemical substances, like mineral acids. Furthermore, their decomposition is often accelerated when they are mixed with certain metallic compounds. Consequently, these compounds should always be isolated from most other compounds, particularly flammable and combustible materials. Ideally, these hazardous materials should be stored alone in a temperature-regulated, ventilated structure with individual containers separated from one another. Because they are unstable, peroxo-organic compounds should never be stored in bulk unless diluted. Sources of ignition should be absent from the vicinity.

When peroxo-organic compounds are involved in a fire, deluging amounts of water should be applied as soon as practical. As with other fires involving oxidizers, the use of smothering-type extinguishers may be ineffective. Firefighting involving peroxo-organic compounds may be more hazardous than fighting other chemical fires. Once such compounds begin burning, the entire amount contained in a given area may deflagrate. Furthermore, some of the commonly available peroxo-organic compounds are toxic by inhalation, ingestion, and skin absorption; almost all of them severely irritate the eyes. These factors necessitate a need for limiting to a minimum the period during which firefighting is accomplished.

USDOT regulates the transportation of all peroxo-organic compounds as organic peroxides. Containers of these compounds are labeled ORGANIC PEROXIDE as illustrated in Fig. 11-14. When required, their transport vehicles are placarded ORGANIC PEROXIDE.

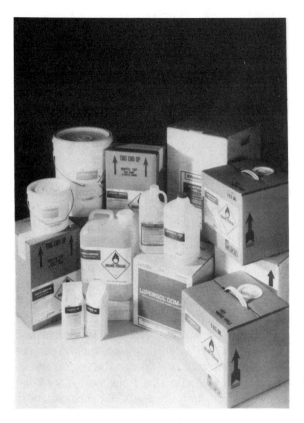

Figure 11-14 Organic peroxides are generally transported and stored in relatively small quantities. In accordance with USDOT regulations, the packages used to transport these substances are labeled ORGANIC PEROXIDE, as indicated. Bulk quantities of organic peroxides are rarely encountered, except when properly diluted. (Courtesy of Pennwalt Corporation, Buffalo, New York)

Benzoyl Peroxide

This is a white, granular, crystalline solid having the chemical formula

In the United States, it is produced in larger amounts than any other peroxo-organic compound. USDOT regulates its transportation as either a paste with various quantities of water or as a dry powder mixed with an inert solid. It is primarily used industrially to induce polymerization in compounds containing the vinyl group, like styrene (Sec. 12-1), but it is also a bleaching agent for flour, fats, oils, and waxes.

Under OSHA, worker exposure to benzoyl peroxide is limited to 5 mg/m^3.

Peracetic Acid

This compound, also known as *acetyl hydroperoxide* and as *peroxyacetic acid,* has the following chemical formula:

$$CH_3-C\underset{O-OH}{\overset{O}{\big<}}$$

It is a colorless liquid, having a strong odor. Peracetic acid is used for the synthesis of certain other organic compounds, as a bactericide, fungicide, and sterilizing agent.

USDOT regulates the transportation of peracetic acid as a solution dissolved in acetic acid. The final composition of this mixture may not contain more than 43% peracetic acid and not over 6% hydrogen peroxide by mass.

Cumene Hydroperoxide

This is a colorless to pale yellow liquid having the following chemical formula:

$$\begin{array}{c} CH_3 \\ | \\ \langle \bigcirc \rangle - C - O - OH \\ | \\ CH_3 \end{array}$$

It is commonly employed for the production of acetone and phenol.

Methyl Ethyl Ketone Peroxide

This is a clear, colorless liquid. While this substance possesses several isomers, one of them has the following molecular structure:

$$\begin{array}{c} H_3C \qquad C_2H_5 \\ \diagdown C \diagup \\ \diagup \qquad \diagdown \\ O \qquad\qquad O \\ | \qquad\qquad | \\ O \qquad\qquad O \\ \diagdown \qquad\qquad \diagup \\ \diagdown C \diagup \\ \diagup \qquad \diagdown \\ H_3C \qquad C_2H_5 \end{array}$$

It is used in the production of polyester acrylics (Sec. 12-5).

USDOT permits the shipment of methyl ethyl ketone peroxide only when it has been diluted to provide a solution having no more than 9% active oxygen by mass.

11-14 CARBON DISULFIDE

 Carbon disulfide is a clear, colorless, relatively dense, water-insoluble liquid. Its chemical formula is CS_2. When pure, it is associated with a pleasant odor; however, industrial grades of carbon disulfide generally have strong, disagreeable odors. Some of its important physical properties are illustrated in Table 11-15.

TABLE 11-15 PHYSICAL PROPERTIES OF CARBON DISULFIDE

Boiling point	115°F (46°C)
Melting point	−169°F (−112°C)
Specific gravity	1.26
Vapor density (air = 1)	2.6
Flash point	−22°F (−30°C)
Autoignition temperature	212°F (100°C)
Lower explosive limit	1%
Upper explosive limit	44%
TLV	20 (skin)

Carbon disulfide is an excellent solvent for rubber, elemental phosphorus, waxes, and resins. Hence, it is used in the manufacture of matches and varnishes. It is also used for the manufacture of rayon, cellophane, and carbon tetrachloride.

Carbon disulfide is not a member of any general class of organic compounds previously noted in Table 11-6. But this substance is introduced in this chapter since several of its properties are cause for concern to firefighters. In particular, carbon disulfide is a flammable liquid, possessing a flash point of −22°F (−30°C); its flammable range extends from 1% to 44% by volume. Thus, carbon disulfide constitutes a major fire and explosion hazard. Its vapor has been known to ignite from as little heat as that radiated from a hot steam pipe or electric light bulb; the vapor may also be ignited by static electricity.

When carbon disulfide burns, sulfur dioxide and carbon monoxide typically form, as the following equation notes:

$$2CS_2(g) + 5O_2(g) \longrightarrow 4SO_2(g) + 2CO(g)$$

Thus, when firefighters attack major fires involving this substance, they must generally use positive-demand, self-contained breathing apparatus to protect against inhalation of these toxic combustion products.

Carbon disulfide itself is also highly toxic by inhalation, ingestion, and skin absorption, causing permanent damage to the central nervous system.

USDOT regulates the transportation of carbon disulfide as a flammable liquid. Containers are labeled FLAMMABLE LIQUID and POISON; when required, their transport vehicles are placarded FLAMMABLE.

Worker exposure to this substance is limited under OSHA to 20 ppm as an 8-hour, time-weighted average.

11-15 FIGHTING FIRES INVOLVING LIQUID ORGANIC COMPOUNDS

Firefighters are often confronted with the following question: Will water successfully extinguish fires involving liquid organic compounds? The answer to this question depends on three properties: water solubility, specific gravity, and flash point.

Let's first consider organic compounds that are water soluble. If such substances catch fire, the addition of water to them causes the flash point of the resulting solutions to rapidly increase. Ultimately, the solutions are no longer capable of supporting combustion. This generally happens when alcohols, ethers, aldehydes, ketones, organic acids, and esters are diluted with water.

Notwithstanding this fact, water may still be ineffective in fighting fires of liquid organic compounds when the burning substance possesses a flash point less than 100°F (38°C). This is particularly true when the water is applied as a stream rather than as a fog. The application of water as a fog is more effective, since the particulates of water can absorb heat more readily and dissipate the burning vapor.

However, consider organic compounds that are water insoluble. The addition of water to these substances results in the formation of a two-phase system that, unless confined, may contribute to the spread of fire. It is in this situation that relative specific gravities need to be recalled. The liquid hydrocarbons, for example, are water-insoluble substances and less dense than water. When water is applied to fires involving such substances, a two-phase system results within its container. The water sinks below the less dense hydrocarbon where it is incapable of absorbing heat; thus, the water is unable to effectively extinguish the fire. By applying more and more water, the hydrocarbon may overflow its container, thereby spreading fire to other nearby areas.

On the other hand, consider fires involving carbon disulfide or the halogenated organic compounds that burn. These substances are more dense than water. When water is added to these substances, a two-phase system again results, but this time the water floats on top of the substance at issue. Hence, the fire extinguishes.

Even when the application of water may effectively extinguish the fires of organic compounds, other fire extinguishers may be preferable. For instance, when properly applied, the use of foam is desirable on fires of water-insoluble organic compounds. NFPA recommends the use of "alcohol" foam on fires of all water-soluble flammable liquids, except those that are only "very slightly" water soluble.

REVIEW EXERCISES

Aliphatic Hydrocarbons

11.1. Classify each of the following formulas as an alkane, alkene, or alkyne: C_5H_{10}; $C_{12}H_{26}$; $C_{13}H_{24}$; and $C_{26}H_{54}$.

11.2. Give the names of the aliphatic hydrocarbons represented by the following carbon skeletons?

(a)
```
C—C—C—C—C—C—C—C
      |
    C—C—C
      |
      C
```

(b)
```
        C
        |
    C—C—C
        |
        C
```

(c)
```
C—C=C—C—C—C—C
        |
        C
```

(d)
```
C=C—C—C
    |
    C
    |
    C
```

11.3. Upon chemical analysis, a sample of low-octane gasoline is found to contain varying amounts of the following aliphatic hydrocarbons: 2,3-dimethylpentane; 3,3-dimethylpentane; 2,2-dimethylpentane; n-hexane; 3-methylpentane; cyclopentane; 2-methylpentane; 2,2-dimethylbutane; n-pentane; isopentane; n-butane; and isobutane. Write condensed formulas for each of these substances.

11.4. Give the condensed formulas of the five compounds having the molecular formula C_6H_{14}. Name each of them using the IUPAC system.

11.5. Using the IUPAC system, name the alkanes having the following condensed formulas:

(a) $CH_3CH_2CH_2CH_2CH_2CH_2CH_2CH_3$

(b)
```
CH3CH2CHCH3
        |
        CH3
```

(c)
```
CH3CH2—CH—CH2CH3
         |
         CH2
         |
         CH3
```

(d)
```
CH3—CH—CH2—CH—CH2CH3
     |          |
     CH3        CH3
```

(e)
```
CH3CH2—CH—CH—CH2CH3
         |   |
         CH3 CH3
```

(f)
```
                       CH3
                       |
CH3—CH—CH2CH2CH2—C—CH3
     |            |
     CH3          CH3
```

11.6. Identify the compounds represented by the following formulas, which possess geometrical isomers:

(a) $$\underset{CH_3}{\overset{CH_3}{\diagdown}}C=C\underset{H}{\overset{CH_2CH_3}{\diagup}}$$

(b) $$\underset{CH_3CH_2}{\overset{CH_3CH_2}{\diagdown}}C=C\underset{CH_3}{\overset{H}{\diagup}}$$

(c) $$\underset{CH_3}{\overset{CH_3CH_2}{\diagdown}}C=C\underset{CH_3}{\overset{CH_2CH_3}{\diagup}}$$

(d) $$\underset{H}{\overset{Br}{\diagdown}}C=C\underset{H}{\overset{Br}{\diagup}}$$

11.7. Using the IUPAC system, name the alkenes having the following formulas:

(a) $$\underset{H}{\overset{CH_3}{\diagdown}}C=C\underset{H}{\overset{H}{\diagup}}$$

(b) $$\underset{H}{\overset{CH_3CH_2}{\diagdown}}C=C\underset{H}{\overset{CH_2CH_3}{\diagup}}$$

(c) $$\underset{CH_3}{\overset{CH_3CH_2CH_2}{\diagdown}}C=C\underset{CH_2CH_3}{\overset{CH_3}{\diagup}}$$

11.8. Using the IUPAC system, name the alkynes having the following formulas:

(a) $H-C\equiv C-CH_3$

(b) $CH_3CH_2-C\equiv C-CH_2CH_3$

(c) $CH_3CH_2CH_2CH_2-C\equiv C-H$

(d) $$CH_3-\underset{\underset{\underset{CH_3}{|}}{\underset{CH_2}{|}}{\overset{|}{CH}}}{CH}-CH_2-C\equiv C-H$$

Aromatic Hydrocarbons

11.9. Name the derivatives of benzene whose formulas follow:

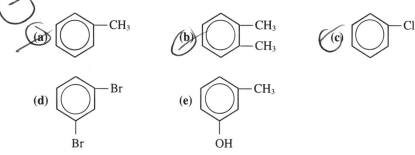

Functional Groups

11.10. Identify the class of organic compounds (alcohol, ester, aldehyde, ketone, organic peroxide, ether, or organic acid) represented by each of the compounds whose formulas follow:

(a)

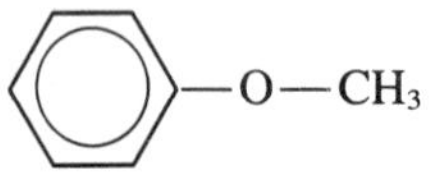

$\langle\!\langle\bigcirc\rangle\!\rangle$—O—CH$_3$

(b)

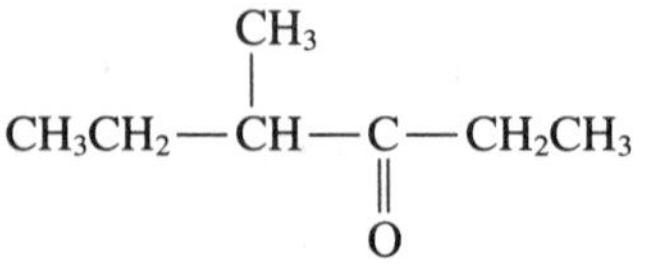

$$CH_3CH_2-\underset{\underset{O}{\overset{\displaystyle CH_3}{|}}}{\overset{}{CH}}-\overset{}{C}-CH_2CH_3$$

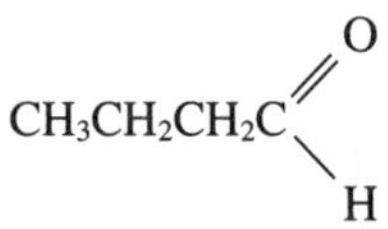

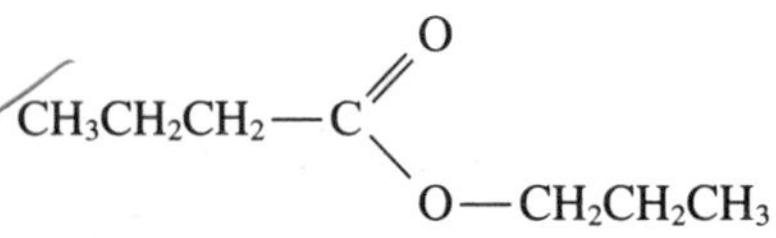

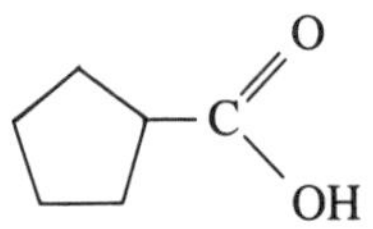

(c) CH$_3$CH$_2$CH$_2$C$\overset{\displaystyle O}{\underset{H}{\diagdown}}$

(d) CH$_3$CH$_2$CH$_2$—C$\overset{\displaystyle O}{\underset{O-CH_2CH_2CH_3}{\diagdown}}$

(e)

$\bigcirc$—C$\overset{\displaystyle O}{\underset{OH}{\diagup}}$

(f)

$$CH_3CH_2-\underset{\underset{CH_2CH_3}{|}}{\overset{\overset{CH_3}{|}}{C}}-OH$$

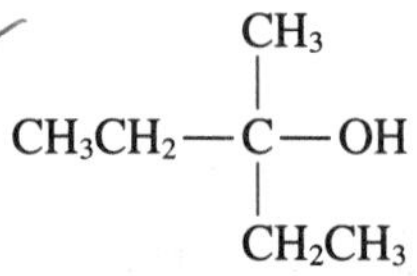

(g) CH$_3$CH$_2$—O—CH(CH$_3$)$_2$

(h) CH$_3$CH$_2$—O—O—CH(CH$_3$)$_2$

Petroleum Products

11.11. Oil refineries often separate crude petroleum into individual mixtures that are commercially sold as oil products.
   (a) Name the products that are typically obtained from crude petroleum.
   (b) In what general way is each product isolated from the others?
   (c) Arrange them according to increasing flash point.

11.12. Asphalt is a black solid or viscous liquid derived from crude petroleum. It is often used for paving or road coating. Asphalt is ordinarily considered to be a relatively inert material, similar to concrete. Yet a major fire may destroy asphalt, but not affect concrete. Why is this so?

Common Hydrocarbons

11.13. Identify three naturally occurring sources of methane.

11.14. Why is it absolutely forbidden to smoke tobacco products inside coal mines?

11.15. What is the most likely reason that USDOT limits the pressure in acetylene cylinders to 250 psig when they are intended for shipment?

11.16. While compressed gases should generally be stored upright when they are confined within steel cylinders, why is such a practice even more essential when storing cylinders of acetylene?

Halogenated Hydrocarbons

11.17. Which of the following substances do not possess flash points?
    (a) 2,3-dimethylpentane   (b) cyclopentane
    (c) *trans*-1,2-dichloroethylene   (d) trichlorofluoromethane
    (e) octafluorocyclobutane

11.18. A common industrial solvent and refrigerant is Freon-113 (also known as Freon TF), whose principal component is 1,1,2-trichloro-1,2,2-trifluoroethane.
    (a) Give the condensed formula of this substance.
    (b) What environmental hazard do large amounts of Freon TF cause when they are allowed to evaporate into the atmosphere?

11.19. Freon-12, a refrigerant, is dichlorodifluoromethane.
    (a) Is this substance flammable or nonflammable?
    (b) When heated to a temperature exceeding approximately 540°C (1000°F), Freon-12 decomposes. Identify its most likely decomposition products.

11.20. What is the most likely reason that USEPA regulates PCBs more than any other single substance or group of substances.

Alcohols

11.21. Why is the cost of denatured alcohol much less than the cost of absolute alcohol?

11.22. What adverse health effects are likely to be experienced by an individual who consumes denatured alcohol?

11.23. In the United States, whiskey, scotch, gin, and similar alcoholic beverages are available commercially in a quantity called a *fifth*, equal to one-fifth of a gallon (about 760 mL). Why may it be fatal for a 150-lb individual to consume a fifth of whiskey within a 15-min period?

11.24. Russian vodka is commercially available from approximately 70 to 90 proof. What is the percentage range by volume of ethyl alcohol in Russian vodka?

Ethers

11.25. Most organic compounds can be safely stored for many years. Yet it is generally advisable to dispose of ethers that were purchased more than a year ago. Why?

11.26. *p*-Dioxane is a cyclic ether having the following formula:

    Its vapor density (Sec. 3-4) is 3.0, and its flash point is 54°F (12°C). Why does this substance constitute a special fire and explosion risk?

11.27. Before their use as solvents, ethers are frequently distilled to remove dissolved water. Why should such distillations never be carried to dryness?

Aldehydes and Ketones

11.28. Why are local exhaust ventilation systems recommended for anatomy laboratories and mortuaries?

11.29. Which USDOT warning label is required on 55-gal steel drums used to transport each of the following substances?
(a) Methyl isobutyl ketone
(b) Formalin, when its flash point is 176°F (80°C)

Organic Acids

11.30. Organic acids do not ordinarily constitute particularly severe fire and explosion risks.
(a) Why?
(b) What is the primary hazard associated with organic acids?

Esters

11.31. Ethyl formate is an ester commonly employed as a solvent for cellulose nitrate and cellulose acetate. Its flash point is −4°F (20°C).
(a) What is the USDOT hazard class for this substance?
(b) What warning label, if any, is affixed on an ethyl formate container when this material is shipped?
(c) What warning placards, if any, are affixed to a truck carrying a total of 900 lb of ethyl formate in 55-gal drums?

Peroxo-organic Compounds

11.32. Why is it a sound procedure to store peroxo-organic compounds only in quantities limited to the minimum amounts required in cooled, ventilated buildings separated by at least 100 ft from areas where other chemical substances are stored?

11.33. Why should a solution of a peroxo-organic compound not be stored at a temperature at which the mixture solidifies or the compound precipitates from solution?

Carbon Disulfide

11.34. What is the most likely reason USDOT forbids the transportation of carbon disulfide on passenger-carrying aircraft, cargo aircraft, and water vessels transporting explosives?

Organic Compounds and Firefighting

11.35. Cumene, also known as isopropylbenzene, is a colorless liquid having a flash point of 96°F (35°C). Its specific gravity is 0.9 at room temperature. Which fire extinguisher is preferable for extinguishing a fire involving 4900 gal of cumene contained in a 5000-gal storage tank: ordinary foam, alcohol foam, or water?

11.36. Di-*n*-butyl phthalate is an oily liquid having the following physical properties: It is water soluble only to the extent of 0.04 g per 100 g of water at 77°F (25°C). Its flash point is 215°F (102°C). Its specific gravity is 1.1 at room temperature. Which fire extinguisher is preferable for extinguishing a fire involving 4500 gal of di-*n*-butyl phthalate contained in a 5000-gal storage tank: ordinary foam, alcohol foam, or water?

11.37. Isopropyl alcohol is used industrially for the production of acetone and its derivatives. Some of its important physical properties are the following: It is a colorless liquid having a flash point of 59°F (15°C). Its specific gravity is 0.7863 at room temperature. Which fire extinguisher is preferable for extinguishing a fire involving 4500 gal of isopropyl alcohol contained in a 5000-gal storage tank: ordinary foam, alcohol foam, or water?

12

Chemistry of Some Common Plastics and Textiles

In the modern world carpets, upholstery, bedding, insulation, drapes, packaging materials, components of domestic appliances of all varieties, and home and office interior decorations are primarily made from materials that are either a plastic or fiber. The chemistry associated with the hazardous nature of these materials comprises the subject matter of this chapter.

It is best to begin the study of plastics and textiles by reviewing several appropriate definitions. A *plastic* is a polymer (defined below), generally synthetic in nature, combined with fillers, reinforcing agents, and other materials and capable of being shaped or molded with or without the application of heat. Examples of some common plastics are polyvinyl chloride, polystyrene, and polymethyl methacrylate.

As a class of materials, plastics have varying physical and chemical properties. Consequently, individual plastics often respond differently when heated. Some soften when exposed to heat and return to their original condition when cooled to room conditions; these materials are said to be *thermoplastic* in nature. Others solidify or set irreversibly when heated; they are *thermosetting* plastics.

Certain natural and synthetic polymers may be drawn into threads or yarns characterized by a high tenacity and an extremely high ratio of length to diameter (several hundred to one); they are called *fibers*. Natural fibers may be derived from vegetable, animal, or mineral sources; examples of these categories of natural fibers are cotton, wool, and asbestos, respectively. Examples of synthetic fibers are nylon and rayon. Fibers differ widely in form, flexibility, durability, and porosity. They are used to make numerous goods, including rope, woven cloth, matted fabrics, brushes, shingles, and other building and insulating materials and as stuffing for pillows and upholstery.

The products produced by weaving fibers are known collectively as *textiles*. The most common examples of textiles are garments, carpets and carpet paddings, towels, curtains, blankets, mattresses, and upholstery fabrics.

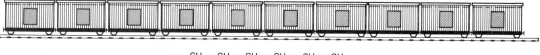

$$-CH_2-CH_2-CH_2-CH_2-CH_2-CH_2-$$

Figure 12-1   The mid-section of a freight train, each of whose railcars is identical. This assemblage resembles that of a polymer's macromolecular structure, each of whose units is also identical. This partial structure is that of polyethylene.

In the modern world, plastics and fibers are even more commonly used than wood and metal. Directly related to their wide occurrence is the fact that plastics and fibers are generally involved in almost all common fires.

12-1 POLYMERIZATION

A *polymer* is the product of a chemical reaction called polymerization. In this reaction relatively large molecules, called *macromolecules,* are formed, each of which is comprised of a number of repeating smaller units or parts. (The word *polymer* is derived from the Greek *poly* meaning many and *meros* meaning parts.) Hence, a polymer may be thought of as a many-unit compound, typically composed of hundreds or thousands of repeating units. The simple compounds from which polymers are made are called *monomers* (*mono* meaning one).

The macromolecular structure of a polymer may be compared to the mid-section of a freight train (that is, missing its locomotive and caboose), each of whose railcars, is identical; this comparison is illustrated in Fig. 12-1. Like a railcar, the repeating unit in the polymer has couplings (chemical bonds) at its opposite ends.

The units of which a polymer is comprised may be nonidentical; in such instances the substance is called a *copolymer*. A molecule of a copolymer may be compared to be section of a freight train without a locomotive or caboose and having alternating units of dissimilar railcars, as Fig. 12-2 illustrates; the individual railcars may alternate either regularly or irregularly.

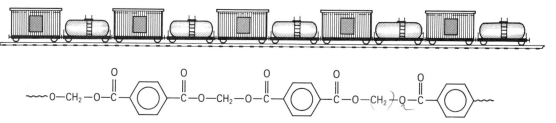

Figure 12-2   The mid-section of a freight train composed of regularly alternating dissimilar railcars. This assemblage resembles that of a copolymer's macromolecular structure, whose alternating units are also dissimilar. The alternation of dissimilar units may occur either regularly or irregularly. This partial structure is that of polyethylene terphthalate.

When the three-dimensional structures of polymers are examined, we find that these chains of repeating units are frequently cross-linked. An example of a cross-linked polymer is shown in Fig. 12-3. Industrially, plastics manufacturers often attempt to increase cross-linking between chains. This causes the polymer to become denser and thus harder and sturdier. Most polymer chains are also folded, coiled, or looped into definite shapes, which, in part, are responsible for the polymer's properties. For our purposes here, however, we need only to be concerned with the one-dimensional sequence by which atoms are bonded to one another in a polymer's macromolecules.

Some substances pose a hazard in that they may spontaneously polymerize under unique conditions. This is a point for concern, as such reactions are highly exothermic. When the substance is confined, the autopolymerization is likely to accelerate and result in explosion. Fortunately, however, autopolymerization may be inhibited by the addition of certain other chemical substances, each of which is unique to an individual monomer. To warn firefighters of the potential for autopolymerization in a substance prone to this phenomenon, a P is sometimes written in the bottom quadrant of the 704 symbol used to summarize the substance's properties.

Industrially, polymer production accounts for a significant activity of the modern chemical industry. Polymers may be chemically produced by either of two mechanisms: *addition* and *condensation*. The resulting polymers are called *addition polymers* and *condensation polymers,* respectively. We shall briefly review these two chemical reactions independently.

Addition Polymerization

Polystyrene is an example of an addition polymer; it is formed by the polymerization of the monomer styrene (more correctly called phenylethene or phenylethylene). The overall polymerization reaction may be represented by an equation like the following:

styrene   polystyrene

The formula on the left of the arrow represents styrene, the monomer. The formula on the right of the arrow represents a fragment of the polystyrene molecule. Note that the portion of this formula represented in brackets is repeated over and over again (n times, where n is a larger integer). The symbol $\sim\!\sim$ denotes the repetition of this unit.

Industrially, polymerization is typically initiated in a controlled fashion. One way by which this is accomplished is to use certain substances that readily form free

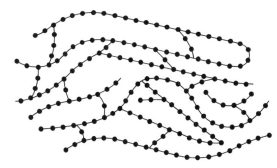

Figure 12-3 A cross-linked polymer. The blackened circles designate an arbitrary monomer, not its atoms. Cross-linking gives the polymer extra strength and durability.

radicals when exposed to heat or light. Such initiators are usually organometallic compounds (Sec. 8-3) or organic peroxides (Sec. 11-12). Once formed, the free radicals combine with neutral monomer molecules to form more complex free radicals, which in turn react with other monomer molecules until the supply of the monomer has been exhausted.

For illustrative purposes, consider the polymerization of styrene induced by free radicals resulting from dissociation of benzoyl peroxide. This polymerization may be envisioned as consisting of independent steps, some of which are represented by the following equations:

In the first step, the oxygen–oxygen bond in benzoyl peroxide is ruptured, resulting in the production of benzoyl free radicals. In the second step, a benzoyl free radical reacts with a styrene molecule to form a more complex free radical. In the third step, this free radical reacts with another molecule of styrene to form a still more complex free radical. Further steps beyond those illustrated successively add more units to the chain in a self-propagating fashion until a long chain of the polymer has formed. In this instance, the repeating unit is the following:

$$\text{C}_6\text{H}_5-\text{CH}-\text{CH}_2-$$

The polymer thus produced is *polystyrene*. It is commercially useful in thermally insulated equipment, electrical insulation, coaxial televison cable, and other products. Several other examples of addition polymer are identified in Table 12-1.

Condensation Polymerization

Condensation polymers are formed from two dissimilar monomers. One means by which a copolymer may be formed consists of a chemical reaction similar to that noted in Sec. 11-12 for the production of an ester from an alcohol and an organic acid. A condensation polymerization reaction requires the interaction between an alcohol with at least two hydroxy (—OH) groups and an acid with two carboxyl groups. This type of alcohol is called a *glycol;* the acid is called a *dicarboxylic acid*. When heated with a catalyst, molecules of these compounds combine, with the simultaneous elimination of water. The resulting polymer is called a *polyester*.

Consider the chemical reaction between the monomers ethylene glycol and succinic acid. The first step in this reaction is illustrated by the following equation:

$$
\begin{array}{ccccc}
\text{OH} & & \text{COOH} & & \text{COOCH}_2\text{CH}_2\text{OH} \\
| & & | & & | \\
\text{CH}_2 & & \text{CH}_2 & & \text{CH}_2 \\
| & + & | & \longrightarrow & | \qquad\qquad + \ \text{H}_2\text{O}(l)\\
\text{CH}_2 & & \text{CH}_2 & & \text{CH}_2 \\
| & & | & & | \\
\text{OH}(l) & & \text{COOH}(s) & & \text{COOH}(s) \\
\text{ethylene glycol} & & \text{succinic acid} & & \text{an intermediate, not yet} \\
& & & & \text{a polymer}
\end{array}
$$

The intermediate compound resulting from this reaction possesses potentially reactive groups at both ends of the carbon–carbon chain. It may react with another molecule of ethylene glycol and another molecule of succinic acid, eliminating two molecules of water and forming an even more complex intermediate, as the following equation notes:

TABLE 12-1 EXAMPLES OF ADDITION POLYMERS

Monomer	Repeating Units

Ethylene

$$H \quad\quad H$$
$$\backslash \quad\quad /$$
$$C = C$$
$$/ \quad\quad \backslash$$
$$H \quad\quad H$$

Polyethylene

$$\begin{array}{c} H \ H \ H \ H \ H \ H \ H \ H \ H \ H \ H \ H \ H \ H \\ |\ |\ |\ |\ |\ |\ |\ |\ |\ |\ |\ |\ |\ | \\ -C-C-C-C-C-C-C-C-C-C-C-C-C-C- \\ |\ |\ |\ |\ |\ |\ |\ |\ |\ |\ |\ |\ |\ | \\ H \ H \ H \ H \ H \ H \ H \ H \ H \ H \ H \ H \ H \ H \end{array}$$

CO
CO₂
H₂O
Cₙ

Vinyl chloride

$$H \quad\quad H$$
$$\backslash \quad\quad /$$
$$C = C$$
$$/ \quad\quad \backslash$$
$$H \quad\quad Cl$$

Polyvinyl chloride

$$\begin{array}{c} H \ H \ H \ H \ H \ H \ H \ H \ H \ H \ H \ H \ H \ H \\ |\ |\ |\ |\ |\ |\ |\ |\ |\ |\ |\ |\ |\ | \\ -C-C-C-C-C-C-C-C-C-C-C-C-C-C- \\ |\ |\ |\ |\ |\ |\ |\ |\ |\ |\ |\ |\ |\ | \\ H \ Cl \ H \ Cl \ H \ Cl \ H \ Cl \ H \ Cl \ H \ Cl \ H \ Cl \end{array}$$

HCl

Acrylonitrile

$$H \quad\quad H$$
$$\backslash \quad\quad /$$
$$C = C$$
$$/ \quad\quad \backslash$$
$$H \quad\quad CN$$

Polyacrylonitrile

$$\begin{array}{c} H \ H \ H \ H \ H \ H \ H \ H \ H \ H \ H \ H \ H \ H \\ |\ |\ |\ |\ |\ |\ |\ |\ |\ |\ |\ |\ |\ | \\ -C-C-C-C-C-C-C-C-C-C-C-C-C-C- \\ |\ |\ |\ |\ |\ |\ |\ |\ |\ |\ |\ |\ |\ | \\ H \ CN \ H \ CN \ H \ CN \ H \ CN \ H \ CN \ H \ CN \ H \ CN \end{array}$$

HCl

Tetrafluoroethylene

$$F \quad\quad F$$
$$\backslash \quad\quad /$$
$$C = C$$
$$/ \quad\quad \backslash$$
$$F \quad\quad F$$

Polytetrafluoroethylene

$$\begin{array}{c} F \ F \ F \ F \ F \ F \ F \ F \ F \ F \ F \ F \ F \ F \\ |\ |\ |\ |\ |\ |\ |\ |\ |\ |\ |\ |\ |\ | \\ -C-C-C-C-C-C-C-C-C-C-C-C-C-C- \\ |\ |\ |\ |\ |\ |\ |\ |\ |\ |\ |\ |\ |\ | \\ F \ F \ F \ F \ F \ F \ F \ F \ F \ F \ F \ F \ F \ F \end{array}$$

Styrene

$$CH = CH_2$$

(benzene ring)

Polystyrene

$$\begin{array}{c} H \ H \ H \ H \ H \ H \ H \ H \ H \ H \ H \ H \ H \ H \\ |\ |\ |\ |\ |\ |\ |\ |\ |\ |\ |\ |\ |\ | \\ -C-C-C-C-C-C-C-C-C-C-C-C-C-C- \\ |\quad |\quad |\quad |\quad |\quad |\quad | \\ H \quad H \quad H \quad H \quad H \quad H \quad H \end{array}$$

xS Cₙ

(benzene rings)

Methyl methacrylate

$$CH_3$$
$$|$$
$$CH_2 = C$$
$$|$$
$$C = O$$
$$|$$
$$O$$
$$|$$
$$CH_3$$

Polymethyl methacrylate

$$\begin{array}{c} H \ CH_3 \ H \ CH_3 \ H \ CH_3 \ H \ CH_3 \ H \ CH_3 \ H \ CH_3 \\ |\quad|\quad|\quad|\quad|\quad|\quad|\quad|\quad|\quad|\quad|\quad| \\ -C-C-C-C-C-C-C-C-C-C-C-C- \\ |\quad|\quad|\quad|\quad|\quad|\quad|\quad|\quad|\quad|\quad|\quad| \\ H \ C=O \ H \ C=O \ H \ C=O \ H \ C=O \ H \ C=O \ H \ C=O \\ \quad |\quad\quad\quad|\quad\quad\quad|\quad\quad\quad|\quad\quad\quad|\quad\quad\quad| \\ \quad O \quad\quad\quad O \quad\quad\quad O \quad\quad\quad O \quad\quad\quad O \quad\quad\quad O \\ \quad |\quad\quad\quad|\quad\quad\quad|\quad\quad\quad|\quad\quad\quad|\quad\quad\quad| \\ \quad CH_3 \quad\ CH_3 \quad\ CH_3 \quad\ CH_3 \quad\ CH_3 \quad\ CH_3 \end{array}$$

TABLE 12-2 EXAMPLES OF CONDENSATION POLYMERS

Monomers	Repeating Unit

Ethylene glycol

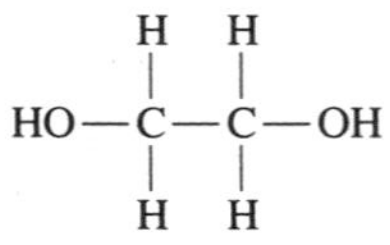

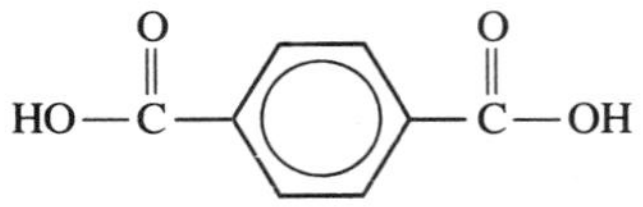

A polyester (polyethylene terephthalate)

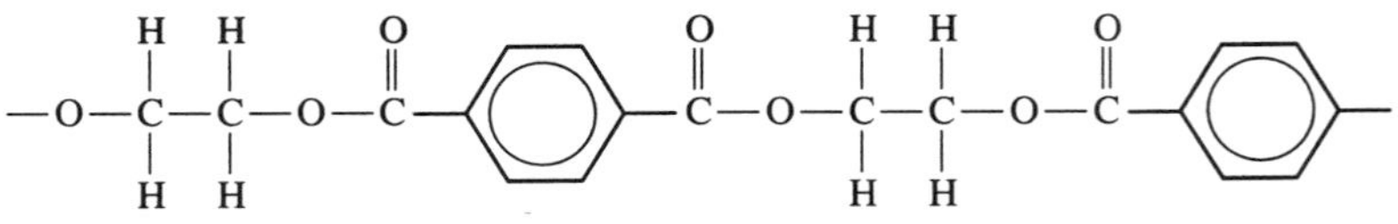

Terephthalic acid

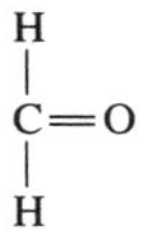

Phenol-formaldehyde resin

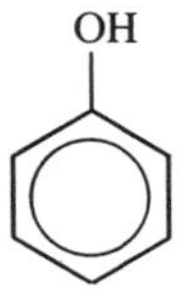

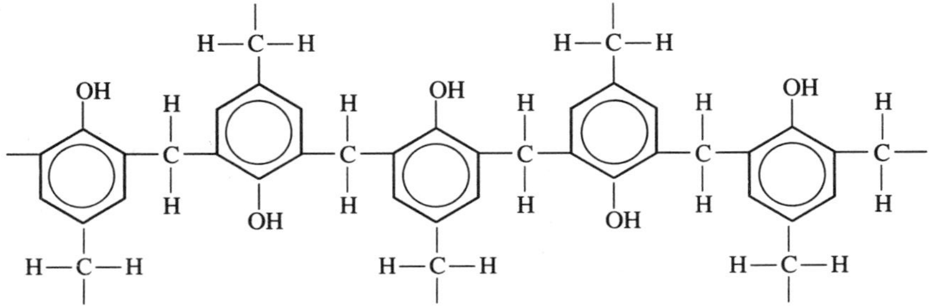

Phenol

Formaldehyde

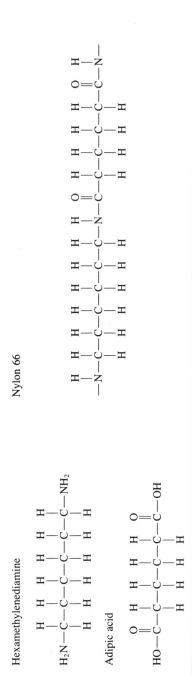

Hexamethylenediamine

Adipic acid

Nylon 66

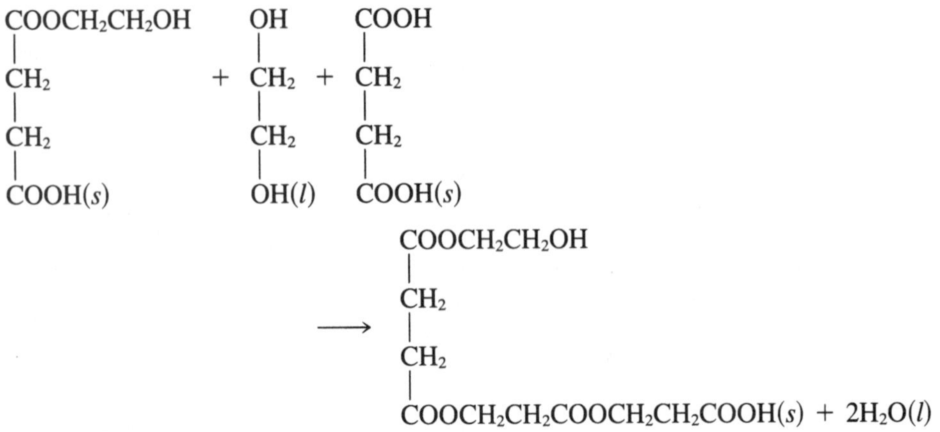

$$COOCH_2CH_2OH \quad OH \quad COOH$$
$$| \qquad\qquad\quad | \qquad\quad |$$
$$CH_2 \qquad + \;\; CH_2 \; + \; CH_2$$
$$| \qquad\qquad\quad | \qquad\quad |$$
$$CH_2 \qquad\qquad CH_2 \quad\; CH_2$$
$$| \qquad\qquad\quad | \qquad\quad |$$
$$COOH(s) \qquad\;\; OH(l) \quad COOH(s)$$

$$COOCH_2CH_2OH$$
$$|$$
$$CH_2$$
$$\longrightarrow \quad |$$
$$CH_2$$
$$|$$
$$COOCH_2CH_2COOCH_2CH_2COOH(s) + 2H_2O(l)$$

This intermediate also possesses reactive groups on both ends of the molecule. Thus, it too may combine with more ethylene glycol and succinic acid, thus increasing the length of the chain. Finally, when either monomer is exhausted, the polyester that remains is described by a formula like the following:

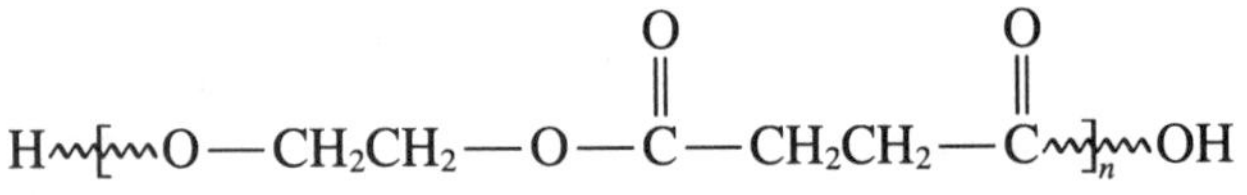

$$\begin{array}{ccc} & O & O \\ & \| & \| \\ H\text{---}\!\{\text{---}O\text{---}CH_2CH_2\text{---}O\text{---}C\text{---}CH_2CH_2\text{---}C\}_n\text{---}OH \end{array}$$

Additional examples of condensation polymers are noted in Table 12-2.

12-2 GENERAL FEATURES REGARDING THE BURNING OF SYNTHETIC POLYMERS

Most polymers used commercially are organic compounds; hence, at elevated temperatures, most products burn that have been made from synthetic polymers. Their combustion is generally associated with three significant features:

1. The product often melts as it burns.
2. The surface of the product tends to char.
3. The burning of polymer products evolves a voluminous amount of smoke, carbon monoxide, and other hazardous fire gases.

In a typical fire environment, the melting of polymeric products is a phenomenon characterized simultaneously with beneficial and detrimental features. Melting usually causes the polymer to drip from its source of origin to elsewhere, as from ceiling tile to the underlying floor. This dripping of a molten polymer closely resembles the dripping of hot candlewax. On the one hand, dripping of molten polymer serves to act as a cooling mechanism, removing heat from the immediate site of combustion. This could prevent the polymer from burning. On the other hand, when the molten polymer is already burning, the dripping is likely to contribute to the spread of fire from one area to another.

Frequently, polymer burning first involves the thermal degradation of the polymer itself into simpler chemical species, usually the monomers from which the polymer was formed. Sometimes, these simple molecules diffuse to the surface of the material, where they mix with atmospheric oxygen and burn. However, under exposure to intense heat, the vapors of these simpler substances may also migrate away from the immediate burning area and accumulate elsewhere, such as near a ceiling in a room, where they again mix with air and ignite.

Heat may also be conducted or radiated through a polymeric material, causing thermal decomposition of the polymer at a different location from where the heat was originally applied. Figure 12-4 illustrates a wall made of wooden support beams on which polymeric paneling has been fastened. Once heat has conducted or radiated through the wall, the polymer in the paneling may decompose, even though the heat source is some distance away. Then a combustible mixture of simple organic substances forms and ignites when exposed to an ignition source. This phenomenon is an example of *flashover;* it frequently contributes to the spread of fire from one room to an adjacent room. Such fires pose special problems in large public buildings constructed in part from plastic materials, since polymeric products burn much faster and hotter than wood and other natural products.

The number of fire-related fatalities usually changes dramatically when polymeric products have been widely used throughout a building that catches fire. These fatalities are usually caused by inhalation of fire gases, as opposed to burns or other means. In small rooms, the concentration of fire gases can increase to life-threatening levels in a matter of seconds. Furthermore, when a fire occurs in such buildings, the greater impact of the fire may occur far from the scene of the fire itself. This is sometimes caused by the convective movement of fire gases and the

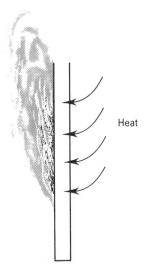

Heat

Figure 12-4   Heat that is conducted or radiated through this section of a wall causes the thermal decomposition of the polymeric paneling on its opposite side. The decomposition of a polymer is associated with the production of simple organic substances, which slowly migrate from their point of origin and mix with surrounding air. When the concentration of this mixture is within its flammable range, an ignition source causes the mixture to ignite.

thermal degradation products through ventilation systems, trash chutes, and other routes from the fire scene to elsewhere. This movement of fire gases spreads the fire, but inhalation of the gases may also be fatal for people far removed from the fire itself.

The burning of products made from synthetic polymers often produces a different mixture of fire gases than that generated from the burning of nonplastic products. Carbon monoxide is still the most prevalent fire gas in the immediate environment of these fires, but other fire gases are typically more abundant than conventionally observed. In particular, lethal concentrations of either hydrogen cyanide or hydrogen chloride are likely to be produced.

In addition to the fire gases, other toxic products are likely to form in fires involving polymeric materials. Most commonly observed is the unsaturated aldehyde commonly known as *acrolein;* its chemical formula is $CH_2 = CHCHO$. It is a pungent smelling, intensely irritating lachrymator whose TLV is only 0.1 ppm. Even a 1-minute exposure to air containing 1 ppm of acrolein causes nasal and eye irritation. Inhalation of air containing as little as 10 ppm of acrolein may be fatal in a few minutes.

The character of the smoke produced in fires involving polymeric materials varies according to the chemical nature of the polymer. More smoke typically develops in fires involving polymers made from aromatic monomers, as opposed to aliphatic monomers. Burning polystyrene, for instance, produces much more soot than burning polyethylene.

Figure 12-5 illustrates how rapidly fire may spread in untreated fabrics produced from synthetic polymers. To improve the fire safety of these textiles, various fire retardants are now incorporated into them, or flameproofing agents are coated on them. There is no doubt that these procedures improve the fire safety of textiles.

Figure 12-5 A grim demonstration. The manikin shown to the far left is sucessively photographed 30 seconds, 60 seconds, and 90 seconds after application of a flame to its blouse. This illustrates how quickly fire may spread when certain synthetic fabrics burn. (Courtesy of I. N. Einhorn, Flammability Research Center, University of Utah, Salt Lake City, Utah)

However, the protection offered by these substances is only partially effective. The technology for treating plastics and textiles to make them more resistant to ignition is complex and has not been perfected. Polymeric materials still burn, even when treated with fire retardants and flameproofing agents, especially at the high temperatures experienced during major fires.

Typical fires retardants and flameproofing agents used to produce modern-day textiles include halogenated hydrocarbons, certain water-insoluble metal salts, antimony trioxide, tricresyl phosphate, and other phosphate esters. One such fire retardant became cause for concern in the 1970s: *tris-*(2,3-dibromopropyl) phosphate, commonly known as TRIS. Its chemical formula is the following:

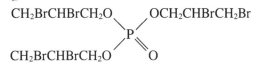

$$CH_2BrCHBrCH_2O \diagdown \qquad \diagup OCH_2CHBrCH_2Br$$
$$P$$
$$CH_2BrCHBrCH_2O \diagup \qquad \diagdown\diagdown O$$

This substance was primarily used as a coating on certain synthetic fabrics like children's sleepwear. While effective as a fire retardant, it was found to be carcinogenic in experimental animals by both oral administration and dermal application. For these reasons, the manufacture and use of *tris-*(2,3-dibromopropyl) phosphate was banned in the United States.

Aside from the voluntary gestures on the part of plastics and textile manufacturers, legal attempts have also helped to make textiles safer from a fire-related perspective. At the federal level, the *Flammable Fabrics Act* responds to public concern about several serious accidents involving brushed rayon high-pile sweaters and children's cowboy chaps, both of which flash-burn when ignited. This statute requires manufacturers to submit apparel fabrics to a specified 45° ignition test and rate-of-burn test. The fact that flash-burning fabrics are no longer on the American market is due directly to implementation of this law.

In 1967 the Flammable Fabrics Act was further amended to include interior furnishings, paper, plastics, and other materials used in wearing apparel and interior furnishings. Regulatory standards were subsequently adopted applying to carpets and rugs, children's sleepwear, and mattresses.

In addition, the Department of Consumer Affairs of various states has adopted flammability regulations applicable to a variety of interior furnishings. Certain specific requirements apply to their use in public buildings. Even though the hazard can never be totally eliminated, the combined efforts of such groups assist in reducing fire-related injuries when plastics and textiles are involved in fires.

12-3 VEGETABLE AND ANIMAL FIBERS

Many commonly encountered textiles have been made from naturally occurring vegetable and animal fibers. Cotton and linen are examples of vegetable fibers, and wool and silk are examples of animal fibers. In addition, these materials may be chemically altered (especially cotton) to produce synthetic fibers, which have experienced a particular popularity in clothing items. We shall note some properties of the more common fibers, with special emphasis on their flammable features.

Cellulose and Cellulosic Derivatives

When the natural binding agent, called *lignin,* is removed from wood, the principal substance that remains is *cellulose*. This polymer serves as the main structural component of the cell walls of all plants and thus is generally regarded as nature's most important polymer. Approximately 50% by mass of wood is cellulose; cotton is almost 100% cellulose. Commercially, cellulose serves as the raw material of the paper, wood, cotton textiles, and cellulose plastics industries.

The chemical formula of cellulose is often abbreviated $(C_6H_{10}O_5)_n$, where n is on the order of 5000. The chemical formula is

The repeating unit in this structure is called *β-glucose*. The β-form of glucose needs some description. As indicated above, the glucose molecules are depicted as planar, hexagonal slabs with the darkened edge projected toward the viewer. The ring carbon atoms are not directly indicated, but are understood to be present where any two lines intersect. By convention, when the hydroxyl group is positioned on the carbon atom indicated as 1 *above* the plane, the corresponding substance is β-glucose. The chemical bonds between the β-glucose units are straight, and the molecules move into a line off the horizontal. These bonds are represented bent in order to keep the chain in the horizontal plane.

When the hydroxy group is positioned *below* the plane with all other positions unaltered, the corresponding substance is *α-glucose*. The two forms of glucose are structural isomers. Once again, the β-isomer occurs in cellulose. The α-isomer is the repeating unit in *starch,* an important natural polymer, but not of concern here.

Cotton comes from any of four species of *Gossypium* that grow in warm climates throughout the world. To produce cotton as a yarn, the raw cotton is first boiled in a dilute solution of sodium hydroxide to remove any wax that naturally forms with the fibers. Then the cotton is bleached with chlorine, sodium hypochlorite, or similar substance. Next it is passed through a vat of dilute sulfuric acid to neutralize any remaining alkaline materials, and, finally, it is washed with water. At this point, the cotton is ready to be spun into a yarn and woven into cloth.

Cotton is sometimes immersed in a concentrated solution of sodium hydroxide. This act forces the cotton to swell. Each of its flattened fibers becomes rounded and more lustrous and takes on additional strength. This process is called *mercerizing,* and the product is called *mercerized cotton* (after the discoverer, John Mercer). Mercerizing allows dyes to penetrate cotton more readily. Cotton textiles may also be chemically treated to produce forms that are wrinkleproof or forms that can be washed and drip-dried with few wrinkles. It may also be blended with synthetic fibers, resulting in blends like 50% cotton, 50% acrylics.

While cotton is a natural fiber, several partially synthetic fibers may be derived from it or from cellulose that originates in wood. For example, cellulose acetate is a fiber produced when cellulose is allowed to chemically react with acetic acid; it is called *rayon,* or artificial silk. Rayon possesses a pronounced strength, but it ignites and burns readily. Hence, it is no longer as popular in fabrics as it once was. Aside from its use in cloth, cellulose acetate is used in motion-picture film, airplane wings, and safety glasses.

Another product derived from cellulose is *nitrocellulose,* or *guncotton* (Sec. 13-7). It is made by reacting cellulose with nitric acid. Nitrocellulose is a chemical explosive. Nonetheless, a form of nitrocellulose may also be dissolved in a solvent and applied to cloth; the product is called *patent leather*.

Linen

Linen is another vegetable fiber. From a chemical viewpoint, linen is nearly pure cellulose. This fiber is derived from the stalk of the flax plant, *Linum usitatissimum*. Linen fibers are among the strongest naturally occurring fibers. Linen also absorbs moisture faster than any other fiber. Decades ago, linen was often used as the fabric for summer garments in very warm climates. Linen also leaves virtually no lint and thus is potentially useful in towels. However, linen lacks resiliency, the ability to spring back when stretched. This lack of resiliency causes linen to wrinkle readily.

The difference between the physical properties of cotton and linen is due to the nature of their fibers. When examined under a microscope, individual cotton fibers are observed to be short and twisted and resemble flattened tubes, while linen fibers are relatively long, with periodic junctions for cross-linking, and resemble transparent tubes. These features combine to make linen fibers stronger than cotton fibers.

When triggered by an ignition source, all cellulosic materials burn; however, they do not melt. Fires involving only cellulosic materials are class A fires, which may be effectively extinguished with water. Chemical analysis of the thermal degradation products found in the smoke of cellulosic fires reveals the presence of nearly 200 different chemical substances. Thus, the mechanism by which cellulose burns is obviously very complicated. Scientists believe that the primary step involves the formation of levoglucosan, whose chemical formula is the following:

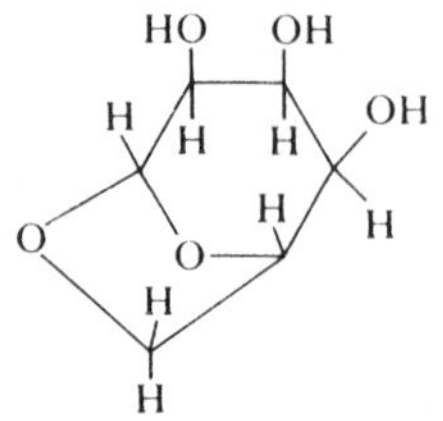

This substance then undergoes further degradation to a variety of simpler compounds.

Wool and Silk

The most common animal fabrics employed in textiles are wool and silk. The macromolecular structure of these materials differs significantly from that of cellulose. Wool is the curly hair of sheep, goats, llamas, and certain other animals. Like all animal hair, wool is a protein, certain complex substances having recurring amide groups ($-NH_2$). A portion of the macromolecular structure of wool is shown in Fig. 12-6. The condensed formula, $C_{42}H_{157}N_5SO_{15}$, has been proposed for wool.

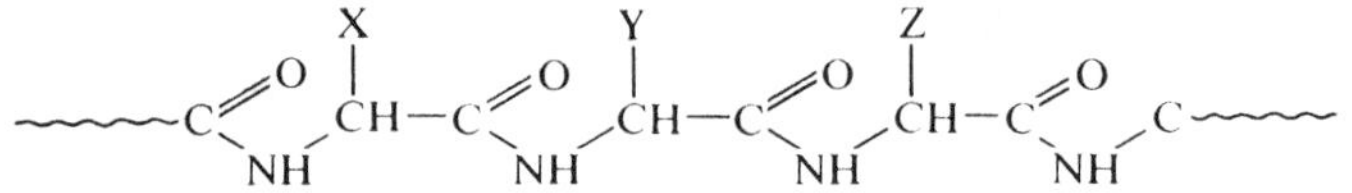

Figure 12-6   A portion of the macromolecular structure of wool. In this protein, X, Y, and Z represent specific substances called *amino acids*, compounds having the chemical formula $R-CHNH_2COOH$. Only 21 amino acids are commonly found in naturally occurring proteins. The interested reader should consult more advanced chemistry texts on this subject.

Under a microscope, wool fibers resemble tiny, overlapping scales, much like the scales on fish. These fibers bend and conform to a variety of shapes. They also possess resiliencey and thus tend to hold their original shape quite well.

Silk is the soft, shiny fiber produced by silkworms to form their cocoons. Silk fibers are very strong, elastic, and smooth. These qualities make silk one of the most useful fibers for the textile market. However, unwinding the long, delicate silk threads of a cocoon is a tedious process; hence, silk tends to be relatively costly.

Chemically, silk consists of a mixture of two relatively simple proteins, *silk fibroin* and *sericin*. Silk fibroin has the general structure illustrated for wool in Fig. 12-6, but X, Y and Z are almost always either $-CH_3$, $HO-C_6H_5-CH_2-$, and $HO-CH_2-$, respectively. Sericin possesses a similar structure, but the primary recurring group in this protein is $HO-CH_2CH-NH_2-$. The condensed formula for silk has been proposed as $C_{15}H_{23}N_5O_6$.

Under a microscope, silk fibers appear semitransparent. This property accounts for their lustrous sheen. Silk is also resilient; its fibers readily spring back to their original position when stretched, bent, or folded.

Since wool and silk are proteins, they possess several similar properties. Each is difficult to ignite, having ignition temperatures in excess of approximately 570°C (1058°F); and when ignited, they generally burn very slowly. This is especially true of rugs and other textiles in which the fibers have been woven tightly. Woolen and silk textiles also tend to smolder and char. However, on the positive side, fires involving woolen and silk textiles are generally easy to extinguish, since these materials are capable of absorbing great quantities of water.

Since the macromolecules of wool have sulfur and nitrogen atoms, burning wool produces sulfur dioxide and nitrogen dioxide. On the other hand, silk macromolecules have nitrogen atoms, but not sulfur atoms; hence, burning silk produces nitrogen dioxide, but not sulfur dioxide. The atmosphere near smoldering woolen and silk textiles often contains ammonia and hydrogen cyanide.

12-4 POLYVINYL POLYMERS

Polyvinyl polymers are those that contain the vinyl group $CH_2{=}CH{-}$. Unlike cellulose derivatives, the polyvinyl polymers are totally synthetic; that is, they are prepared from monomers that do not occur naturally.

Many polyvinyl polymers may be considered as derivatives of polyethylene, although polyethylene itself is the only polyvinyl polymer that is directly prepared from ethylene.

$$\text{\textasciitilde}CH_2CH_2\text{\textasciitilde}(CH_2CH_2)_n\text{\textasciitilde}CH_2CH_2\text{\textasciitilde}$$
polyethylene

For example, polystyrene may be imagined as a derivative of polyethylene in which a hydrogen atom has been substituted with a phenyl group in each repeating unit.

Polyvinyl polymers are ordinarily produced industrially as addition polymers from their respective monomers. The following equations illustrate the formation of several important polyvinyl polymers:

ethylene   polyethylene

propylene   polypropylene

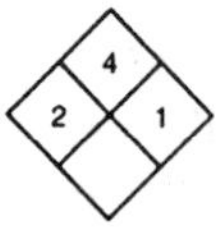

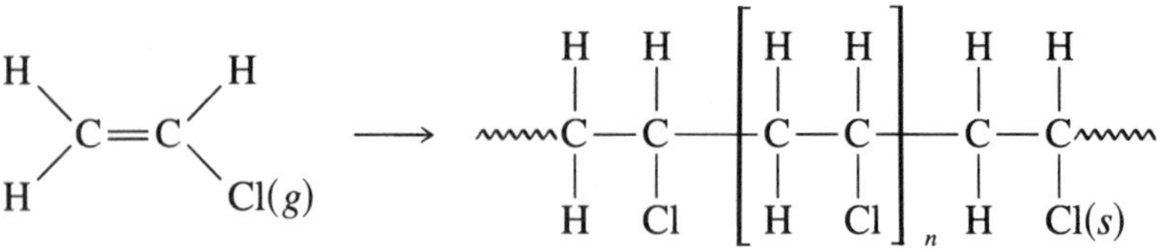

vinyl chloride

polyvinyl chloride

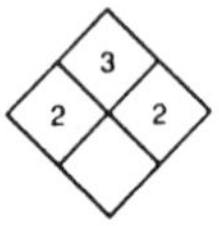

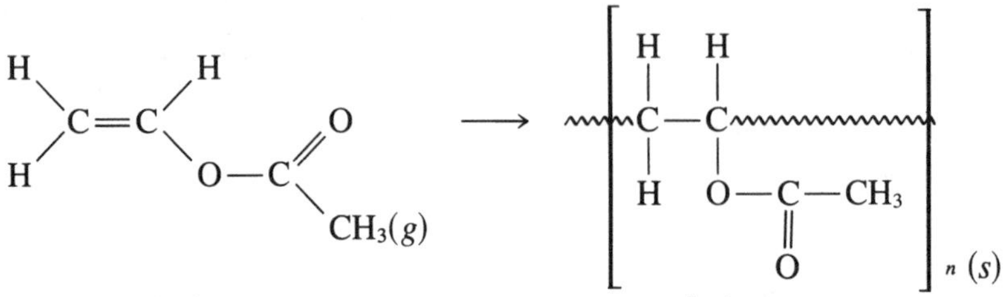

vinyl acetate

polyvinyl acetate

Polyethylene and polyvinyl chloride are usually regarded as the most important of these polyvinyl polymers.

The referenced 704 diamond symbols are those of the monomers from which these polyvinyl polymers are produced. These substances are hazardous materials for several reasons. All of them burn, typically at relatively low ignition temperatures. Furthermore, several of these monomers are prone to autopolymerization. USDOT regulates the transportation of methyl acrylate, methyl methacrylate, and styrene, when inhibited, as flammable liquids.

Certain copolymers may also be produced from a combination of the monomers used to prepare addition polymers. For instance, vinyl chloride (chloroethene or chloroethylene) and vinylidene chloride (1,1-dichloroethene) form the copolymer industrially known as *Saran*, as the following equation illustrates:

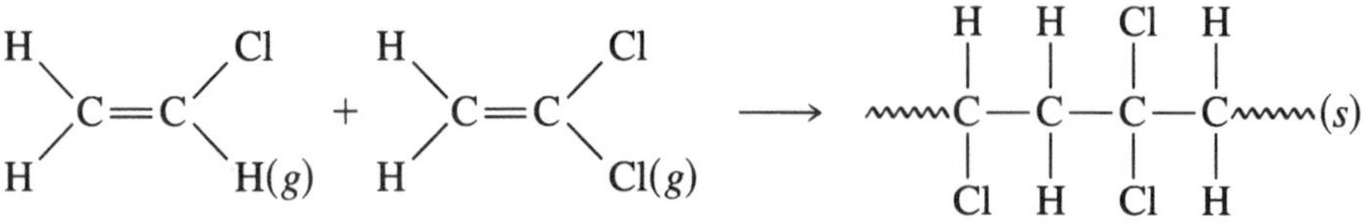

vinyl chloride   vinylidene chloride   saran

Saran is used in the form of sheets, tubes, rods, fibers, and other molded items. The fibers are used to produce a number of textiles like carpets, curtains, and upholstery fabrics.

Note that some polyvinyl polymers contain only carbon, hydrogen, and oxygen, whereas polyvinyl chloride and the vinyl chloride–vinylidene chloride copolymer contain chlorine; polyacrylonitrile contains nitrogen. As we shall note in more detail shortly, these features of the polymer structures are important factors for understanding the nature of the gases that form when polymers are exposed to intense heat sources.

TABLE 12-3 PHYSICAL PROPERTIES OF SEVERAL POLYVINYL POLYMERS

	Polyethylene	Polypropylene	Polystyrene	Polyvinyl chloride
Melting point	241°–351°F (116°–177°C)	286°–309°F (141°–154°C)	194°F (90°C)	—
Temperature corresponding to maximum vaporization	878°–972°F (470°–522°C)	880°–898°F (471°–481°C)	622°–844°F (328°–451°C)	—
Temperature corresponding to onset of thermal degradation	847°–896°F (453°–480°C)	682°–837°F (361°–447°C)	790°–842°F (421°–450°C)	505°–889°F (263°–476°C)
Heat of combustion (cal/g)	10,008–11,176	11,112	9480–10,425	9952

Some polyvinyl polymers first soften and melt when they are involved in fires. Polystyrene, polyethylene, and polypropylene begin to soften at 194°, 185°, and 248°F (90°, 85°, and 120°C), respectively. Table 12-3 lists some other physical properties of these three polyvinyl polymers, as well as for polyvinyl chloride. In specific temperature ranges, these polymers undergo thermal degradation, often forming the monomer from which they were produced, as well as methane, ethane, ethylene, benzene, toluene, ethylbenzene, and other comparatively simple substances. It is the combination of these latter substances that actually catches fire and burns.

The fact that polyvinyl polymers thermally decompose to their respective monomers is important when considering the adverse health effects these substances are likely to pose during fire-related incidents. Table 12-4 identifies the threshold limit values of several selected monomers that polymerize to form polyvinyl polymers.

The following brief overview summarizes the important fire-related features of some selected polyvinyl polymers.

TABLE 12-4 THRESHOLD LIMIT VALUES
OF SOME COMMON MONOMERS

Monomer	TLV (ppm)
Acrylonitrile	20
1, 3-Butadiene	1000
Chloroprene	25
Ethyl acrylate	25 (skin)
Ethylene oxide	50
Methyl acrylate	10 (skin)
Styrene	100
Vinyl chloride	500
Vinylidene chloride	5

Polyethylene

This is one of the most commonly used polyvinyl polymers. It is encountered in either of three forms: cross-linked, linear, and low molecular weight. All three forms are white solids, but cross-linked polyethylene is thermosetting, whereas the linear and low-molecular-weight forms are thermoplastic. The cross-linked form is produced by irradiating the linear form with γ-radiation (Sec. 14-2). Low-molecular-weight polyethylene possesses a molecular weight from 2000 to 5000, while the molecular weight of the linear form approximates 6000 or more.

Linear polyethylene is used in products like those shown in Fig. 12-7. A film may be produced from polyethylene from which dozens of products are produced, like plastic bags for packaging foods, soft goods, and hardware; the film is also used in the building industry as a vapor and moisture barrier and in agriculture for mulching, silage covers, greenhouse glazings, pond liners, and animal shelters; and the

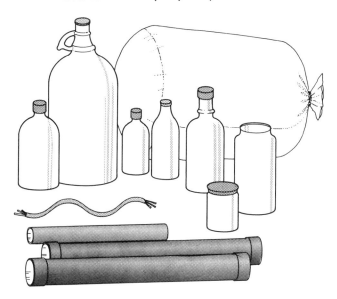

Figure 12-7   Examples of some common products made from polyethylene.

film is employed as liners in drums and other shipping containers. Low-molecular-weight polyethylene is used in coatings and polishes, while cross-linked polyethylene is largely used for making molded products like toys and containers of all sizes.

On the positive side, polyethylene burns. Since it is composed of only carbon and hydrogen, the combustion products resulting from burning polyethylene are carbon monoxide, carbon dioxide, and water. A characteristic associated with burning polyethylene film is that products made from it tend to disintegrate during combustion into many burning, molten globules.

Polyvinyl Chloride

Polyvinyl chloride has received considerable attention from fire safety personnel. This polymer has largely replaced cotton as an insulating material in all electrical wiring. It is also commonly employed for production of phonograph records, floor tile, piping, imitation leather, and shower curtains. Polyvinyl chloride is produced by polymerizing vinyl chloride.

On the positive side, products made of polyvinyl chloride do not easily burn, particularly when compared to wood; hence, it provides an element of safety in home furnishing products. However, when strongly heated, polyvinyl chloride forms a mixture of vapors and gases, including the toxic gas hydrogen chloride (Sec. 7-7). When inhaled, this gas may cause serious destructive damage to the mucous membranes. Furthermore, carbon particulates in smoke adsorb hydrogen chloride. When smoke is inhaled, these particulates are likely to bypass the body's upper respiratory mucous membranes and become lodged in the lungs, where hydrogen chloride may cause pulmonary edema.

Exposure of polyvinyl chloride to elevated temperatures may also form vinyl chloride. This substance is a human carcinogen. Thus, when involved in fires, polyvinyl chloride products may potentially pose a life-threatening situation in multiple ways.

Polyacrylonitrile

This polyvinyl polymer is popularly known by the trade names *Orlon, Acrilan, Creslan,* and *Zefran*. It was the first of a number of synthetic polymers that became popular as *acrylic fibers*. Today, polyacrylonitrile is largely used in textiles, since it possesses an outstanding resistance to sunlight, is quick drying, and is easy to launder.

Polyacrylonitrile is produced from the monomer acrylonitrile (vinyl cyanide). Thus, polyacrylonitrile possesses the repeating unit

$$-CH_2CH-$$
$$|$$
$$CN$$

When heated under fire conditions, products made from polyacrylonitrile evolve the fire gas hydrogen cyanide. Exposure to heat also causes polyacrylonitrile to decompose to its monomer, acrylonitrile. This substance burns very easily. Hence, once ignited, untreated polyacrylonitrile products usually burn rapidly.

Polymethyl Methacrylate

Several polyvinyl polymers are produced from the esters of acrylic acid and methacrylic acid. The methyl esters of these acids have the following chemical formulas:

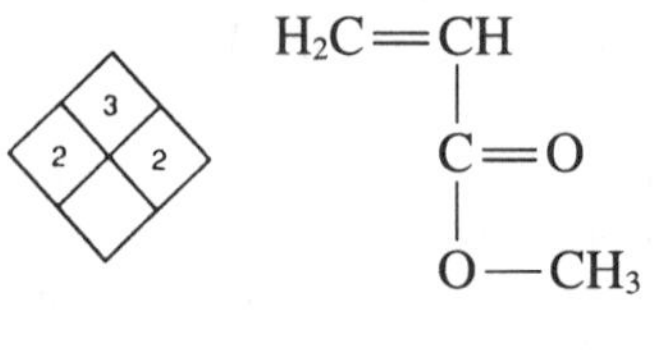

methyl acrylate

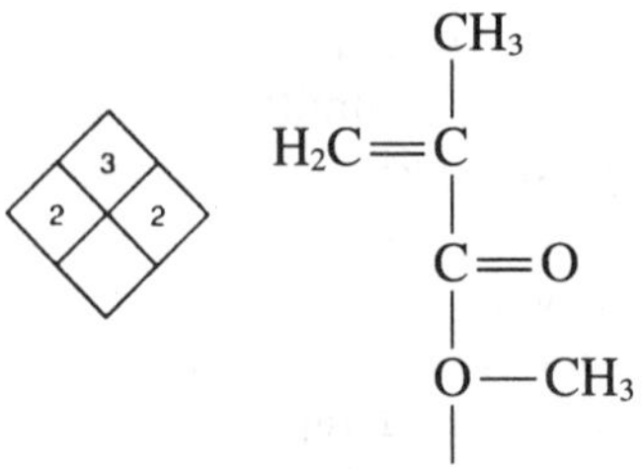

methyl
methacrylate

Each of these substances polymerizes, as the following equations illustrate:

methylacrylate   polymethacrylate

methyl methacrylate   polymethyl methacrylate

Polymethyl methacrylate is popular as a plastic known commercially as *Plexiglas*. It is a clear substance and may be manufactured as a transparent, translucent, or opaque material. It is virtually unbreakable; thus, it is used in windshields, windows, and other products where strength and transparency are simultaneously important factors. When heated, polymethyl methacrylate undergoes thermal degradation at approximately 320°F (160°C), forming the monomer, methyl methacrylate, which readily catches fire and burns.

12-5 POLYACETALS, POLYETHERS, AND POLYESTERS

Under appropriate conditions, several aldehydes, ethers, and esters undergo polymerization to form polymers with recurring carbon-to-oxygen units. These substances are generally condensation polymers.

Aldehydes having low molecular weights readily autopolymerize. This is evidenced by the polymerization of the simplest aldehyde, formaldehyde. When aqueous solutions of formaldehyde are permitted to stand for some time, they slowly deposit a white, unstable polymeric solid called *paraformaldehyde* whose molecular formula follows:

$$HO -\!\!\!\left[\!\!\left.\rule{0pt}{9pt}\sim\!\!\!\sim CH_2 - O \sim\!\!\!\sim \right]\!\!\right]_{n}\!\!- H$$

The interger n equals 8 to 100. Paraformaldehyde is most commonly used as a bactericide and fungicide, since when heated or exposed to moisture it decomposes, forming gaseous formaldehyde.

Paraformaldehyde is an example of a polymer called a *polyacetal*. These polymers contain repeating units of the type $-O-CH_2-O-CH_2-O-CH_2-$. Another example of a polyacetal is a thermoplastic polymer known by the trade name *Delrin*. Its molecular formula is the following:

$$CH_3-\underset{\underset{O}{\|}}{C}-O\sim\left[-CH_2-O\sim\right]_n-CH_2-O-\underset{\underset{O}{\|}}{C}-CH_3$$

It is used to make door handles, piping, and mechanical and automotive parts.

Formaldehyde also forms a number of copolymers with substances like phenol and urea. These copolymerizations are represented by the following equations:

$HCHO(g) + $ OH(*l*) phenol ⟶

$HCHO(g) + CO(NH_2)_2(l)$ urea   ⟶

The phenol–formaldehyde copolymer is commonly known by the trade name *Bakelite*. It is historically significant in that it was the first thermosetting plastic to be successfully manufactured. Today it is used in products requiring hardness, moisture-resistance, and durability, like binding agents for wood, protective coatings, table-tops, furniture panels, and even grinding wheels.

A urea–formaldehyde copolymer was one of the first of such substances to be made in white, pastel, and colored products. It is widely used in lamp shades, dinnerware, and other decorative household applications. Similar copolymers may be made from formaldehyde and more complex amines, organic compounds containing the —NH_2 functional group; they are commonly called *melamines* and manufactured under the trade name *Melamine*. Melamine products are used as molding compounds with cellulose and mineral powders as fillers and also for laminating, textile treatment, and leather processing. They are also used to manufacture dinnerware and other plastic items.

These formaldehyde copolymers burn at elevated temperatures. When exposed to heat, they typically evolve formaldehyde as a thermal degradation product. Since urea–formaldehyde polymers contain nitrogen atoms, they form the fire gas, nitrogen dioxide (Sec. 9-9), as a combustion product.

A similar group of polymers are the *polyethers;* they contain the recurring unit —CH_2CH_2—O—. Polyethers have the unusual property of being completely miscible with water. An example of a polyether is polyethylene oxide; it is produced by the polymerization of ethylene oxide, as the following equation notes:

ethylene oxide   polyethylene oxide

Polyethylene oxide is used to make water-soluble packaging films.

Finally, recurring carbon-to-oxygen bonds also occur in polymers called *polyesters*. The recurring unit is the following:

$$-O-(CH_2CH_2)_n-O-C-(CH_2CH_2)_n-C-$$

These substances are generally condensation polymers; as noted earlier, they are produced from glycols and dicarboxylic acids.

A common polyester fiber is polyethylene terephthalate. It is also known by its trade names *Dacron* in the United States and *Terylene* in England. Production of polyethylene terephthalate is represented by the equation at the top of page 436. Polyethylene terephthalate is commonly employed in textiles, since it resists wrinkling and fading and may be readily combined with natural fibers like wool. Other than its common use as the fabric from which curtains, drapes, and other home furnishings are made, polyethylene terephthalate is also to make fire hoses.

Polyesters can be very hazardous when employed in textiles, since they readily burn. Some polyesters begin to soften at 256°C (493°F) and drip from one point to

$$\underset{\substack{\text{terephthalic}\\\text{acid}}}{\overset{\text{COOH}}{\underset{\text{COOH}(l)}{\bigcirc}}} + \underset{\substack{\text{ethylene}\\\text{glycol}}}{\overset{\text{OH}}{\underset{\text{OH}(l)}{\overset{\text{H}-\text{C}-\text{H}}{\text{H}-\text{C}-\text{H}}}}} \longrightarrow$$

$$HO\left[\overset{O}{\overset{\|}{C}}-\bigcirc-\overset{O}{\overset{\|}{C}}-O-CH_2-CH_2-O\right]_n H(s)$$

polyethylene terephthalate

another. For this reason, most polyesters intended for use in fabrics are treated with a flame retardant to effectively raise their ignition temperatures.

12-6 NITROGENOUS POLYMERS: POLYAMIDES AND POLYURETHANES

Polyamide polymers are substances characterized by the presence of the amide functional group —CONH. The most common polyamides are members of a family of such polymers called *nylon*. For instance, the polyamide known as nylon-66 is produced from adipic acid and hexamethylene diamine, as the following equation illustrates:

$$\underset{\substack{\text{hexamethylene}\\\text{diamine}}}{\overset{\text{NH}_2}{\underset{\text{NH}_2(l)}{(\text{CH}_2)_6}}} + \underset{\substack{\text{adipic acid}}}{\overset{\text{COOH}}{\underset{\text{COOH}(l)}{(\text{CH}_2)_4}}} \longrightarrow$$

$$HOOC(CH_2)_4-\overset{O}{\overset{\|}{C}}\left[\overset{H}{\overset{|}{N}}-(CH_2)_6-\overset{H}{\overset{|}{N}}-\overset{O}{\overset{\|}{C}}-CH_2CH_2CH_2CH_2-\overset{O}{\overset{\|}{C}}\right]_{\sim60}\overset{H}{\overset{|}{N}}-(CH_2)_6NH_2(s)$$

Most of us think of the use of nylon polymers in hosiery and other wearing apparel, but it is also used to make many other items like tire cord, bristles for toothbrushes, hairbrushes and paint brushes, fish nets and lines, turf for athletic fields, parachutes, tennis rackets, film, and automotive upholstery. Nylon polyamides are often selected for these varied uses because they are the toughest, strongest, and most elastic of the common polymers.

Molded items made of polyamides are generally difficult to ignite. However, nonmolded polyamides ignite readily, especially fabrics. When polyamides burn, nitrogen dioxide (Sec. 9-9) is a product of combustion. Figure 12-8 illustrates that polyamides generally possess the lowest ignition point of the common polymers used in textiles. Nevertheless, when a source of heat, like a lighted cigarette, is placed on

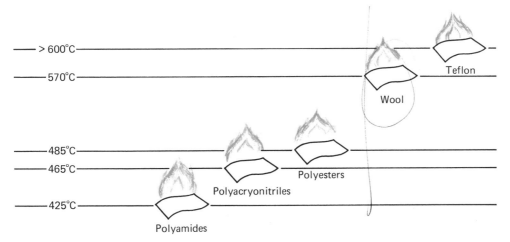

Figure 12-8   Some average ignition temperatures (not to scale). A polymer typically ignites within some range of temperature, which depends in part on the density of the textile. These ignition temperatures can often be achieved for some common materials by the heat evolved from an ordinary flat iron.

a nylon fabric, it generally only burns a hole in the fabric, but does not consume the entire fabric.

A *polyurethane* is a substance produced from a glycol and organic diisocyanate. Simple isocyanates have the general chemical formula R—N=C=O; diisocyanates are compounds having two isocyanate groups (—NCO). The following equation illustrates the production of a polyurethane from tetramethylene glycol and hexamethylene diisocyanate:

$$
\begin{array}{cc}
\text{OH} & \text{N=C=O} \\
| & | \\
\text{(CH}_2\text{)}_4 \quad + \quad \text{(CH}_2\text{)}_6 & \\
| & | \\
\text{OH}(l) & \text{N=C=O}(l)
\end{array} \qquad \longrightarrow
$$

tetramethylene   hexamethylene
glycol   diisocyanate

$$
\left[\text{---O}-(\text{CH}_2)_4-\text{O}-\overset{\overset{\text{O}}{\|}}{\text{C}}-\overset{\overset{\text{H}}{|}}{\text{N}}-(\text{CH}_2)_6-\overset{\overset{\text{H}}{|}}{\text{N}}-\overset{\overset{\text{O}}{\|}}{\text{C}}\text{---} \right](s)
$$

polyurethane

Polyurethanes are available commercially as rigid and flexible plastics. The rigid foam of polyurethane is used in certain forms of insulation, as well as in other structural components. This foam may be conveniently sprayed directly on concrete. The flexible foams of polyurethane are used in bedding and furniture cushioning and in carpet padding. The cushioning in car seats is frequently made of polyurethane.

Untreated polyurethane products burn, although the burning pattern may be altered by special designing, as Fig. 12-9 illustrates. Such products pose a pronounced

Figure 12-9 Foamed rigid plastic materials made from polyurethane are recognized for their insulating quality and light weight. The polyurethane rigid foam in this "Corner Test" at the Factory Mutual Test Center has been specially formulated and protected, so its contribution to the burning is minimal. (Courtesy of Factory Mutual Engineering & Research, Norwood, Massachusetts)

hazard when encountered in fires, since their combustion produces unusually voluminous amounts of nitrogen dioxide (Sec. 9-9). So much nitrogen dioxide is formed that the smoke from polyurethane fires is generally red, denoting the characteristic color of this fire gas. Not all polyurethane products burn with ease. Some have been treated with a fire retardant to make them relatively resistant to combustion. Nonetheless, even these treated polyurethane products burn at the elevated temperatures achieved during a major fire.

The potential fire-related hazard of polyurethance products has led to the issuance of warning labels, particularly on polyurethane insulation. A typical warning is the following:

Fire hazard when used inside buildings. Use only on exterior applications
or when confined between walls of materials considered acceptable when
used in construction.

While fires involving polyurethane may be effectively extinguished with water, it is essential to check for total fire extinguishment. Polyurethane foam retains considerable heat, even when its fire has been physically extinguished. This retained heat often causes polyurethane to again catch fire and burn.

12-7 NATURAL AND SYNTHETIC RUBBER

The term *rubber* refers to any of the natural or synthetic polymers having two main properties: deformation under strain and elastic recovery after vulcanization. Natural rubber is one of our most commercially important polymers. Its principal source is the rubber tree, which is indigenous to Brazil but common to countries of the Far East as well. Natural rubber is present in the sap of the rubber tree, called *latex*. This sap is a white, milky fluid that contains about 30% to 35% by weight of the polymer whose structure is $(C_5H_8)_n$; the remainder is primarily water. This abbreviated formula may be represented as follows:

$$\underset{\text{\tiny wwCH}_2}{\overset{\text{H}}{\diagdown}}C=C\overset{\text{CH}_3}{\underset{\text{CH}_2\text{CH}_2}{\diagup}}\overset{\text{H}}{\diagdown}C=C\overset{\text{CH}_3}{\underset{\text{CH}_2\text{CH}_2}{\diagup}}\overset{\text{H}}{\diagdown}C=C\overset{\text{CH}_3}{\underset{\text{CH}_2\text{ww}}{\diagup}}$$

When strongly heated, natural rubber decomposes into its monomer, *isoprene*, or 2-methyl-1,3-butadiene. The equation that illustrates the thermal degradation of natural rubber is the following:

$$(C_5H_8)_{1500-6000}(l) \longrightarrow \underset{\text{H}}{\overset{\text{H}}{\diagdown}}C=C\underset{\text{CH}_3}{\overset{\text{H}}{-}}C=C\underset{\text{H}(l)}{\overset{\text{H}}{\diagup}}$$

         natural rubber   isoprene

Isoprene is a raw material used for the production of many products, such as those shown in Fig. 12-10.

While isoprene is the principal component of natural rubber, it may also be synthesized from certain petrochemical products. It is a clear, flammable liquid that possesses an especially low flash point, −65°F (−54°C). Since isoprene is likely to autopolymerize, USDOT requires it to be mixed with an inhibitor when transported. The transportation of inhibited isoprene is regulated as a flammable liquid. Isoprene is used to manufacture certain forms of synthetic rubber.

Natural rubber becomes sticky, especially in warm climates. On the other hand, vulcanized rubber retains a firm shape over a relatively wide range of temperature. *Vulcanized rubber* is a natural rubber that has been heated with sulfur. During vulcanization, sulfur atoms bond between chains of polyisoprene, resulting in a structure like the following:

$$\begin{array}{c}\text{wwwC}-\underset{\text{S}}{\overset{\text{CH}_3}{\text{C}}}=\underset{\text{H}}{\overset{\text{H}}{\text{C}}}-\underset{\text{H}}{\overset{\text{H}}{\text{C}}}-\underset{\text{S}}{\overset{\text{H}}{\text{C}}}-\underset{\text{H}}{\overset{\text{CH}_3}{\text{C}}}=\overset{\text{H}}{\text{C}}-\overset{\text{H}}{\text{C}}\text{www}\end{array}$$

With only 3% sulfur by mass, the rubber becomes soft and still elastic, but with ad-

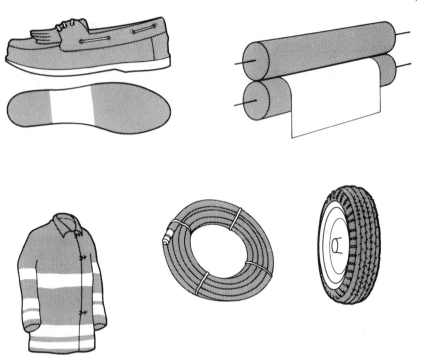

Figure 12-10 Examples of some common products either made directly from rubber or from polymers with rubber additives.

ditional sulfur, it becomes hard and loses its elasticity. When rubber has been vulcanized to the extent of 68% by mass, it becomes a black solid called *ebonite* or *hard rubber*. It is used for making black piano keys and the casings for automobile storage batteries.

Other ingredients besides sulfur are generally added to vulcanized rubber. For instance, carbon may be added to make it stronger, and some amount of old rubber is generally added when producing a new batch of rubber. Depending on the projected use of the rubber, air may be whipped into the latex or ammonium carbonate may be added. When heated, the ammonium carbonate decomposes to ammonia and carbon dioxide, which thereafter become trapped in the complex rubber structure. This product is called *foam rubber*. It is commonly used in cushions and mattresses.

The rubber industry attempted for decades to reproduce the molecular structure of natural rubber from isoprene, but their early attempts failed. The form of rubber that nature selectively produces is the *cis*-isomer of polyisoprene, in which all the methyl groups are oriented in one direction about the carbon–carbon double bonds. While chemists were able to produce a form of polyisoprene, its methyl groups were randomly oriented in all direction, *cis* and *trans*. It was not until the 1950s that rubber chemists were finally successful in developing a synthetic rubber whose molecular structure matched that made by nature. To reproduce the natural form, special organometallic compounds were employed as stereochemical catalysts. However, the synthetic preparation of the counterpart of natural rubber is still a comparatively costly venture.

Other synthetic forms of rubber have also been prepared. The first such form was *Thiokol,* the trademark for a series of polysulfide rubbers having the recurring unit $-S-S-CH_2-CH_2-S-S-$. This substance was produced from 1,2-dichloroethane and sodium polysulfide. An outstanding feature of Thiokol is its impermeability to attack by gasoline. Thiokol products were commonly used throughout World War II when it was difficult for the United States to obtain natural rubber from the Far East. However, production costs for Thiokol have never competed favorably with natural rubber from foreign sources. Hence, it is no longer commonly encountered.

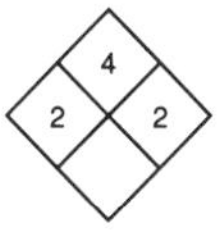

The first commercially valuable synthetic rubber was *neoprene.* This substance may be produced by polymerizing *chloroprene,* or 2-chloro-1,3-butadiene, a process accomplished simply by heating. The polymerization of chloroprene is illustrated by the following equation:

$$CH_2=CH-\underset{\underset{Cl}{|}}{C}=CH_2(g) \longrightarrow \left(\sim\!CH_2-\underset{\underset{Cl}{|}}{C}=C-CH-CH_2\!\sim\!\right)_{\!n}(s)$$

$$\text{chloroprene} \qquad\qquad\qquad \text{neoprene}$$

When vulcanized with zinc oxide, single strands of neoprene cross-link with oxygen atoms, replacing the chlorine atoms. Vulcanized neoprene has the following structure:

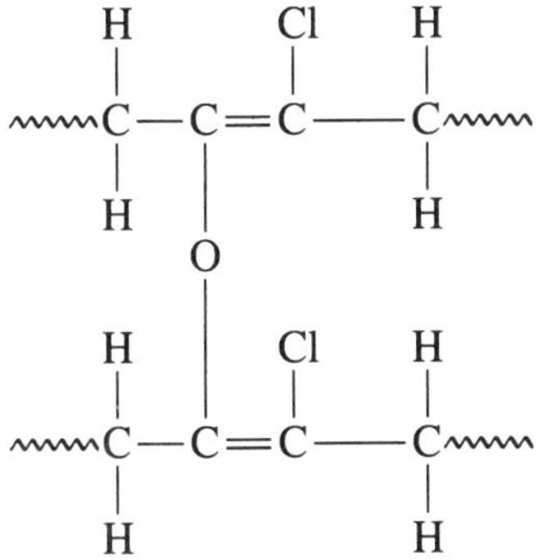

$$
\begin{array}{ccccc}
& H & Cl & H & \\
& | & | & | & \\
\sim\!\!\!\sim\!\!\!\sim C & - C & = C & - C & \sim\!\!\!\sim\!\!\!\sim \\
& | & | & | & \\
& H & | & H & \\
& & O & & \\
& | & & & \\
& H & Cl & H & \\
& | & | & | & \\
\sim\!\!\!\sim\!\!\!\sim C & - C & = C & - C & \sim\!\!\!\sim\!\!\!\sim \\
& | & & | & \\
& H & & H &
\end{array}
$$

Like Thiokol, neoprene rubber is not chemically attacked by gasoline. It is even used in gasoline hoses at filling stations.

Other attempts have been made to produce synthetic rubber economically. Of all such attempts, none begins to compete with the production of a butadiene–styrene copolymer known commercially as either *GRS* (government rubber styrene) or *SBR* (styrene butadiene rubber). The associated production is illustrated by the following equation:

$$CH_2{=}CH{-}CH{=}CH_2(g) \; + \quad CH{=}CH_2(g) \quad \longrightarrow$$

1,3-butadiene   styrene

$$\sim\!\!\sim\!\!\sim CH_2{-}CH{=}CH{-}CH_2CH_2{-}CH{-}\overset{\displaystyle CH_2\!\sim\!\!\sim}{\underset{\displaystyle \bigcirc}{CH}}CH{=}CH_2(s)$$

(GRS or SBR)

GRS is vulcanized using elemental sulfur.

GRS resists wear more than any other synthetic rubber. Consequently, it is used in lieu of natural rubber for a variety of purposes, mostly associated with the production of tires for motor vehicles.

To understand the combustion of different rubbers, it is important to examine the nature of their component molecules. The combustion of untreated natural rubber, for instance, produces only carbon monoxide, carbon dioxide, and water vapor. The combustion of vulcanized rubber, on the other hand, produces each of these gases as well as the toxic fire gas sulfur dioxide. The suffocating odor of sulfur dioxide is almost always detected when stockpiles of tires burn. These fires also generally yield a particularly dense smoke.

Since the molecular structure of neoprene is comprised of chlorine-containing units, its combustion products include the toxic gas hydrogen chloride. Furthermore, since it has been vulcanized with metallic oxides, particulates of compounds of these metals are likely to be present in the smoke accompanying the combustion.

The atmosphere near burning rubber is almost always likely to consist of a mixture of toxic gases. Consequently, fires involving rubber products should always be attacked from the upwind side or at right angles. When exposure to the smoke and fumes from such fires is essential, the use of positive-demand, self-contained breathing apparatus is essential to prevent fatalities.

REVIEW EXERCISES

General Features of Polymerization

12.1. Draw the partial structure of the polymer macromolecule that results from the successive addition of four recurring styrene units.

12.2. An egg is a mixture of various proteins, certain complex polymers essential for the life process. Does a cooked egg represent a mixture of thermoplastic or thermosetting polymers?

12.3. Camphor has long been used to increase the hardness of certain plastics. Speculate on the manner by which camphor contributes hardness to plastics.

12.4. When transporting ethyl acrylate, USDOT require the use of the following proper shipping name on papers accompanying the shipment: "Ethyl acrylate, inhibited." What does the term inhibited imply about ethyl acrylate?

12.5. Illustrate by means of a chemical equation how the addition polymer polyvinylidene chloride is produced from its monomer.

12.6. Write the equation that illustrates the condensation of two molecules of 1,6-hexamethylene glycol and two molecules of adipic acid.

General Features Concerning the Burning of Synthetic Polymers

12.7. When a new car is parked in the sun for some time with the windows closed, it acquires the odor of "plastic."
  (a) Speculate on the general nature of the substances that give rise to this odor.
  (b) How does this incident resemble that of a polymeric product that has been exposed to heat?

12.8. Identify the fires gases that are likely to form during the high-temperature combustion of the following polymers:
  (a) Polypropylene, $(\!\sim\!\!C_3H_8\!\sim\!)_n$
  (b) Polyvinylidene chloride, $(\!\sim\!\!CH_2CCl_2\!\sim\!)_n$
  (c) Polytetrafluoroethylene, $(\!\sim\!\!CF_2CF_2\!\sim\!)_n$, commonly known as *Teflon*

12.9. Some polymeric products are manufactured in an *expanded* foam. This is accomplished by blowing a solvent into the polymer (thus causing its expansion), molding it into desirable shapes, and allowing the solvent to evaporate. An example is an item made of *Styrofoam;* such items have been manufactured from expanded polystyrene. Why do such expanded products burn more rapidly than forms of the same polymer that have not been expanded?

12.10. What substance in the smoke of many plastic fires is most likely responsible for causing the eyes to tear?

12.11. Pound for pound, which substance yields more carbon monoxide when it burns: a polymer or the monomer from which it was made?

Cellulose and Its Derivatives

12.12. Cotton and linen are primarily composed of the same substance, cellulose. However, when ironing garments made from these fabrics, a very hot iron must be used to remove wrinkles from linen fabrics, although a moderately hot iron is normally adequate for removing wrinkles from cotton. Why are these different heat strengths required to remove wrinkles from fabrics that are chemically composed of the same substance?

12.13. Given similar surface areas, which ignites easier and burns more readily: fabrics made from vegetable fibers or from animal fibers?

Polyvinyl Polymers

12.14. Why do fire marshals often discourage the use of decorating items, shower curtains, and upholstery fabrics made for use in a high-rise public building from either polyacrylonitrile or polyvinyl chloride?

12.15. Acrilan is an acrylic fiber produced as a copolymer from acrylonitrile and methyl acrylate. It may be used in woven and knitted clothing fabrics, carpets, drapes, upholstery, and electrical insulation.
   (a) Write an equation illustrating the formation of Acrilan from its monomers.
   (b) Identify the most likely chemical nature of the mixture of gases formed when Acrilan products are exposed to an intense source of heat.
   (c) Identify the chemical nature of the gases that comprise the mixture of combustion products generated when Acrilan burns.

12.16. A copolymer known by the tradename *Dynel* is produced from vinyl chloride and acrylonitrile. Dynel is used in textiles.
   (a) Write an equation illustrating the formation of Dynel from its monomers.
   (b) Identify the most likely chemical nature of the mixture of gases formed when Dynel products are exposed to an intense source of heat.
   (c) Identify the chemical nature of the gases that comprise the mixture of combustion products generated when Dynel burns.

Polyacetals, Polyethers and Polyesters

12.17. Paraformaldehyde was formerly used as a fungicide in mushroom houses prior to planting. Why is the use of paraformaldehyde for this purpose now discouraged?

12.18. Polyesters are often blended with cotton to produce a fabric that is a blend of both polymers. What is the principal fire-related benefit achieved from blending polyesters with cotton?

Polyamides and Polyurethanes

12.19. Like other organic polymers, nylon burns when ignited. However, burning nylon is not ordinarily considered as hazardous as the burning of many other nontreated polymeric products. Why is this so?

12.20. Arriving at the scene of a fire in a warehouse, a fire chief notes that the smoke issuing from the fire is red with broad streaks of black.
   (a) Which fire gas is probably present in the smoke?
   (b) Which polymeric product in the warehouse is likely to give rise to this combustion gas?

(c) What special directions should be provided to firefighters who must attack this fire?

Natural and Synthetic Rubber

12.21. Why is the burning of untreated natural rubber associated with virtually no detectable odor, whereas the burning of vulcanized rubber is associated with a suffocating odor?

12.22. Why does vulcanization cause a polymer to be simultaneously elastic and "rubbery"?

12.23. Which is likely to burn at a faster rate: foam rubber or sheet rubber textiles?

13

Chemical Explosives

A *chemical explosive* is a substance that detonates spontaneously or as the result of friction, mechanical impact, or heat. It is ordinarily perceived as a substance whose primary intended purpose is to accomplish some act, like demolition, when it detonates. In this latter regard, chemical explosives are distinguished from certain other substances, like gasoline or flammable gases, which when confined in containers and ignited are sometimes said to explode. They are also distinguished from nuclear explosives, which detonate because of the occurrence of certain nuclear phenomena.

Chemical explosives have been utilized for centuries in two main ways. During wartime, they are used in artillery and other weapons as ammunition to destroy cities, sink ships, and kill the enemy. In peacetime, they help to mine ore, drill oil wells, tunnel through mountains, clear land, and loosen underground rock formations; they are also components of ammunition used for sporting and other purposes. Chemical explosives also constitute a tool for fighting unusually large, uncontrolled fires in certain areas, such as in oil fields, prairies, forests, or segments of a metropolitan area. In these situations, an explosive detonation serves as the mechanism for creating a firebreak and thus preventing further spread of fire.

Fortunately, most individuals appreciate the degree of danger associated with chemical explosives. Hence, explosives are typically stored in special areas under lock and key and are rarely encountered in the ordinary disasters to which firefighters are called. Nevertheless, chemical explosives may be encountered, however rarely, when responding to fire-related incidents, including transportation mishaps.

Explosives experts maintain that the majority of accidents involving explosives could have been prevented. However, such accidents can be prevented only when safe procedures have been implemented for transporting, storing, and handling explosives. Special training in their use should always be undertaken by individuals who intend to employ a chemical explosive for some purpose. Such specialized training is beyond the scope of this chapter.

A chemical explosive is a highly reactive substance. When not intended for future use, it should be disposed of. Under RCRA (Sec. 1-5), an explosive must be treated, stored, and disposed of as a hazardous waste in accordance with regulatory procedures. We have previously observed several ways by which a hazardous waste may exhibit the characteristic of reactivity (see introduction to Chapters 8 and 9). Reactivity may also be exhibited by properties that relate to a waste's potential explosive nature. This occurs when a representative sample of the waste has any of the following properties:

1. It is normally unstable and readily undergoes violent change without detonating.
2. It is capable of detonation or explosive reaction when subjected to a strong initiating source or when heated under confinement.
3. It is readily capable of detonation or explosive decomposition or reaction at standard temperature and pressure.
4. It is a forbidden explosive (Sec. 13-3) or a division 1.1, 1.2, or 1.3 explosive (Sec. 13-3).

13-1 CHARACTERISTICS AND CLASSIFICATION OF EXPLOSIVES

When an explosive detonates, it *suddenly* undergoes a very rapid chemical transformation into gases and vapors of relatively simple substances, with the simultaneous production of unusually large amounts of energy. Some common simple substances into which an explosive ordinarily transforms are carbon monoxide, carbon dioxide, nitrogen, oxygen, and water vapor. These substances absorb some of the energy that simultaneously evolves during explosive transformations. This causes them to be rapidly heated from the prevailing ambient temperature to a highly elevated temperature, typically 5430°F (3000°C). The increase in temperature causes them to simultaneously expand so rapidly that they exert an exceedingly high pressure on the surroundings.

Let's consider the explosive substance nitroglycerin (Sec. 13-5). When it explodes, nitroglycerin is transformed into a mixture of carbon dioxide, nitrogen, oxygen, and water vapor. One gram of nitroglycerin is completely decomposed in less than one-millionth of a second. The phenomenon is represented by the following equation:

$$\begin{array}{c} CH_2ONO_2 \\ | \\ 4CHONO_2 \longrightarrow 12CO_2(g) + 10H_2O(g) + 6N_2(g) + O_2(g) \\ | \\ CH_2ONO_2(l) \end{array}$$

Note that atmospheric oxygen is unnecessary for this detonation. On the other hand, oxygen atoms are typically present in the molecules of most chemical explosives.

The detonation of a chemical explosive is an example of an oxidation–reduction reaction (Sec. 10-2) in which the explosive substance is both the oxidizing agent and the reducing agent.

At the elevated temperature accompanying a detonation, the products of the explosive reaction expand to occupy nearly 10,000 times their original volume. The simultaneous increase in pressure accompanying the expansion causes the resulting physical effect of an explosion, like that noted in Fig. 13-1. Generally, this expansion propagates at rates that exceed the speed of sound, often resulting in the development of shock waves and an unusually loud noise. These shock waves cause the explosive's shattering power, called its brisance. The brisance is an important factor when selecting a chemical explosive for a specific purpose, as clearing rock for a roadway.

Figure 13-1 The detonation of a chemical explosive may be initiated in several ways, such as by heat or mechanical impact. This detonation was induced by exposing the material to pulsed, high-energy x rays. The x-ray machine is located in the domed building and is part of the Dynamic Testing Division at Los Alamos National Laboratory, Los Alamos, New Mexico. The machine is used to study explosions and the nature of explosive materials. (Courtesy of the U.S. Department of Energy, Washington, D.C.)

Some commercially available chemical explosives are actually mixtures of substances that include an oxidizer, like ammonium nitrate (Sec. 10-6). These mixtures of substances often detonate by means of a combination of phenomena; but typically, the oxidizer thermally decomposes, thus providing an additional supply of oxygen to the explosive.

Individual chemical explosives may be classified into either of the following two groups, which relate to the rapidity and sensitivity with which they detonate: *high,* or *detonating*; and *low,* or *deflagrating.* There are no clear demarcations in rate of speed or shock sensitivity between high and low explosives; instead, these terms are useful only in a relative sense. High explosives undergo chemical transformations quite rapidly, whereas low explosives transform hundreds of times more slowly and typically with less ease. The rate of detonation may be as great as 4 miles/s (6000 m/s) for a high explosive, but only 900 ft/s (270 m/s) for a low explosive. On the other hand, detonating some substances that are considered high explosives requires the use of a blasting cap or similar activating device. This classification of explosives as high or low is not entirely satisfactory, since some substances may be categorized into either class, depending on their state of subdivision. Smokeless powder, for instance, has been typically considered a low explosive; but under proper conditions, its detonation resembles that of a high explosive.

Chemical explosives may also be classified as either primary or secondary explosives. *Primary explosives* are unstable substances that are extremely sensitive to heat, mechanical shock, and friction; hence, they develop shock waves in an extremely short time period. Typical primary explosives are lead azide and mercury fulminate (Sec. 13-13). By comparison, *secondary* explosives are unstable substances that are relatively insensitive to heat, mechanical shock, and friction. They typically require the use of a booster to bring about detonation. Typical secondary explosives are cyclonite (Sec. 13-9), tetryl (Sec. 13-10), and PETN (Sec. 13-11).

Low explosives should not be associated with the potential for a lesser degree of hazard; neither should the insensitivity associated with secondary explosives be taken as a sign of stability. In fact, low explosives often *deflagrate;* that is, they burn furiously and persistently. However, low explosives *may* explode, causing considerable damage at the explosion site. Furthermore, both primary and secondary explosives possess considerable brisance.

From a chemical viewpoint, the unstable nature of an explosive may often be understood by examining the oxidative potential of its atoms. Many explosives contain one or more pyrotechnically active groups of atoms within their structures. These groups of atoms are called *explosophores*. The presence of explosophores within the structural framework of a substance is likely to give it an explosive potential. Many organic compounds are innately unstable when they contain any explosophore whatsoever. In certain instances, *all* organic compounds containing a certain explosophore are unstable. For instance, organic azides are compounds containing the azide group, —N_3; all organic azides are potentially explosive.

Why are some substances more prone to explode than others? Examination of the molecular structure of explosives illustrates that nitrogen atoms are frequently components of explosives. In fact, the most commonly encountered explosophore in commercial explosives is the *nitro* group, —NO₂. When multiple nitro groups are present in a single molecule of an organic compound, the substance is often chemically unstable. This is exemplified by such commercial explosives as nitroglycerin (Sec. 13-5), nitrocellulose (Sec. 13-7), trinitrotoluene (Sec. 13-8), PETN (Sec. 13-11), and picric acid (Sec. 13-12).

Artillery Ammunition

The sensitivity of individual chemical explosives may be greatly appreciated by examining the particular function each component of a round of artillery ammunition plays, as the round is activated. A typical round is illustrated in Fig. 13-2. Its components are classified into six parts, each of which plays a unique burning or explosive function with a different order of magnitude.

1. Primer: In a round of artillery ammunition, the primer is generally a mixture of an oxidizer, like potassium chlorate, and some other arbitrary substance that readily burns. This material is usually confined in a cap or tube at one end of the round and is activated by a percussion action, as from a firing pin.

2. Igniter: In a round of ammunition, black powder (Sec. 13-4) almost always plays the role of the igniter. The oxidation–reduction reaction between the components of black powder is initiated by the heat evolved from the burning of the primer.

3. Propellant: This is a material like smokeless powder [nitrocellulose (Sec. 13-7)] that deflagrates and forms a relatively large volume of gases and vapors. In a round of artillery ammunition, the quantity is intentionally limited to prevent a major explosion, but sufficient to generate an adequate volume of vapors and gases to propel the projectile forward. The explosive that plays the role of the propellant is physically separated by means of a thin wall from the other chemical explosives in the round.

4. Detonator: This component of the round explodes from the mechanical shock

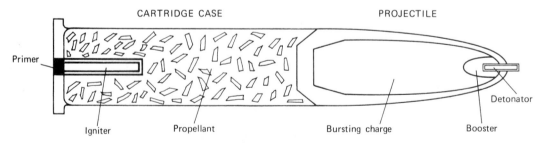

Figure 13-2   A typical round of artillery ammunition, illustrating its component parts. Each part plays a specific role in the proper functioning of the round. Chemical explosives are sometimes identified by the function that they perform in such a round (for example, primer or booster).

received when the propelling round encounters its target. In a round of artillery, it may be present in only a limited quantity.

5. Booster: This component of the round is induced to explode from the heat generated by the explosion of the detonator. As with the detonator, the quantity of the booster may be limited to that which generates a prescribed brisance. The function of the booster is to intensify the brisance resulting from explosion of the detonator.

6. Bursting charge: This is typically a high explosive that detonates from the shock resulting from explosion of the booster. It is the main charge and is present in a round of artillery ammunition in sufficient quantity to produce the brisance that causes the intended demolition.

The chain of events from initiation of the primer to detonation of the bursting charge is called an *explosive train.* Each component of an explosive train must properly actuate to result in the effectiveness of the entire round.

All other types of ammunition may be regarded as modifications of the round of artillery ammunition. A rifle cartridge, for instance, contains only a primer and propellant, or a primer, igniter, and propellant; the projectile is generally a mass of metal, such as lead or steel pellets. An ordinary bomb is also a modification of a round of artillery ammunition, but generally lacks a propellant, since the detonator explodes from the force achieved by dropping the bomb from a high altitude.

When explosives are used for nonmilitary blasting purposes, the propellant is also ordinarily unnecessary. In these instances, a blasting cap holds a prescribed amount of a high explosive, which is generally detonated by means of a booster. The assembly containing the bursting charge and booster is connected to several hundred yards of electric wire and is activated by a plunger-type magneto apparatus from a safe distance.

13-2 ENCOUNTERING CHEMICAL EXPLOSIVES WHEN FIREFIGHTING

Federal, state, and local ordinances and regulations require chemical explosives to be stored in a specially constructed area called a *magazine,* such as that shown in Fig. 13-3. OSHA regulates the manner by which magazines are to be constructed and maintained, as well as the actual manner by which explosive materials are stored within them.

As a general rule, high explosives should not be stored with blasting caps, fuses, primers, and detonators and should only be stored in the minimum quantities needed for their intended purpose within a defined time frame. Whenever possible, magazines should be separated by hundreds of yards from other nearby buildings or by whatever distance is physically feasible. These storage recommendations concur with the sound precautionary measure that advises explosive users to take appropriate measures to prevent explosives from being directly involved in fires. While a fire could cause an explosive to burn, the heat of the fire could also serve as the initiating mechanism by which it detonates.

Figure 13-3 Explosives are stored in specially designed rooms or bunkers called *magazines*. The magazine noted in the foreground of this photograph is a bunker of masonry construction. All components are bullet, weather, and fire resistant. The entire roof is covered with sand, except structures that are required for interior ventilation. Inside the magazine, packages of explosives are stored flat with the top sides up and in stable configurations. The interior is ventilated to prevent dampness and heating of stored explosives. The land surrounding the magazine is kept clear of all combustible materials for a distance of at least 25 ft (7.6 m). (Courtesy of the Los Alamos National Laboratory, Los Alamos, New Mexico)

As a general rule, fires likely to involve explosives should not be actually fought, although every effort should be taken to assure that fire does not reach magazines. An actual explosion in a typical situation could demolish an area as large as a city block and adversely affect individuals even farther away. Thus, to limit the number of possible fatalities, an area extending to at least 2000 ft (610 m) from the fire should be cleared of all spectators and fire personnel.

When explosives are involved in a transportation incident, they should not be removed from their transportation vehicle until an explosives expert can supervise their removal. Figure 13-4 illustrates a truck-emergency card, which provides instructions on recommended procedures in the event of a truck fire involving explosives. When it becomes essential to destroy explosives, assistance is generally sup-

DON'T FIGHT EXPLOSIVES FIRES!

(But you _can_ prevent fire from reaching the explosives... and save lives)

1 IDENTIFY THE CARGO!
Look for these signs on the truck or trailer

| EXPLOSIVES A | OXIDIZERS |
| EXPLOSIVES B | |

2 ACT FAST!
Take the right action ... use proper extinguishers on tires, engine, cab or body

CARGO FIRE
- STOP ALL TRAFFIC AND CLEAR THE AREA FOR 2000 FEET IN ALL DIRECTIONS.
- DON'T FIGHT FIRE! (CARGO MAY EXPLODE).
- WHEN TRACTOR-TRAILER IS INVOLVED, SEPARATE TRACTOR FROM TRAILER IF POSSIBLE.

TIRE FIRE
- WHEN TRACTOR-TRAILER IS INVOLVED, SEPARATE TRACTOR FROM TRAILER IF POSSIBLE.
- USE PLENTY OF WATER—DOUSE IT. IF WATER IS NOT AVAILABLE USE DRY CHEMICAL FIRE EXTINGUISHER OR DIRT.
- CAUTION: FIRE MAY START AGAIN. STAND BY WITH EXTINGUISHER READY.
- CONSERVE DRY CHEMICAL—USE IN SHORT BURSTS.
- GET TIRE OFF AND AWAY FROM VEHICLE.

ENGINE OR CAB FIRE
- WHEN TRACTOR-TRAILER IS INVOLVED, SEPARATE TRACTOR FROM TRAILER IF POSSIBLE.
- USE DRY CHEMICAL FIRE EXTINGUISHER, WATER OR FOAM.
- DISCONNECT ONE BATTERY CABLE.

BODY FIRE
- CLEAR AREA BEFORE FIRE REACHES CARGO.
- WHEN TRACTOR-TRAILER IS INVOLVED, SEPARATE TRACTOR FROM TRAILER IF POSSIBLE.
- USE DRY CHEMICAL EXTINGUISHER, WATER OR FOAM.
- DO NOT FIGHT FIRE WHEN IT REACHES CARGO.

Figure 13-4 A safety message to law enforcement and fire protection personnel. EXPLOSIVE A and EXPLOSIVE B refer to two classes of chemical explosives formerly designated by the USDOT. (Courtesy of the Institute of Makers of Explosives, Washington, D.C.)

plied by explosives manufacturers to fire departments, law enforcement agencies, inspection and regulatory bodies, and the users of the explosives.

13-3 CHEMICAL EXPLOSIVES AND USDOT REGULATIONS

As first noted in Sec. 5-2, USDOT regulates the transportation of chemical explosives and articles containing chemical explosives. For regulatory purposes, USDOT recognizes five divisions of class 1 (explosives): 1.1, 1.2, 1.3, 1.4 and 1.5. These divisions are defined as follows:

Division 1.1: Substances and articles that have a mass explosion hazard.

Division 1.2: Substances and articles that have a projection hazard, but not a mass explosion hazard.

Division 1.3: Substances and articles that have a fire hazard and either a minor blast hazard or a minor projection hazard or both, but not a mass explosion hazard. This division comprises explosive articles and substances that give rise to considerable radiant heat or that burn one after another, producing minor blast or projection effects, or both.

Division 1.4: Substances and articles that present no significant hazard.

Division 1.5: Very insensitive substances that have a mass explosion hazard.

For transportation purposes, shippers of explosives must also evaluate the *compatibility group* of a given explosive substance or explosive article. There are 12 compatibility groups, represented by one of the following capital letters: A, B, C, D, E, F, G, H, J, K, L, or S. Assignment of the compatibility group is based on the nature of the explosive substance or article according to the appropriate classification provided in Table 13-1. When classifying an explosive substance or article for transportation, shippers readily identify its compatibility group by reviewing the entries in column 3 of the Hazardous Materials Table. USDOT requires the appropriate compatibility group to be displayed on the EXPLOSIVE 1.1, 1.2, 1.3, 1.4, or 1.5 label that is affixed to the packaging (see, for example, Fig. 5-10).

For certain chemical explosives and explosive articles, the word "forbidden" appears in column 3 of the Hazardous Materials Table. As with other hazard classes, forbidden means that USDOT prohibits the particular explosive from being offered or accepted for transportation. These latter explosives, called *forbidden explosives,* are identified in Table 13-2.

Finally, special markings are displayed on the packaging used to ship chemical explosives and explosive articles. USDOT requires packages to be marked with the net quantity of material and its proper shipping name, supplemented by additional descriptive text to indicate the material's commercial or military name. Table 13-3 provides some examples of such names. In addition, shippers mark packages containing chemical explosives and explosive articles with specific precautionary statements. For instance, when detonating fuses are transported, their packages are ap-

TABLE 13-1 COMPATIBILITY GROUP ASSIGNMENT

Explosive article or substance	Division	Compatibility group
Primary explosive substance	1.1	A
Article containing a primary explosive substance and not containing two or more effective protective features	1.1, 1.2, 1.4	B
Propellant explosive substance or other deflagrating explosive substance or article containing such explosive substance	1.1, 1.2, 1.3, 1.4	C
Secondary detonating explosive substance or black powder or article containing a secondary detonating explosive substance, in each case without means of initiation and without a propelling charge or article containing a primary explosive substance and containing two or more effective protective features	1.1, 1.2, 1.4, 1.5	D
Article containing a secondary detonating explosive substance, without means of initiation, with a propelling charge (other than one containing a flammable or hypergolic liquid)	1.1, 1.2, 1.4	E
Article containing a secondary detonating explosive substance, with its own means of initiation, with a propelling charge (other than one containing a flammable or hypergolic liquid) or without a propelling charge	1.1, 1.2, 1.3, 1.4	F
Pyrotechnic substance, or article containing a pyrotechnic substance, or article containing both an explosive substance and an illuminating, incendiary, lachrymatory, or smoke-producing substance (other than a water-activated article or one containing white phosphorus, phosphide, or flammable liquid or gel)	1.1, 1.2, 1.3, 1.4	G
Article containing both an explosive substance and white phosphorus	1.2, 1.3	H
Article containing both an explosive substance and a flammable liquid or gel	1.1, 1.2, 1.3	J
Article containing both an explosive substance and a toxic chemical agent	1.1, 1.3	K
Explosive article or substance containing an explosive substance and presenting a special risk needing isolation of each type	1.1, 1.2, 1.3	L
Article or substance so packed or designed that any hazardous effects arising from accidental functioning are confined within the package, unless the package has been degraded by fire, in which case all blast or projection effects are limited to the extent that they do not significantly hinder or prohibit firefighting or other emergency response efforts in the immediate vicinity of the package	1.4	S

propriately marked as follows: HANDLE CAREFULLY—DO NOT STORE OR LOAD WITH ANY HIGH EXPLOSIVES. This combination of markings conveys unique warning messages to firefighters and others who encounter these packages when responding to emergency incidents involving explosives.

TABLE 13-2 FORBIDDEN EXPLOSIVES

Explosive compounds, mixtures, or devices that ignite spontaneously or undergo marked
   decomposition when subjected to a temperature of 167°F (75°C) for 48 consecutive hours.

New explosive compounds, mixtures, or devices, except as otherwise provided.

Explosive mixtures or devices containing an ammonium salt and a metallic chlorate.

Explosive mixtures or devices containing an acidic metal salt and a metallic chlorate.

Leaking or damaged packages of explosives.

Nitroglycerin, diethylene glycol dinitrate, or other liquid explosives not otherwise authorized.

Loaded firearms.

Fireworks that combine an explosive and a detonator or blasting cap.

Fireworks containing yellow or white phosphorus.

Toy torpedoes, the maximum outside dimension of which exceeds 7/8 in., or toy torpedoes
   containing a mixture of potassium chlorate, black antimony, and sulfur with an average weight
   of explosive composition in each torpedo exceeding four grains.

13-4 BLACK POWDER

The explosive known as *black powder* is an intimate mixture of charcoal, sulfur and
either potassium or sodium nitrate. It is a low explosive, sometimes employed as a
blasting agent; it is also used as a component of fireworks and certain forms of ammunition. A variation of black powder, called *black gunpowder,* is a mixture of 15
parts charcoal, 10 parts sulfur, and 75 parts potassium nitrate by mass.

When initiated, black powder is generally said to "explode." Actually, when
black powder explodes, its components undergo the oxidation–reduction reaction illustrated by the following equation:

$$32KNO_3(s) + 3S_8(s) + 16C(s) \longrightarrow 16K_2CO_3(s) + 16N_2(g) + 24SO_2(g)$$

In other words, black powder does not truly detonate: It deflagrates. Nevertheless,
the phenomenon occurs so furiously that it often resembles an explosion. Furthermore, black powder is such a reactive mixture of substances that it should always be
stored, transported, handled, and used as if it actually were an explosive.

USDOT regulates the transportation of black powder as an explosive material.
Packages of granular and compressed black powder are labeled EXPLOSIVE 1.1D;
their transport vehicles are placarded EXPLOSIVES 1.1D.

13-5 NITROGLYCERIN

Pure nitroglycerin is a heavy, oily liquid whose appearance resembles water. The
commercial product, however, is generally pale yellow and viscous. Some other
physical properties of nitroglycerin are noted in Table 13-4.

Chemically, the proper name of glycerin is glyceryl trinitrate. It is an ester, not
a nitro-organic compound. It is prepared by dropping glycerol, a trihydroxy alcohol,

TABLE 13-3 EXAMPLES OF COMMERCIAL OR MILITARY NAMES OF EXPLOSIVE
ARTICLES

Actuating cartridges, explosive, fire extinguisher	Explosive mine
Actuating cartridges, explosive, valve	Explosive projectile
Ammunition for cannon with empty projectiles	Explosive release devices
Ammunition for cannon with explosive projectiles	Explosive rivets
	Explosive torpedo
Ammunition for cannon with gas projectiles	Flexible linear-shaped charges, metal clad
Ammunition for cannon with incendiary projectiles	Hand signal devices
Ammunition for cannon with inert-loaded projectiles	Igniter cord
	Initiating explosive
Ammunition for cannon with smoke projectiles	Instantaneous fuse
	Low blasting explosive
Ammunition for cannon with solid projectiles	Mild detonating fuse, metal clad
Ammunition for cannon with tear gas projectiles	Propellant explosives
Ammunition for cannon without projectiles	Rifle grenades
	Rocket ammunition with explosive projectiles
Ammunition for small arms with explosive projectiles	Rocket ammunition with gas projectiles
Ammunition for small arms with incendiary projectiles	Rocket ammunition with incendiary projectiles
	Rocket ammunition with illuminating projectiles
Black blasting powder	Rocket ammunition with smoke projectiles
Black pellet powder	Rocket motors
Black powder	Safety fuse
Black rifle powder	Signal flares
Boosters (explosive)	Small-arms ammunition
Bursters (explosive)	Smoke candles
Common fireworks	Smoke grenades
Cordeau detonating fuse	Smoke pots
Detonating primers	Smoke signals
	Special fireworks
Explosive bomb	Supplementary charges (explosive)
Explosive cable cutters	Toy caps

into a cooled mixture of nitric acid and sulfuric acid. The synthesis is represented by
the following equation:

$$\begin{array}{ccc}
CH_2OH & & CH_2O-NO_2 \\
| & & | \\
CHOH & + \ 3HNO_3(l) \ \longrightarrow & CHO-NO_2 \quad + \ 3H_2O(l) \\
| & & | \\
CH_2OH(l) & & CHO-NO_2(l)
\end{array}$$

The sulfuric acid participates catalytically.

Nitroglycerin is extremely sensitive to shock. It is a high explosive. When nitroglycerin explodes, the resulting brisance is approximately 3 times that of an
equivalent amount of gunpowder and proceeds roughly 25 times faster. A slight jar-
ring, dropping, or throwing against a hard surface is certain to make nitroglycerin

TABLE 13-4 SOME PHYSICAL PROPERTIES OF NITROGLYCERIN

Density	1.59 g/mL
Melting point	55°F (13.2°C)
Detonation velocity	7.8 km/s
Sensitivity	Very high (almost a primary explosive)

detonate. Due to its high degree of sensitivity, it is rarely used directly as an explosive. When it explodes, nitroglycerin decomposes into carbon dioxide, water vapor, nitrogen, and oxygen, as noted in Sec. 13-1.

When heated to approximately 350°F (177°C), nitroglycerin spontaneously explodes. The —NO_2 explosophores chemically react with the remainder of each nitroglycerin molecule. Self-detonation may be initiated in a second way. Upon exposure to atmospheric moisture, nitroglycerin tends to hydrolyze, forming a mixture of glycerol and nitric acid. This mixture is particularly susceptible to spontaneous decomposition. Thus, repeated exposure of nitroglycerin to the atmosphere is likely to produce a highly treacherous mixture that may explode at the slightest provocation.

Nitroglycerin is also toxic by ingestion, inhalation, and skin absorption. It causes the dilation of small veins, capillaries and coronary blood vessels, which leads to severe headaches, flushing of the face, and a drop in blood pressure. However, this property of nitroglycerin may also be used medicinally for treatment of heart and certain blood-circulation diseases. For this latter purpose, nitroglycerin is dissolved in a solvent like alcohol or acetone and is called *spirits of nitroglycerin*. These spirits are also used to produce nitroglycerin tablets.

Table 13-1 indicates that nitroglycerin is a forbidden explosive. However, when desensitized, USDOT regulates the transportation of liquid nitroglycerin as a division 1.1D explosive. Packages of desensitized nitroglycerin are labeled EXPLOSIVE 1.1D; and their transport vehicles are placarded EXPLOSIVES 1.1D. USDOT regulates the transportation of spirits of nitroglycerin containing from 1% to 10% by mass nitroglycerin in alcohol as a flammable liquid. Containers of spirits of nitroglycerin are labeled FLAMMABLE LIQUID, and their transport vehicles are placarded FLAMMABLE.

13-6 DYNAMITE

In 1867, a Swedish engineer, Alfred Nobel, discovered that nitroglycerin could be absorbed into a porous material, such as siliceous earth. The resulting material was found to be much safer to handle than liquid nitroglycerin; it became known as *dynamite*. Nobel acquired a fortune from this discovery and the subsequent manufacture of dynamite, which he used in part to establish a fund for the world-famous Nobel prizes.

Dynamite is not only safer to handle than nitroglycerin, it may also be transported and used with less hazard of spontaneous decomposition. In fact, dynamite is so safe to handle that a detonating cap is required to explode it. Notwithstanding this fact, dynamite is a high explosive and can be sensitive to heat, shock, and friction.

In contemporary times, dynamite is produced by absorbing nitroglycerin in a mixture of wood pulps, sawdust, flour, starch, and similar carbonaceous materials. Calcium carbonate is often added as an antacid to neutralize the nitric acid that forms by spontaneous decomposition. Ethylene glycol dinitrate is usually added as an antifreeze, and oxidizers are also normally added. This mixture of substances is then packed into cylindrical cartridges made of waxed paper, which vary in size from $\frac{7}{8}$ to 8 in. (2 to 20 cm) in diameter and from 4 to 30 in. (10 to 76 cm) in length. A typical stick of dynamite, like that shown in Fig. 13-5, may be a cylinder roughly 1.5 in. × 8 in. (3.8 cm × 20 cm) and weigh 0.5 lb (230 g).

Figure 13-5 Cylindrical cartridges into which dynamite has been packed are generally referred to as *sticks* of dynamite. These sticks are commercially available in a number of sizes. (Courtesy of IRECO, Inc., Salt Lake City, Utah)

Three forms of dynamite are commonly encountered today: *ammonia dynamite, straight dynamite,* and *gelatin dynamite.* All three forms contain the mixture identifed above, but ammonia dynamite and straight dynamite contain ammonium nitrate (Sec. 10-6) and sodium nitrate (Sec. 10-9), respectively, as oxidizers; gelatin dynamite contains nitrocellulose in a concentration of about 1% by mass, which

TABLE 13-5 PHYSICAL PROPERTIES OF SOME FORMS OF DYNAMITE

	Straight dynamite	Ammonia dynamite	Gelatin dynamite
Density (g/mL)	1.3	0.8–1.2	1.3–1.6
Chemical composition	20%–60% nitroglycerin depending on grade, sodium nitrate, carbonaceous material, antacid, and moisture	20%–60% nitroglycerin depending on grade, ammonium nitrate, carbonaceous material, sulfur, antacid, and moisture	20%–60% nitroglycerin depending on grade, gelatinized in nitrocellulose
Detonation velocity (km/s)	4–6	0.8–1.2	0.75–1.1
Sensitivity	High	High	High

Figure 13-6 The loading of a borehole with a water-gel cartridge. (Courtesy of E. I. Du Pont de Nemours & Co., Wilmington, Delaware)

serves to thicken the nitroglycerin and give dynamite additional brisance when it explodes. Table 13-5 provides additional information about the compositional nature of these forms of dynamite. Gelatin dynamite is less sensitive to shock and friction than ammonia dynamite and straight dynamite; also note that it may contain as much as 30% more nitroglycerin by mass than either of the other forms.

Notwithstanding the relative safety of dynamite compared to nitroglycerin, dynamite *does* contain nitroglycerin. Thus, when it explodes, dynamite produces the same brisance as an equivalent amount of nitroglycerin. When it is involved in fires, dynamite occasionally only burns; but the heat from a fire may also cause dynamite to explode. For these reasons, fires involving dynamite should not be fought.

From the 1920s through the 1930s, dynamite was the most popularly used explosive for peacetime purposes. But accidental dynamite detonations often occurred, particularly when handling old dynamite. New chemical explosives became known that were far safer to transport, store, and use than dynamite. These safer explosives largely replaced dynamite; an example is the water-gel cartridge illustrated in Fig. 13-6. Even though safer explosives are commercially available, some explosives experts still prefer the use of dynamite for unique demolition assignments.

USDOT regulates the transportation of dynamite as a division, 1.1D explosive. The proper shipping name is "Explosive, blasting, type A." Containers of dynamite are labeled EXPLOSIVE 1.1D, and their transport vehicles are placarded EXPLO-SIVES 1.1D.

13-7 NITROCELLULOSE

As noted in Sec. 12-3, nitrocellulose is an explosive produced by reacting cotton with nitric acid. The substance is a polymer* having a chemical structure resembling the following:

In appearance, nitrocellulose is a white solid, resembling cotton prior to being physically altered. For use as an explosive, it may be blocked, gelled, flaked, granulated, or powdered, and it may be either dry or wettened with water or ethyl alcohol.

Dry nitrocellulose is generally regarded as a low explosive:[†] that is, instead of detonating, it often deflagrates. Its flash point is only 55°F (13°C). When nitrocellulose burns, virtually no smoke is produced; its combustion products are carbon monoxide, carbon dioxide, water vapor, and nitrogen, not nitrogen dioxide. Nitrocellulose mixed with nitroglycerin is often called *smokeless powder*, since each of these combustion products is colorless. When dry, nitrocellulose is extremely sensitive to friction and poses a severe fire and explosion risk. Nitrocellulose burns more rapidly than virtually any other solid. Tons of nitrocellulose may be completely consumed by fire within a matter of minutes.

Nitrocellulose is probably most commonly encountered as a component of small-arms ammunition, but is also used frequently as the propellant in artillery ammunition. Nitrocellulose gels in organic solvents like acetone and ethyl alcohol. This feature is advantageous in that gelled nitrocellulose may be shaped into a variety of forms which solidify as the solvent evaporates. Munitions containing nitrocellulose are manufactured in this fashion. When intended for use as a chemical explosive in demolition work, nitrocellulose is almost always combined with a second explosive, like nitroglycerin.

USDOT regulates the transportation of nitrocellulose in several ways, as summarized in Table 13-6.

---

*Cellulose may be nitrated to varying degrees. The structure noted here is that of completely nitrated cellulose, which most closely resembles the structure of nitrocellulose that is a chemical explosive. The form of nitrocellulose that is intended for use in plastics is not as highly nitrated.

[†] Since cellulose may be nitrated to varying degrees, different forms of nitrocellulose may be produced. Each form typically acts as a low explosive. Nevertheless, to promote safety, USDOT regulates the transportation of dry nitrocellulose as a high explosive.

TABLE 13-6   USDOT-REGULATED FORMS OF NITROCELLULOSE

Form	Hazard class
Nitrocellulose, dry or wetted with less than 20% water or alcohol by mass	Division 1.1D explosive
Nitrocellulose, plasticized with not less than 18% plasticizing substance by mass	Division 1.3C explosive
Nitrocellulose solution, flammable, with not more than 12.6% nitrogen by mass, and not more than 55% nitrocellulose, flash point less than 73°F (23°C)	Flammable liquid
Nitrocellulose solution, flammable, with not more than 12.6% nitrogen by mass, and not more than 55% nitrocelleulose, flash point not less than 73°F (23°C), but not more than 140°F (60.5°C)	Flammable liquid
Nitrocellulose, unmodified or plasticized with less than 18% plasticizing substance by mass	Division 1.1D explosive
Nitrocellulose, wetted with not less than 25% alcohol by mass	Division 1.3C explosive
Nitrocellulose with alcohol, not less than 25% alcohol by mass, and not more than 12.6% nitrogen by dry mass	Flammable solid
Nitrocellulose with plasticizing substance not less than 18%, plasticizing substance by weight, and not more than 12.6% nitrogen by dry mass	Flammable solid
Nitrocellulose with water, not less than 25% water by mass	Flammable solid

13-8 TRINITROTOLUENE

2,4,6-Trinitrotoluene is a pale yellow solid, although the commercial grade is generally yellow to dark brown. It is prepared by the nitration of toluene, using a mixture of nitric acid and sulfuric acid; the latter acts catalytically. The production of trinitrotoluene is noted by the following equation:

$$\text{C}_6\text{H}_5\text{—CH}_3(l) + 3\text{HNO}_3(l) \longrightarrow \text{C}_6\text{H}_2(\text{CH}_3)(\text{NO}_2)_3 + 3\text{H}_2\text{O}(l)$$

Some important physical properties of trinitrotoluene are noted in Table 13-7.

     Trinitrotoluene is a high explosive. When used as an explosive, it is more commonly referred to as TNT. During World Wars I and II, TNT was used on a very large scale as a military explosive. During peacetime, it has also been used considerably for such purposes as mining.

     TNT possesses the following three desirable features that make it valuable as a commercial chemical explosive:

TABLE 13-7 SOME PHYSICAL PROPERTIES OF TRINITROTOLUENE

Density (g/mL)	1.59 (cast)
	1.45 (pressed)
	0.8 (grained)
Detonation velocity (km/s)	5.1–6.9
TLV	1.5 ppm (skin)
Melting point	178°F (81°C)
Sensitivity	Low
Detonating temperature	878°F (470°C)

1. *For a high explosive, it is unusually insensitive to heat, shock, and friction.* In fact, relatively small quantities of TNT typically burn when involved in fires. TNT does not ordinarily detonate unless unusually large amounts are activated simultaneously. Figure 13-7 demonstrates the detonation of 500 tons (454,000 kg) of TNT.

2. *It does not react with atmospheric moisture.* Hence, unlike nitroglycerin, it is not prone to hydrolysis.

3. *It is not susceptible to spontaneous decomposition,* even after years of storage. The solid may even be melted with steam with little fear of explosive decomposition.

Figure 13-7 The detonation of 500 tons (454,000 kg) of trinitrotoluene. The massiveness of this explosion may be noted by comparing its size to those of the objects in the foreground. (Courtesy of the U.S. Department of Energy, Las Vegas, Nevada)

The latter feature is an advantageous one for preparing explosive mixtures containing TNT, since molten TNT may be mixed with other explosives or oxidizers and cast into the shape of blocks or poured into ammunition shells. One such mixture consists of 80 parts of ammonium nitrate and 20 parts of TNT by mass; it is called *amatol*. It has found wide use as a military and industrial explosive. Other such mixtures are *cyclonite* [composition B (Sec. 13-9)] and *tetrytol* (Sec. 13-10).

Trinitrotoluene is highly toxic by ingestion, inhalation, and skin absorption. The adverse health effects resulting from exposure to trinitrotoluene are similar to those noted earlier for exposure to nitroglycerin.

USDOT regulates the transportation of trinitrotoluene as an explosive material. Dry trinitrotoluene and trinitrotoluene containing at least 10% water by mass are regulated as explosive materials. Packages containing either dry trinitrotoluene or trinitrotoluene wettened with less than 30% water by mass are labeled EXPLOSIVE 1.1D; and their transport vehicles are placarded EXPLOSIVES 1.1D.

13-9 CYCLONITE

The explosive known commercially as either *cyclonite, RDX* or *hexogen* is chemically named either cyclotrimethylenetrinitramine or hexahydro-1,3,5-trinitro-*s*-triazine. It is prepared from hexamethylenetetramine, nitric acid, ammonium nitrate, and acetic anhydride, as the following equation illustrates:

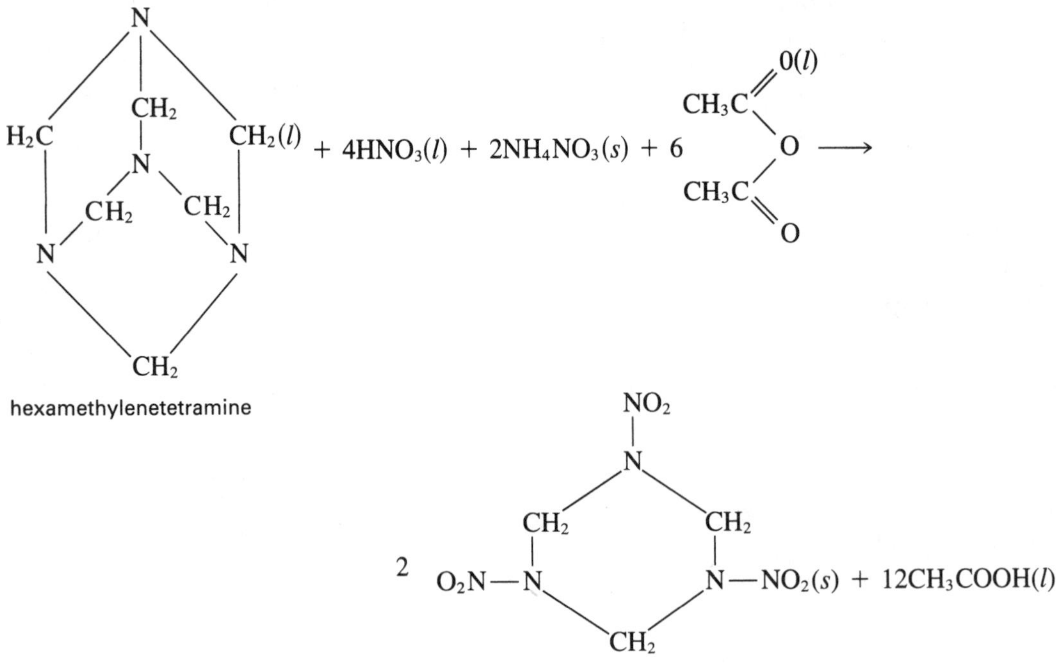

hexamethylenetetramine

$$+ \; 4HNO_3(l) \; + \; 2NH_4NO_3(s) \; + \; 6 \; \text{(acetic anhydride)} \longrightarrow$$

$$2 \quad \text{(cyclonite)} \quad + \; 12CH_3COOH(l)$$

cyclonite

TABLE 13-8 SOME PHYSICAL PROPERTIES OF CYCLONITE

Density (g/mL)	1.2 (loose)
	1.6 (pressed)
Detonation velocity (km/s)	6.8–8.0
Sensitivity	High
Melting point	396°F (202°C)
Detonating temperature	386.6°F (197°C)

Cyclonite is a white solid; some of its other physical properties are noted in Table 13-8.

As an explosive, cyclonite is approximately one and one-half times as powerful as TNT. As the pure substance, it is extremely sensitive to explosive decomposition. However, when cyclonite has been mixed with beeswax, it acquires thermal stability, even when exposed to relatively high temperatures. The beeswax serves to desensitize trinitrotoluene.

Mixed with beeswax in varying proportions, cyclonite was used widely through World War II as the bursting charge in bombs. Two compositions are still employed. *Composition A* is a mixture of 91% cyclonite and 9% beeswax by mass. From the standpoint of the speed with which it detonates, this composition ranks first among popular commercial explosives. A charge of composition A in a 1-ton bomb detonates in approximately 1/4 millisecond. This relatively high rate of detonation yields tremendous brisance. *Composition B* is a mixture of 60% cyclonite, 40% trinitrotoluene, and 1% beeswax by mass. This mixture has largely replaced composition A in artillery shells.

Cyclonite is the principal component of *plastics explosives*. These explosives may be easily molded into puttylike shapes and attached directly to structures that are to be demolished. Like the compositions of cyclonite with beeswax, cyclonite in plastics explosives retains the brisance of pure cyclonite, but is less sensitive to heat, shock, and friction. The fact that cyclonite is thermally stable makes it potentially useful when explosives are needed for firefighting. The main disadvantage associated with plastics explosives is their tendency to become brittle in cold climates.

Dry cyclonite is a forbidden explosive; but USDOT regulates the transportation of desensitized cyclonite and cyclonite containing not less than 15% water by mass as explosive materials. Packages of such materials are labeled EXPLOSIVE 1.1D; and their transport vehicles are placarded EXPLOSIVES 1.1D.

13-10 TETRYL

The explosive known commercially as *tetryl* is chemically named 2,4,6-trinitrophenylmethylnitramine. It is produced by nitrating the product resulting from reacting chloro-2,4-dinitrobenzene with methylamine, as the following equation illustrates:

tetryl

Tetryl is a yellow solid; some of its other physical properties are noted in Table 13-9.

Since World War II, tetryl has been the standard chemical explosive used as the booster in artillery ammunition by the U.S. military; that is, tetryl acts as the initiating agent for less sensitive explosives. Tetryl is very sensitive to heat, shock, and friction. It detonates at the rate of approximately 25,000 ft/s (7500 m/s).

TABLE 13-9 SOME PHYSICAL PROPERTIES OF TETRYL

Density (g/mL)	1.45 (pressed)
Detonation velocity (km/s)	7.0
Sensitivity	High
Melting point	266°F (130°C)
Detonating temperature	500°F (260°C)

Tetryl may be mixed with molten trinitrotoluene and a small amount of graphite to produce another explosive, called *tetrytol*. This explosive is occasionally used as the bursting charge in artillery ammunition.

USDOT regulates the transportation of tetryl as an explosive material. Packages containing tetryl are labeled EXPLOSIVES 1.1D, and their transport vehicles are placarded EXPLOSIVES 1.1D.

13-11 PETN

The explosive known commercially at *PETN* (pronounced "pettin") is chemically named *pentaerythritol tetranitrate*. It is produced by the nitration of pentaerythritol, as the following equation illustrates:

$$
\underset{\text{pentaerythritol}}{\text{HOCH}_2 - \overset{\displaystyle \overset{\text{OH}}{|}\ \overset{\text{CH}_2}{|}}{\underset{\displaystyle \underset{\text{CH}_2}{|}\ \underset{\text{OH}}{|}}{C}} - \text{CH}_2\text{OH}(s)} \ + \ 4\text{HNO}_3(l) \ \longrightarrow
$$

$$
\text{NO}_2 - \text{O} - \text{CH}_2 - \overset{\displaystyle \overset{\text{ONO}_2}{|}\ \overset{\text{CH}_2}{|}}{\underset{\displaystyle \underset{\text{CH}_2}{|}\ \underset{\text{ONO}_2}{|}}{C}} - \text{CH}_2 - \text{ONO}_2(s) \ + \ 4\text{H}_2\text{O}(l)
$$

PETN is a white solid; some of its other physical properties are noted in Table 13-10.

As an explosive, PETN is nearly as powerful as cyclonite. Like tetryl, it is sometimes used as the booster in artillery ammunition, but it is probably most commonly encountered as a form of *primacord*, a detonation fuse consisting of a core of PETN wrapped in fabric.

TABLE 13-10   SOME PHYSICAL PROPERTIES OF PETN

Density (g/mL)	1.6 (pressed)
Detonation velocity (km/s)	7.92
Sensitivity	High
Detonation temperature	410°F (210°C)

Dry PETN is a forbidden explosive; but USDOT regulates the transportation of wettened or desensitized PETN as an explosive material. Packages containing such forms of PETN are labeled EXPLOSIVE 1.1D, and their transport vehicles are placarded EXPLOSIVES 1.1D.

13-12 PICRIC ACID

Picric acid is the common name for 2,4,6-trinitrophenol. It is produced by nitrating phenol, as the following equation illustrates:

$$\text{C}_6\text{H}_5\text{OH}(s) + 3\text{HNO}_3(l) \longrightarrow \text{C}_6\text{H}_2(\text{OH})(\text{NO}_2)_3(s) + 3\text{H}_2\text{O}(l)$$

It is a yellow solid at room temperature. Picric acid is used industrially as an explosive and for dyeing textiles.

As an explosive, picric acid is just slightly more sensitive to explosive decomposition than TNT; it is also about as stable as TNT. On the other hand, picric acid reacts with certain metallic compounds to form salts called *picrates*. The picrates of copper, iron, lead, nickel, and zinc are considered dangerously sensitive to explosive decomposition; for instance, dry iron(III) picrate is similar to PETN in explosive sensitivity. Thus, while picric acid itself is relatively safe to handle, the presence of metallic picrates in impure picric acid is likely to initiate an accidental detonation.

Another picrate potentially useful as a chemical explosive is *ammonium picrate*. This substance is about as stable as picric acid, requiring a primary explosive to detonate it.

Picric acid is highly toxic. In humans, the ingestion of picric acid may cause nausea, vomiting, diarrhea, abdominal pain, itching, and skin disorders.

USDOT regulates the transportation of dry picric acid and picric acid wettened with less than 30% water by mass as explosive materials. Packages containing such forms of picric acid are labeled EXPLOSIVE 1.1D, and their transport vehicles are placarded EXPLOSIVES 1.1D. USDOT regulates the transportation of picric acid wettened with more than 30% water by mass as a flammable solid. Packages containing this form of picric acid are labeled FLAMMABLE SOLID, and their transport vehicles are placarded FLAMMABLE SOLID.

13-13 PRIMARY EXPLOSIVES

Three chemical explosives are frequently used in detonators and fuses to produce the detonation wave that initiates the booster or bursting charge. They are mercury fulminate, lead azide, and lead styphnate; each is a high explosive. Their physical properties are noted in Table 13-11. Each is highly sensitive to heat and shock. They are used in percussion caps, shells, cartridges, and detonators. Since these primary explosives are compounds of lead and mercury, they are highly toxic substances.

Mercury fulminate is the common name for mercury(II) cyanate; its chemical formula is $Hg(CNO)_2$. It is produced by pouring a nitric acid solution of mercury(II) nitrate into ethyl alcohol, but the chemical reaction is not entirely understood. Mercury fulminate is a white to gray solid.

Lead azide has the chemical formula $Pb(N_3)_2$. It is prepared by reacting

TABLE 13-11 SOME PHYSICAL PROPERTIES OF SEVERAL PRIMARY EXPLOSIVES

	Density (g/mL)	Detonation temperature	Detonation velocity (km/s)
Mercuric fulminate	3.6	356°F (180°C)	4.7
Lead azide	4.0	662°F (350°C)	5.1
Lead styphnate	2.5	513°F (267°C)	4.8

sodium azide and lead acetate, as the following equation illustrates:

$$2NaN_3(aq) + Pb(C_2H_3O_2)_2(aq) \longrightarrow Pb(N_3)_2(s) + 2NaC_2H_3O_2(aq)$$

It is a colorless solid.

Lead styphnate is the legal label name for lead trinitroresorcinate. It is produced by reacting lead acetate with trinitroresorcinol (styphic acid), as the following equation notes:

lead styphnate

It is a yellow-orange solid.

Lead azide and lead styphnate are specifically identified by USDOT as forbidden explosives. When these primary explosives have been wettened, however, US-DOT regulates their transportation as explosive materials. Packages of either mercuric fulminate, lead azide, or lead styphnate wettened with not less than 20% water by mass or mixed with alcohol and water are labeled EXPLOSIVE 1.1A, and their transport vehicles are placarded EXPLOSIVES 1.1A.

REVIEW EXERCISES

General Characteristics of Chemical Explosives

13.1. Indicate three differentiating features associated with the detonation and deflagration of chemical explosives.

13.2. Under certain conditions, ammonium nitrate is a material that will "explode"; yet the transportation of ammonium nitrate is regulated by USDOT as an oxidizer rather than an explosive. Why is this most likely so?

13.3. Write the chemical equation that illustrates the detonation of ordinary dynamite.

13.4. Small-arms ammunition consists of a metallic cartridge case, the primer and propelling charge, and an explosive projectile with or without a detonating fuse, all in one assembly. Describe the function of each component.

Firefighting Involving Chemical Explosives

13.5. Certain chemical explosives are known to be stored in a locked shed at a construction site located downwind from a burning two-story building. The shed and burning building are approximately 90 m (100 yd) apart. What appropriate actions should firefighters take to most effectively safeguard lives and property?

13.6. A truck is placarded EXPLOSIVES 1.1D. It becomes involved in a transportation mishap along with two other motor vehicles, a passenger-carrying bus and an automobile. No fire is evident at the scene of the accident, but the odor of rubber can be detected. What actions should members of emergency response forces take to safeguard lives and property?

USDOT Regulations Regarding the Transportation of Chemical Explosives

13.7. Dry cyclotetramethylenetrinitramine, commonly called *HMX,* is a "forbidden explosive."
  (a) What does this term mean?
  (b) If HMX is wettened with not less than 15% water by mass, USDOT regulates its transportation as an explosive material. Why is it unsafe to transport this chemical explosive in its dry state, but relatively safe to transport it when wettened?

13.8. An explosive mixture or device containing an acidic metal salt and a chlorate is an example of a forbidden explosive. What is the most likely reason that USDOT forbids the transportation of such a mixture or device?

13.9. Type A fireworks are pyrotechnic devices that pose a risk of mass explosion when packed for transport. What hazard division and compatibility group are assigned to type A fireworks intended for shipment?

13.10. A rocket motor is a device designed to propel a rocket or missile. One type of rocket motor consists of a charge of solid propellant in a metal cylinder fitted with one or more nozzles and fueled by a flammable liquid. The nature of the propellant is such that it generally represents a projection hazard, but not a mass explosion hazard. What hazard division and compatibility group are assigned to rocket motors intended for shipment?

13.11. A branch of the military service intends to transport bombs by air that are assembled with a high explosive and a poisonous gas. The high explosive potentially represents a projection hazard, but not a mass explosion hazard.
  (a) What labels does USDOT require to be affixed to packages containing such cargo?
  (b) What special markings does USDOT require on these packages?

Black Powder

13.12. Why is the explosive potential of black powder eliminated when it is drenched with water?

13.13. Why does the deflagration of black powder not necessarily require atmospheric oxygen?

13.14. USDOT regulates black powder as a division 1.1D explosive, even though explosives experts normally classify it as a low explosive. What is the most likely reason USDOT takes this position?

Nitroglycerin and Dynamite

13.15. Why is old nitroglycerin more susceptible to explosion by shock and heat than freshly prepared nitroglycerin?

13.16. Liquid nitroglycerin is less susceptible to explosive decomposition by shock when it has been absorbed into a porous material; yet it retains nearly the same degree of explosive effectiveness. Why is this so?

13.17. Why is the addition of an antacid to the formulation of dynamite likely to result in a product that is less sensitive to shock and heat?

13.18. Why are the production and manufacture of nitroglycerin typically accomplished in well-ventilated areas?

Miscellaneous Chemical Explosives

13.19. Write a balanced equation illustrating the deflagration of nitrocellulose, the chief component of guncotton. Its chemical formula may be approximated as $C_6H_7O_2(ONO_2)_3$.

13.20. When used in an explosive device, USDOT requires that a desensitized explosive component may not freeze at temperatures above $-10°F$ ($-22°C$). Why are such explosives more likely to detonate if they have been frozen?

13.21. Cyclonite is less susceptible to explosive decomposition by shock when it has been mixed with beeswax, yet it retains its usefulness as a chemical explosive. Why is this so?

13.22. Two trucks are involved in a transportation incident in which the cargo of the second truck has caught fire. The first truck is placarded DANGEROUS, but the second truck has not been placarded. Upon arriving at the scene of the accident, a fire chief discovers from shipping papers that the first truck contains 150 lb (68 kg) of low explosives. Based on this minimal information, what directions should be given to firefighters to best safeguard lives and property?

13.23. Which situation is potentially the more hazardous: a fire involving primary explosives or a fire involving an equal quantity of secondary explosives?

13.24. When PETN detonates, a mixture results of carbon dioxide, carbon monoxide, nitrogen, and water vapor. Write a balanced chemical equation that illustrates this chemical reaction. The chemical formula of PETN may be condensed as $C(CH_2ONO_2)_4$.

14

Radioactive Materials

In today's world, radioactive materials are used somewhat regularly. In the nuclear power industry, for instance, either uranium or plutonium is required to generate electrical energy, albeit indirectly. In the field of medicine, various radioactive materials are used to help diagnose certain diseases, as well as to aid in their treatment. Some radioactive materials are employed industrially as sources of radiation for the sterilization of products against the possible presence of bacteria. Others are used to detect the presence of buried pipeline and to gauge the thickness of plastic film. In research, results derived from the use of specific radioactive materials have greatly influenced our understanding of phenomena in agriculture, medicine, biology, and such diverse fields as astrophysics, art, and archeology.

Notwithstanding these beneficial features, certain radioactive materials may also be used in ways that could adversely affect life on our entire planet. One such way is associated with the use of certain radioactive materials in nuclear weapons. The awesome force of nuclear weapons was first exposed to the world in 1945 toward the end of World War II, when atomic bombs were dropped on the Japanese cities of Hiroshima and Nagasaki. This act served as the impetus for ending the war; but we must recognize that there are individuals and their offspring living today who still suffer from cancer and other health problems resulting from the initial blast of the atomic bomb or from exposure to its radioactive fallout products.

In contemporary times, many people throughout the world fear that nuclear weapons may again be used to maintain the current worldwide balance of political power. Their fear is based on the possibility of a major nuclear confrontation between the superpowers. At its very worst, such an act could result in abruptly ending our civilization.

Such fear of exposure to radioactive materials is not confined to war zones. The operation of nuclear reactors, for instance, involves our primary peaceful use of radioactive materials. Even with rigid controls, there is a considerable basis for fear-ing the effects from exposure to the radioactive materials associated with reactors. Nuclear fuel, for instance, is highly radiotoxic. Even in the tiniest quantities, plutonium is searingly radiotoxic; it ranks with *botulinum* toxin (Table 11-14) as one of the world's most poisonous substances.

Ever since the nuclear industry first became operative, the public has expressed concern that radioactive waste products could be inadvertently released to the environment. Such events are no longer regarded as mere hypothetical possibilities. There have been well-documented instances of airborne releases of radioactive materials from reactors in the United States, the Soviet Union, and elsewhere. The worst such disaster of this type occurred in 1986 at the Soviet Union's Chernobyl nuclear power plant. A series of operator errors unleashed a power surge that triggered an explosion and partial meltdown of the reactor's fuel. Tons of radioactive fuel and by-products were subsequently discharged. Using limited data, scientists estimate that the risk of acquiring cancer increased from 100 to 1000 times the norm for the thousands of individuals directly exposed to the radioactive fallout associated with this one incident.

The growth of the nuclear power industry during the past several decades has also caused the evolution of a related fear among some segments of the population. This fear is associated with the intensely radioactive nuclear waste that is generated from the operation of nuclear reactors. Year after year, the quantity of such wastes has grown to staggering amounts; but what can be done with it? Considerable apprehension has been voiced about the magnitude of harm that may result from disposing of it in landfills; some citizens also worry about the adverse affect that a transportation mishap involving nuclear wastes would have on the residents of nearby communities.

While grounds exist for these fears, we should note that regulatory bodies have taken steps to minimize the hazards associated with radioactive materials. In the United States, under the auspices of the Atomic Energy Act of 1954, the U.S. Nuclear Regulatory Commission (USNRC) licenses and oversees any facility that uses intensely radioactive materials. Furthermore, the construction and operation of commercial nuclear reactors are regulated by the USNRC, as well as the construction and operation of nuclear waste disposal sites. In the workplace, employers are responsible for assuring that individual workers are protected against unnecessary exposure to radiation or radioactive materials through regulations promulgated under OSHA. Finally, when radioactive materials are shipped, their transportation is regulated by USDOT. In combination, these various regulations provide an element of protection against the potential hazards that could arise from the inadvertent exposure of the public to radioactive materials.

Nonetheless, when compared to the other classes of hazardous materials, radioactive materials are generally regarded as having the very highest degree of hazard. What properties do they possess that give rise to this particularly hazardous nature? What can be done to minimize the adverse effects from radiation exposure? These are among the questions that will be answered in this final chapter.

14-1 SOME FEATURES OF ATOMIC NUCLEI

In Sec. 3-4, we noted that there are two primary constituents of the atomic nucleus, protons and neutrons. The nuclei of all atoms of the same element possess the same number of protons, but they may differ by the number of neutrons they possess. These different nuclei of the same element are called the *isotopes* of that element.

The number of protons found in the nucleus of an atom is called the *atomic number*. The number of protons equals the number of electrons in the neutral atom. The number of protons or electrons may be readily identified for any atom of an element by reference to a periodic table (Sec. 3-5) or other tables that list atomic numbers. The total number of protons and neutrons possessed by a particular isotope is called its *mass number*. The mass number corresponds to the atomic mass of an isotope, rounded off to the nearest whole number.

Hydrogen has an atomic number of 1; this means that each and every hydrogen atom possesses one proton. Hydrogen atoms exist in either of three isotopic forms having the unique names and compositions that follow:

1. *Protonium.* This is simplest of the hydrogen isotopes and, in fact, the simplest of all atoms. Its nucleus has one proton, but no neutrons.
2. *Deuterium.* The nucleus of this second hydrogen isotope is composed of one proton and one neutron.
3. *Tritium.* The nucleus of the third hydrogen isotope is composed of one proton and two neutrons.

Every neutral hydrogen atom has one electron, but when a hydrogen atom is ionized, it is stripped of its electron. Only its nucleus remains. Thus, when protonium atoms are ionized, only protons remain. When deuterium atoms are ionized, each nucleus that remains is composed of one proton and one neutron; these nuclei are called *deuterons*. When tritium atoms are ionized, the remaining nuclei are called *tritons;* each is composed of one proton and two neutrons.

Isotopes are designated by the symbol $_Z^A X$, where Z is the atomic number of an element having a chemical symbol X, and where A is the mass number. Hence, the three hydrogen isotopes are designated by the symbols $_1^1 H$, $_1^2 H$, and $_1^3 H$, respectively. For each, the symbol Z equals 1, the number of protons. The number of neutrons may be obtained by difference: For protonium, the number is 0; for deuterium, the number is 1; and for tritium, the number is 2.

Only the isotopes of hydrogen have unique names. The isotopes of other elements are named by simply identifying the element and the mass number of the isotope at issue. Thus, nuclei designated as $_6^{12}C$, $_{92}^{235}U$, and $_{19}^{40}K$ are named carbon 12, uranium 235 and potassium 40, respectively.

All elements have from 3 to approximately 25 isotopes. Many isotopes are stable; that is, they retain whatever structure they have and do not undergo spontaneous changes. On the other hand, many other nuclei are subject to spontaneous transformations. These nuclei are said to *disintegrate* or *decay*. The phenomenon is called *radioactivity*. The unstable nuclei at issue are said to be *radioactive;* they are called *radioisotopes*. Two hydrogen isotopes, protonium and deuterium, are stable nuclear species, but tritium is a radioisotope.

Radioactivity is not affected by the physical or chemical changes that occur in a substance. Hence, when radioactive materials are subjected to changes in pressure, volume, temperature, or chemical nature, their spontaneous disintegration is not altered in any way.

When a radioisotope undergoes a change, it usually emits a particle; less commonly, it may absorb an electron; and either process may be accompanied by the simultaneous emission of energy. When the transformation occurs, the radioisotope has been converted into a new nucleus, which may be either stable or radioactive itself. Frequently, radioisotopes undergo several transformations before they are converted into stable nuclei.

These transformations occur during time periods. The time during which an arbitrary number of nuclei is reduced to half this value is called the *half-life* of that radioisotope. For instance, suppose we have 1 million atoms of tritium and that they are set aside for 12.3 years. After the elapse of this time period, only 500,000 tritium atoms remain. After another 12.3 years elapses, only 250,000 atoms remain. Consequently, the half-life of tritium is 12.3 years.

Half-life periods vary appreciably. The half-life of uranium 238, the principal naturally occurring radioisotope of uranium, is 4.5 billion years; but the half-life of astatine 216, a radioisotope of an artificially produced halogen, is only 0.0003 s.

Each element has at least one radioisotope. The hundred or so elements collec-

TABLE 14-1   NATURALLY OCCURRING RADIOISOTOPES

Isotope	Type of disintegration	Half-life in years	Relative isotopic abundance
$^{3}_{1}$H	β	12.26	0.00013
$^{14}_{6}$C	β	5570	
$^{40}_{19}$K	β, EC[a]	1.2×10^{9}	0.012
$^{87}_{37}$Rb	β	6.2×10^{10}	27.8
$^{115}_{49}$In	β	6×10^{14}	95.8
$^{138}_{57}$La	β, EC	$\sim 2 \times 10^{11}$	0.089
$^{144}_{60}$Nd	α	$\sim 5 \times 10^{15}$	23.9
$^{147}_{62}$Sm	α	1.3×10^{11}	15.1
$^{176}_{71}$Lu	β	4.6×10^{10}	2.60
$^{187}_{75}$Re	β	$\sim 5 \times 10^{10}$	62.9
$^{190}_{78}$Pt	α	$\sim 1 \times 10^{12}$	0.012
$^{226}_{88}$Ra	α	1622	
$^{232}_{90}$Th	α	1.4×10^{10}	100
$^{235}_{92}$U	α	7.13×10^{8}	0.72
$^{238}_{92}$U	α	4.5×10^{9}	99.28

[a] EC = electron capture.

tively have nearly 1200 known radioisotopes. Since the age of Earth is estimated at some 3.6 billion years, few radioisotopes occur naturally, as Table 14-1 notes. If radioisotopes were present when Earth was formed, which most likely was true, most of them disappeared long ago. Many radioisotopes encountered today are produced by artificial means in nuclear reactors or particle accelerators.

14-2 TYPES OF RADIATION

The transformations of radioisotopes are associated with three types of radiation: *alpha radiation* (α), *beta radiation* (β), and *gamma radiation* (γ). These different mechanisms of radioactive transformation are illustrated in Fig. 14-1. They are examples of *ionizing radiation,* since the passage of either type through matter results in its ionization. Each type is discussed independently.

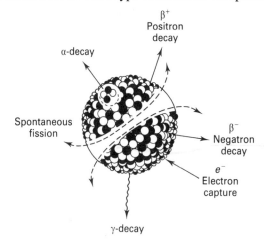

Figure 14-1 The modes by which a radioisotope may decay (open circles are protons, solid black circles are neutrons). Of these modes, the rarest is spontaneous fission. The most common modes are those associated with the production of beta and gamma radiation.

Alpha Radiation

Many isotopes having atomic numbers greater than 83 tend to disintegrate by emitting particles consisting of two protons and two neutrons. The particles are called *alpha particles*. These particles are the nuclei of doubly ionized helium atoms. They may be symbolized as $^4_2$He, but more generally we write α. When the nuclei of many such atoms decay, many alpha particles are correspondingly emitted. This combination of alpha particles is called *alpha radiation*. It is associated with a relatively large amount of energy, ranging from 4 to 8 MeV per particle;* but since alpha particles are doubly ionized, this energy is readily dissipated when it passes through only a few centimeters of air. It may also be absorbed by a thin piece of

---

*The energy of alpha, beta, and gamma radiation is typically quoted in multiples of a unit called an *electron volt* (eV). One electron volt is equivalent to 1.602×10^{-19} J. One million electron volts (MeV) is equivalent to 1.602×10^{-13} J.

most forms of matter; for instance, alpha radiation may be absorbed by the thickness of this page.

When a radioisotope emits an alpha particle, its atomic number correspondingly decreases by 2, while its mass number decreases by 4. An example of a radioisotope that disintegrates by alpha particle emission is uranium 238. This transformation is designated by the following equation:

$$^{238}_{92}\text{U} \longrightarrow ^{234}_{90}\text{Th} + ^{4}_{2}\text{He} \quad (\text{or } \alpha)$$

The equation notes that the uranium 238 nucleus changes into another one, thorium 234, by emitting an alpha particle. The particle is written to the right of the arrow, designating that it has been emitted from the uranium 238 nucleus.

Beta Radiation

The second mode of radioactive disintegration is asociated with three different processes. The first such process involves the emission of an electron from the nucleus of a radioisotope. When electrons are encountered in nuclear phenomena, they are called *negatrons* and designated as β^- or $_{-1}^{0}e$. When the nuclei of many such atoms decay, many negatrons are correspondingly emitted. This is one form of *beta radiation*.

When negatron emission occurs, the mass numbers of the associated nuclei remain unchanged; but the atomic number increases by 1.* An example of a radioisotope that disintegrates by emitting a negatron is thorium 234. This is the nucleus produced when uranium 238 disintegrates. Upon emitting a negatron, each thorium 234 nucleus becomes protoactinium 234, as the following equation designates:

$$^{234}_{90}\text{Th} \longrightarrow ^{234}_{91}\text{Pa} + _{-1}^{0}e \quad (\text{or } \beta^-)$$

The emission of a negatron from the nucleus raises a fundamental question: How may electrons be emitted from nuclei when they are not components of nuclei? This question suggests that the process of negatron emission is more involved than an equation is capable of representing. Such equations summarize overall nuclear phenomena. More pointedly, however, each neutron within the unstable nucleus transforms into a proton and an electron; the proton then becomes part of the new nucleus, and the electron is simultaneously emitted. This conversion of a neutron ($_{0}^{1}n$) into a proton and an electron is designated by the equation

$$_{0}^{1}n \longrightarrow _{1}^{1}\text{H} + _{-1}^{0}e$$

The second process associated with beta decay involves emission of a *positron* from the nucleus. This particle has all the features typically associated with the electron, except one: A positron is positively charged. It is symbolized as either $_{+1}^{0}e$ or β^+.

Radioisotopes emitting positrons retain their mass numbers but decrease in

---

* Beta decay processes are also associated with the production of neutrinos and antineutrinos, neutral particles having a particularly small mass. These particles are of no interest to us here.

atomic number by 1. The pertinent nuclear event consists of the conversion of a proton into a neutron and positron, as the following equation indicates:

$$\ce{^1_1H} \longrightarrow \ce{^1_0}n + \ce{^0_{+1}}e$$

An example of a radioisotope that disintegrates by positron emission is sodium 22. This nucleus is converted into neon 22, which is represented by the following equation:

$$\ce{^{22}_{11}Na} \longrightarrow \ce{^{22}_{10}Ne} + \ce{^0_{+1}}e \quad (\text{or } \beta^+)$$

Positrons represent another form of beta radiation.

Finally, the third process associated with beta radiation involves the union of an unstable nucleus with extranuclear electrons; it is called *electron capture*. A radioisotope that transforms by electron capture decreases its atomic number by 1, but retains the identity of its mass number. Each electron captured by such nuclei reacts with a proton, thereby forming a neutron. The neutron then becomes part of the structure of the product nucleus. This individual nuclear event is represented by the following equation:

$$\ce{^1_1H} + \ce{^0_{-1}}e \longrightarrow \ce{^1_0}n$$

An example of a radioisotope that disintegrates by electron capture is oxygen 15. The overall nuclear process is represented as follows:

$$\ce{^{15}_8O} + \ce{^0_{-1}}e \longrightarrow \ce{^{15}_7N}$$

Gamma Radiation

Nuclear transformations are frequently accompanied by the simultaneous emission of the third form of radiation: *gamma radiation*. This is a form of electromagnetic radiation, like x radiation, infrared radiation, and ultraviolet radiation, none of which possesses mass or charge. The electromagnetic spectrum is illustrated in Fig. 14-2. The various forms of radiant energy are characterized by their wavelengths. Ultraviolet, infrared, and radio waves have long wavelengths; gamma radiation and x radiation have short wavelengths. The components of the electromagnetic spectrum with short wavelengths are very energetic, so much so that they ionize matter through

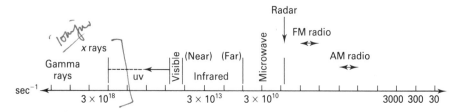

Figure 14-2 The components of the electromagnetic spectrum as a function of frequency. Gamma and x radiation, shown to the far left, are forms of radiant energy associated with short wavelengths and high frequencies. Such forms are sufficiently energetic to ionize the matter through which they pass.

which they pass. However, the components with long wavelengths are relatively nonenergetic; they are forms of nonionizing radiation.

The individual components of gamma radiation are called *gamma rays* or *photons,* represented as γ. Since they are so energetic, they are also extremely penetrating and only absorbed by dense forms of matter, like blocks of lead.

When gamma rays are emitted from a radioisotope, no change occurs in either the atomic number or mass number. Instead, some fraction is removed of the energy of excitation that causes the nucleus to be unstable. Imagine a radioisotope that exists in only two energy states. The more energetic form, called the *excited state,* may change into the other one; but when this change occurs, gamma rays are emitted from the nucleus. The phenomenon may be illustrated by the following equation, where the excited state is identified by an asterisk:

$$(_Z^A X)^* \longrightarrow \ _Z^A X + \gamma$$

The emission of gamma radiation is often represented diagrammatically as illustrated below:

Each line represents a discrete energy state of the atomic nucleus.

As a specific example of gamma radiation, consider the radioisotope cobalt 60. When it decays, cobalt 60 first changes to an excited state of nickel 60; in turn, this state emits two gamma rays having energies of 1.173 MeV and 1.332 MeV. This process is represented as follows:

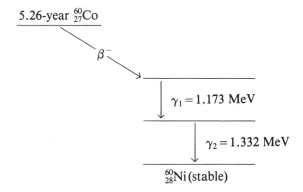

14-3 MEASUREMENT OF RADIOACTIVITY

The most commonly encountered instrument for the detection of radiation is the *Geiger counter,* illustrated in Fig. 14-3. Like all other such radiation-detection instruments, it operates on the fact that radiation creates ionization in certain media.

Figure 14-3 A portable Geiger counter, often used for surveying an area when monitoring for the presence of beta and gamma radiation. The operating voltage is supplied by batteries. The indicating meter is calibrated to record a radiation intensity unit, like milliroentgens per hour or counts per minute. For accurate interpretation, the instrument must be calibrated with radiation of the same type and energy whose intensity is to be measured. (Courtesy of Fisher Scientific Company, Pittsburgh, Pennsylvania)

As these instruments were developed, a number of units of radiation measurement were simultaneously defined, the most common of which are reviewed here.

The oldest unit of measuring radiation is the *roentgen*. It is the international unit of radiation quantity for gamma radiation, but the unit may be used to measure alpha and beta radiation as well. One roentgen is the amount of radiation that produces sufficient ion pairs in 1 mL (0.001293 g) of dry air measured at 32°F (0°C) and 1 atm to carry one electrostatic unit of electrical charge of either sign.

Another unit of radiation measurement is the *rad*. This is the quantity of ionizing radiation that results in the absorption of 0.00001 J of energy by 1g of a material. This unit is commonly employed for measuring the quantity of radiation absorbed by the body or the body's tissues upon exposure to radioactive materials.

Radioactivity is very commonly measured in an amount called the *activity*. One unit used to measure activity is the *curie* (Ci). One curie is the amount of radiation corresponding to 37 billion disintegrations per second (3.7×10^{10} disintegrations/s). The SI unit of activity is the *becquerel* (Bq), which is defined as 1 disintegration per second. One curie equals 3.7×10^{10} Bq. The *specific activity* of a radioactive material is the activity per unit mass of that substance.

Finally, a measure of the quantity of any ionizing radiation to the body or the body's tissues may be measured in terms of its *estimated biological effect*. While the roentgen or rad could be directly used for this measurement, it is often measured by use of a unit called the *rem (roentgen equivalent man)*. The relation of the rem to other radiation units depends on the biological effect under consideration and on the conditions for radiation. For instance, OSHA considers each of the following exposures to the indicated types of ionizing radiation to be equivalent to an exposure of 1 rem:

1. Exposure of 1 roentgen due to x radiation or gamma radiation
2. Exposure of 1 rad due to x radiation, gamma radiation, or beta radiation.
3. Exposure of 0.1 rad due to high-energy protons
4. Exposure to 0.05 rad due to particles heavier than protons and with sufficient energy to reach the lens of the eye.

The SI unit of radiation exposure is the coulomb per kilogram (c/kg), which equals approximately 3876 R. The SI unit of dose-equivalent is the *sievert* (Sv), which is equal to 100 rem.

The amount of radiation to which an individual has been exposed in the workplace is often estimated through the use of appropriate personal monitoring equipment, such as the film badges or pocket dosimeters shown in Fig. 14-4. OSHA requires employers to provide radiation-monitoring devices to those employees who are likely to be exposed to radiation and further requires the employees to use them.

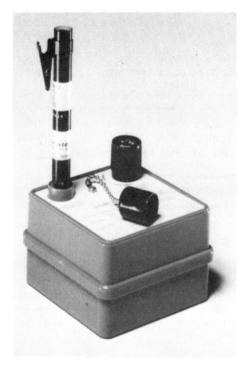

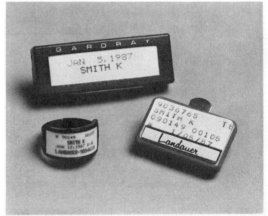

Figure 14-4 A pocket dosimeter on the left and three types of radiation-monitoring badges (film body, thermoluminescent body, and thermoluminescent ring) on the right. For individuals who work near radiation sources or operating x-ray equipment, these devices provide an estimate of the total amount of beta, gamma, and x radiation to which they have been exposed. (Courtesy of Lab Safety Supply Co., Janesville, Wisconsin (pocket dosimeter), and R.S. Landauer, Jr. and Co., Glenwood, Illinois, radiation-monitoring badges).

Film badges usually consist of photographic emulsions that are very sensitive to radiation. They are worn for a specified period, usually one each day, and then professionally developed. The radiation exposure is ascertained by comparing the developed film against those previously exposed to known amounts of radiation.

Pocket dosimeters are small ionization chambers, typically calibrated to read between 0 and 200 milliroentgens. After exposure to radiation, they are read through use of an auxiliary reader instrument to determine the amount of exposure to ionizing radiation.

What amount of radiation exposure is considered safe? The answer to this question is not straightforward, but as a guideline, it is appropriate here to consider the position taken by OSHA in the workplace. The amount of radiation to which a worker may be exposed in segments of the workplace is regulated by OSHA to certain specified amounts. One such segment is called a *restricted area;* it is any area the access to which is controlled by the employer for purposes of protecting individuals against exposure to radiation or radioactive materials. OSHA requires the following:

"(a) Except as indicated below in (b), no employer may possess, use or transfer sources of ionizing radiation in such a manner as to cause any individual in a restricted area to receive during a calendar quarter a dose in excess of the limits specified in Table 14-2.

(b) An employer may permit an individual in a restricted area to receive doses to the whole body greater than those indicated in Table 14-2 only as long as during any calendar quarter, the dose to the whole body does not exceed 3 rems; furthermore, the dose to the whole body, when added to the accumulated occupational dose to the whole body, does not exceed $5 \times (N - 18)$ rems, where N equals the individual's age in years at his last birthday.

(c) No employer may allow any employee under 18 years of age to receive in any calendar quarter a dose in excess of 10% of the limits specified in Table 14-2."

TABLE 14-2 AMOUNTS OF RADIATION EXPOSURE LIMITED TO WORKERS UNDER OSHA

Area of the body	Rems per calendar quarter
Whole body; head and trunk; active bloodforming organs; lens of the eyes; or gonads	1.25
Hands and forearms; feet and ankles	8.75
Skin of whole body	0.50

14-4 ADVERSE HEALTH EFFECTS FROM RADIATION EXPOSURE

Each of us is constantly exposed to a certain amount of inescapable low-level ionizing radiation from naturally occurring radioisotopes. This radiation is part of our natural environment. Furthermore, we are periodically exposed to ionizing radiation when x-ray patterns are taken to trace defects in bones and teeth or to detect the presence of tumors. Such intentional exposure does not generally result in the development of health problems, since the intensity of the radiation is relatively low. But suppose an individual is exposed to high-intensity ionizing radiation, for example, in concentrations exceeding those listed in Table 14-2. What adverse health effects are likely to be caused from this exposure?

To answer this question, let's examine what occurs when radiation passes through matter. The subsequent chemical effects may be profound. As we have noted, the primary action of alpha, beta, or gamma radiation is to ionize materials through which they pass. This ionization may be represented for a molecule of an arbitrary substance A by the equation that follows:

$$A \rightsquigarrow A^+ + e^-$$

The wiggly arrow indicates that the reaction has been induced by radiation; e^- symbolizes an electron.

As many such events occur, the number of ions and electrons correspondingly increase. When this occurs, several secondary actions of radiation result. In particular, the ions may combine with any of the electrons to form an excited state of A, as the following equation notes:

$$A^+ + e^- \longrightarrow A*$$

Here, $A*$ refers to a molecule of A that possesses excess energy. $A*$ is unstable. Possessing excessive energy, it may dissociate entirely into molecules of new substances, as the following equation illustrates:

$$A* \longrightarrow B + C$$

The ions that form from the initial action of radiation may also react with neutral molecules. Such reactions produce new ions and free radicals. The direct exposure of a substance to radiation may also result in the production of free radicals. As we noted before, these are highly reactive chemical species, which can initiate a variety of chemical reactions.

Let's consider specifically what is likely to occur when water is exposed to ionizing radiation. First, primary ionization occurs:

$$H_2O(l) \rightsquigarrow (H_2O)^+(l) + e^-$$

Then the $(H_2O)^+$ ion may react with a neutral water molecule to form a hydroxyl radical:

$$(H_2O)^+(l) + H_2O(l) \longrightarrow (H_3O)^+(aq) + \cdot OH(aq)$$

Exposure of water to radiation may also result in the production of hydrogen atoms and hydroxyl free radicals.

$$H_2O(l) \longrightarrow H \cdot (aq) + \cdot OH(aq)$$

When the concentration of hydroxyl radicals has increased substantially, two hydroxyl radicals are certain to interact. This may result in their union to form hydrogen peroxide.

$$2 \cdot OH(aq) \longrightarrow H_2O_2(aq)$$

Much research has been devoted to examining the nature of the chemical species that result when water is exposed to radiation. The results of such research have an important implication when evaluating the effects of radiation on humans. Water constitutes three-fourths of all the body's tissues. Every living cell in the body contains water. Thus, when foreign substances are formed in cells due to the radiolysis of water, the cellular biochemistry may be significantly altered, causing damage or death to the affected cells.

Any body tissue can be exposed to a certain amount of radiation without experiencing deleterious effects. In fact, unless the radiation is unusually intense, it is not accompanied by any physical sensation. Ultimately, however, increased exposure to radiation causes body tissue to become necrotic or ulcerating. The amount of radiation that causes this effect depends on the type of radiation absorbed, its energy, the specific area of the body that has been exposed to the radiation, and the age of the individual. The chronic effect on the body's tissues depends on the number and nature of the cells whose biological function has been altered. Burns frequently develop superficially, followed by the development of tumors.

Radioisotopes that have been ingested and incorporated into the body may be difficult, if not impossible, to eliminate. Radioisotopes of heavy-mass elements decay primarily by emitting alpha and gamma radiation; they also tend to deposit in the bones. Once incorporated in the structure of bone matter, they may induce the destruction of the molecular components of its cells and cause cancer. Radioisotopes of elements found naturally in the body tend to accumulate in the structures where the element localizes. For instance, when radioactive calcium 45 is ingested, it deposits in bone matter just as nonradioactive calcium isotopes do.

The nature of radiation is extremely important when ascertaining the biological damage likely to be caused by exposure to radiation. Alpha radiation is very energetic, but it is easily absorbed externally by the epidermis, the outer layer of the skin (see Fig. 9-2). Since the epidermis consists of dead skin cells, external exposure to alpha radiation is likely to represent only a minor hazard. Ingested, however, the energetic alpha radiation may localize in a minute area of bone matter where the resulting biological damage may be severe.

Beta radiation penetrates deeper than alpha radiation. A 3-MeV negatron, for instance, travels through 0.6 in. (15 mm) of tissue before being absorbed. Thus, beta radiation passes through the epidermis into the dermis, a layer of skin tissue containing living cells. For this reason, external and internal exposure to beta radiation may cause biological damage.

Gamma and x radiation cause the most severe deleterious effects on body tissues. Neither is easily absorbed. Hence, gamma and x radiation easily pass through tissue, causing ionization of the various substances they encounter. This means that external and internal exposure to gamma and x radiation causes more severe biological damage than exposure to either alpha or beta radiation.

The relative abilities of alpha, beta, and gamma radiation to penetrate matter are illustrated in Fig. 14-5.

The most severe cases of radiation exposure for large numbers of people oc-

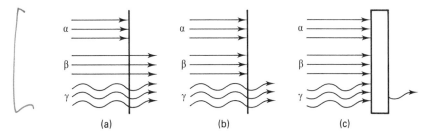

Figure 14-5 The relative penetrating power of alpha, beta, and gamma radiation: (a) sheet of paper, (b) sheet of aluminum, and (c) block of lead.

curred during the World War II bombings of Hiroshima and Nagasaki and in the Marshall Islands following bomb testing at Bikini Atoll in 1954. Medical studies of the victims who survived have yielded information regarding the biological effect resulting from exposure to radiation, part of which is summarized in Table 14-3 for x and gamma radiation.

TABLE 14-3 X-RAY AND GAMMA-RAY DOSES REQUIRED TO PRODUCE VARIOUS SOMATIC EFFECTS

Dose (rads)	Effect
0.3 weekly	No observable effect
60 (whole body)	Reduction of lymphocytes (white blood cells formed in lymphoid tissues as in the lymph nodes, spleen, thymus, and tonsils)
100 (whole body)	Nausea, vomiting, fatigue
200 (whole body)	Reduction of all blood elements
400 (whole body)	50% of an exposed group would probably die
500 (gonads)	Sterilization
1000 (skin)	Erythema (reddening of the skin)

14-5 FISSIONABLE RADIOISOTOPES

Scientists have been able to synthetically produce numerous radioisotopes not found naturally. One method of accomplishing this feat involves irradiating stable or long-lived isotopes in specially designed machines, like cyclotrons and bevatrons. Samples of such isotopes are generally irradiated with α particles, protons, deuterons, and other particles that have been accelerated to relatively high velocities. When the velocity of such particles is increased, so are their energies. When these high-energy particles bombard nuclei, nuclear reactions often occur. A nuclear reaction results in the artificial transmutation of one nucleus into another. Many artificially produced radioisotopes that are used in research studies have been prepared in this fashion.

A second method of producing radioisotopes involves the use of neutrons formed in nuclear reactors. Every nuclear reactor is functional because of a special nuclear reaction called *nuclear fission,* illustrated in Fig. 14-6. This is a process

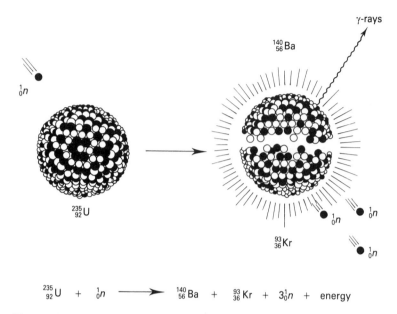

$$\underset{92}{\overset{235}{}}U \; + \; \underset{0}{\overset{1}{}}n \; \longrightarrow \; \underset{56}{\overset{140}{}}Ba \; + \; \underset{36}{\overset{93}{}}Kr \; + \; 3\underset{0}{\overset{1}{}}n \; + \; energy$$

Figure 14-6 The fission of a uranium-235 nucleus. A low-velocity neutron (upper left) is first absorbed by this nucleus, which, in this instance, fissions to produce barium 140, krypton 93, three neutrons, and energy.

whereby a nucleus splits into two others, accompanied by the release of a relatively large amount of energy and several neutrons. In the nuclear power industry, engineers harness this energy and convert it into electrical energy.

The neutrons generated in fission may be used to bombard other nuclei. This serves as a means of artificially producing radioisotopes that do not exist naturally. A commonly encountered radioisotope is cobalt 60. When it decays, cobalt 60 releases gamma radiation, which may be used industrially to induce cross-linking in polyethylene (Sec. 12-4), to vulcanize rubber, or in other ways. Cobalt 60 does not occur naturally. Instead, it is produced by irradiating cobalt 59, the only naturally occurring stable isotope of cobalt, with low-velocity neutrons. The production of cobalt 60 in this manner is illustrated by the following equation:

$$\underset{27}{\overset{59}{}}Co \; + \; \underset{0}{\overset{1}{}}n \; \longrightarrow \; \underset{27}{\overset{60}{}}Co \; + \; \gamma$$

Nuclear fission is not only important in connection with the operation of nuclear reactors. It is also the reaction that occurs during the detonation of an atomic bomb. Such fission processes occur spontaneously as a mode of decay for some nuclei (see Fig. 14-1), but such events are relatively rare; more commonly, they must be artificially induced. Even then, only certain nuclei can be induced to undergo fission; they are said to be *fissile*.

One such nucleus is uranium 235. It is induced to undergo fission by exposing it to thermal (that is, low velocity) neutrons. When uranium 235 captures thermal neutrons, uranium 236 first forms. But uranium 236 is unstable; to achieve stability, each uranium 236 nucleus fissions into two other nuclei, simultaneously releasing energy and several neutrons. One such event is represented by the following equation:

$$^{235}_{92}U + ^{1}_{0}n \longrightarrow ^{236}_{92}U \longrightarrow ^{89}_{35}Br + ^{145}_{57}La + 2^{1}_{0}n + 192 \text{ MeV}$$

There are many such fission events that occur when uranium 235 fissions, resulting in the formation of approximately 200 fission products. These nuclei are distributed among the elements from zinc ($Z = 30$) to terbium ($Z = 65$) and have mass numbers from $A = 72$ to 161. In a reactor, they comprise its waste products, a mixture of radioisotopes like those indicated in Table 14-4. In an atomic bomb, they comprise the components of fallout. In either event, they constitute a highly radiotoxic mixture.

Only 0.72% by mass of naturally occurring uranium is uranium 235. The more abundant isotope, uranium 238, is not fissile. The thermal neutron-induced fission of uranium 235 is a particularly important phenomenon, since more neutrons are formed for each neutron that are consumed by reaction. These neutrons that are formed can be utilized to induce the fission of other nuclei. In order words, a *chain*

TABLE 14-4 PRINCIPAL RADIOACTIVE FISSION PRODUCTS IN NUCLEAR WASTE

Radioisotope	Half-life	Emitted radiation
$^{85}_{36}Kr$	4.4 h	β, γ
$^{95}_{40}Zr$	65 d	β, γ
$^{89}_{38}Sr$	54 d	β
$^{90}_{38}Sr$	27 yr	β
$^{95}_{41}Nb$	90 h	β, γ
$^{99}_{43}Tc$	5.9 h	β
$^{103}_{44}Ru$	39.8 d	β, γ
$^{103}_{44}Rh$	57 min	e^{-} [a]
$^{106}_{44}Ru$	1 yr	β
$^{106}_{45}Rh$	30 s	β, γ
$^{129}_{52}Te$	34 d	β, γ
$^{129}_{53}I$	1.7×10^{7} yr	β, γ
$^{131}_{53}I$	8 d	β, γ
$^{133}_{54}Xe$	2.3 d	β, γ
$^{137}_{55}Cs$	33 yr	β, γ
$^{140}_{56}Ba$	12.8 d	β, γ
$^{140}_{57}La$	40 h	β, γ
$^{141}_{58}Ce$	32.5 d	β, γ
$^{144}_{58}Ce$	590 d	β, γ
$^{143}_{59}Pr$	13.8 d	β, γ
$^{144}_{59}Pr$	17 min	β
$^{147}_{61}Pm$	2.26 yr	β

[a] e^{-} denotes internal conversion electrons. Internal conversion is a relatively rare process in which an unstable nucleus gives its excess energy directly to an orbital electron, instead of undergoing de-excitation by means of gamma-ray emission. The electron is subsequently ejected from the nucleus.

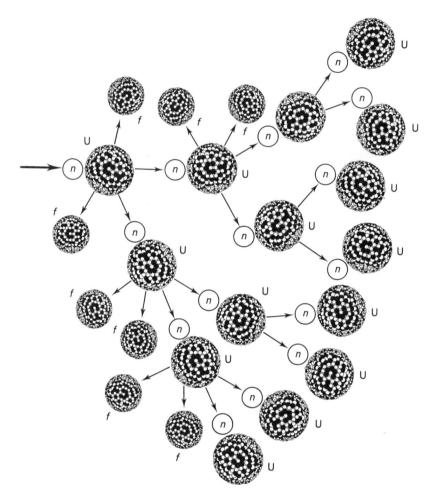

Figure 14-7 A nuclear chain reaction. The neutron (*n*) adjacent to the arrow is absorbed by a uranium 235 nucleus, which subsequently undergoes fission. Fission by-products (*f*) and additional neutrons are thus produced. These neutrons induce the fission of other uranium 235 nuclei. The chain reaction is self-perpetuating until either the fuel is depleted or the neutrons are absorbed by matter other than the fuel.

reaction may be initiated like that in Fig. 14-7, which under proper conditions results in the production of tremendous energy. In a nuclear reactor the chain reaction proceeds in a controlled fashion. By contrast, in the atomic bomb the chain reaction builds up energy at an explosive rate.

May a nuclear reactor explode like an atomic bomb? The answer to this question requires knowing that, for either instance, a certain minimum mass of fissionable material must be present. This is called the *critical mass*. When the mass of the fissionable material is less than the critical mass, most of the neutrons that are formed during fission escape from the mass and do not induce the fission of other nuclei. The perpetuation of a chain reaction requires that at least as many neutrons be retained by the system, as reactants for further fission events, as were consumed in the fission reaction from which they were produced.

In the atomic bombs shown in Fig. 14-8, two pieces of fissionable material, each of whose mass was less than the critical mass, were driven together by explosive charges into a small volume. This required only 1 μs. In this tiny instant, an

Figure 14-8   Top: "Little Boy," the atomic bomb detonated over Hiroshima, Japan, on August 6, 1945. This bomb contained fissile uranium 235, was 28 in. in diameter, 120 in. long, weighed about 9000 lb, and killed more than 200,000 people. Bottom: "Fat Man," the atomic bomb detonated over Nagasaki, Japan, on August 9, 1945. This bomb contained fissile plutonium 239, was 60 in. in diameter, 128 in. long, weighed about 10,000 lb, and killed more than 74,000 people. (Courtesy of the Los Alamos National Laboratory, Los Alamos, New Mexico)

enormous amount of energy was simultaneously produced from numerous fission events. This caused the bombs to detonate.

By contrast, in a nuclear reactor, one assembly of fissile material is present in an amount just slightly larger than its critical mass. There is far less fissionable material per unit volume in any nuclear reactor than there was in either atomic bomb detonated over Japan. Furthermore, the number of neutrons produced can be restricted in a reactor by using *control rods,* long rods of cadmium that can be inserted into the reactor to effectively absorb extra neutrons. Thus, there is little fear that a reactor will explode. Given all normal conditions, even if the chain reaction in a nuclear reactor became uncontrollable, the reactor would most likely melt before it would detonate.

14-6 USDOT REGULATIONS REGARDING RADIOACTIVE MATERIALS

When radioactive materials are transported, their radioactive nature generally poses a greater potential risk to public health and the environment than any of the substance's chemical properties. Thus, for transportation purposes, it is generally this radioactive nature that must be readily identifiable. This is accomplished by reviewing shipping papers to determine the specific radioisotopes being transported and their activities and by recognizing the shipper's use of RADIOACTIVE labels and placards (see Fig. 5-16).

USDOT regulates the transportation of all radiosotopes whose specific activities equal or exceed 0.002 μCi/g (74 Bq/g). To minimize worker exposure to radioactive materials, the materials are required to be contained in either of two forms of USDOT-approved packaging, identified as type A and type B packaging. Some typical forms of these packaging types are illustrated in Figs. 14-9 and 14-10.

Radioactive materials are designated in the Hazardous Materials Table under only the following entries, which, as appropriate, serve as components of their proper shipping names:

> Radioactive material, excepted package, articles manufactured from natural or depleted uranium *or* natural thorium
>
> Radioactive material, excepted package, empty packaging
>
> Radioactive material, excepted package, instruments *or* articles
>
> Radioactive material, excepted package, limited quantity of material
>
> Radioactive material, fissile n.o.s., *Class I, II or III*
>
> Radioactive material, low specific activity, *LSA*, n.o.s.
>
> Radioactive material, n.o.s.
>
> Radioactive material, special form, n.o.s.
>
> Thorium metal, pyrophoric
>
> Thorium nitrate, solid

PACKAGE MUST WITHSTAND NORMAL CONDITIONS
OF TRANSPORT ONLY WITHOUT LOSS OR DISPERSAL OF THE
RADIOACTIVE CONTROL CONTENTS.

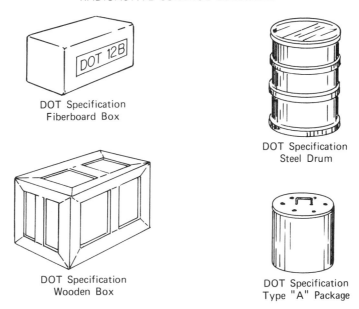

DOT Specification
Fiberboard Box

DOT Specification
Steel Drum

DOT Specification
Wooden Box

DOT Specification
Type "A" Package

Figure 14-9 Typical forms of type A packaging for containment of radioactive materials during transportation. These forms of packaging are designed to withstand the normal conditions of transport, as demonstrated by retention of the integrity of containment and shielding, to the extent required by USDOT.

Uranium hexafluoride (*fissile containing more than 1% U-235*)

Uranium hexafluoride (*fissile excepted or nonfissile*)

Uranium metal, pyrophoric

Uranyl nitrate hexahydrate solution

Uranyl nitrate, solid

Aside from use of these general descriptions, the proper shipping name provides the identity of the radioisotopes shipped in each package. Abbreviations are commonly employed for this purpose, for example, ^{99}Mo. For nonfissile radioisotopes, the physical or chemical form of the material, its activity, and the category of label are also provided. Thus, a shipper provides the following proper shipping name for a type A package containing 400 μCi/g of cadmium 115 labeled cadmium chloride: "Radioactive material, n.o.s., (^{115}Cd, 400 μCi/g as cadmium chloride, RADIOACTIVE WHITE–I), 7, UN2982, PG Type A." When a fissile material is transported, the shipper also provides the transport index.

During transport, one of three types of labels may be required by USDOT to be affixed to the exterior surface of packages containing radioactive materials. Shippers determine the proper label to affix by ascertaining the radiation level of the

PACKAGE MUST STAND BOTH NORMAL AND
ACCIDENT TEST CONDITIONS WITHOUT LOSS OF
CONTENTS.

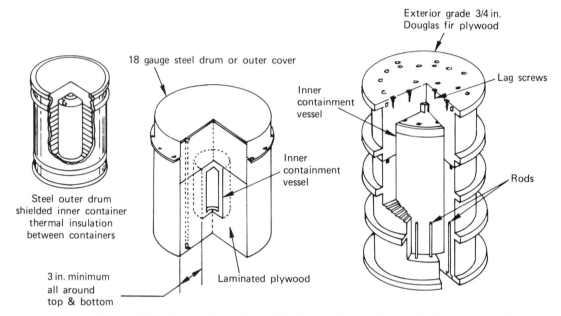

Figure 14-10 Typical forms of type B packaging for containment of radioactive materials during transportation. These forms of packaging are designed to withstand the damaging effects of a transportation accident, as demonstrated by the retention of the integrity of containment and shielding, to the extent required by USDOT using specified test procedures.

contents at the surface of the package, the fissile class of the material, and, if appropriate, the transport index. These three factors are described as follows:

1. *Radiation level:* The external radiation level is determined through use of an appropriate instrument or by arithmetic calculation when the physical properties of the radiosotope are known.
2. *Fissile class:* When the radioactive material is fissile, shippers are required to assure that the consignment does not attain criticality under both normal and accident conditions of transport. For such purposes, packages containing radioactive materials are classified into one of the following three groups:
   a. *Fissile class I:* Packages that may be transported in unlimited number, and in any arrangement, and that require no nuclear criticality safeguards during transport. A transport index is not assigned to fissile class I packages for the purposes of nuclear criticality safety control, but the external radiation levels may require a transport index.
   b. *Fissile class II:* Packages that may be transported together in any arrangement but in numbers that do not exceed an aggregate transport index of 50. For the purposes of nuclear criticality safety control, individual packages

may have a transport index of not less than 0.1 and not more than 10. However, the external radiation levels may require a higher transport index. These shipments require no nuclear criticality safety control by the shipper during transportation.

c. *Fissile class III:* Packages that do not meet the requirements of fissile class I or II and that are controlled during transportation by special arrangements between the shipper and carrier.

3. *Transport Index:* The transport index for a package means either of the following:

a. The number expressing the maximum radiation level in millirems per hour at 40 in. (1 m) from the external surface of the package (or microsieverts per hour divided by 10 at 1 m).

b. For a fissile class II or III package, the larger of the following numbers: (1) the number expressing the maximum radiation level under (a) or (2) the number obtained by dividing 50 by the allowable number of such packages.

Based on knowledge of these three parameters, the appropriate radioactive label is affixed to a package containing a radioactive material according to the following criteria:

1. A RADIOACTIVE WHITE-I label is used when the radiation level originating from the package at any time during normal transport does not exceed 0.5 mrem/h (5 μSv/h) at any point on the external surface of the package, the package does not belong to fissile class II or III, and it is not transported under special arrangement.

2. A RADIOACTIVE YELLOW-II label is used when the package does not belong to fissile class III, is not being transported under special arrangement, and the radiation level exceeds 0.5 mrem/h (5 μSv/h), but not 50 mrem/h (0.5 mSv/h), provided the transport index at any time during normal transportation does not exceed 1.0.

3. A RADIOACTIVE YELLOW-III label is used when the package belongs to fissile class III, the surface radiation level exceeds 50 mrem/h (0.5 mSv/h), and the transport index at any time during normal transport does not exceed 10.

Each radioactive label provides spaces in which the shipper inscribes the name of the contents, specific activity, and, for RADIOACTIVE II and RADIOACTIVE III labels, the transport index.

14-7 ENCOUNTERING RADIOACTIVE MATERIALS WHILE FIREFIGHTING

In the workplace, OSHA requires signs to be conspicuously posted, like either of those illustrated in Fig. 14-11(a), (b), or (c), to warn individuals that a given area is either of the following:

1. A *radiation area,* that is, one in which radiation exists at a level from which the body could receive in any hour a dose in excess of 5 millirem, or in any five consecutive days a dose in excess of 100 millirem.

2. A *high radiation area,* that is, one in which radiation exists at such levels that a major portion of the body could receive in any one hour a dose in excess of 100 millirem.

3. An *airborne radioactivity area,* that is, one in which airborne radiation is likely to exist.

These signs bear a magenta or purple three-bladed propeller on a yellow background, along with the word CAUTION and either of the words RADIATION

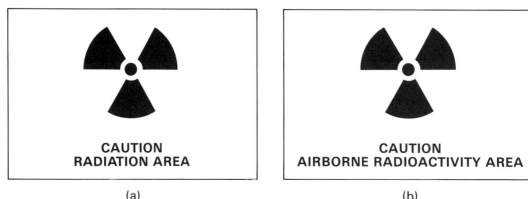

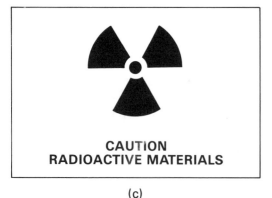

Figure 14-11 OSHA requires one or more of these caution signs to be conspicuously posted in areas where radioactive materials are stored or otherwise located. The three-bladed symbol is the conventional radiation caution sign. OSHA also requires that labels be affixed to certain radioactive material storage and transport containers; these labels resemble the caution sign illustrated in (c).

AREA, HIGH RADIATION AREA, or AIRBORNE RADIOACTIVITY AREA. OSHA also requires the conspicuous posting of one or more of the signs shown in Fig. 14-10 in any area or room in which radioactive material is stored or used, as well as on containers used to transport or store radioactive materials.

Encountering either of such signs while responding to a building fire makes one aware of the likelihood that radioactive materials are in the nearby vicinity. But what should be done to protect oneself against unnecessary exposure to ionizing radiation while responding to an emergency? The answer to this question lies in applying three basic principles to assure that the radiation exposure is minimized: shielding, time, and distance.

During firefighting, the use of shielding is ordinarily unnecessary since, when encountered, radioactive materials are most likely to be properly contained and stored so as to minimize exposure to the accompanying radiation. Nevertheless, it is wise to note that a barrier of any type between an individual and a radioactive material minimizes the radiation dose.

When it is essential to be exposed to ionizing radiation, the very best method of protection is to limit the time of exposure and to maintain a position as far removed from the source as is practical. Limiting the time of exposure and staying far removed from the radiation, but still in a position that allows one to accomplish a given task, assures that only the minimum dose is received.

The intensity of radiation decreases as the inverse square of the distance from the source. This is the *inverse square law of radiation,* which may be arithmetically expressed as follows:

$$I = \frac{I_0}{r^2}$$

In this equation, I_0 is the original intensity of a source of radiation, and I is the intensity at a distance r. Thus, if a radioactive material registers 1000 r on a Geiger counter held 0.3 m (1 ft) from the source, it registers only 250 r when held 0.6 m (2 ft) from the source. Thus, doubling the distance between an individual and the source of radiation reduces one's exposure to one-fourth the original value.

When radioactive materials are involved in a transportation mishap where there is suspected radioactive contamination from leakage or damaged containers, the carrier is required to notify the regional office of the U.S. Department of Energy, or appropriate state or local radiological authorities. When firefighters and other emergency response personnel respond to such incidents, care must be exercised to avoid possible inhalation, ingestion, or contact with the radioactive materials. Loose radioactive materials and associated packaging materials should be segregated in an area pending disposal instructions from responsible radiological authorities. If a fire has occurred, it should be extinguished from a distance. All measures should be taken to prevent the spread of radioactivity by keeping the quantity of runoff water to a minimum.

14-8 RESIDENTIAL RADON

Radioisotopes of radium are continuously forming from the decay of naturally occurring uranium and thorium. The concentration of these radium radioisotopes is fairly low in nature, except in areas where these metals are present in the metal-bearing ore or in granite or black shale. Such geological formations are relatively common below the upper strata of Earth throughout specific areas of the United States.

Buried deep beneath Earth's surface, radium ordinarily presents little problem. However, when radium isotopes decay, they change into a colorless, odorless gas called *radon*. For instance, when radium 226 decays, it is converted into radon 222, as the following illustrates:

$$^{226}_{88}\text{Ra} \longrightarrow \, ^{222}_{86}\text{Rn} + \alpha$$

Thus formed, radon slowly percolates through the soil and frequently enters the atmosphere, where it is thought to dissipate innocuously. The average atmospheric radon concentration is 4 pCi/L.

However, radon may also sneak through cracks in building foundations, wells, drainpipes, and cinder-block walls. Here, in buildings where there is little exchange between inside and outside air, the concentration of radon may increase to pose a health hazard.

All radon isotopes are radioactive. They emit ionizing radiation and become new radioisotopes. Radon 222 decays as follows:

$$^{222}_{86}\text{Rn} \longrightarrow \, ^{218}_{84}\text{Po} + \alpha$$

Polonium 218 decays by a combination of alpha and negatron emission, forming lead 214 and astatine 218, respectively.

$$^{218}_{84}\text{Po} \longrightarrow \, ^{214}_{82}\text{Pb} + \alpha$$

$$^{218}_{84}\text{Po} \longrightarrow \, ^{218}_{85}\text{At} + \beta^-$$

These products of radioactive decay are themselves radioisotopes.

When radon is inhaled, its decay products may become lodged in the lungs where they could irradiate the lung tissues for great lengths of time, even an entire lifetime. This means that breathing an atmosphere enriched in radon increases the risk of acquiring lung cancer. Some experts say that radon is the leading cause of lung cancer among nonsmokers, but such remarks are mostly conjectural. Credible scientific information is currently unavailable; hence, no one knows with certainty the precise risk that results from inhaling atmospheres enriched in radon. Nevertheless, this much is probably true: The risk of acquiring lung cancer from exposure to residential radon is proportional to the amount of radon that enters a home and the length of time it remains in living areas. Furthermore, the effectiveness by which radon is able to cause cancer is increased when the exposed person is a smoker. Toxicologists estimate that smoking may increase the carcinogenic potential of radon by 15 times.

Unfortunately, little can be done to improve the structure of a home against concentrating excessive levels of radon. USEPA suggests increasing the air flow into and through the home by opening windows and using fans. When homes have crawl spaces, vents in these areas should be kept open year round. Implementation of these suggestions results in a reduction of the atmospheric radon concentration, which in turn, reduces the risks likely to be caused by radon exposure.

REVIEW EXERCISES

Features of Atomic Nuclei

14.1. How many protons and neutrons are there in each of the following nuclei: (a) lithium 6; (b) carbon 14; (c) iodine 130; (d) plutonium 239?

14.2. The principal radioisotopes that are likely to pose an environmental hazard from nuclear power plants are krypton 85, xenon 133, and xenon 135.
(a) Why is this so?
(b) How many protons and neutrons are there in each of these nuclei?

14.3. Phosphorus 32 has a half-life of 15 days. If 2 million atoms of ^{32}P-labeled phosphoric anhydride are set aside for 30 days how many atoms will remain? How many remain after 45 days?

14.4. The nucleus is not the ordinary domain for electrons. How is it that an electron may be emitted during the spontaneous decay of certain radioisotopes?

14.5. A particular radioisotope, X, is only safe at concentrations of less than 1 mg/mile^2. A nuclear explosion site has a concentration of 10 mg/mile^2 just after the explosion of a bomb has occurred. The main contaminant of the area is X, whose half-life is 24 h. Using the above criterion of safety, will the site be safe in 4 days? If not, how many days must elapse before the concentration of X reaches a safe level?

Types of Radiation

14.6. Identify the product nucleus in each of the following instances:
(a) $^{24}_{11}Na \rightarrow$ ____ $+ \beta^-$
(b) $^{22}_{11}Na \rightarrow$ ____ $+ \beta^+$
(c) $^{65}_{30}Zn \rightarrow$ ____ (electron capture)
(d) $^{95}_{40}Zr \rightarrow$ ____ $+ \beta^-$
(e) $^{131}_{53}I \rightarrow$ ____ $+ \beta^-$
(f) $^{210}_{84}Po \rightarrow$ ____ $+ \alpha$

14.7. Identify the particle emitted during each of the following nuclear transformations:
(a) $^{105}_{45}Rh \rightarrow ^{105}_{46}Pd +$ ____
(b) $^{248}_{97}Bk \rightarrow ^{248}_{98}Cf +$ ____
(c) $^{233}_{87}Fr \rightarrow ^{229}_{85}At +$ ____
(d) $^{140}_{59}Pr \rightarrow ^{140}_{58}Ce +$ ____

14.8. In what fundamental way is gamma radiation similar to microwave radiation? In what fundamental way is it different?

Measurement of Radioactivity

14.9. Why is it generally recommended that persons under 18 years of age accumulate no radiation?

14.10. Fifteen grams of carbon 14 labeled sodium carbonate have an activity of 5000 pCi. What is the specific activity of this radioactive material?

14.11. Under regulations promulgated by OSHA, what is the maximum accumulated radiation dose in rems recommended for a 48-year-old male employee?

Adverse Health Effects from Radiation Exposure

14.12. Strontium 90 is a component of the radioactive waste generated during nuclear fission. Suggest a reason why ingestion of substances containing this radioisotope is likely to pose a health hazard.

14.13. When is exposure to alpha radiation considered to be particularly dangerous to one's health?

14.14. Two laboratory technicians are periodically exposed to low levels of gamma radiation. One is a cigarette smoker, while the other is a nonsmoker. Which is exposed to the higher risk of acquiring lung cancer?

Nuclear Fission

14.15. Identify the nucleus that completes each of the following equations:
   (a) $^{235}_{92}U + ^{1}_{0}n \rightarrow$ _____ $+ ^{93}_{36}Kr + 3^{1}_{0}n$
   (b) $^{235}_{92}U + ^{1}_{0}n \rightarrow$ _____ $+ ^{147}_{57}La + 3^{1}_{0}n$
   (c) $^{235}_{92}U + ^{1}_{0}n \rightarrow$ _____ $+ ^{144}_{54}Xe + 2^{1}_{0}n$

14.16. Nuclear fission occurs in both nuclear reactors and atomic bombs. Yet the detonation of an atomic bomb is unlike the normal operation of a nuclear reactor. Why are these two phenomena so different in degree of hazard?

14.17. USDOT requires the carrier of a fissile material to assure that "criticality" cannot be reached under any foreseeable circumstances of transport. Describe what would most likely occur if criticality was achieved.

USDOT Requirements Regarding Radioactive Materials

14.18. A carrier specializing in the transportation of radioactive materials desires to transport 0.05 μCi of cobalt 60 intended for industrial use. When the activity is measured 1 m from the external surface of the package, a Geiger counter records 0.4 mrem/h.
   (a) Which USDOT-approved label should be affixed to the package?
   (b) What information should be entered on it?

14.19. The number of packages bearing RADIOACTIVE YELLOW-II or -III labels stored in any one temporary transit area is limited by USDOT so that the sum of the transport indexes in any individual group of packages does not exceed 50. Groups of these packages must be stored so as to maintain a spacing of at least 6 m (20 ft) from other groups of packaging containing radioactive materials. What is the most likely reason USDOT promulgated his particular regulation?

Encountering Radioactive Materials While Firefighting

14.20. During a transportation mishap, a motor vehicle overturns; it is placarded RADIOAC-TIVE. A firefighter discovers a ruptured steel drum with the exterior marking TYPE B and a RADIOACTIVE YELLOW-III label. Neither the vehicle nor the steel drum is involved in fire. What should be done to best safeguard lives and property?

Radon

14.21. To minimize exposure to radium 226, an engineer transfers the material to a test tube and locks it securely in a lead-walled safe. However, sometime later, it is discovered that the walls of the interior of the safe are slightly radioactive. How did the safe acquire this radioactivity?

14.22. Two buildings having identical structural features are situated over different geological formations, both of which contain radium-bearing rock. In one instance, the rock is interladen with porous soil; in the other, with clay. Assuming all other factors to be the same, which building is likely to pose the greater health risk from accumulation of radon?

Index

TABLE OF ELEMENTS AND ATOMIC WEIGHTS

in GRAMS

	Symbol	Atomic Number	Atomic Weight		Symbol	Atomic Number	Atomic Weight
Actinium	Ac	89	227*	Mercury	Hg	80	200.59
Aluminum	Al	13	26.9815	Molybdenum	Mo	42	95.94
Americium	Am	95	243*	Neodymium	Nd	60	144.24
Antimony	Sb	51	121.75	Neon	Ne	10	20.183
Argon	Ar	18	39.948	Neptunium	Np	93	237*
Arsenic	As	33	74.9216	Nickel	Ni	28	58.71
Astatine	At	85	210*	Niobium	Nb	41	92.906
Barium	Ba	56	137.34	Nitrogen	N	7	14.0067
Berkelium	Bk	97	245*	Nobelium	No	102	253*
Beryllium	Be	4	9.0122	Osmium	Os	76	190.2
Bismuth	Bi	83	208.980	Oxygen	O	8	16
Boron	B	5	10.811	Palladium	Pd	46	105.4
Bromine	Br	35	79.909	Phosphorus	P	15	30.9738
Cadmium	Cd	48	112.40	Platinum	Pt	78	195.09
Calcium	Ca	20	40.48	Plutonium	Pu	94	242*
Californium	Cf	98	248*	Polonium	Po	84	210*
Carbon	C	6	12.01115	Potassium	K	19	39.102
Cerium	Ce	58	140.12	Praseodymium	Pr	59	140.907
Cesium	Cs	55	132.905	Promethium	Pm	61	145*
Chlorine	Cl	17	35.453	Protactinium	Pa	91	231*
Chromium	Cr	24	51.996	Radium	Ra	88	226*
Cobalt	Co	27	58.9332	Radon	Rn	86	222*
Copper	Cu	29	63.54	Rhenium	Re	75	186.2
Curium	Cm	96	245*	Rhodium	Rh	45	102.905
Dysprosium	Dy	66	162.50	Rubidium	Rb	37	85.47
Einsteinium	Es	99	247*	Ruthenium	Ru	44	101.07
Erbium	Er	68	167.26	Samarium	Sm	62	150.35
Europium	Eu	63	151.96	Scandium	Sc	21	44.956
Fermium	Fm	100	254*	Selenium	Se	34	78.96
Fluorine	F	9	18.9984	Silicon	Si	14	28.086
Francium	Fr	87	223*	Silver	Ag	47	107.870
Gadolinium	Gd	64	157.25	Sodium	Na	11	22.9898
Gallium	Ga	31	69.72	Strontium	Sr	38	87.62
Germanium	Ge	32	72.59	Sulfur	S	16	32.064
Gold	Au	79	196.967	Tantalum	Ta	73	180.948
Hafnium	Hf	72	178.49	Technetium	Tc	43	99*
Helium	He	2	4.0026	Tellurium	Te	52	127.60
Holmium	Ho	67	164.930	Terbium	Tb	65	158.924
Hydrogen	H	1	1.00797	Thallium	Tl	81	204.37
Indium	In	49	114.82	Thorium	Th	90	232.038
Iodine	I	53	126.9044	Thulium	Tm	69	168.934
Iridium	Ir	77	192.2	Tin	Sn	50	118.69
Iron	Fe	26	55.847	Titanium	Ti	22	47.90
Krypton	Kr	36	83.80	Tungsten	W	74	183.85
Lanthanum	La	57	138.91	Uranium	U	92	238.03
Lawrencium	Lw	103	259*	Vanadium	V	23	50.942
Lead	Pb	82	207.19	Xenon	Xe	54	131.30
Lithium	Li	3	6.939	Ytterbium	Yb	70	173.04
Lutetium	Lu	71	174.97	Yttrium	Y	39	88.905
Magnesium	Mg	12	24.312	Zinc	Zn	30	65.37
Manganese	Mn	25	54.9381	Zirconium	Zr	40	91.22
Mendelevium	Md	101	256*				

*Mass number of isotope of longest known half-life

Cigarettes   EMF

Sociology

Political

Economic.

medical

Public Health

Scientific